Film Formation in Waterborne Coatings

ACS SYMPOSIUM SERIES **648**

Film Formation in Waterborne Coatings

Theodore Provder, EDITOR
ICI Paints

Mitchell A. Winnik, EDITOR
University of Toronto

Marek W. Urban, EDITOR
North Dakota State University

Developed from a symposium sponsored
by the Division of Polymeric Materials:
Science and Engineering, Inc.

American Chemical Society, Washington, DC

Library of Congress Cataloging-in-Publication Data

Film formation in waterborne coatings / Theodore Provder, editor
Mitchell A. Winnik, editor, Marek W. Urban, editor.

    p.   cm.—(ACS symposium series, ISSN 0097–6156; 648)

    "Developed from a symposium sponsored by the Division of Polymeric
Materials: Science and Engineering, Inc. at the 210th National Meeting
of the American Chemical Society, Chicago, Illinois, August 20–24,
1995."

    Includes bibliographical references and indexes.

    ISBN 0–8412–3457–4

    1. Emulsion paint—Congresses.

    I. Provder, Theoder, 1939– . II. Winnik, Mitchell A.
III. Urban, Marek, W., 1953– . IV. American Chemical Society.
Division of Polymeric Materials: Science and Engineering. V. American
Chemical Society. Meeting (210th: 1995: Chicago, Ill.): Science and
Engineering. VI. Series.

TP934.F465   1996
667'.9—dc20   96–31614
                                           CIP

This book is printed on acid-free, recycled paper.

Advisory Board

ACS Symposium Series

Foreword

THE ACS SYMPOSIUM SERIES was first published in 1974 to provide a mechanism for publishing symposia quickly in book form. The purpose of this series is to publish comprehensive books developed from symposia, which are usually "snapshots in time" of the current research being done on a topic, plus some review material on the topic. For this reason, it is necessary that the papers be published as quickly as possible.

Before a symposium-based book is put under contract, the proposed table of contents is reviewed for appropriateness to the topic and for comprehensiveness of the collection. Some papers are excluded at this point, and others are added to round out the scope of the volume. In addition, a draft of each paper is peer-reviewed prior to final acceptance or rejection. This anonymous review process is supervised by the organizer(s) of the symposium, who become the editor(s) of the book. The authors then revise their papers according to the recommendations of both the reviewers and the editors, prepare camera-ready copy, and submit the final papers to the editors, who check that all necessary revisions have been made.

As a rule, only original research papers and original review papers are included in the volumes. Verbatim reproductions of previously published papers are not accepted.

ACS BOOKS DEPARTMENT

Contents

MORPHOLOGY AND FILM STRUCTURE

ix

INDEXES

Preface

NEW COATINGS TECHNOLOGIES such as high solids, powder, water-borne, and radiation-curable coatings have been developed during the past 15 years to meet the challenges of (a) governmental regulations in the area of ecology (emission of volatile organic compounds), (b) long-term increasing costs of energy and petroleum-based solvents, (c) more-active public consumerism, and (d) the continual need for cost-effective, high-performance coatings in a highly competitive and global business environment. These new coatings technologies require the use of water as the major solvent with water-soluble or high-molecular-weight latex polymers or the use of strategically designed, low-molecular-weight polymers, oligomers, and reactive additives that, when further reacted, produce high-molecular-weight and cross-linked polymers. This has led to a need for improved methods of materials characterization in diverse areas such as molecular-weight distribution analysis, particle-size distribution assessment and characterization, rheology of coatings, film-formation and cure-process characterization, morphological characterization, and spectroscopic analysis, as well as a need for improved methods for modeling and predicting materials properties and processes.

The film-formation process is key to the development of the ultimate physical and chemical properties of waterborne systems and the consequent end-use properties. Thus, commercially successful, cost-effective waterborne coating products require practical application of the knowledge that underpins our understanding of the film-formation process. In recent years significant advances have been made in our knowledge and understanding of the film-formation process.

This book covers significant advances recently made in our understanding of the film-formation process. These advances have been brought about by collateral advances in instrumentation technology and its application to studying the stages involved in the film-formation process. Data obtained from the instrumental methods have allowed the confirmation of some aspects of the mechanism of film formation and have furthered the understanding of the subtleties and complexities of commercially relevant waterborne coatings. The first section of this book focuses on the mechanism of film formation and the uses of advanced instrumental methods such as fluorescence spectroscopy, small-angle neutron scattering, and dielectric spectroscopy to study the film-formation process. The second section of the book focuses on measuring film mechanical properties and relating them to the film-formation process. The third section focuses on relevant morphology and film structure, which is a consequence of the physics and chemistry of the film-formation process. The fourth section focuses on the application of novel chemistry and processes to develop unique film structures for a variety of coating systems. We hope this book will encourage and foster additional investigations to further our understanding of the commercially relevant and scientifically challenging issue of film formation in waterborne coatings.

Acknowledgments

We are grateful to the authors for their effective oral and written communications and to the reviewers for their critiques and constructive comments. We also gratefully acknowledge the American Chemical Society Division of Polymeric Materials: Science and Engineering, Inc., for its financial support of the symposium on which this book is based, which was presented at the 210th National Meeting of the ACS in Chicago, Illinois, August 20–25, 1995.

THEODORE PROVDER
Strongsville Research Center
ICI Paints
16651 Sprague Road
Strongsville, OH 44136–1739

MITCHELL A. WINNIK
Department of Chemistry
University of Toronto
Toronto, Ontario M5S 1A1
Canada

MAREK W. URBAN
Department of Polymers and Coatings
North Dakota State University
Fargo, ND 58105

July 22, 1996

Mechanism Studies

Chapter 1

Film Formation of Acrylic Copolymer Latices: A Model of Stage II Film Formation

S. T. Eckersley[1,3] and A. Rudin[2]

[1]Department of Chemical Engineering and [2]Department of Chemistry, University of Waterloo, Waterloo, Ontario N2L 3G1, Canada

A model of latex film formation is proposed, where capillary and interfacial forces act in tandem to promote latex film formation. The radius of the contact region between the deformed particles is predicted by the model and is determined by scanning electron microscopy. Model parameters include capillary water surface tension, polymer surface tension, polymer/water interfacial tension, polymer viscoelastic properties ($G*$ and $\eta*$) and elapsed drying time. The model was initially evaluated as a function of particle size for poly(methyl methacrylate-*co*-butyl acrylate) latexes. Experimental results validated the model, despite the estimation of several parameters. In the subsequent analysis of the model, the viscoelastic nature of the copolymer was varied by the addition of molecular weight modifiers (CBr_4 chain transfer agent and ethylene glycol dimethacrylate crosslinking monomer) during synthesis. In contrast to the initial model evaluation, experimentally determined values of the parameters were employed. Film drying kinetics, the surface energetics of the system and hydroplasticization of the copolymer were investigated. Results showed that the degree of film formation ranged from complete to superficial. Comparison of the model predictions and experimental observations supported the model.

As a film-forming latex dries, it is transformed from a colloidal dispersion into a continuous polymer film having mechanical integrity. The quality of the fused film, in combination with the bulk polymer character, determines the ultimate coating properties. Consequently, an understanding of the process of film formation is critical for the development of latex coatings.

[3]Current address: Emulsion Polymers Research, Dow Chemical Company, 1604 Building, Midland, MI 48674

The mechanism of film formation has been studied for nearly half a century and remains the subject of active debate. Several alternative analyses of the film formation process are presented in this volume. We have published a number of articles related to the second stage of latex film formation. This chapter is a compilation of much of that work into a unified whole.

Background

The process of film coalescence is considered to occur in three stages, as depicted in Figure 1. During the first stage, bulk water evaporation occurs at a constant rate. The second stage begins when sufficient water has evaporated that the particles pack in an ordered array. Water evaporates from the interstitial voids at a reduced rate and the latex particles deform. The third stage begins once the film is macroscopically dry. Polymer diffuses across the residual particle boundaries as the film is aged. The research presented here is concerned with the second stage of film coalescence. In this chapter, we use the terms film formation, fusion, and deformation synonymously to identify the second stage of latex coalescence. The term coalescence is used to refer to the overall process (stages I through III).

The earliest mechanism of film formation was proposed by Dillon *et al* (*1*). These authors suggested that deformation occurred in the dry state, subsequent to water evaporation. They proposed a dry sintering mechanism where the polymer underwent viscous flow, driven by the tendency to reduce surface energy. A mechanism based on capillary forces due to the presence of interstitial water was proposed by Brown (*2*). Vanderhoff *et al* (*3*) proposed a wet sintering mechanism where particle fusion resulted from the polymer / water interfacial tension. Subsequent models (*4,5,6,7*) were also based on surface energetics. Sheetz (*5*) proposed that capillary forces initiated the coalescence process and that subsequent fusion resulted from compaction of the film by the fused surface layer during further water evaporation. Kendall and Padget (*7*) proposed that particle deformation was elastic.

In an earlier article (*8*), we presented evidence suggesting that previous models did not fully account for experimental observations. We proposed a model where the capillary force and interfacial forces of earlier models were complementary (*8*). The model predicts the radius 'a' of the contact area between latex particles as shown in the schematic of Figure 2. Assuming that the polymer is a linear viscoelastic material (as was suggested by Lamprecht (*6*)), the deformations due to each force are additive. The contact radius is the sum of the contributions from the capillary and interfacial forces and is given by the general expression:

$$a = a_{capillary} + a_{interfacial} \qquad (1)$$

The resulting equation gives the radius of the contact area (a) as a function of particle radius (R), water surface tension (σ), interfacial tension (γ), elapsed drying time (t), and the polymer viscoelastic properties (G* (complex modulus) and η* (complex viscosity)) according to:

$$a = \left(\frac{2.80R^2\sigma}{G*} \right)^{1/3} + \left(\frac{3\gamma Rt}{2\pi\eta*} \right)^{1/2} \qquad (2)$$

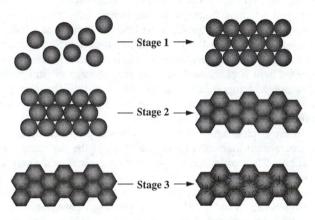

Figure 1. Illustration of the Film Formation Process

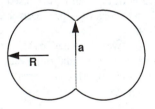

Figure 2. Radius of the Circle of Contact 'a' of Two Polymer Spheres

The interfacial tension term can be either polymer / water interfacial tension or polymer / air surface tension, as will be discussed later. The present chapter describes the evaluation of the model given by equation 2.

Experimental

Emulsion Polymerization. Latexes were synthesized by semi-continuous emulsion polymerization. To eliminate the effects of composition, the proportions of all ingredients (except water) remained constant between recipes. Seeded reactions were used to produce monodispersed latexes with a range of particle sizes. Further details of the emulsion polymerizations can be found in (9). Sodium dodecyl benzene sulphonate (Siponate DS-10, Alcolac Inc.) was used as the emulsifier in this series. Typical recipes and particle size data are given in Table I. The seed latex used in the seeded recipe is the product of the unseeded polymerization. Latex characteristics are given in Table II. Minimum film temperature is expressed as the average ± the 95% confidence interval.

Table I: Emulsion Polymerization Recipes (Particle Size Series)

	Latex P4 (Unseeded)	Latex P6 (Seeded)
Reactor charge		
deionized water (g)	268	180
ammonium persulphate initiator (g)	1.35	0.85
P4 seed latex (g)	-	131
Monomer emulsion		
methyl methacrylate (g)	105	65.6
butyl acrylate (g)	105	65.6
methacrylic acid (g)	2.55	1.60
sodium dodecyl benzene sulphonate (g)	0.16	0.10
deionized water (g)	70	43
Particle size distribution		
number average diameter, D_n (nm)	501	788
weight average diameter, D_w (nm)	509	798
polydispersity index, D_w/D_n	1.02	1.01

A second series of latexes was made with molecular weight modifiers. Carbon tetrabromide and carbon tetraiodide chain transfer agents were used. Ethylene glycol dimethacrylate (EGDM) was employed as the crosslinking monomer. These polymerizations were surfactant-free. A general recipe is given in Table III. The quantities of molecular weight modifiers used (x) and the resulting latex and polymer characteristics are given in Table IV.

All reactions were performed in a one liter kettle reactor equipped with an overhead condenser and a jacketed mechanical stirrer. The agitation rate was maintained at 250 rpm throughout the reaction.

The water and initiator were charged to the reactor and maintained at a temperature of 80°C. The monomer emulsion (or monomer mixture) was fed to the

Table II: Latex Characteristics (Particle Size Series)

Latex	D_n (nm)	D_w/D_n	$\overline{MFT}$ (°C)
P1	148	1.05	11.5 ± 1.0
P2	318	1.02	11.7 ± 1.4
P3	413	1.04	13.0 ± 1.4
P4	501	1.02	13.1 ± 1.6
P5	571	1.03	15.1 ± 0.9
P6	788	1.01	14.7 ± 0.9
P7	816	1.02	15.9 ± 1.0
P8	1125	1.03	14.8 ± 1.7
P9	1234	1.00	16.4 ± 1.2

reactor at a constant rate of ~ 1 mL·min^{-1}. No monomer accumulation was observed at any time. Therefore, it was assumed that the reaction was starved-fed and that the composition of the polymer was uniform throughout the latex particle. Following monomer addition, the reaction was continued for 1 h. The latex was then gradually cooled to ambient temperature. Finally, the latex was filtered through a 100-mesh screen to remove the minimal amount of grit that formed during the polymerization. The pH of the latexes was adjusted to 9 by the addition of aqueous ammonia solution. Particle size measurements were obtained using an ICI-Joyce Loebl Disk Centrifuge according to the method of (*10*).

Table III: Surfactant Free Emulsion Polymerization Recipes
(Molecular Weight Modified Series)

Reactor charge
deionized water (g)   210
ammonium persulphate initiator (g)   1.35
Monomer mixture
methyl methacrylate (g)   101.4
butyl acrylate (g)   101.4
methacrylic acid (g)   2.55
molecular weight modifier (g) x

Surface Tensions. The surfactant-free latexes synthesized for these studies were subject to settling under the influence of gravity. Isolation of the continuous phase was desired. Hence, the liquid phase was allowed to separate and was decanted. The liquid was centrifuged at 2700 rpm for ~ 2 h. The supernatant was decanted and the procedure was repeated. The continuous phases of several latexes (M1, M6, M12 and M13) were obtained in this manner. Five measurements of the continuous phase surface tensions were obtained using a calibrated ring tensiometer (*11*). The differences between the four latexes were negligible. Therefore, the average was calculated to be 48.8 dyne·cm^{-1} with a 95% confidence interval of 0.8 dyne·cm^{-1}. The surface tension of latex M6 as a function of post-added NP-40 (nonyl phenol ethylene oxide adduct (40 mols EO)) was also obtained using the ring tensiometer.

Table IV: Latex Characteristics (Molecular Weight Modified Series)

Latex	Modifier	x (g)	D_n (nm)	D_w/D_n	$\overline{MFT}$ (°C)	G* (0.1 rad·s^{-1}) at 22°C dyne·cm^{-2}(10^7)
M1	CBr_4	5.00	671	1.010	8.0 ± 0.5	2.2
M2	CBr_4	5.00	890	1.010	12.0 ± 1.8	2.2
M3	CBr_4	2.50	680	1.030	9.9 ± 0.1	12
M4	CBr_4	0.96	606	1.010	11.5 ± 0.4	32
M5	CI_4	0.10	647	1.005	11.9 ± 0.9	-
M6	none	-	580	1.010	11.0 ± 0.3	7.6
M7	EGDM*	0.40	582	1.020	12.6 ± 0.9	-
M8	EGDM	0.50	588	1.010	12.7 ± 1.0	-
M9	EGDM	1.50	438	1.020	14.0 ± 0.5	-
M10	EGDM	3.00	699	1.010	14.0 ± 0.9	18
M11	EGDM	3.00	1002	1.010	15.8 ± 1.5	18
M12	EGDM	8.00	899	1.007	18.6 ± 1.6	71
M13	EGDM	12.00	984	1.006	21.4 ± 1.6	162

*ethylene glycol dimethacrylate

Minimum Film Temperature (MFT).   An apparatus similar to that used by Protzman and Brown (*12*) was used to determine the MFT's of the various latexes. An insulated stainless steel bar replaces an aluminum bar in the original apparatus. Cooling at one end of the bar is achieved by two 12-V ceramic thermoelectric cooling modules. The cooling rate was maintained by means of a feedback control device. Heat was not applied at the opposite end of the bar, since all the MFT's were below room temperature. The temperature gradient along the bar was determined by eight thermocouples installed at intervals along the bar. The thermocouples were connected to a digital temperature indicator which had an accuracy of ±0.1°C.

A glass plate that permitted visual observation of the drying films covered the stainless steel bar. Prior to application of the latexes, the cooling mechanism was activated and nitrogen gas flow from the cold to hot end of the bar at a rate of 2000 mL·min^{-1} was started. The nitrogen gas minimized condensation of water at the cold end of the bar and maintained the humidity at a constant level. The temperature of the bar was allowed to equilibrate for about six hours. The glass plate was then removed. Approximately equal volumes of the latexes were applied to the channels down the length of the bar and the glass plate was quickly replaced. Drying of the latexes took approximately four hours.   During this time, five replicate measurements of the temperature gradient along the bar were obtained and subsequently averaged.   The MFT was determined as the temperature at which clarity of the dry film was observed.

The deformation of the dried latex cast films was examined by scanning electron microscopy (SEM).   Electron microscopy specimens were obtained by drying a thin film of latex on the aluminum sample stub for 24h at room temperature (22°C). Prior to exposure to the electron beam, the films were gold sputtered to a thickness of 1.6 (10^{-4}) m to prevent charging of the film surface.

Dynamic Mechanical Spectroscopy (DMS)

Sample Preparation. It was thought that the film integrity might effect the response of the material. Therefore, all dynamic mechanical tests were performed with completely fused polymer samples. The latexes were dried in a convection oven at 60°C then ground in a Wiley mill. All samples were then thoroughly fused by pressing at elevated temperature and pressure. The actual pressing conditions were varied since the materials had very different viscoelastic properties. Typical conditions were T=100°C, P=16psi for 60s (polymer M1), and T=100°C, P=570psi for 900s (polymer M13). All samples were pressed several times to ensure complete fusion.

The fused polymer sheets obtained were approximately 2mm thick. For the water plasticization experiments, circular samples of about 25mm in diameter were cut from the sheets and immersed in deionized water at room temperature. The specimens were exposed to water for two months. In the case of the surfactant plasticization experiments, either DS-10 (sodium dodecyl benzene sulphonate) or NP40 (nonyl phenol ethylene oxide adduct (40 mols EO)) was added to a concentration of 0.041 g / g polymer prior to drying of the latex.

Measurements of the polymer moduli were made using a Rheometrics Model 605 mechanical spectrometer. Both torsion rectangular and parallel plate geometries were employed. In the case of the parallel plate geometry, 8mm diameter and 25mm diameter plates were used.

Strain / Temperature Sweeps. The initial strain experienced by the sample was chosen by preliminary experiments. Strain sweeps were performed at the lowest temperature in the experimental range and a frequency of 1.0 rad$\cdot$s^{-1}. Strain sweeps were performed to determine an initial strain value that fulfilled several criteria. Most importantly, the torque on the transducer had to be within the recommended operating limits of the rheometer (this set an upper limit for the strain). Also, the measurement of tan delta had to be within the measurement sensitivity of the rheometer (this set a lower limit for the strain). Finally, to allow comparison of the materials at different deformations and different geometries, it was necessary that the polymer behave in a linear viscoelastic manner at the strain chosen.

Forced oscillation measurements were obtained at two frequencies (0.1 and 1.0 rad$\cdot$s^{-1}) for each temperature in all of the temperature sweep experiments. At the lower end of the temperature range, torsion rectangular geometry was used because of the stiffness of the polymers. When the torque became unacceptably low, the torsion rectangular geometry was exchanged for the small (8mm diameter) parallel plate geometry. At the higher temperatures, the larger (25mm) parallel plates were required. For all geometries, as the signal diminished with increasing sample temperature, the percent strain experienced by the specimen was increased. This procedure is only acceptable in the linear viscoelastic strain region, where the dynamic mechanical response of the material is not a function of the degree of deformation. Linear viscoelasticity was confirmed by performing a strain sweep at the terminal experimental temperature for each geometry. As with the low temperature strain sweep, the test was done at a frequency of 1.0 rad$\cdot$s^{-1}.

Results and Discussion

The model of equation 2 was investigated by varying the latex particle size, surface energetics (via changes in surfactant concentration) and polymer architecture (through the addition of molecular weight modifiers during polymerization). In addition, latex drying kinetics were investigated as a function of polymer architecture.

Particle Size Effect. A preliminary evaluation of the model focused on the effect of particle size on latex film formation properties (8). The minimum film temperature (MFT) was found to be a weak function of particle size, a result supported by other research (13). Equation 2 suggests that this should be the case. As particle size increases, equation 2 predicts that the degree of coalescence 'a/R' will be reduced. A particle size increase can be compensated for by increasing the film formation temperature, since an increase in temperature causes a reduction in the polymer modulus and viscosity. Therefore, the relationship between MFT and particle size lends qualitative support to the model of equation 2, as well as earlier models (1,2,4,6).

In (8), the model was evaluated quantitatively for the latexes of Table II. The contact radius as a function of particle radius was determined from the scanning electron micrographs of films dried at ambient temperature (22°C). The variation in the degree of film deformation is shown in the micrographs of Figure 3. The smaller particle size latexes were completely deformed and the large size particles showed only superficial deformation. Contact radii were measured directly from the micrographs for latexes P5 - P9. The films prepared from the smaller diameter latexes were too highly fused to allow determination of the contact radii from the micrographs. Model predictions for ambient drying conditions were made using equation 2. Low and high estimates of the contact radii were made. These were based on drying times of one and two hours respectively. The experimental contact radii and those predicted by equation 2 were compared and showed good agreement.

Figure 4 demonstrates that there was reasonable agreement between the predicted and experimental values of the contact radii. However, several approximations and assumptions were made in the model evaluation. It was assumed that water was present throughout the film formation process. That is, it was assumed that particle deformation was negligibly slow once water evaporation was complete and that any deformation due to the polymer surface tension could be neglected. An approximate value of the polymer / water interfacial tension (27 dyne·cm^{-1}) was calculated using the method of Owens and Wendt (14). This value is likely an overestimate for a carboxylated latex containing emulsifier. A value of 30 dyne·cm^{-1} was assumed for the surface tension of water. Approximate film drying times of one and two hours were assumed. It was assumed that the emulsifier did not plasticize the polymer, which was confirmed by DMS. However, the possibility of plasticization by water was not investigated. Linear viscoelasticity was assumed and was confirmed by DMS strain sweeps. In the subsequent work described here, we concentrated on a more rigorous testing of the model.

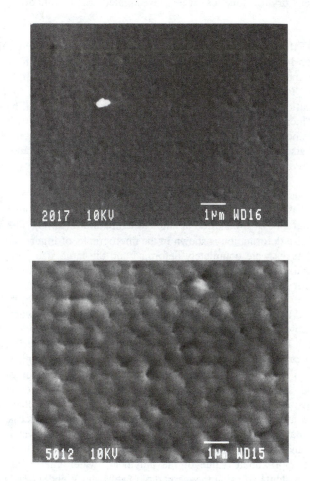

Figure 3. Scanning Electron Micrographs of Latex Films of Varying Particle Size: (a) Latex P2, D_n = 318 nm; (b) Latex P5, D_n = 571 nm; (c) Latex P8, D_n = 1125nm

c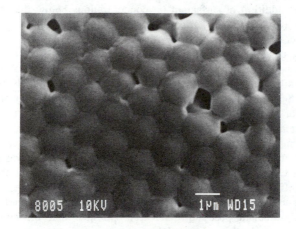

Figure 3. Continued.

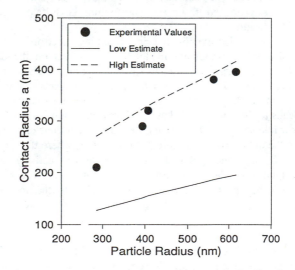

Figure 4. Contact Radius as a Function of Particle Radius for Drying Times of One Hour (Low Estimate) and Two Hours (High Estimate). Data from (*8*).

The Effect of Surfactant Type and Concentration. The contact radius of equation 2 is dependent on the magnitude of polymer interfacial tension. The model predicts that reducing water surface tension and polymer / water interfacial tension should decrease the degree of particle deformation. It was postulated that the post-polymerization addition of surfactant (at concentrations below the critical micelle concentration (CMC)) would affect latex film forming behaviour. This was investigated for latex M6.

Two typical coatings surfactants were studied (*11*): an anionic (sodium dodecyl benzene sulphonate, DS-10) and a nonionic (nonyl phenol ethoxylate (40 mols EO), NP40). The CMC's were determined to be approximately 10^{-3} g DS-10/(g polymer) and $8(10^{-3})$ g NP40/(g polymer) for latex M6. The MFT's were found to be independent of surfactant concentration. This is contrary to the model, which predicts a decrease in the degree of deformation with reduced interfacial tension. An increase in emulsifier concentration reduces the interfacial tension (provided the concentration of surfactant is less than the CMC). Therefore, it was expected that the surfactant concentration would have a pronounced effect on film formation below the CMC. This was not observed, despite a clear effect of surfactant concentration on latex surface tension. The latexes were not dialyzed to replace the serum before surfactant post-addition. It is possible that the concentration of pseudosurfactant and other water-soluble species may have masked any effect of post-added surfactant. The sulfate ion terminated oligomeric or polymeric pseudosurfactant is produced from the ammonium persulfate used during the emulsion polymerization.

The films were also examined using scanning electron microscopy. The concentration of anionic surfactant did not have an effect on the degree of film deformation. However, increasing the level of the nonionic surfactant had a marked effect, as shown in Figure 5. The interstitial regions are smeared, giving a superficial appearance of a more fully fused film. However, closer examination of the two micrographs leads to the conclusion that the increased fusion is localized only in the region at the particle surfaces. The particles clearly retain their original shape. This is in contrast to the micrograph of Figure 3 (D_n=318 nm) where the original spherical particle cannot be detected. Because of their surface activity, the emulsifier molecules will tend to reside at the particle / water interface. It was postulated that the polymer was locally plasticized at the particle surface by the nonionic surfactant. The plasticization of polymer M6 by NP40 was confirmed by dynamic mechanical spectroscopy, as shown in Figure 6.

The Drying Process. Visual observation indicates that the physical processes of film coalescence and drying occur simultaneously. The presence of water during film formation is implicit in the published models of film formation (with the exception of the dry sintering model of (*1*)). Evidently, water is present (at least initially) during the film formation process and will have an impact on the parameters in equation 2.

The concurrent processes of film fusion and drying were explored using environmental scanning electron microscopy (ESEM) (*15*). This technique is unique in that it allows imaging of latexes in an aqueous environment. The drying and film formation of latexes M1 and M13 were followed in real time. In the case of

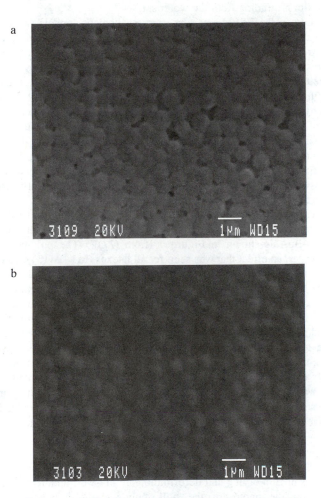

Figure 5. Scanning Electron Micrographs of Latex M6 Films Containing NP40 Nonionic Surfactant: (a) 0.0035 g NP40/g polymer; (b) 0.0199 g NP40/g polymer.
Reproduced with permission from reference (*11*).

latex M1 (synthesized with an excess of CBr_4), a skin formed at the sample surface before water evaporation was complete. This indicated that film formation and drying were occurring simultaneously. In the case of latex M13 (synthesized with an excess of EGDM) the water front was observed to recede from the particles, leaving a barely fused film protruding from the water surface.

In addition to the ESEM imaging, two mechanistic models (16,17) of film drying were evaluated. In the model proposed by Vanderhoff et al (16), a coalesced skin of latex is formed once the particles are closest-packed. Further water loss occurs by diffusion through the fused layer. In the model proposed by Croll (17), a drying front retreats through the film, leaving a 'dry' layer at the film surface which increases in thickness over time. The film continues to dry by water vapor percolation through the channels between particles. Gravimetric water loss studies in (15) showed that drying rate was independent of the polymer architecture. For example, latexes M1 and M13 dried at the same rate, despite vast differences in polymer character. This result supports the percolation drying mechanism of Croll. However, the ESEM observation of M1 latex skin formation is inconsistent with this model. It was postulated that the polymer skin remained sufficiently porous (or hydrophilic) to allow unhindered water flux.

The presence of water in the drying film has a direct impact on polymers which are hydrophilic in nature. The effect of hydroplasticization on the modulus of bulk pMMA-co-BuA is shown in Figure 7 (from (11)). The polymer is clearly plasticized by water, an effect that was seen for all latexes (M1 - M13) studied.

The model proposed here assumes that the polymer composition is uniform throughout the latex particle and hence, that hydroplasticization occurs uniformly. This is not an unreasonable assumption for low levels of methacrylic acid comonomer. However, in general, this may be an unrealistic assumption. The carboxylic acid(s) used for stabilization, along with residual ionic initiator fragments will have the tendency to locate at the particle surface. There is likely to be localized hydroplasticization at the particle / water interface, regardless of the hydrophobicity of the bulk polymer. This effect has not been accounted for in this research, but warrants further attention. In addition, an equilibrium quantity (<2 mass %) of water persists in the latex film, long after it is macroscopically dry at ambient relative humidity. The effect of this residual water is not clear.

Rigorous Model Evaluation

The relative rates of film drying and deformation are not well understood as yet and are dependent on the nature of the bulk polymer, the polymer architecture, environmental conditions, and more subtle factors such as particle inhomogeneity. Given the wide range of conditions for film formation, it is unlikely that a single mechanism (i.e., wet versus dry sintering) is satisfactory for all conditions. Consequently, we chose to evaluate the model of equation 2 at the extremes possible for latex film formation (18). At one extreme, deformation follows drying. That is, the capillary force initiates the film formation process in the presence of water. Once the contact between the particles has been established, film formation continues by a dry sintering mechanism. The polymer / air interfacial tension and the dry viscoelastic properties then apply. At the other extreme, it is assumed that the

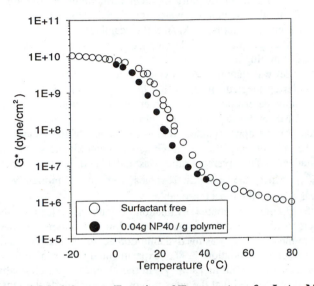

Figure 6. Modulus as a Function of Temperature for Latex M6 and Latex M6 with NP40 Nonionic Surfactant Post-Added. Data from (11).

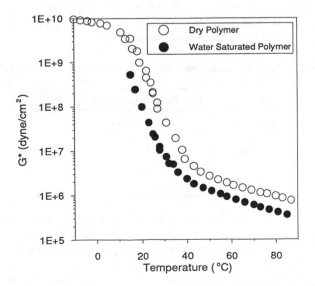

Figure 7. Hydroplasticization of *p*MMA-*co*-BuA (Polymer M6). Data from (11).

polymer retains a significant quantity of water throughout the second stage of film formation. Then, the polymer / water interfacial tension and hydroplasticized viscoelastic parameters apply. Again, the capillary force initiates the initial deformation. After particle / particle contact is established, further deformation is driven by a wet sintering mechanism.

The model was quantitatively evaluated with the capillary force and either a wet or dry sintering mechanism occurring in conjunction. This was accomplished by experimentally determining the parameters in equation 2. The moduli were measured by DMS under dry and wet (hydroplasticized) conditions (11). The wet moduli could only be approximated, due to the nature of the measurement. It was found that the ratio of log G^*_{dry} to log G^*_{wet} was approximately constant for all the polymers studied. The average ratio was calculated, yielding the relationship $G^*_{wet} = G^{*0.90}_{dry}$. Since the individual values were not completely reliable, this approach was deemed the most reasonable.

In the preliminary model evaluation, calculated values of the interfacial tensions were used. The polymer surface tension and polymer / water interfacial tensions were measured (19) using dynamic contact angle analysis and are used in the rigorous model evaluation. The experimentally determined values of the interfacial tensions are given in Table V. In the absence of an alternative approach, it was necessarily assumed that the latex polymer / water interfacial tension could be measured from the contact angle between dry pMMA-co-BuA and water. This approach assumes that a clearly defined interface exists. Water plasticization of the hydrophilic particle surface was disregarded. Experimental evidence suggests that that this assumption has questionable validity: Measurements of the dynamic polymer / water contact angle in (19) indicated a strong interaction between the polymer films (M4, M6, and M10) and water. The films imbibed water upon immersion, indicating that water plasticization did occur.

Table V: Experimentally Determined Values of Interfacial Tensions

	Interfacial Tension (dyne·cm^{-2})
$\sigma_{water\ phase}$	48.8
$\gamma_{polymer/air}$	24
$\gamma_{polymer/water}$	28

The latexes of Table IV were chosen to evaluate the model. It was felt that these materials represented an exacting test since the moduli ranged over two orders of magnitude. In addition, the polymers had varying levels of crosslinking. Table IV shows the effect that the molecular architecture had on the minimum film temperature. An increase in the concentration of chain transfer agent (either CBr_4 or CI_4) caused a reduction in MFT from the control latex (M6). Use of EGDM crosslinking monomer yielded a corresponding increase in MFT. The effect of latex particle size on the MFT was confirmed by the latex pairs M1/M2 and M10/M11 where M1 and M10 were used as seed latexes in the polymerizations of M2 and

M11, respectively. In the case of both pairs of latexes, there was a corresponding increase of the MFT with particle diameter.

Scanning electron micrographs of films prepared from the latexes of Table IV are shown in Figure 8, which shows the clear relationship between the degree of film formation and the polymer architecture. The model of equation 2 was again evaluated by comparing the predicted and experimental contact radii for films dried at room temperature (22°C). In order to represent the data in two dimensions, equation 2 was condensed into the simplified expression shown in the ordinate which is in terms of the particle radii and the dry polymer modulus, where η^* is replaced by G^*/ω. For the fully fused latexes (M1-M4), the radii are calculated on the following basis: Two spheres having radii R deform to a single sphere having radius R'. The relationship between the radii is then R'=1.26R.

Figure 9 shows that the experimentally measured contact radii fell within the range predicted by the model. As Table I shows, the particle size was maintained over a fairly narrow range. Therefore, the changes in the degree of deformation were primarily determined by variation of the viscoelastic properties of the materials. The best agreement between the model and experiment was for the dry sintering mechanism. The wet sintering process predicted the films to be far more deformed than was actually measured by experiment. This suggests that our model overestimates the forces promoting particle deformation. This may due to the assumption that the particles are homogeneously plasticized by water. Recent studies on a vinyl acetate / butyl acrylate system showed relatively fast film formation that appeared to be dominated by the surface regions of the particles (20).

Figure 9 indicates that the model predicts the degree of film fusion for a wide range of polymer characteristics. However, the model has several shortcomings. The model is highly sensitive to the values of the viscoelastic parameters. It is valid only for small strains (a criterion that is clearly violated for some of the latexes). In order to choose the appropriate values for G^* and η^*, the time scales of the deformation and flow processes must be known accurately (they are not). Some of these shortcomings could be addressed by a more sophisticated approach to the problem. For instance, the process of film formation could be modeled in three dimensions, accounting for polymer viscoelasticity, surface energetics, the geometry of deformation, and drying kinetics. Clearly, this would be a daunting task.

As mentioned earlier, hydrophilic polymer will reside in a shell at the particle surface. The assumption of uniform plasticization is likely an oversimplification. We assume a hard sphere model for latex particles. In actuality, the character of the hydrophilic shell at the particle surface is poorly defined. The true nature of the surface region is obviously very important to the ongoing study of film formation.

Recently, several research groups have addressed the question of the special role that water plays in the film coalescence process (21,22,23). Dobler *et al* (22) have demonstrated that particle coalescence can occur in a wet environment under the influence of polymer / water interfacial tension. The same authors studied film formation under standard conditions (*i.e.*, with water evaporation) (23). They found that neither polymer / water interfacial tension (under wet conditions) nor polymer / air interfacial tension (under dry conditions) contributed to the film coalescence process under limiting conditions of temperature and relative humidity. These

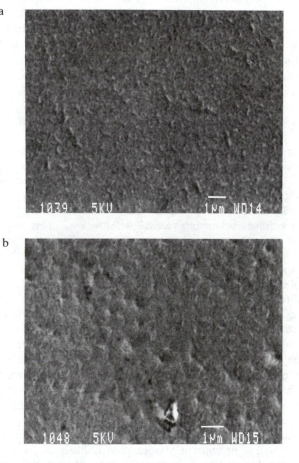

Figure 8. Scanning Electron Micrographs of Latex Films
Synthesized with Molecular Weight Modifiers:
(a) Latex M1, 2.4% CBr_4 (based on polymer)
(b) Latex M4, 0.5% CBr_4; (c) Latex M10, 1.4% EGDM
(d) Latex M11, 1.4% EGDM; (e) Latex M13, 5.5% EGDM
Reproduced with permission from reference (18).

c

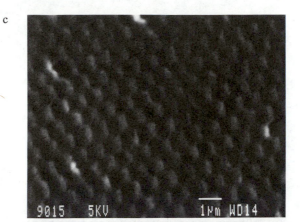

d

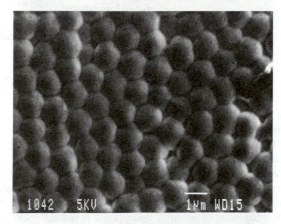

e

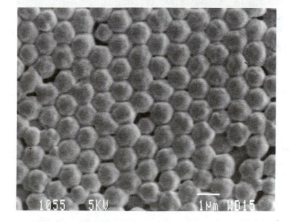

Figure 8. Continued.

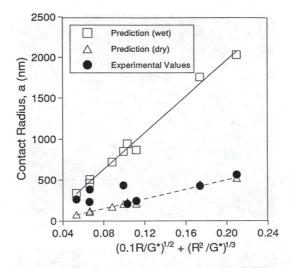

Figure 9. Contact Radius as a Function of the Model Parameters for Film Formation Under Wet and Dry Conditions. Data from (18).

authors felt that polymer modulus alone determined the fusion process and that the mechanism proposed by Sheetz (5) was most reasonable.

More recently, other researchers have revisited the concept of film formation under the influence of polymer / air interfacial tension (*i.e.*, a dry sintering mechanism). Sperry *et al* (24) suggest that water is not necessary for film formation to occur. In a series of simple experiments, they found that hydrophobic polymers will form a film by a dry sintering mechanism. These authors suggest that the only effect of water is hydroplasticization of hydrophilic polymers. Similarly, Mazur and Plazek (25) invoke a dry coalescence mechanism in a recent publication. These authors essentially focus on a dry sintering mechanism appropriate for powder coatings. The initial response is found to be purely elastic with the viscous component only contributing at very long times.

In general, the current research that is published in this volume and elsewhere (22, 23, 24) recognizes the need to develop a latex film formation mechanism that is appropriate to realistic film formation conditions.

Conclusions

We have studied the process of film formation with respect to the nature of the latex polymer (particle size, viscoelasticity, hydroplasticity) and the kinetics of drying and deformation. We have extended earlier models of film formation to a comprehensive model where capillary forces and sintering mechanisms both contribute to film coalescence. We recognize that film formation does not occur under unique conditions (i.e., by either a 'wet' or 'dry' mechanism). In fact, the degree of film fusion will depend on the nature of the polymer (e.g., hydrophilicity and molecular weight) as well as the environmental conditions of film formation (relative humidity and temperature). Consequently, we have evaluated the model at the limits that would be expected at realistic film formation conditions. In particular, we have focused on the effects of changing latex particle size and modulus. We have demonstrated that the model is consistent with experimental observation over a wide range of conditions, despite some limitations.

Acknowledgments

The authors would like to thank the Natural Sciences and Engineering Research Council of Canada for financial support of this research.

Literature Cited

1.   W. E. Dillon, D.A. Matheson and E.B. Bradford, *J. Coll. Sci.*, **1951**, *6*, 108.
2.   G. L. Brown, *J. Polym. Sci.*, **1956**, *22*, 423.
3.   J. W. Vanderhoff, H.L Tarkowski, M.C. Jenkins and E.B. Bradford, *J. Macromol. Chem.*, **1966**, *1*, 361.
4.   G. Mason, *Br. Polym. J.*, **1973**, *5*, 101.
5.   D.P. Sheetz, *J. Appl. Poly. Sci.*, **1965**, *9*, 3759.
6.   J. Lamprecht, *Coll. Polym. Sci.*, **1980**, *258*, 960.
7.   K. Kendall and J.C. Padget, *Int. J. Adhes. Adhes.*, **1982**, *2*, 149.
8.   S.T. Eckersley and A. Rudin, *J. Coat. Tech.*, **1990**, *62(No. 780)*, 89.
9.   S.T. Eckersley, G. Vandezande and A. Rudin, *JOCCA*, **1989**, *7*, 273.
10.   M.J. Devon, T. Provder, and A. Rudin In *Particle Size Distribution II. Assessment and Characterization*; Editor, T. Provder; ACS Symposium   Series 472; American Chemical Society: Washington, DC, 1991; p. 135.
11.   S.T. Eckersley and A. Rudin, *J. Appl. Polym. Sci.*, **1993**, *48*, 1369.
12.   T.F. Protzman and G.L. Brown, *J. Appl. Poly. Sci.*, **1960**, *4*, 8.
13.   D.P. Jensen and L.W. Morgan, *J. Appl. Poly. Sci.*, **1991**, *42*, 2845.
14.   D.K. Owens and R.C. Wendt, *J. Appl. Poly. Sci.*, **1969**, *13*, 1741.
15.   S.T. Eckersley and A. Rudin, *Prog. Org. Coat.*, **1994**, *23*, 387.
16.   J.W. Vanderhoff, E.B. Bradford and W.K. Carrington, *J. Poly. Sci., Symp.*, **1973**, *41*, 155.
17.   S. Croll, *J. Coat. Tech.*, **1986**, *58(No. 734)*, 41.
18.   S.T. Eckersley and A. Rudin, *J. Appl. Polym. Sci.*, **1994**, *53*, 1139.
19.   S.T. Eckersley, R. O'Daiskey and A. Rudin, *J. Coll. Int. Sci.*, **1992**, *152*, 455.
20.   G. del Rio, *Ph.D. thesis*; University of Waterloo: Waterloo, Ont., 1995.
21.   D.J. Meier and F. Lin, *Langmuir*, **1995**, *11*, 2726.
22.   F. Dobler, T. Pith, M. Lambla and Y. Holl, *J. Coll. Int. Sci.*, **1992**, *152*, 1.
23.   F. Dobler, T. Pith, M. Lambla and Y. Holl, *J. Coll. Int. Sci.*, **1992**, *152*, 12.
24.   P.R. Sperry, B.S. Snyder, M.L. O'Dowd, and P.M. Lesko, *Langmuir*, **1994**, *10*, 2619.
25.   S. Mazur and D.J. Plazek, *Prog. Org. Coat.*, **1994**, *24*, 255.

Chapter 2

Mechanisms of Particle Deformation During Latex Film Formation

F. Dobler[1] and Y. Holl[2]

[1]Centre de Recherches Rhône-Poulenc, 52 rue de la Haie Coq,
93308 Aubervilliers Cedex, France
[2]Institut Charles Sadron, Centre National de la Recherche Scientifique
and University of Strasbourg, 4 rue Boussingault, 67000 Strasbourg,
France

The aim of this article is to review the literature dealing with the deformation step in the phenomenon of film formation from a latex. It will be shown that the situation of this problem in the literature is quite confusing. The article will be divided in two parts. The first will recall the classical theories proposed in the past to account for the deformation of latex particles in the film formation process, namely the dry sintering theory by Dillon et al., Brown's capillary theory, the wet sintering theory by Vanderhoff et al., and the surface layer theory by Sheetz. The second part will deal with following works which were all attempts to experimentally verify and/or improve the main theories.

One can distinguish three steps in the process of film formation from a latex (*1, 2*).
i) In the first step, water evaporates at a constant rate until a stage is reached where particles form a dense packing of spheres. The solid volume fraction, for a latex monodispersed in size, is then close to 0.74. An additional condition for high packing fraction is sufficient colloidal stability. Otherwise, a less ordered structure with lower polymer volume fraction will result.
ii) At the beginning of the second step, particles show at the surface of the latex, and the rate of water evaporation decreases. Forces start to act which ensure the deformation of the particles in such a way that polymeric material fills all the space. Acting forces have to surmount the mechanical resistance of the particles against deformation. The spheres are transformed into rhombic dodecahedra. A rhombic dodecahedron has 12 rhombic faces, and 14 vertices of two types. Six vertices correspond to 4-fold axes, and 8 to 3-fold axes. It is not easy to draw this polyhedron. A schematic representation can be found in reference 3. They can also be observed in beautiful freeze fracture micrographs of poly(butyl methacrylate) latex films (*4*). At this stage, interfaces between particles still exist (*5*). For latexes, the compaction/deformation step is favorable from a thermodynamical point of view

0097–6156/96/0648–0022$15.50/0

because of the strong decrease of the total area of the particle-water or particle-air interfaces.

There are other processes which show some analogies with the deformation step in latex film formation, for example the drainage of foams and the sintering of metallic powders. There are two kinds of foams, wet and dry ones. In wet foams, the bubbles are spherical, the volume fraction of liquid is high, around 40 %. By drainage, the liquid volume fraction is decreased, the cells become polyhedral, a dry foam is formed. An important problem, still nowadays, concerns the shape of the cells in dry foams. What shape leads to the minimum surface for a given volume? A solution to this problem was proposed by Kelvin in the 19th century (*6*), for monodisperse cells disposed in a cubic centered lattice, the "tetrakaidecahedron". This "barbaric" polyhedron possesses 14 faces, 6 squares and 8 hexagons. It took 63 years to recognize that this solution was erroneous (*7*). Recently (*8*), another solution was proposed, based on computer simulations, and confirmed experimentally. It consists in an association of two kinds of polyhedral cells, a tetradecahedron (2 hexagonal and 12 pentagonal faces) and a dodecahedron with pentagonal faces. Obviously, there are marked differences between foams and latexes. In latexes, the viscosity of the particles is enormously higher and the interfacial tension is generally lower than in foams. However, one can ask the questions : "Can the study of foam structure be helpful in the field of latexes?" "Are other shapes than the rhombic dodecahedron possible in monodisperse latex films?"

Sintering of metallic powders is a process well known in metallurgy (*9*). It consists in the compression of a metallic powder in a mold at a temperature well below the melting temperature, leading to a continuous, or at least with low porosity, metallic part. The sintering mechanisms are discussed in the simple case of two spherical particles in contact (Fig. 1) (*10, 11*). Three mechanisms have been proposed. The metal vapor pressure over a convex surface (P_1, Fig.1) is higher than the vapor pressure over a flat surface whereas it is lower over a concave surface. P_1 is thus higher than P_2. A "vaporization-condensation mechanism" can then transport metallic atoms from the convex surfaces to the particle contact zone, thus increasing the size of the contact zone. The second mechanism is analogous to the first one for vacancies. The vacancy concentration is lower near a convex surface than near a concave one (Thomson Freundlich law), $C_1 < C_2$. Vacancies migrate in order to equalize the concentrations. This, again, corresponds to a transport of matter towards the contact zone. The third mechanism is based on Laplace's equation which states that the pressure near the center of one particle [P(o)] is higher than the pressure in the contact zone [P(x)] [see Vanderhoff's mechanism (*12, 13*) in the following part]. This pressure gradient is also the cause of an increase of the size of the contact zone. With all these mechanisms : $X^p/R^q = Kt$; with X = radius of the contact zone, R = radius of the particles, p,q depending on the particular mechanism considered, and K= function of temperature, t = time. In the latex film formation process, only the third mechanism can be operative and we shall see latter on that it was proposed indeed (*12, 13*).

iii) The third step of the latex film formation corresponds to the evolution of the interfaces between particles. They tend to disappear by interdiffusion of the macromolecules from one particle to the neighbors. It is sometimes called maturation or further gradual coalescence or autohesion. Film properties like mechanical strength and permeability are altered during this step (*14, 15*). Some authors introduce more steps in the latex film formation. For instance, Joanicot et al. (16) split the third step in two parts which they call coalescence and interpenetration. However, the description of the process in three steps is the most widely accepted and will be retained in this paper.

The aim of this article is to review the literature dealing with the second step of the latex film formation process, i.e. the deformation step. It will be shown that the situation of this problem in the literature is still quite confusing, despite the fact that it has been tackled for the first time many years ago, in the early fifties (17). Yet, it is very important from both scientific and practical points of view. To illustrate the practical importance of the deformation step mechanisms, let us give the following example. It will be shown in the second part of this paper that, currently, one of the most controversial points is to know whether deformation occurs in the presence or in the absence of water. This seems a quite academic question. It is not, because it determines the mechanisms by which the surfactant will be distributed in the final latex film. And surfactant distributions are extensively studied nowadays because of their strong influence on properties of latex films (18-20).

The subject of deformation of latex particles is often briefly reviewed in the introduction of papers dealing with latex film formation but without enough details (21, 22); or the review can be more extensive but incomplete. For instance, both Eckersley and Rudin (23) and Rios (24) do not even quote Sheetz's contribution (25) which we consider as a major one. Thus, it seemed to us that the deformation step deserved a proper review paper. The objectives of the article are the following.

- Present a review of the literature from the most classical contributions to the most recent ones. However, being perfectly exhaustive is not among our goals.
- Propose some personal views on a certain number of points, on capillary pressure, on deformation criteria, on the role of the particle-air interfacial tension, and so on. Doing this, we expose ourselves to the criticism of subjectivity. We assume the risk.
- Arouse a renewed interest of the scientific community working in the field of latex films for the deformation step and stimulate further work on this problem.

The article is divided in two main parts. The first one presents the classical theories proposed in the past to account for the deformation of latex particles in the film formation process, namely the dry sintering theory by Dillon et al. (17), Brown's (26) capillary theory, the wet sintering theory by Vanderhoff et al. (12, 13), and the surface layer theory by Sheetz (25). The second part will deal with following works which were all attempts to experimentally verify and/or improve the main theories.

The Main Classical Theories

Dry Sintering. This theory was proposed in 1951 by Dillon et al. (17).Its key points are the following.
- The latex dries before particles loose their shape.
- Particles are deformed by the particle-air interfacial tension, estimated at 30 mN/m. Particles are submitted to a pressure P given by $P = C. \gamma_{PA} / R$, C being a constant, γ_{PA} the particle-air interfacial tension, and R the radius of the particles.
- Particles loose their shape by viscous flow according to Frenkel's law (27) :
  $\theta^2 = 3\gamma_{PA} t / 2\pi\eta R$ θ defined in figure 2, t = time, η = viscosity of the polymer.

According to Dillon et al. (*17*), the pressure P is important enough to ensure the deformation of particles with radius less than 100 nm.

A drawback of this approach is that it ignores the viscoelastic character of the polymer. Brown (*26*) criticized this theory. He stated that deformation can also occur for slightly crosslinked particles where viscous flow is impossible. However, his main criticism of the dry sintering theory was against the fact that the particles deform in air rather than in water. Brown considered as well established that evaporation of water and deformation are concurrent phenomena and that water evaporation plays a major role in film formation.

Capillarity. The capillary theory was proposed by Brown in 1956 (*26*). Brown stated that in given conditions some latexes form films whereas latexes of other characteristics do not form films. Similarly, from a given latex a film can be obtained under certain sets of conditions and not under different ones, for instance at lower temperatures. Thus, there are forces which promote particle deformation and forces which resist it. Deformation will occur as long as promoting forces are stronger than resisting ones. Promoting forces are, according to Brown, Van der Waals attraction (F_{VW}), gravity (F_G), surface tension forces (F_S), and capillary forces (F_C). Resisting forces are electrostatic repulsion (F_{EL}) and elastic resistance of the particle against deformation (F_R). Deformation will take place when the following condition is always fulfilled :

$$F_{VW} + F_G + F_S + F_C > F_{EL} + F_R$$

Brown considered F_{VW}, F_G, F_S, and F_{EL} as negligible. The condition reduces then to :

$$F_C > F_R$$

The capillary force was calculated via the capillary pressure. For this purpose, Brown took into consideration the small circle of radius r included in the space between three contacting particles (Fig. 3). The radius of curvature of the water - air interface was taken equal to r and thus the pressure over this concave surface was $P_C = 2 \gamma_{WA} / r$, with γ_{WA} = water-air interfacial tension. Simple geometrical considerations allow to relate r to R (the radius of the particles) in the following way : $r = 0.155R$ (see reference 23 for explanation). The capillary pressure becomes $P_C = 12.9 \gamma_{WA} / R$. This expression is based on the incorrect and unnecessary assumption that the radius of curvature of the water - air interface equals r (see next section). It was used by several authors after Brown, including Mason (*28*), Lamprecht (*29*) and Eckersley and Rudin (*23*). The capillary force was calculated by multiplying the capillary pressure by the area of the surface of contact between particles. This is obviously erroneous, as was emphasized by Mason (*28*) several years later. On the other hand, the elastic resistance of the particles was determined by assuming a purely elastic behavior for the polymer. A quantitative condition for deformation was inferred in the following form :

$$G < 35 \gamma_{WA} / R \qquad \text{G being the shear modulus of the particle.}$$

This theory was and probably still is the most popular one. However, it suffers from many weaknesses. i) The first is the erroneous calculation of capillary pressures and forces. We shall come back to this point in the second section of this chapter. ii) The second weakness is the assumption of elastic spheres, when everybody knows the viscoelastic character of polymers. Lamprecht (*29*) calculated a new quantitative criterion for deformation taking into account the viscoelasticity of the particles. And Lamprecht's calculation was further improved by Eckersley and Rudin (*23*).

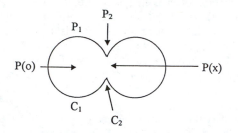

Figure 1: Sintering of two spherical metallic particles. $P_{1,2}$ = metal vapor pressures. P(o) = pressure at the center of the particle. P(x) = pressure in the contact zone. $C_{1,2}$ = vacancy concentrations.

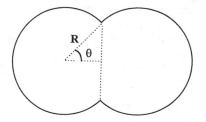

Figure 2: Dry sintering of two polymeric particles according to Dillon et al. (17). R = radius of the particle. θ = half angle of the contact zone.

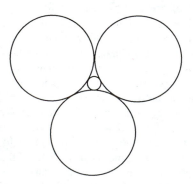

Figure 3: Top view of the drying latex showing three contacting particles with a capillary full of water in between.

iii) Another problem, as pointed out by Sheetz (*25*), arises from the assumption of a zero polymer-water contact angle. In general, the capillary force acting on a particle along what is called the triple line, the line where three phases (air, polymer, water) are in contact, has a vertical (normal) and a horizontal (parallel) component (Fig. 4). When the contact angle is not equal to zero, the parallel component of the capillary force increases whereas the normal component decreases. The case where the level of water is beneath the equatorial plane of the particles is not considered here. If the contact angle is higher than 45°, the parallel component is always larger than the normal component. iv) The main criticism formulated by several authors (*23, 25, 30*) against Brown's theory is that capillary forces are not strong enough to ensure complete deformation of the particles, even when the radius is small. This question is still controversial nowadays (*31*), as will be discussed later on.

Wet Sintering. The wet sintering theory was developed by Vanderhoff et al. (*12, 13*) in an attempt to explain deformation of large particles. In this case, the main driving force for deformation is the particle-water interfacial tension. It is the deformation of two spheres in contact which was considered (Fig. 5). Laplace equation was used to calculate the pressure gradient $P_2 - P_2'$ (see figure 5 for definition of symbols) in the following way :

$P_2 - P_1 = 2\gamma_{PW} / r$ with γ_{PW} = particle-water interfacial tension, r = radius of the particle;

$P_2' - P_3 = \gamma_{PW}(1/r_2 - 1/r_1)$ r_1, r_2 defined in figure 5;

$P_3 = P_1$;

$\Delta P = P_2 - P_2' = \gamma_{PW}(1/r_1 - 1/r_2 + 2/r)$

The radius of curvature r_1 being extremely small, in the range of a few nanometers, ΔP is positive and can be quite high. This pressure gradient pushes polymeric material from the particles towards the contact zone, thus increasing the size of this zone. Vanderhoff et al. then considered two cases, model I where r_1 is constant during coalescence of the two particles and model II where r_1 progressively increases (Fig. 6). Results for model I are shown in figure 7. Figure 7 indicates that the pressure gradient ΔP increases when coalescence proceeds but tends towards a limit which is even quicker reached when the radius of the particle increases. The opposite is obtained in model II (Fig. 8) : the pressure gradient decreases when θ increases and this decrease is even faster when the radius of the particles is more important.

When spheres come in touch during evaporation of water, there is only one point of contact and the mechanism described above is not yet operative. In order to explain the start of the deformation, Vanderhoff et al. invoked the presence of a thin layer of water around the particles (Fig. 9). In this situation, the water-air interfacial tension is the driving force for deformation, and the expression of the pressure gradient can be written like this :

$\Delta P = \gamma_{WA}[1/r_1 - 1/r_2 + 2/(r +s)]$ with s = thickness of the water layer.

Two remarks can be made about the wet sintering theory. i) Experimental evidences exist (*21, 25*) that particle-water interfacial tension is indeed able to ensure particle deformation but only to a limited extend. ii) Precise experimental testing of Vanderhoff's models is difficult because, as pointed out by the authors themselves,

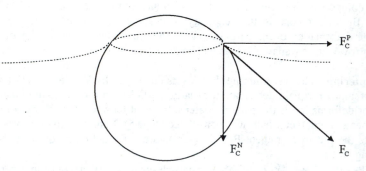

Figure 4: Capillary force F_C acting on a particle. F_C^P and F_C^N are the parallel and nornal components of the capillary force, respectivelly. The dashed line represents the water level.

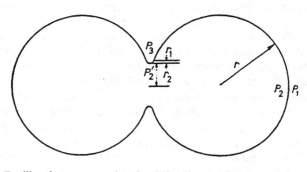

Figure 5: Radii of curvature involved in the coalescence of two spheres. Reproduced with permission from reference 13. Copyright 1970 SCI.

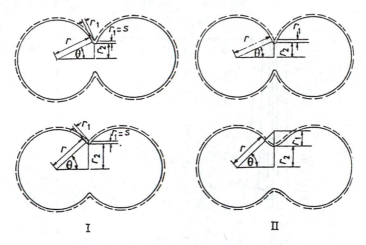

Figure 6: Models for the coalescence of two spheres. (I) r_1 constant; (II) r_1 increasing with θ. Reproduced with permission from reference 13. Copyright 1970 SCI.

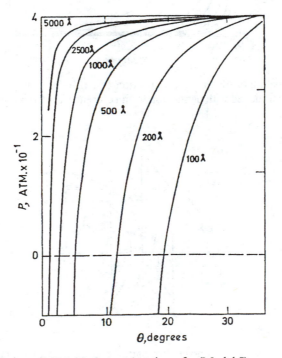

Figure 7: Variation of ΔP with θ as a function of r (Model I). r_1 = constant = 25 Å; γ_{PW} = 10 mN/m. Reproduced with permission from reference 13. Copyright 1970 SCI.

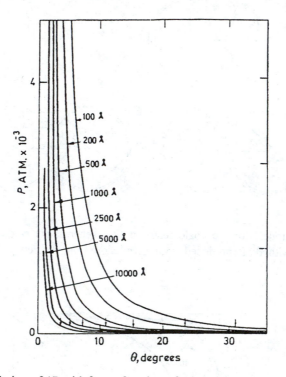

Figure 8: Variation of ΔP with θ as a function of r (Model II, r_1 = variable); γ_{PW} = 10 mN/m. Reproduced with permission from reference 13. Copyright 1970 SCI.

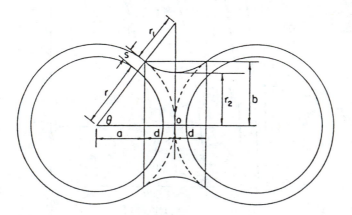

Figure 9: Two spheres in contact after partial evaporation of water. Reproduced with permission from reference 13. Copyright 1970 SCI.

separating the effects of water-air and particle-water interfacial tensions is tricky. Such a verification was only rarely tried.

Surface Layer. The last classical explanation for particle deformation was proposed by Sheetz in 1965 (*25*). In his article, Sheetz first emphasized the importance of the horizontal component of the capillary force acting on a particle. He then stated that the capillary forces, especially the horizontal components, provoke the deformation of the particles of the top layer in such a way that a continuous polymeric film covers the drying latex. In order to continue to leave the latex, water has to diffuse through this membrane. This gives rise to a normal compressive work which compresses the whole array of particles, leading to deformation. The magnitude of this work, in an isothermal and reversible process, is given by

$$W = - PdV$$

with P = vapor pressure of water over the film and dV = decrease of the volume of the latex due to loss of water. The energy is supplied as heat from the surroundings. Sheetz (*25*) performed a rapid calculation for 10 g of a hypothetical sample which properties are : density = 1.00 g / cm^3; particle diameter = 0.2 µm; solids = 75%; γ_{PA} = 30 mN/m; γ_{PW} = 5 mN/m. The free energy changes associated with the decrease of the particle water interface, with the decrease of the particle-air interface, and with the evaporation of water are 1.13 J, 6.75 J, and 344 J, respectively. The later contribution is the most important and sufficient to ensure deformation of the particles.

Sheetz also described an experience which supports his views. A MMA/n-BA copolymer latex was cast on a support. A portion of this wet film was covered directly with a thin water permeable acrylic film. Another portion of the same wet latex was also covered with the thin film but an air gap was maintain between the two films. Both portions of the latex dried at the same rate (10h at 24°C). The latex film directly covered was continuous and clear whereas the other part, not in direct contact with the thin film, was cloudy and discontinuous.

Surprisingly, Sheetz 's theory aroused little interest in the scientific community. In the early seventies, only Yeliseeva (*32*) claimed that she agreed with Sheetz's description of the film formation process.

Table I summarizes the key features of the four classical theories.

Following Works

As indicated in the introduction, the articles published after the main classical theories presented above were all attempts to verify and/or improve them. No really new theory was proposed after the sixties. The works which will be discussed now are all more or less related to one of the four main theories. Relation with a particular theory will be the classification criterion we shall adopt in this second part. Thus, we shall present first the articles connected to the dry sintering theory, then to the capillary theory, and so on. This presentation will be interrupted by a couple of more personal remarks.

Dry Sintering. The sintering of particles obtained from a freeze-dried latex was studied by dynamic scanning calorimetry (*33, 34*). The loss of particle-air interfacial area was observed by an exotherm in the DSC diagram. Stewart and Johnson (*35*)

Table I : The main classical theories. Summary.
In bold characters, the driving force for deformation.

- **Bradford 1951 : DRY SINTERING**

  LATEX DRIES BEFORE PARTICLES DEFORM
  PARTICLE-AIR INTERFACIAL TENSION
  PARTICLE = VISCOUS LIQUID

- **Brown 1956 : CAPILLARY THEORY**

  CAPILLARY PRESSURE
  PARTICLE = ELASTIC SOLID
  FIRST DEFORMATION CRITERION

- **Vanderhoff 1966 : WET SINTERING**

  PARTICLE-WATER INTERFACIAL TENSION
  (and water -air interfacial tension)

- **Sheetz 1965 : SURFACE LAYER**

  CLOSURE OF THE SURFACE LAYER (CAPILLARY FORCES)
  EVAPORATION OF WATER → COMPRESSION WORK

have also studied the dry sintering of a freeze-dried polystyrene latex by DSC. Sintering occurs at 102° and is completed within 1.2 min. On the other hand, they calculated the total time required for sintering, using the equation derived by Vanderhoff et al. (*12, 13*) for their wet sintering theory, and found 0.95 min. It was claimed that this result supported Vanderhoff's equation when the particle-water interfacial tension is replaced by the polymer surface tension.

In this section, a very important paper is the one by Sperry and coworkers (*22*). These authors studied the particle deformation step by minimum film formation temperature (MFT) measurements. The main objective of the paper was to address the question of the role of water in the deformation step. Does water play a special role in the process, as was claimed by Brown, Vanderhoff or Sheetz? Several kinds of polymers were used, rather hydrophobic ones like ethyl acrylate/styrene and butyl acrylate/styrene copolymers or a butyl methacrylate homopolymer and more hydrophilic ones like ethyl acrylate/methyl methacrylate or butyl acrylate/methyl methacrylate copolymers. All those systems also contained 1 wt% of acrylic acid and a small amount of surfactant. MFT were measured versus nature of the polymer, time, particle size, and water content of the deposited film and in the drying environment. Latexes were cast on the MFT bar in the conventional wet form (condition W, as noted by the authors). Or they were cast and dried on the gradient bar well below the glass transition temperature of the polymer, this resulting in a dry array of non deformed particles and then only the bar was powered (condition D). The test environment was either dry , a high flow of dry air in a large volume above the film (noted D again), or humid, open to room air, 65 % relative humidity at 25 °C (noted H). Four sets of conditions were then used, ranging from very dry, DD (no water effects) to very wet, WH (conventional wet conditions). The main findings were that, for hydrophobic polymers, the wet MFT, measured under WH conditions, is essentially identical to the dry MFT, measured under DD conditions. The authors concluded that water does not play a special role in the process. For them, film formation can be separated in two independent components, namely water evaporation and particle deformation in the dry stage. Deformation is ensured by particle-air interfacial tension and Van der Waals attraction between particles. It was also claimed that the results of dry MFT versus time measurements supported the Johnson-Kendall-Roberts model (*36*) of particle adhesion and deformation. On the other hand, for hydrophilic polymers, the wet MFT was found lower than the dry MFT by around 10°C. According to the authors, this result could be rationalized in terms of plasticization by water. "These effects have been found to coincide with earlier reports on the changes in polymer modulus and Tg as a function of water content."

Size effects on MFT measurements were also studied for one polymer, poly(butyl methacrylate). A few literature results are incidentally recalled. Some authors do not observe any MFT versus particle size effect (*37*), some authors observe a small one (*23*), and finally, some can see a significant one (*38*). Sperry et al. report a rather important effect, 31°C for 350 nm particles to around 40°C for 700 nm ones. This was only detected in dry MFT measurements. "The increase of time-dependent dry MFT with increasing particle size correlates with a simple viscous flow model expressing the longer time required for the larger interstitial voids to shrink to a size that results in visual transparency of the film."

Sperry et al.'s paper (22) is certainly a major one in the field of latex film formation mechanisms. However, two important criticisms can be raised against it. The first one is that for the hydrophobic polymers considered in this study, MFT is rather high, above 40°C. Even under what the authors call humid conditions, the drying kinetics could be quite high, giving not enough time to possible water effects to be operative. It would then not be surprising that dry and wet MFT values became identical. The butyl methacrylate homopolymer is expected to behave like the hydrophobic copolymers. However, its dry and wet MFT values are different, rather like the hydrophilic polymers. This could be due to a drying rate effect as mentioned by the authors themselves, because the Tg of the polymer is closer to ambient temperature. This casts some doubt on the conclusions drawn for hydrophobic polymers. Clearly, possible drying rate effects on MFT should be considered. This is another controversial point in MFT measurements, one more. Some authors observe a decrease of MFT when water evaporation is slowed down (39, 40) and some do not (37, 41). The second criticism arises from the work by Lin and Meier (31) on capillary forces operating in a latex film. These authors claim that capillary pressure can be very large even when the water content in a latex film is very low. The question is "Are conditions DD in Sperry et al.'s work dry enough to avoid condensation of a small amount of water in the film during particle deformation, this water giving rise to a considerable capillary pressure?"

In the introduction of this article, we mentioned that we consider the question of the presence or absence of liquid water at the precise place where deformation occurs as crucial. This point is raised in the discussion of Sperry et al.'s paper. The authors give a clear answer :"We propose that this is an environment in which liquid water is absent." There are some rather convincing experimental evidences supporting this view, under certain circumstances. Keddie et al. (42, 43) have studied the film formation process by multiple-angle-of-incidence ellipsometry and environmental scanning electron microscopy. They were looking for kinetics and rate limiting steps in the phenomenon. According to their results, the important parameter is the position of the Tg with respect to the temperature of the film formation process. When the film is formed well above Tg, evaporation of water and deformation of particles occur concomitantly and evaporation of water is the rate limiting step. When Tg and film formation temperature become close, there is no longer the evaporation of water which is the rate limiting step but the deformation of the particles. In other words, water evaporates first and then a polymer surface tension driven deformation of the particles occurs. Furthermore, the results in the latter case are in agreement with a model based on the viscous flow of polymers. When water evaporates first, there is a water front receding from the top of the drying latex towards the support, leaving an array of air surrounded almost undeformed particles. Those findings are supported by preliminary results of Reffner et al. (44). Sperry et al.'s call (22): "Critical experiments that define the particle environment at the onset of deformation (...) are certainly called for." seems to have been heard. The answer to the question "Presence or absence of water?" seems to be: "It depends on the drying conditions with respect to the particle characteristics." More work is required to confirm and precise this point.

Capillary Theory. An attempt was made in 1966 to verify the effect of the water-air interfacial tension on the film formation process (45). A poly(vinyl acetate) latex was dried at 6°C, below MFT, extensively washed to get rid of the surfactant, dried again and then redispersed either in pure water (γ_{WA} = 70 mN/m) or in a 1% surfactant solution (γ_{WA} = 26 mN/m). A film was formed at 21°C, above MFT, with both latexes. The film obtained from the dispersion in pure water had better mechanical properties than the other. This work was a first proof of the importance of the water-air interfacial tension.

Coalescence Criteria. Coalescence criteria, the first having been proposed by Brown, have aroused much interest in the latex community. Brodnyan and Konen (*37*) tried a verification of Brown criterion in the following way. In limit conditions for film formation, according to Brown's criterion, the term $G(MFT).R / \gamma_{WA}$ should be constant; G(MFT) is the shear modulus at MFT, R is the radius of the particles. The authors have measured G(MFT) for different values of R and γ_{WA}. Unfortunately, results were not convincing because of a lack of precision in the measures of G(MFT).

Mason (*28*) recalculated the capillary forces, and by taking into account the deformation of the particles, inferred a new form of deformation criterion :

$$G < 266 \, \gamma_{WA} / R$$

Lamprecht (*29*) was the first to take into account the viscoelastic character of the polymer. He distinguished two cases : a film formation process taking place at constant stress or at constant strain rate. In the first case, he calculated the radius a of the contact zone as the following :

$$a^3(t) = \frac{3R}{16} \int_{-\infty}^{t} J(t-\tau) \frac{dF_C}{d\tau} \cdot d\tau \quad \text{where J(t) is the creep function and Fc the}$$

capillary force taken from Mason (*28*). The deformation condition is then :

$$a^3 (t^*) > 0.0294 R^3 \quad t^* \text{ being the time necessary for a complete evaporation of}$$

water;
and the deformation criterion becomes :

$$1 / J(t^*) < 95 \, \gamma_{WA} / R$$

In the second case, a process at constant strain rate, Lamprecht calculated the force which resists deformation :

$$F_R(t) = \frac{16}{3R} \int_{-\infty}^{t} G(t-\tau) \frac{da^3}{d\tau} \cdot d\tau \quad \text{with}$$

$a(t) = R^3 (v_0 t / 3d)^{3/2}$ where d is the thickness of the array of close packed undeformed particles and v_0 the evaporation rate by unit area of water / air interface. The deformation condition being $F_R(t) < F_C$, one ends with :

$$1.33 (v_0/d)^{3/2} \cdot \int_{0}^{t^*} G(t-\tau) \tau^{1/2} d\tau \, \langle \, P_C \quad \text{where } t^* = 0.283 \, d / v_0 \text{ and}$$

$P_C = 12.9 \, \gamma_{WA} / R$ the expression given by Brown (*26*) for the capillary pressure.

Lamprecht's viscoelastic approach was used by Eckersley and Rudin (*23*) with a slight modification. According to these authors, Lamprecht's deformation condition, a^3 (t*) > 0.0294R^3, is erroneous and should be written $a^3 (t^*) > 0.0836R^3$. The deformation criterion then becomes

$$1 / J(t^*) < 34 \, \gamma_{WA} / R$$

Again, Brown's expression for the capillary pressure was used by Eckersley and Rudin. This expression is based on incorrect assumptions. It should be calculated in the following way.

Capillary Pressure. The calculation starts with the determination of the capillary force acting vertically on a particle partially showing at the surface of a drying latex (see figure 4). The force acts along the triple line, where air, water and polymer are in contact. The driving force is the difference between the particle-air (γ_{PA}) and the particle-water (γ_{PW}) interfacial tensions. When γ_{PA} is higher than γ_{PW}, which is the case for latex particles, the system tends to pull the particle downwards in order to minimize interfacial energies. The magnitude of the force is proportional to the length of the triple line and to the difference $\gamma_{PA} - \gamma_{PW}$, it can then be written

$Fc = 2\pi r(\gamma_{PA} - \gamma_{PW})$ with r = radius of the circle made by the triple line. By using Young's equation

$\gamma_{PA} - \gamma_{PW} = \gamma_{WA} \cos\theta$ with γ_{WA} = water-air interfacial tension, θ = contact angle of water on the polymer; the capillary force can be written as

$Fc = 2\pi r \gamma_{WA} \cos\theta$

This expression corresponds to the exact solution of the capillary rise problem (*46*). In our case, there is an additional complication due to the curvature of the sphere. One has to add a term $\cos\alpha$, α being the angle between Fc and F_C^N (see figure 4), in order to get the vertical component of the capillary force. The final equation is

$F_C^N = 2\pi r \gamma_{WA} \cos\theta\cos\alpha$

This equation could easily be checked experimentally using the Wilhelmy technique (*47-49*).

The calculation of the pressure, now, consists in choosing an appropriate surface where the total magnitude of the capillary forces is easy to determine and to divide this magnitude by the area of the surface. Let us consider the simple situation where water is at the level of the equatorial plane of the first layer of particles ($\cos\alpha = 1$), before deformation of the particles has started. The surface is a rectangle (Fig. 10) which sides are 2R and $2\sqrt{3}$ R, R being the radius of the particles. Inside of this rectangle, one finds the equivalent of two particles. The magnitude of the capillary forces is thus

$Fc = 2 \times 2\pi R \gamma_{WA} \cos\theta$. The capillary pressure is

$Pc = 4\pi R \gamma_{WA} \cos\theta / 4\sqrt{3}R^2$.

If $\cos\theta = 1$, $Pc = 1.8 \gamma_{WA} / R$. The numerical coefficient is quite different from the value of 12.9 given by Brown. If θ and α are different from zero, it is even lower. The consequence is that all deformation criteria using Brown's equation for capillary pressure cannot be right.

Remark Concerning Deformation Criteria. Deformation criteria were all established in the framework of the capillary theory and were all of the general form

$G < K \gamma_{WA} / R$ K being a numerical coefficient ranging from 34 to 266 depending on the particular assumptions made by the different authors. These criteria are difficult to establish for several reasons. It is difficult to precisely take into account and characterize the deformation of the particle. The viscoelastic approaches by Lamprecht (*29*) and Eckersley and Rudin (*23*) are probably oversimplified and non linear effects should probably also be taken into account (*50, 51*). Furthermore, quantification of deformation driving forces is not yet really possible. On the other hand, the deformation criteria are difficult to verify because this requires a precise measure of moduli at MFT, a region where they are highly temperature dependent. Another difficulty arises from the problem of plasticization of the polymers by water. This point was raised by several authors (*22, 35, 41, 52*). It is often meaningless to use moduli measured on dry films in order to check a deformation criterion. For all those reasons, efforts committed to establish and verify deformation criteria are of little interest at the present stage.

Let us now come back to the papers dealing with Brown's capillary theory. The ones by Eckersley and Rudin are well known and important. In their first article (*23*), they calculated the highest values of moduli at which deformation was still possible according to Brown's or Mason's models. Those values are significantly lower than the experimentally measured ones. They also measured the radius of the contact zone between particles by SEM and compared the values with calculated ones based on capillary (see above) and interfacial forces. Again, they observed that radii determined on the basis of capillary forces alone were well below experimental values. The agreement became better when they added a contribution from particle-water interfacial tension. Their conclusions were that: "The capillary force acting alone is insufficient to effect film coalescence. However, if the capillary force is accompanied by an interfacial driving force, there is sufficient coalescing force to cause the observed degree of film fusion." In subsequent work, they concentrated on a more rigorous testing of their model (*53*). They carefully considered the effect of hydroplasticization, thus improving the accuracy of the measurements of particle moduli, either in a wet or in a dry environment (*54*). They also studied drying mechanisms (*55*) showing, like Keddie et al. (*42*), that, roughly speaking, soft particles are deformed in the presence of water and harder particles remain almost undeformed when water has left. Consequently, they considered either a wet or a dry sintering process as the additional contribution to capillary forces. The various interfacial tensions involved in the deformation process were more accurately determined (*56*). A new testing of their model involving capillary and interfacial forces was performed (*57*). The conclusion was that the fit with experience was satisfactory. Our opinion is that, despite of all improvements, it is hard to be convinced by Eckersley and Rudin's model. Among the most important criticisms which can be raised, we would like to mention the followings. i) The parameter time was not clearly discussed. Was the radius of the contact zone considered as a constant after a certain period of time? Or was the time between latex casting and SEM observation kept constant? Those points should be clarified. ii) All is based on an observation of the surface of the latex. Is the surface really representative of what occurs in the bulk of the film? Many authors, including ourselves, doubt it. iii) There is such an uncertainty concerning the actual deformation conditions, whether rather dry or rather wet, that an approximate fit between extreme conditions (dry and wet, see figure 2 in reference 57) is hardly convincing.

Another argument against capillary forces can be found in reference 30. Dobler et al. have established a limit film formation diagram for model latexes (*58*) in terms of temperature and relative humidity (Fig. 11). The model particles have a well defined core shell structure (*59*) with 10, 15 or 25 wt% of methacrylic acid in the shell. The latexes are called CS10, CS15 and CS25, CS standing for core-shell. The important point here is that all latexes fall on the same line, regardless of the composition of the

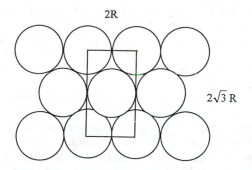

Figure 10: Top view of the first layer of particles in a drying latex. Rectangular surface used for the calculation of the capillary pressure.

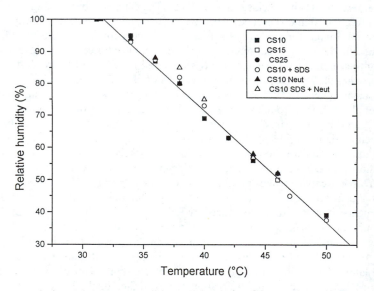

Figure 11: Limit coalescence conditions (T/RH) for CS10, CS15, CS25 purified latexes and for CS10 with SDS adsorbed or neutralized by NaOH or both (neutralized and SDS adsorbed). CS stands for core-shell. The core is a styrene / butyl acrylate copolymer and the shell a styrene / butyl acrylate / methacrylic acid terpolymer. In CS10, CS15 and CS25 latexes, the shells contain 10, 15 and 25 wt% of methacrylic acid, respectively. In the upper part of the diagram, latexes do form films, in the lower part they do not. Reproduced with permission from reference 30. Copyright 1992 Academic Press.

shell, the presence or absence of adsorbed surfactant, the neutralization of the methacrylic acid in the shell. When SDS is added to the CS10 latex at a concentration of 1.6 g / liter, one can calculate (*30*) that the capillary force is divided by a factor of 2. Under the assumption that capillary forces alone ensure deformation of the particles and on the basis of the variation of the modulus of the particles with temperature (*60*), it is possible to calculate that, at a given relative humidity, the limit temperature for deformation should be higher for the latex containing SDS by 4°C. This difference is significant and is not observed experimentally. This seems to indicate that capillary forces alone are not strong enough to deform the particles, a conclusion similar to that of Eckersley and Rudin.

Above, we presented a way to calculate the capillary pressure when the water phase is continuous in the drying latex. Another approach was proposed by Lin and Meier (*31*) in the case where the water phase becomes discontinuous and only forms rings of water around the contact zones between particles. Lin and Meier's paper elegantly renews the vision of the role of capillary pressure in the latex particle deformation process. The authors used a model latex of poly(iso-butyl methacrylate), a highly hydrophobic polymer, with a Tg of 65°C. They determined Tg of both wet and dry latexes by DSC and could not detect any difference in the measured values. From this, they concluded that poly(iso-butyl methacrylate) was not at all plasticized by water. The deformation kinetics was established via AFM measurements of peak-to-valley distances (corrugation heights) on the top of a monolayer of latex particles deposited on mica. The corrugation height progressively decreases as deformation proceeds. Particles deformed either in a completely dry or in a wet environment. The wet situation was obtained by exposing the dry particles to a water saturated atmosphere. Lin and Meier observed that deformation takes place 10 times faster for the wet than for the dry particles. This difference was attributed to the capillary pressure arising from the presence of water. The authors calculated capillary pressures and forces versus water contents and polymer-water contact angles. They showed that capillary pressures, and forces to a lesser extend, increase when water content decreases and can reach very high values with seemingly insignificant amounts of water in the interstitial regions of spherical particles. "Furthermore, (...) capillary pressure in wet systems can be of sufficient magnitude to explain the observed deformation of the latex particles at 50°C, i.e., 15°C below Tg." The authors conclusion was affirmative. "We maintain that capillary pressure typically is the dominant driving force for film formation."

Lin and Meier (*31*) also raised a geometrical consideration concerning particle deformation. They showed by AFM that, at least at the surface, the deformation process does not change the particle center-to-center spacing. A similar conclusion was drawn by Winnick and coworkers (*61*). The question is always the same. To what extent is the surface of a latex film representative of what happens in the bulk?

We are presently in a situation where some authors contest the capillary theory whereas some others strongly support it. The question of the presence or absence of water at the point where particles deform becomes even more critical.

Wet Sintering.   The first who verified the possibility of deforming latex particles under the sole influence of particle-water interfacial tension was Sheetz (*25*). Latexes were agglomerated in dialysis bags to a solids content of 62.3 % and then placed in deionized water and maintained for 2 hours at 36°C. The solids content, determined gravimetrically, raised to 84.8 %. Further heating at 50°C for 2 more hours only increased the solids content to 87.2 %. When the agglomerated latex was put in a 1 % surfactant solution instead of pure water for 2 hours at 36°C, the solids content increased to 66.2 %.

Mason (28) criticized Vanderhoff's wet sintering theory (12, 13) on the basis that forces and pressures were confused. We shall not develop this point.

It was already mentioned above that Stewart and Johnson (35) used Vanderhoff's approach to study the dry sintering of latex particles. Their experiments and calculations seemed to validate Vanderhoff's model.

The more extensive experimental study of wet sintering was performed by Dobler et al. (21). They followed the solids content increase of arrays of model latex particles in water as a function of time, temperature, surfactant adsorption and methacrylic acid neutralization. The model latexes had core-shell particles with the same diameters and the same bulk properties but different surface characteristics. It was confirmed that particle-water interfacial tension was indeed able to provoke deformation of the particles, even at temperatures below the glass transition temperature. The kinetics of deformation was closely related to interfacial tension. It was shown that the rate decreased when interfacial tension decreased, i.e., when the amount of methacrylic acid increased in the shell or when sodium dodecyl sulfate was adsorbed on the particles or when the methacrylic acid was neutralized by sodium hydroxide. Activation energies of deformation in water were measured. They are equal to the activation energy of the motions of the polymer at the glass transition as determined by dynamic mechanical spectrometry (60).

This work allowed Dobler et al. to verify Kendall and Padget theory of deformation of elastic spheres (62). According to Kendall and Padget, the diameter of the contact zone between particles, d, is given by

$$\frac{4d^3}{D^2} \times \frac{E}{\gamma(1-\nu^2)} = 18\pi$$

E being the Young's modulus of the polymer, ν the Poisson's ratio, γ the surface tension and D the diameter of the particles. The latex film is fully formed when d=D. For Eckersley and Rudin (23), this latter condition is rather $d^3=0.0836D^3$. For the CS10 latex at 32°C, the shear modulus is 10^8 Pa (60). If Kendall's equation and Rudin's conditions are used, it is found that total deformation would only be possible for γ values above 600 mN/m. Such values are higher than usual particle-water interfacial tensions by a factor of around 100 and impossible to reach. Thus, the theory proposed by Kendall and Padget is not applicable to particle deformation in water.

Surface Layer. Sheetz's theory of surface layer (25) has aroused only little interest in the literature. Even if some authors report observations which seem to support Sheetz's theory, they do not relate them to Sheetz's work. For instances : "It was found that the latex synthesized with excess CBr_4 formed a coalesced skin at the film surface while water was still present in the drying film." (53); or "This fits in with the idea of a skin of partly-coalesced latex near the surface existing above a reservoir of water near the substrate interface." (43). It is only in Dobler et al. (30) that one can find an explicit support to Sheetz theory.

Following their work with model latexes in water (21), Dobler et al. have studied the film formation process in standard conditions, i.e., with evaporation of water, always using the same latexes. They compared film formation kinetics in water to those when water simultaneously evaporates and established the limit conditions for film formation in terms of temperature and relative humidity (Fig 11). It was demonstrated that in limit conditions, particle-water interfacial tension has a negligible contribution to the mechanism of latex film formation when film formation and water evaporation are concurrent. Evaporation of water seems to be at the origin of forces which ensure deformation of particles. It was shown that the origin of particle deformation is probably not the capillary forces which can develop at the surface of the latex when a solid volume fraction around 75% is reached. This conclusion was drawn after preliminary research and it would be important to further investigate this point in order to establish it more firmly. Dobler et al.'s results seem to support Sheetz's theory of film formation. A simple visual observation of the latices during drying was used as a source of information. Whatever the latex, with or without SDS, one could observe the apparition of iridescent spots at the surface when the solids content reached 25-30%. These spots then grew in size until the whole surface became iridescent. If, during the extension period of the spots, the relative humidity was raised to 100% in order to stop evaporation of water, the spots stopped growing and started to decrease in size from the edges. A central part always remained, the size of which was larger when water evaporation had been stopped later in the course of the drying process. If relative humidity was decreased to allow water to evaporate again, the spots grew again. However, once the whole surface had become iridescent, the iridescence was irreversible. In Sheetz's theory, closing of the surface takes place when the particles are already close packed. This seems not to be the case in Dobler's system. The surface of the latices becomes iridescent far before the stage of close packing is reached. This iridescence shows that particles organize regularly at the surface. When it covers the entire surface, iridescence is irreversible. This shows that particles come into contact in an irreversible manner. When in contact, they deform and form a superficial polymeric film. Drying then can occur by diffusion of water through the surface layer and, when the solid volume fraction reaches 74%, deformation of the whole array of particles can start. The iridescence is observed at a solid volume fraction always less than 40% for all latices. Thus, the compression phenomenon is identical for all latices, which explains that the limit conditions are identical in all cases. Further experimental evidence for surface closure is desirable. Fluorescence could be a valuable technique to use.

Conclusion

The problem of particle deformation mechanisms in the phenomenon of film formation from a latex is very important and interesting from both scientific and practical points of view. However, despite several major recent contributions, the present situation is still extremely confusing. One can find articles supporting all existing main theories : dry and wet sintering, capillarity, surface layer. And some authors propose to combine several different driving forces. No group has yet succeeded in fully convince the whole scientific community involved in polymer colloid science. There is at least one point on which everybody should agree : mechanisms are complex and depend on the nature of the latex and on the film formation conditions.

It appears that one of the key points, currently, is to know whether particles deform in the presence or in the absence of water. This question is related to the problem of drying mechanisms of latexes and emulsions in general. Although an old problem (*63-65*), it is not clear and still under investigation (*55, 67*).

Another crucial issue is to consider the latex film formation mechanisms in filled systems (*68*). In spite of its practical importance, this problem seems still in its infancy in scientific literature.

As one can see, there are lots of stimulating questions remaining open for future research work.

Literature Cited

1. Bradford, E.B.; Vanderhoff, J.W. *J. Macromol. Chem.* **1966**, *1*, 335.
2. Kast, H. *Makromol. Chem., Macromol. Symp.* **1985**, *10/11*, 447.
3. Roulstone, B.J.; Wilkinson, M.C.; Hearn, J.; Wilson, A.J. *Polym. Internat.* **1991**, *24*, 87.
4. Wang, Y.; Kats, A.; Juhué, D.; Winnick, M.A. *Langmuir* **1992**, *8*, 1435.
5. Distler, D.; Kanig, G. *Colloid Polym. Sci.* **1978**, *256*, 1052.
6. Thomson, W.(Lord Kelvin) *Phil. Mag.* **1887**, *25*, 503.
7. Matzke, E. *Bulletin of the Torrey Botanical Club* **1950**, *77*, 222.
8. Weaire, D.; Phelan, R. *Phil. Mag. Letters* **1994**, *69*, 107.
9. Benjamin, J.S.; Bomford, M.J. *Metall. Trans.*, **1977**, *8A*, 1301.
10. Coble, R.L. *J. Appl Phys.* **1970**, *41*, 4798.
11. Johnson, D.L.; Cutler, I.B. *J. Am. Ceram. Soc.* **1963**, *46*, 541.
12. Vanderhoff, J.W.; Tarkowski, H.L.; Jenkins, M.C.; Bradford, E.B. *J. Macromol.Chem.* **1966**, *1*, 131.
13. Vanderhoff, J.W. *Br. Polym. J.* **1970**, *2*, 161.
14. Voyutskii, S.S. *J. Polym. Sci.* **1958**, *23*, 528.
15. Voyutskii, S.S.; Ustinova, Z.M. *J. Adhesion* **1977**, *9*, 39.
16. Joanicot, M.; Wong, K.; Maquet, J.; Chevalier, Y.; Pichot, C.; Graillat, C.; Lindner, P.; Rios, L.; Cabane, B. *Prog. Colloid Polym. Sci.* **1990**, *81*, 175.
17. Dillon, R.E.; Matheson, L.A.; Bradford, E.B. *J. Colloid Sci.* **1951**, *6*, 108.
18. Kientz, E.; Holl, Y. *Colloid Surface* **1993**, *78*, 255.
19. Niu, B.J.; Urban, M.W. *J. Applied Polym. Sci.* **1995**, *56*, 377.
20. Charmeau, J.Y.; Kientz, E.; Holl, Y. *Prog. Org. Coating* **1995**, in press.
21. Dobler, F.; Pith, T.; Lambla, M.; Holl, Y. *J. Colloid Interface Sci.* **1992**, *152*, 1.
22. Sperry, P.R.; Snyder, B.S.; O'Dowd, M.L.; Lesko, P.M. *Langmuir* **1994**, *10*, 2619.
23. Eckersley, S. T.; Rudin, A. *J. Coating Technol.* **1990**, *62 (780)*, 89.
24. Rios-Guerrero, L. *Makromol. Chem., Macromol. Symp.* **1990**, *35/36*, 389.
25. Sheetz, D.P. *J. Appl. Polym. Sci.* **1965**, *9*, 3759.
26. Brown, G.L. *J. Polym. Sci.* **1956**, *22*, 423.
27. Frenkel, J. *J. Phys. (USSR)* **1943**, *9*, 385.
28. Mason, G. *Br. Polym. J.* **1973**, *5*, 101.
29. Lamprecht, J. *Colloid Polym. Sci.* **1980**, *258*, 960.
30. Dobler, F.; Pith, T.; Lambla, M.; Holl, Y. *J. Colloid Interface Sci.* **1992**, *152*, 12.
31. Lin, F.; Meier, D.J. *Langmuir* **1995**, *11*, 2726.
32. Yeliseeva, V.I. *Polymerization of film forming materials;* Khimiya Press: Moscow, 1971.
33. Mahr, T.G. *J. Phys. Chem* **1970**, *74*, 2160.
34. Bertha, S.L.; Ikeda, R.M. *J. Appl. Polym. Sci.* **1971**, *15*, 105.
35. Stewart, C. W.; Johnson, P.R. *Macromolecules* **1970**, *3*, 755.

36. Johnson, K.L.; Kendall, K.; Roberts, A.D. *Proc. Roy. Soc. London* **1971**, *A 324*, 301.
37. Brodnyan, J.G.; Konen, T.J. *J. Appl. Polym. Sci.* **1964**, *8*, 687.
38. Jensen, D.P.; Morgan, L.W. *J. Appl. Polym. Sci.* **1991**, *42*, 845.
39. Protzman, T.F.; Brown, G.L. *J. Appl. Polym. Sci.* **1960**, *4*, 81.
40. Myers, R.R.; Schultz, R.K. *J. Appl. Polym. Sci.* **1964**, *8*, 755.
41. Powell, E.; Clay, M.J. *J. Appl. Polym. Sci.* **1968**, *12*, 1765.
42. Keddie, J.L.; Meredith, P.; Jones, R.A.L.; Donald, A.M. *Macromolecules* **1995**, *28*, 2673.
43. Keddie, J.L.; Meredith, P.; Jones, R.A.L.; Donald, A.M. *Proceedings of the American Chemical Society, PMSE* **1995**, *73*, 144.
44. Reffner, J.R.; Fu, Z.; Boczar, E.M.; Lesko, P.M.; Kirk, A.B. *Proceedings of the American Chemical Society, PMSE* **1995**, *73*, 364.
45. Redknap, E.F.; *J. Oil Color Chem. Assoc.* **1966**, *49*, 1023.
46. Adamson, A.W. *Physical Chemistry of Surfaces;* John Wiley & Sons, New York, Chichester, Brisbane, Toronto, Singapore, 1990; p 14.
47. Wilhelmy, L.; *Ann. physik* **1863**, *119*, 117.
48. Andrade, J.D.; Smith, L.M.; Gregonis, D.E. In *Surface and Interfacial Aspects of Biomedical Polymers*; Andrade, J.D., Ed.; Plenum Press: New York and London, 1985, Vol. 1; p. 249.
49. Vergelati, C.; Perwuelz, A.; Vovelle, L.; Romero, M.A.; Holl Y. *Polymer* **1994**, *35*, 262.
50. Kuczynski, G.C.; Neuville, B.; Toner, H.P. *J. Appl. Polym. Sci.* **1970**, *14*, 2069.
51. Mazur, S.; Plazek, D.J. *Prog. Organic Coating* **1994**, *24*, 225.
52. Kan,C.S. *TAPPI Notes, Adv. Coat. Fundam.* **1993**, 101.
53. Eckersley, S.T.; Rudin, A. *Proceedings of the American Chemical Society, PMSE* **1995**, *73*, 3.
54. Eckersley, S.T.; Rudin, A. *J. Appl. Polym. Sci.* **1993**, *48*, 1369.
55. Eckersley, S.T.; Rudin, A. Prog. Org. Coat. **1994**, *23*, 387.
56. Eckersley, S.T.; Rudin, A.; O'Daiskey, R. *J. Colloid Interface Sci.* **1992**, *152*, 455.
57. Eckersley, S.T.; Rudin, A. *J.Appl. Polym. Sci.* **1994**, *53*, 1139.
58. Dobler, F.; Pith, T.; Holl, Y.; Lambla, M. *J. Appl. Polym. Sci.* **1992**, *44*, 1075.
59. Dobler, F.; Affrossman, S.; Holl Y. *Colloid Surface* **1994**, *A89*, 23.
60. Dobler, F. *PhD dissertation* **1991**, University of Strasbourg, France.
61. Goh, M.C.; Juhue, D.; Leng, O.M.; Wang, Y.; Winnick, M.A. *Langmuir* **1993**, *9*, 1319.
62. Kendall, K.; Padjet, J.C. *Int. J. Adhes. Adhes.* **1982**, *2*, 149.
63. Vanderhoff, J.W.; Bradford, E.B.; Carrington, W.K. *J. Polym. Sci. Symp.* **1973**, *41*, 155.
64. Croll, S. *J. Coating Technol.* **1986**, *58 (734)*, 41.
65. Croll, S. *J. Coating Technol.* **1987**, *59 (751)*, 81.
66. Chevalier, Y.; Pichot, C.; Graillat, C.; Joanicot, M.; Wong, K.; Maquet, J.; Lindner, P.; Cabane, B. *Colloid Polym. Sci.* **1992**, *270*, 806.
67. Feng, J.; Winnick, M.A. *Proceedings of the American Chemical Society, PMSE* **1995**, *73*, 90.
68. Joanicot, M.; Cabane, B.; Wong, K. *Proceedings of the American Chemical Society, PMSE* **1995**, *73*, 143.

Chapter 3

Geometric Considerations in Latex Film Formation

Edwin F. Meyer III

ICI Paints, Strongsville Research Center, 16651 Sprague Road, Strongsville, OH 44136

The theoretical study of latex film formation involves the transformation of a sphere into a polyhedron which, along with its neighbors, fills all space. The type of polyhedra formed depends upon the manner in which the latex particles are packed together after the film is dried. If the lattice is known, the polyhedron, and hence its geometric parameters, can be determined. These geometric parameters can then be used to model the degree of particle deformation and to calculate the interparticle forces, pressures and contact areas at various stages of the film formation process. Of the two lattices that close-pack spheres, the face-centered cubic is by far the most popular in the literature. However, hexagonal close packed is another close-packing possibility that should be considered.

The theoretical study of latex film formation involves the transformation of a sphere into a polyhedron which, along with its neighbors, fills all space. The type of polyhedra formed depends upon the manner in which the latex particles are packed together after the film is dried. If the lattice is known, the polyhedron, and hence its geometric parameters, can be determined. These geometric parameters can then be used to model the degree of particle deformation and to calculate the interparticle forces, pressures and contact areas at various stages of the film formation process. Of the two lattices that close-pack spheres, the face-centered cubic is by far the most popular in the literature. However, hexagonal close packed is another close-packing possibility that should be considered.

0097–6156/96/0648–0044$15.00/0
© 1996 American Chemical Society

The process of latex film formation has been divided into three[1], four[2], and five[3] steps. These steps involve particle ordering, evaporation of the water, particles sticking to each other, deformation of the particles, coalescence of the particles and finally interpenetration of the cores of the particles. It is understood that these steps overlap somewhat. That is, there is particle deformation before the evaporation is complete and there is some interdiffusion while the particles are deforming. To characterize the film formation process researchers have developed models to represent the physical phenomena involved. Virtually all the models use the rhombic dodecahedron (RDH) as the shape to which the latex spheres ultimately conform. This shape is shown in figure 1. The RDH has 12 equivalent rhombi as its sides. Using this shape as a model, researchers can calculate contact areas, the length of contact edges, lattice spacing and various other geometric parameters as a function of the deformation and then relate these back to physical parameters such as surface tension, contact pressure, and evaporation rate. However, the RDH will be the ultimate shape of the spherical latex particles only if the particles are packed in a face-centered cubic (FCC) lattice. The FCC lattice is one of two lattices that close-pack spheres. The other is the hexagonal close-packed (HCP) lattice. If the latex particles are arranged in a hexagonal close packed (HCP) lattice then the final shape is a trapezo-rhombohedron (TRH), which is shown in figure 2. The TRH has six trapezoidal sides and six rhombic sides. In actual latex films the structure into which the particles ultimately settle after the evaporation process depends upon many parameters. The final packing structure is often a mixture of more than one lattice. To get an understanding of the geometry of the film formation process it would be helpful to examine each possible lattice individually. For simplicity we will focus only on the two lattices that close pack spheres: FCC and HCP. Both of these lattices are characterized by 12 nearest neighbors and a packing fraction of $\pi\sqrt{2}/6$ which is about 74%. These two lattices can be thought of as multiple layers of triangular nets. A net is the two-dimensional counterpart of a lattice. Two layers of triangular nets will not determine the nature of the three dimensional packing. The differentiation comes in the third layer. If the third layer is placed so the spheres are directly atop those in the first layer then we have a section of an HCP lattice. If the spheres in the third layer are placed over the intersticies of the first layer so that none of the spheres in the three layers are directly atop each other, then we have a

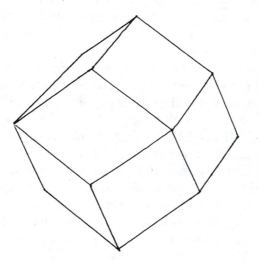

Figure 1. A rhombic dodecahedron

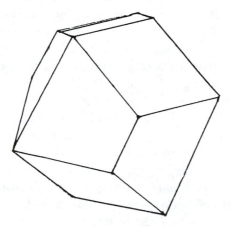

Figure 2. A trapezo-rhombohedron

section of an FCC lattice. If the initial layer is not a triangular net but a square net, then a HCP structure is not possible and thus the FCC is the only way to get close packing. However, a mono-layer of spheres will naturally conform to a triangular net [4] so it is reasonable to assume that surface at which latex film formation begins is a triangular net of spherical latex particles.

To compare the geometric characteristics of the RDH and the TRH it is helpful to have a length metric. The metric used here is the perpendicular distance from the center of the polyhedra to any of the twelve sides. Since this length is a characteristic pseudo-radius of the polyhedra we will call this length R. With this definition we can calculate the characteristics of the sides of the shapes (table 1).

Next we can calculate the edge length, surface area and volume of these two shapes as a function of R, the center to side distance (table 2).

So, an RDH and a TRH with the same center-to-side distance have the same total edge length, the same surface area and the same volume. Using the calculated volume we can compare the lattice spacing of the dodecahedra with the lattice spacing of the close-packed spheres before they were deformed. Obviously the center to center distance of the spheres is $2r$, where r is the radius of the sphere. The center to center distance of the dodecahedra in the lattice is $2R$, simply from the way we defined R. To compare R with r we have to make an assumption regarding the density change of the particles upon deformation and the isotropy of the deformation. It is common practice to assume that the mass density of the particle does not change upon deformation and that the deformation is isotropic. Since we are assuming no density change, we set the volume of the two geometrical shapes equal to the volume of the original latex particle. Setting the expressions for the volumes equal,

$$\left(\frac{8}{\sqrt{2}}\right) R^3 = \left(\frac{4\pi}{3}\right) r^3$$

and solving for the ratio r/R gives us,

$$\frac{r}{R} = \sqrt[3]{\frac{6}{\sqrt{2}\pi}} = 0.9047$$

So, if the deformation is isotropic, the lattice spacing of the dodecahedra in the final lattice is less than the lattice spacing of close-packed latex particles. This means that as the spheres deform their centers must get closer together.

Table   1: The characteristics of the sides of the rhombic dodecahedron (RDH) and the trapezo-rhobmohedron (TRH) with a center-to-side distance of R

	RHOMBIC SIDE		
SHAPE	long diag	short diag	edge
RDH	$2R$	$\sqrt{2}R$	$(\sqrt{6}/2)R$
TRH	$2R$	$\sqrt{2}R$	$(\sqrt{6}/2)R$
	TRAPEZOIDAL SIDE		
SHAPE	long edge	short edge	skew sides
TRH	$(2\sqrt{6}/3)R$	$(\sqrt{6}/3)R$	$(\sqrt{6}/2)R$

Table   2: The total edge length, surface area and volume of a rhombic dodecahedron (RDH) and a trapezo-rhombohedron (TRH) with a center-to-side distance of R

	Edge Length	**Surface Area**	**Volume**
RDH	$12\sqrt{6}R$	$\left(24/\sqrt{2}\right)R^2$	$\left(8/\sqrt{2}\right)R^3$
TRH	$12\sqrt{6}R$	$\left(24/\sqrt{2}\right)R^2$	$\left(8/\sqrt{2}\right)R^3$

As we have seen the rhombic dodecahedron and the trapezo-rhombohedron share many geometric features. An RDH and a TRH that have the same volume also have the same number of sides, number of edges (24), and number of corners (14). Of these 14 corners, each shape has eight corners where three edges come together and six corners where four edges come together. If the distance from the center of the RDH to the center of any side is R, the distance from the center to any of the eight three-edge corners is $\sqrt{3/2}R$ and the distance from the center to any of the six four-edge corners is $\sqrt{2}R$. The same thing is true of the TRH. Also, the interior angle between opposite sides of any four-edge corner is $90°$ and the interior angle between an edge of the three-edge corner and the line bisecting the opposite side is $125.26^{\circ}(90^{\circ} + \sin^{-1}(1/\sqrt{3}))$ for both shapes. The four-edge corners will be the location of the greatest deformation of the latex particle because the four edge corners are the furthest points from the surface of the original spherical latex particle and because the four-edge corners are sharper than the three-edge corners.

With all these similarities there are a couple of important difference between the two shapes. One involves the directionality of the deformation. In the symmetric rhombic dodecahedron the six four-edge corners are orthogonal. That is, they can be found along the x, y, and z axes of a cartesian coordinate system whose origin is at its center. This positioning spaces the four edge corners as far away from each other as possible. In the trapezo-rhombohedron the six four-edge corners are spaced, not at 90° from each other, but at 70.5° $((2\sin^{-1}(1/\sqrt{3}))$ as measured from the center of the TRH. The second difference involves the angle between the edges of the four-edge corners. In the rhombic dodecahedron the angle between the edges and the angle between the sides of the four-edge corners are all 90°. However, in the trapezo-rhombohedron the angle between the edges of the four-edge corners is 109.47°, the tetrahedral angle. So, the four-edge corners of the RDH are sharper than the four-edge corners of the TRH. These differences are due to the asymmetry of the trapezo-rhombohedron shape. The six trapezoids form a ring whose plane of symmetry is parallel to the original layer of latex particles. The consequence of this asymmetry is that hexagonal close packed spheres will have to deform to a greater extent in the direction along the layers of latex particles and to a lesser extent in the direction perpendicular to these layers. The symmetry of the face centered cubic lattice results in a uniform deformation of the particle in all six orthogonal directions.

The two main experimental techniques used to provide insight into the packing of latex particles are microscopy and neutron scattering. In the literature there are micrographs of a freeze fracture surface showing hexagonally packed particles each exhibiting three rhombi on the exposed surface[5]. These are being used as evidence supporting the prescence of rhombic dodecahedra. However, a lattice of trapezo-rhombohedra will give the same structure if fractured along its dominant fracture plane. Some micrographs do indeed provide conclusive evidence for the presence of the RDH shape[3], [5]. However, it seems the detection of a section of FCC packing is being cited as evidence that all latex particles are packed in an FCC lattice. The small angle neutron scattering results[3], which are cited as evidence of the FCC structure, are inconclusive with respect to the FCC vs HCP question. Indeed, the authors never address HCP packing; they support the face-centered cubic structure over the body-centered cubic (BCC) structure.

The rhombic-dodecahedron may indeed be the geometric destiny of the spherical latex particle but the current literature provides no conclusive evidence on this matter.

[1] "The Transport of Water Through Latex Films," Vanderhoff, J. W., Bradford, E. B., Carrington, W. K., J. Polym. Sci., Symposium No. 41, 155-174 (1973)

[2] "Film Formation with Latex Particles," Chevalier Y., Pichot C., Graillat C., Joanicot M., Wong K., Maquet J., Linder P., and Cabane B., Colloid Polym Sci 270: 806-821 (1992)

[3] "Ordering of latex particles during film formation," Joanicot M., Wong K., Maquet J., Chevalier Y., Pichot C., Graillat C., Linder P., Rios L. and Cabane B., Progr Colloid Polym Sci 81: 175-183 (1990)

[4] "Mecahnism of Formation of Two Dimensional Crystals from Latex Particles on Substrates." Denkov N. D., Velev O. D., Kralchevsky P. A., Ivanov I. B., Yoshimura H. and Nagayama, K., Langmuir 8, 3183-3190 (1992)

[5] "Freeze Fracture Studies of Latex Films Formed in the Abscence and Presence of Surfactant" Yoncai Wang, Aviva Kats, Didier Juhue and Mitchell A Winnik: Langmuir, 8, 1435-1442 (1992)

Chapter 4

Influence of Polar Substituents at the Latex Surface on Polymer Interdiffusion Rates in Latex Films

Mitchell A. Winnik

Department of Chemistry, University of Toronto, Toronto, Ontario M5S 1A1, Canada

We investigate the influence of surface properties of several different types of poly(n-butyl methacrylate) [PBMA] latex particles on the kinetics of polymer diffusion across the interface in their latex films. A series of core-shell PBMA microspheres were prepared, containing different amounts of methacrylic acid groups in their shell, or with poly(ethylene oxide) [PEO] chains at the surface. These latex are compared to PBMA latex with a surface containing only sparsely spaced $-OSO_3^-$ groups. Interdiffusion, examined by energy transfer measurements on labeled latex, was found to be significantly retarded by the presence of the acid groups or their salts in the latex shell. PEO chains at the latex surface were found to promote interdiffusion only in the early stages of film formation and aging. By contrast, PEG-containing non-ionic surfactant was found to act as a traditional plasticizer for PBMA to promote interdiffusion.

Of the various aspects of latex film formation and aging, the aspect that has seen the greatest progress over the past decade has been in measures of polymer interdiffusion across the interparticle boundary. Originally suggested by Voyutski (1) as the source of the growth in mechanical properties of latex films with time, the first demonstration of interdiffusion was provided by Hahn and coworkers (2) at BASF using small angle neutron scattering [SANS] to study PBMA latex films in which a fraction of the particles were deuterated. This technique has been used to good effect by Klein and Sperling (3) at Lehigh for the case of melt-pressed polystyrene latex films, and more recently by Joanicot et al (4) at Rhône-Poulenc for styrene-butyl acrylate films. Our approach (5) to this problem involves direct non-radiative energy transfer experiments, in which some (typically half) of the particles are labeled with a donor dye (D, e.g. phenanthrene, Phe) and the rest with an acceptor dye (A, e.g. anthracene, An). Interdiffusion brings D and A into proximity, making energy transfer possible when the film sample is excited with light absorbed selectively by D. In this chapter, I review DET results on interdiffusion in latex films, focusing on the effect on the interdiffusion rate of polar groups or substances attached or adsorbed to the latex surface.

An essential characteristic of latex microspheres is that polar groups are introduced at the particle surface during emulsion polymerization. These groups provide colloidal stability for the particle dispersion. When the dispersion is applied to a substrate and allowed to dry, compression forces deform the spherical particles into space filling cells, and the polar groups originally at the particle surface sit at the interface between adjacent cells. When these groups are few in number and well separated, they presumably remain relatively isolated in the interface. If present in larger amounts, as in core-shell latex materials, they then form an interconnected membrane in the film. This hydrophilic membrane gives the nascent film a morphology similar to that of a foam (6).

Here we are interested in the influence of this membrane on the rate of interdiffusion of polymer molecules between adjacent cells. At one extreme, if the membrane coponent is immiscible with the core polymer, and if its glass transition temperature Tg is very high, it might serve as a barrier to prevent interdiffusion. Mechanical strength in such films would come only from interactions within the membrane phase. Interdiffusion could occur only if the system were heated well above this Tg, which would allow for break-up of the membrane phase. Extensive investigations of poly(acrylic acid) [PAA] membranes in latex films have been reported by the Rhône-Poulenc group (4,6). Alternatively, polymers may be able to diffuse through the membrane. In certain instances, the polar membrane might be composed of a substance that might soften the interface and *promote* interdiffusion. Examples of these processes will be given below.

Following Interdiffusion by DET

We prepare latex films on small quartz plates from dispersions containing an equal number of Phe- and An-labeled particles. These samples were annealed for various periods of time in a temperature-controlled oven, removed and cooled to room temperature for donor fluorescence decay $[I_D(t')]$ measurements, and then returned to the oven for further annealing. Decay profile measurements are carried out using the single photon timing technique (5). Decays are first fitted to the equation (7)

$$I(t') = A_1\left[\exp\left(-\frac{t'}{\tau_D} - P\left(\frac{t'}{\tau_D}\right)^\beta\right)\right] + A_2\exp\left(-\frac{t'}{\tau_D}\right) \tag{1}$$

with $\beta = 0.5$, where first term approximates the contribution to the donor decay of energy transfer in regions of the film where polymer interdiffusion has taken place. A better measure of the extent of interdiffusion is obtained from the areas under the decay profiles, calculated by integrating $I_D(t')$, using parameters obtained in eq.(1) after normalizing at t'=0. Note that t' refers to the fluorescence decay time scale and t, to the sample annealing time.

$$\text{Area} = I = \int_0^\infty I_D(t')dt' \tag{2}$$

The decrease in area is proportional to the increase in extent of energy transfer, which in turn is related to the fraction of mixing, f_m.

$$f_m = \frac{\text{Area}(t) - \text{Area}(0)}{\text{Area}(\infty) - \text{Area}(0)} \tag{3}$$

The term f_m in fact measures the quantum efficiency of mixing, which is different from, but proportional to, the mass fraction of mixing (7-9). Since the proportionality constant depends primarily on the An content of the A-labeled latex (10), which is kept constant in all experiments reported here, one can use f_m values to calculate apparent diffusion constants D_{app} for the polymer by reference to a spherical Fickian diffusion model. Simulations (10,11) suggest that values of D_{app} differ from "true" values of the center-of-mass diffusion coefficient D_{cm} by as much as a factor of 2 to 5. In the case of systems in which the level of acceptor labeling and the mixture of D- and A-labeled particles remain constant, changes in D_{app} exactly parallel changes in D_{cm}, and thus provide a powerful means to evaluate changes in polymer diffusion rates (5c,8,9). Here the external variables which cause changes in the diffusion rate are temperature and the surface composition of otherwise identical latex.

Effect of Surface Carboxyl Groups on Interdiffusion

These experiments involve latex films prepared from core-shell latex particles in which PBMA is the core polymer and the shell contains methacrylic acid [MAA] as a comonomer [P(BMA-co-MAA)]. This series of core-shell PBMA latex particles containing different amounts of MAA groups in their shell were prepared by three-stage emulsion polymerization under monomer starved conditions (7). The latex dispersions were purified using a mixed bed ion-exchange resin to remove low molecular weight electrolytes including surfactant. The surface charge density (Q_S) was determined by potentiometric titration of freshly ion-exchanged latex. Films prepared from the ion-exchanged latex have a carboxylic-acid-group-rich phase as an interparticle membrane, whereas films prepared from the same particles at high pH form an ionomer phase in the membrane. We will see that these structures retard but do not prevent interparticle polymer diffusion.

The characteristics of the latex particles and their constituent polymers are presented in Table I. Note that the core-shell samples are similar in size, in molecular weight and molecular weight distribution, but differ only in the content of carboxyl groups. The latex samples referred to as MA2, MA4, and

Table I. Characteristics of PBMA Latex

Latexes	Diameter [nm]	$10^{-5}M_W$	M_W/M_n	$10^5 Q_s$ a) [eq/g]	S. A b) [A^2/acid]
An-MA0	129	4.8	3.6	1.0	730
Phe-MA0	128	4.1	3.2	0.99	750
An-MA2	152	4.8	2.9	4.52	140
Phe-MA2	147	4.4	3.2	4.98	130
An-MA4	150	4.2	2.9	7.98	79
Phe-MA4	150	4.0	2.7	6.54	96
An-MA6	149	4.9	3.0	11.6	55
Phe-MA6	146	4.3	2.6	12.1	54

a) Total surface charge density of -OSO3H and -COOH groups.

b) The average surface area occupied by an acid group.

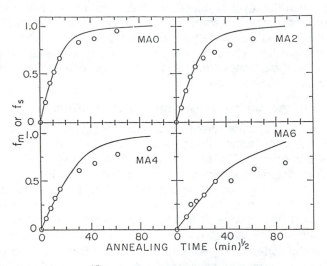

Figure 1. Plots of f_m or f_s vs $t^{1/2}$ for four latex films samples. The solid lines represent f_s, the mass fraction of mixing, calculated from a spherical Fickian diffusion model with diffusion coefficient values ($cm^2 s^{-1}$) of MA0, 6.5 x 10^{-16} ; MA2, 6.0 x 10^{-16}; MA4, 2,7 x 10^{-16}; MA6, 0.9 x 10^{-16}

MA6 have increasing amounts of MAA in their shell. Films were prepared from mixtures of donor- and acceptor-labeled latex, and, once dry, were annealed in an oven. In Figure 1 we present f_m values, plotted against square-root of annealing time, for this series of latex films. Two features of this graph are important. First, we note that the rate of polymer diffusion depends sensitively on the carboxyl group content of the latex: f_m values at the same annealing time decrease significantly in films prepared from latex with increasing carboxyl group content in the shell.

The second important observation about the data in Figure 1 is that interdiffusion occurs even at the early stages of annealing time, in the all films. This result is different from that reported by Joanicot et al (4) for a different latex film. They found that polyacrylic acid [PAA] at the surface of a poly(styrene-co-butyl acrylate) latex effectively suppressed interdiffusion until the film temperature exceeded the T_g of the PAA, at which point the polar membranes ruptured. The essential difference in the two systems is one of carboxyl content in shell. In their system, the shell polymer was extremely rich in PAA. In the latex particles we prepared, the composition of the shell-polymer did not exceed 9 mol% MAA, but this phase comprised about one-third of the polymer in latex (7). Thus our particles have a thicker shell composed of a materials likely to be more miscible with the core polymer.

A deeper analysis of the data is possible through the calculation of D_{app} values. These are in fact cumulative diffusion coefficients, since we use the diffusion model to find the value of D_{app} that best describes f_m at any annealing time. Since the polymers have polydispersity in both molecular weight and composition, these D_{app} values decrease as a function of f_m. The fastest diffusing species dominate the mixing process at early times. D_{app} values calculated from the data in Figure 1 are presented in Figure 2 as a function of f_m.

For each of the samples, we see that the D_{app} values are initially constant, and then decrease sharply once a certain extent of interdiffusion is attained. This crossover point where the diffusion begins to slow down shifts to lower f_m for films richer in -COOH groups. This effect is even more pronounced for samples in which the latex is pretreated with inorganic base (8,9). Neutralization involved one equivalent of base corresponding to the number of titratable acid groups in the latex. In Figure 3, where we compare the behavior of an unneutralized MA6 film with those prepared from dispersions that were neutralized with NH_4OH, $NaOH$, and $Ba(OH)_2$. The first feature of interest here is that D_{app} for the polymers with acid groups present as the ammonium salt are essentially identical at early annealing times with those with their acid groups in the protonated form. Some differences are apparent at later times, but the absence of a significant effect of NH_3 on polymer D_{app} value points to a tendency for the ammonium groups to dissociate upon drying or annealing of the film (8).

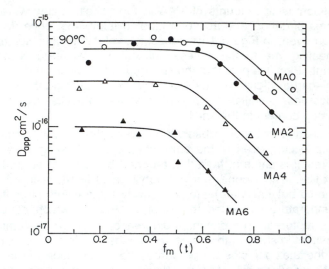

Figure 2. Plot of D_{app} values calculated from the data in Figure 1 as a function of f_m. The symbols are the same as in Figure 1.

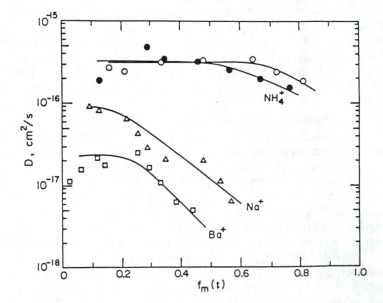

Figure 3. Plot of D_{app} values of latex films of MA6 annealed at 100 °C as a function of f_m. The surface ionizable groups of MA6 latex particles were neutralized by $NH_4OH(m)$, $NaOH(D)$, and $Ba(OH)_2(q)$. The top curve (m) refers to a films prepared from unneutralized MA6, plotted for comparison.

When the MA6 film is neutralized by sodium hydroxide, one observes not only a substantially smaller initial value of D_{app}, but also a very pronounced decrease in D_{app} throughout the interdiffusion process. Moreover, this diffusion is retarded even further in the case of the barium carboxylate film. Here the initial D_{app} value is ca 10^{-17} cm^2/s, and appears to increase before decreasing. One possible reason for the initial increase of D_{app} with f_m may be related to the mutual attraction of ion pairs to form a segregated, ion-rich microphase. This kind of phase in ionomer materials has been detected by small-angle X-ray scattering (12). The major conclusion to be drawn from this data is that neutralization of the carboxylic acid groups at the latex surface with NaOH and Ba(OH)$_2$ induces a pronounced decrease in the interdiffusion rate, and that the divalent Ba^{2+} salt is more effective than Na$^+$ at retarding interdiffusion.

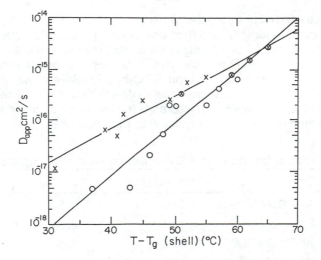

Figure 4. Plot of D_{app} vs (T - Tg) for D_{app} values obtained at $f_m = 0.5$. (x) refers to data obtained from latex with -COOH groups at the surface, examined at different T; (o) refers to latex with -COO$^{(-)}$ groups at the surface, with Na$^+$ or Ba^{2+} as the counterion, examined at 100°C.

It is possible to estimate the Tg of the membrane phase in these various latex samples (9). When we examine values of D_{app} as a function of (T - Tg), where T is the annealing temperature, a surprisingly simple result is obtained. Here we are careful, since D_{app} depends upon f_m, to compare values at identical f_m. As seen in Figure 4, the data from all of our experiments (for $f_m = 0.5$) fall on two relatively closely spaced straight lines, one representing data for the -COOH latex in which T was the variable. The other line represents experiments carried out at 100°C, in which the extent of neutralization and the metal counterion affect Tg. In all of these samples, the annealing temperature T is well above Tg of the membrane copolymer, and thus our results are consistent with those reported by the Rhône-Poulenc group (4,6) for films of latex containing PAA at the surface.

Effect of PEO Chains at the Latex Surface

A second type of PBMA, prepared from a styrene end-capped poly(ethylene oxide) [PEO] macromonomer **1** as a steric stabilizer, has a surface rich in PEO chains (13). PEO is miscible with PBMA and has the potential to act as a plasticizer. Here we find that the membrane polymer actually accelerates interdiffusion in the early stages of the interdiffusion process (14).

$$CH_3\text{-}O\text{-}(CH_2CH_2O)_n\text{-}(CH_2)_7\text{-}\underbrace{\hspace{2cm}}\qquad \textbf{(1)} \qquad \begin{array}{l} n=53 \\ M_w/M_n=1.06 \end{array}$$

These PBMA latex particles were prepared by dispersion polymerization in methanol-water (13), and then transferred to a purely aqueous phase. The characteristics of these microspheres are listed in Table II. The diameters of the two PBMA/PEO particles are very similar to one another and are only slightly larger than those of the PEO-free PBMA sample composed of similar molecular weight polymer. The PBMA/PEO particles contain 14 wt% PEO. Since the PEO was introduced as a macromonomer, it is present in the latex in the form of a PBMA copolymer with PEO branches.

Table II. Characteristics of PBMA Latex with PEO at the Surface

Latex	Diameter/nm	$10^{-5}M_w$	M_w/M_n
An-PBMA[a]	115	2.44	3.6
Phe-PBMA[b]	117	2.86	3.4
An-PBMA/PEO[b]	129	2.64	2.9
Phe-PBMA/PEO[b]	132	2.35	2.7

a) SDS surfactants were removed by ionic exchanged resin.
b) PEO macromonomer is 14.4 wt%.

Films were prepared from the two sets of samples described in Table II. One immediately apparent difference is that the PBMA/PEO latex formed transparent films at 20°C, whereas the PBMA dispersion required higher temperature (ca 30°C). The surface PEO chains appear to lower the minimum film forming temperature [MFT] of the latex. Interdiffusion measurements were carried out on these two films, and f_m values were calculated, Figure 5. These results are quite surprising. One notices first that the surface PEO chains promote interdiffusion at early times. This result is consistent with the PEO also lowering the MFT of the system, and suggests that PEO acts as a plasticizer for the particle surface.

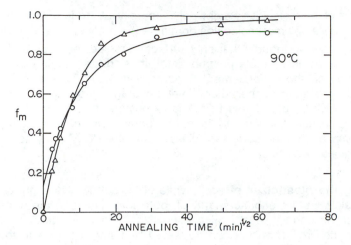

Figure 5. Plot of f_m vs the square-root of annealing time for PBMA(D) and PBMA/PEO (m) latex films at 90 °C.

At later stages of the interdiffusion, the two curves cross, and the PEO-containing latex film experiences a smaller diffusion rate. These experiments are still in their very early stages, and thus it is premature to draw firm conclusions. Our current explanation for this retardation is that it corresponds to the diffusion of the PBMA-PEO copolymer. Its diffusion is slower either because of its enhanced molecular weight or because the branches interfere with the diffusion.

Effect of Non-ionic Surfactant at the Latex Surface

Because of the unusual features of the influence of surface-grafted PEO chains on interdiffusion, we wished to carry out comparison experiments with non-ionic surfactants on ordinary PBMA latex. Here results are reported for NP-20, a nonylphenol surfactant containing on average 20 EO units, along with some preliminary results for NP-100 (15).

$$C_9H_{19}\text{—}\langle\ \rangle\text{—}O(CH_2CH_2O)_n\text{-H} \qquad (2) \text{ NP-n, n= 20, 100}$$

In Figure 6, the values of D_{app} for samples containing various concentrations of NP-20, are plotted against f_m. One sees that there is a marked increase in the polymer diffusion rate by addition of NP-20 surfactants, and the extent of enhancement increases with increasing NP-20 concentration. It is also apparent in Figure 6 that D_{app} values decrease with increasing f_m for all four sets of samples. In the absence of NP-20, this decrease can be explained in terms of the molecular weight polydispersity of the latex polymer, since the most mobile polymer molecules dominate the early stages of mixing.

Originally we thought that only surfactant molecules adsorbed onto the latex surface would promote polymer interdiffusion. According to this idea, we would expect the enhancement of polymer diffusion to level off when the amount of surfactant reached monolayer coverage of the latex. Monolayer coverage of NP-20 onto these PBMA latex occurs at about 6 wt%. Interestingly, D_{app} values continue to increase above this concentration. This strongly implies that the surfactant dissolves in the PBMA polymer and acts as a traditional plasticizer.

Modeling the plasticizer effect. While NP-n surfactants might contribute in several ways to enhancement of polymer diffusion, the most likely mechanism is one in which, as a plasticizer, it acts to increase the free volume in the film. This idea can be tested by fitting the data to the Fujita-Doolittle equation (16),

$$\left[\ln\frac{D_p(T,\Phi_a)}{D_p(T, 0)} \right]^{-1} = f_p(T, 0) + \frac{f_p^2(T, 0)}{F_a \ \beta(T)} \tag{4}$$

Here $f_p(T, 0)$ is the fractional free volume of the polymer in the absence of surfactant, and Φ_a is the volume fraction of the surfactant. $\beta(T)$ represents the difference in fractional free volume between the surfactant ($f_s(T)$) and the polymer ($f_p(T,0)$) at temperature T.

In Figure 7, the values of D_{app} calculated from f_m obtained in the presence of NP-20, are superimposed on corresponding data obtained in the absence of surfactant (15). In this way we obtain a shift factor which corresponds to the inverse of the term on the left hand side of equation (4). One sees that all data can be superimposed onto a single curve, to afford a master curve of D_{app} and f_m. From this result we conclude that NP-20 acts as a traditional plasticizer for PBMA.

In the case of NP-100, where our data are less complete, we still obtain full agreement with eq. (4), but with a significantly smaller value of $\beta(T)$. NP-100 is a less effective plasticizer of PBMA toward polymer diffusion. Thus the influence of PEO-type non-ionic surfactant molecules on polymer interdiffusion in PBMA latex films is relatively simple. The surfactant acts as a plasticizer in complete accord with the Fujita-Doolittle model. As the dispersion dries, the surfactant diffuses into the latex polymer. Understanding the timing and rate of this diffusion process is an important topic for future research.

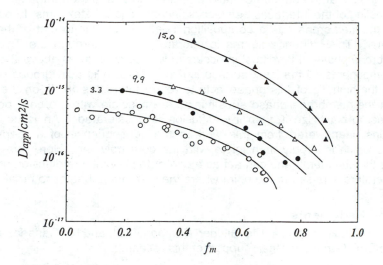

Figure 6. Plot of D_{app} vs f_m for PBMA latex containing various amounts of the non-ionic surfactant NP-20. From top-to-bottom (wt%): 15.0, 9.9, 3.3 , 0.

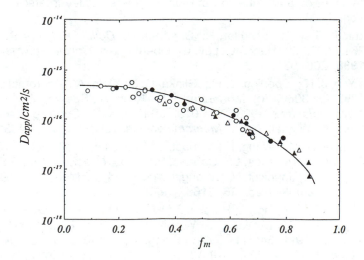

Figure 7. Master curve generated by fitting the data to the Fujita-Doolittle expression, eq. (4), from which a value of b(T) is obtained.

Summary

Polar groups at the surface of latex particles have an important influence on the healing of the interfaces between adjacent cells in latex films. In our systems the membrane has a composition relatively similar to that of the core polymer. This influence seems to operate primarily through the Tg of the membrane phase. Interdiffusion occurs only at temperatures above the Tg of all components. If the latex surface is rich in carboxylic acid groups or their salts, the high Tg of this phase acts to retard interdiffusion. If, on the other hand, the membrane phase substituents are miscible with the core polymer and act to plasticize it, then interdiffusion is accelerated. An issue which remains unanswered from this work is the consequence of a membrane phase which is largely immiscible with the core polymer. Here one anticipates that the membrane can act as a barrier to prevent interdiffusion until the temperature is raised to the point where the membrane begins to break up.

Acknowledgments

The authors thank The Glidden Company, ICI, and ICI Canada, as well as NSERC Canada for their support of this research.

Literature Cited

1. (a) Voyutskii, S.S. *J. Polym. Sci.* **1958**, 32, 528; (b) Voyutskii, S. S., "Autohesion and Adhesion of High Polymers," Wiley-Interscience, New York, 1963.

2. Hahn, K; Ley, G; Schuller, H; Oberthur, R; *Colloid Polymer Sci.*, **1986**, 264, 1092; (b) Hahn, K; Ley, G; Oberthur, R; *Colloid Polymer Sci.*, **1988**, 266, 631.

3. (a) Yoo, J. N.; Sperling, L. H.; Glinka, C. J.; Klein, A., Macromolecules, **1990**, 23, 3962; (b) Mohammadi, N; Klein, A.; Sperling, L. H., *Macromolecules*, **1993**, 26, 1019; (c) Kim, K. D.; Sperling, L. H.; Klein, A.; Wignall, G. D., *Macromolecules*, **1993**, 26, 4624-4631.

4. (a) Joanicot, M.; Wong, K.; Maquet, J.; Chevalier, Y; Pichot, C; Graillat, C; Lindner, P; Rios, L.; Cabane, B., *Prog. Colloid Polym. Sci.*, **1990**, 81, 175; (b) Joanicot, M.; Wong, K.; Richard, J.; Maquet, J.; Cabane, B., *Macromolecules*, **26**, 3168, (1993).

5. (a) Zhao, C.L.; Wang, Y.; Hruska, Z.; Winnik, M. A., *Macromolecules*, **1990** 23, 4082 (b) Wang, Y., Winnik, M. A., Haley F., *J. Coatings Technol.*, **1992**, 64(811), 51. (c) Wang, Y.; Winnik, M. A., *J.Phys. Chem.*, **1993**, 97, 2507; (d) Wang, Y.; Zhou, C. L., Winnik, M. A., *J. Chem. Phys.*, **1991**,95, 2143; (e) Boczar, E. M.; Dionne, B. C.; Fu, Z.; Kirk, A. B.; Lesko, P. M.; Koller, A. D., *Macromolecules*, **1993**, 26, 5772.

6. Chevalier, Y; Pichot, C; Graillat, C; Joanicot, M; Wong, K; Lindner, P; Cabane, B., *Colloid Polym. Sci.*, **1992**, 270, 806.

7.   Kim, H-B.; Wang, Y.; Winnik, M.A.; *Polymer*, **1994**, 35, 1007.

8.   Kim, H-B.; Winnik, M.A. *Macromolecules* ,**1994**, 27, 1779.

.9   Kim, H-B.; Winnik, M.A. *Macromolecules* ,**1995**, 28, 2033.

10.   Dhinojwala, A; Torkelson, J. M., *Macromolecules* ,**1994**, 27, 4817;

11.   (a) Liu, Y. S.; Feng, J.; Winnik, M. A.. *J. Chem. Phys.*, **1994**, 101, 9096; (b) Liu, Y.S., Winnik, M.A., *Makromol. Chem. Symp.*, **1995**, 92, 321; (c) Farinha, J. P.; Martinho, J. M. G.; Yekta, A.; Winnik, M. A., *J. Phys. Chem.* , **1995**, in press.

12.   Eisenberg A., "Ion-Containing Polymers. Physical Chemistry and Structure". Academic Press, NY, 1977.

13.   (a) Kawaguchi S., Winnik M.A., Ito K., *Macromolecules* , **1995**, 28, 1159; (b) Kawaguchi S., Winnik M.A., Ito K., *Macromolecules* , **1995**, in press.

14.   Winnik, M. A.; Kim, H.-B.; Kawaguchi, S., *Prog. Pacific Polym. Soc. 3*, Ghiggino, K. P. ed., Springer-Verlag, Berlin, 1994, pp 247 - 257.

15.   Kawaguchi S., Odrobina, E.; Winnik M.A., *Makromol. Chem. Rapid Comm.*, **1995**, in press.

16.   Fujita, H. *Fortschr. Hochpolym.-Forsch.* **1961**, Bd.3, S.1.

17.   Juhué, D.; Wang, Y.; Winnik, M. A., *Makromol. Chem. Rapid Comm.*, **1993**, 14, 345;

Chapter 5

Steady-State Fluorescence Method To Study Film Formation from Latex Particles Having a High Glass-Transition Temperature

Ö. Pekcan and M. Canpolat

Department of Physics, Istanbul Technical University, Maslak, 80626 Istanbul, Turkey

Direct energy transfer (DET) method conjunction with steady state fluorescence (SSF) technique were used to study interdiffusion of polymer chains across the particle-particle junction, during film formation from high-T latex particles. The latex films were prepared from pyrene (P) and naphthalene (N) labeled poly (methyl methacrylate) (PMMA) particles and annealed in elevated time intervals above glass transition (T_g) at 180°C. Monte Carlo simulations were performed to model the N and P fluorescence intensities (I_N and I_P) using photon diffusion theory. Number of N and P photons (N_N and N_P), emerging from the front surface of the latex film are calculated when only N is excited. A novel correction method was suggested and employed to eliminate the P intensity due to the optical variation in latex film. P intensity solely from the energy transfer processes were monitored versus annealing time and was used to measure the polymer chain diffusion coefficient (D), which was found to be 5.9 x 10^{-13} cm^2/sec at 180°C.

Interdiffusion processes during latex film formation have been widely investigated over the past few years. These phenomena can be considered as special cases of crack healing or polymer welding processes. Latex films are formed from small polymer particles produced initially as a colloidal dispersion, usually in water. The term "latex film" normally refers to a film formed from soft or low-T latex particles (T_g below room temperature) where the forces accompanying the evaporation of solvent are sufficient to deform the particles into a transparent, void-free film. Latex films can also be obtained by compression molding of a dried latex powder composed of a polymer such as polystyrene (PS) or poly (methyl methacrylate) (PMMA) that has T_g above drying temperature. These latexes are generally called high-T particles. During drying process high-T latex particles remain essentially discrete and undeformed. The

0097–6156/96/0648–0064$15.00/0

mechanical properties of such films can be evolved by annealing which is called sintering of latex powder.

Recently, freeze fracture TEM (FFTEM) has been used to study the structure of dried latex films (*1,2*). Small-angle neutron scattering (SANS) has been used to study latex film formation at the molecular level. Extensive studies using SANS have been performed by Sperling and co-workers (*3-5*) on compression-molded PS films. Using time-resolved fluorescence (TRF) measurements in conjunction with dye-labeled particles, interdiffusion during latex film formation has been studied by Winnik and co-workers (*6-9*). Direct non-radiative energy transfer (DET) method was employed to investigate the film formation processes in dye-labeled PMMA (*6,9*) and PBMA (*7,8*) systems. The DET method is particularly sensitive to the early stages, whereas the SANS technique becomes more sensitive to later stages of the interdiffusion processes. The steady state fluorescence (SSF) technique combined with DET was recently used to examine healing and interdiffusion processes in dye-labeled PMMA latex films (*10-12*).

Because of photon diffusion, radiative and non-radiative energy transfer processes, special attention has to be paid during SSF measurements to study the evolution of latex film formation. Film thickness and annealing time intervals are very critical for the quality of the film. In that sense photon diffusion, radiative and non radiative processes compete and may play important role during SSF measurements. In this work, SSF measurements were performed to study interdiffusion processes during film formation from PMMA particles, using DET method.

The PMMA particles prepared by non-aqueous dispersion (NAD) polymerization. These particles were labelled with appropriate donor (naphthalene, N) and acceptor (pyrene, P) chromophores (*13*). The 1-3 µm dia particles were used having two components; the major part, PMMA, comprises 96 mol % of the material and the minor component, poly-isobutylene (PIB) (4 mol %), forms an interpenetrating network through the particle interior (*14,15*) very soluble in certain hydrocarbon media. A thin layer of PIB covers the particle surface and provides colloidal stability by steric stabilization. (These particles were prepared in Professor M.A. Winnik's laboratory in Toronto)

In this paper two sets of isothermal experiments were performed by annealing PMMA latex film samples at 180°C, in elevated time intervals. In the first set; after annealing, only P was excited at 345 nm and emission intensity (I_{op}) was observed versus annealing time. This set of experiments were performed to detect the evolution of the quality (transparency) of the film samples. During the second set of experiments, after annealing the film samples, N and P intensities (I_N and I_P) were monitored against annealing time, where the films were excited at 286 nm. In this set of experiments we aimed to observe the chain interdiffusion across the particle-particle junction. Monte Carlo simulations were carried out to model the I_{op}, I_N and I_P intensities emitted from latex film, using photon diffusion theory (PDT) (*16*). A novel correction method was suggested and employed to separate the P intensity (I'_P) due to nonradiative and radiative energy transfer, from the total P intensity I_P, which produced the characteristics of chain interdiffusion.

Experimental

Latex film preparation were carried out thus; the same weights on N and P particles were dispersed in heptane in a test tube; after complete mixing, a large drop of the dispersion was placed on a round silica window plate with dia of 2 cm. Heptane was allowed to evaporate and the silica window was placed in the Solid Surface Accessory of Model LS-50 fluorescence spectrometer of Perkin-Elmer. All measurements were carried out in the front face position at room temperature. Slit widths were kept at 2.5 mm. The N-P film sample was excited at 286 nm in order to maximize naphthalene absorbance while minimizing pyrene absorbance. Film samples were illuminated only during the actual fluorescence measurements and, at all other times, were shielded from the light source. The film of latex particles was annealed above T_g of PMMA for various periods of time at 180°C. The temperature was maintained within ±2° during the annealing. The variation in optical density of N-P film was controlled by only exciting pyrene at 345 nm for each measurement.

Theorical Models

Photon Diffusion and DET.   The journey of an exciting or emitted photon to or from a dye molecule in a film formed from annealed latex particles can be modelled by photon diffusion theory. The collision probability, 3p of a travelling photon with any scattering center in a film is given by

$$P = 1 - \exp\left(-r / <r> \right) \tag{1}$$

where, r is the distance of a photon between each consecutive collision and $<r>$ is defined as the mean free path of a photon. Here the film is taken as a plane sheet with a thickness of d and the direction of incident photons is taken perpendicular to the film surface (for example in z the direction). In Monte Carlo simulations, after each collision, d is compared with the z component of the total distance

$$S_z = \sum_i^n r_{iz} \tag{2}$$

where i labels the successive collisions during the journey of a photon. Photons emerging from the back and front surfaces of the film without interacting with a pyrene molecule have to satisfy the conditions given below

$$S_z > d \quad \text{and} \quad S_z < 0 \tag{3}$$

respectively. The total number of photons emerging from the front surface is then represented by N_{sc}, which is assumed to be proportional to the intensity, I_{sc} of the light scattered from latex film during SSF measurements.

When the interaction of a photon by a dye molecule is considered, mainly two types of Monte Carlo simulations can be carried out; At first, number of photons (N_{op}) emitted from the front surface of the latex film is calculated when only pyrene molecules are excited which models the direct emission intensity from pyrene (I_{op}). In the second part, number of photons emitted from the latex film is calculated when only naphthalenes are excited. In this part of the simulation non-radiative and radiative energy transfer processes are considered and, naphthalene and pyrene emissions (I_N and I_p) were modeled by counting N and P photons (N_N and N_P) emitted from the front surface of the film.

In order to drive the relation for the fluorescence intensity, I_{op} emitted from the latex film, we defined the probability of a photon encountering a pyrene molecule to be,

$$q = 1 - \exp\left(-s/l\right). \tag{4}$$

Here, s represents the total distance (optical path) and l is the mean distance the photon travels in the film, before it finds a pyrene molecule. If s is large, the probability of the photon encountering a pyrene is high, or vice versa. After collision with a pyrene the photon travels again according to equation 1. Then d is compared with the z component of the total distance, s_z as in equation 3, to find if the photon is emitted from the film surfaces. The number of photons emitted from the front surface of the film is given by N_{op}, which is assumed to be proportional to I_{op}, the fluorescence intensity. In Monte Carlo simulations l and d were taken to be fixed parameters with the values of 150 and 50 respectively. Here we have to note that l is considered to be inversely proportional to the pyrene concentration in latex film which is assumed to be constant during the film formation processes. The mean free path, $<r>$ was varied between 1 to 100 for a given d and for each $<r>$, the number of incident photons was taken to be 3×10^4 during the simulations. The number of collisions, n is varied so that the conditions in equation 2 are satisfied.

In the second part of Monte Carlo simulations, the probability of a photon encountering a naphthalene molecule can be defined as

$$P_N = 1 - e^{-s_N/l_N}. \tag{5}$$

Here, s_N is the total distance and l_N is the mean distance, the photon travels in the film, before it finds a naphthalene molecule. After collision with a naphthalene, photon travels again by satisfying equation 1 and emitted from the film surfaces according to equation 3. The number of naphthalene photons emitted from the front surface of the film is given by N_N and assumed to be proportional to I_N, the naphthalene fluorescence intensity. There is always a certain probability of encountering a pyrene molecule by a naphthalene photon, which is defined by

$$P_P = 1 - e^{-s_P/l_P}. \tag{6}$$

This process is called "radiative energy transfer" where, s_p represent the total distance and l_p is the mean distance, the naphthalene photon travels in the film, before it finds a pyrene molecule. In equations 5 and 6 l_N and l_P are considered to be inversely proportional to the naphthalene and pyrene concentrations in latex film. Here, again after collision photon travels by satisfying equation 1 and emitted from the film surfaces according to equation 3. After the radiative energy transfer is taken place, the number of pyrene photons emitted from the front surface of the film is given by N_{PR}. During simulations l_N and l_P are taken to be fixed parameters with the values of 100 and 1000 respectively.

When the distance between N and P molecules is short enough they can interact and excited N molecule can transfer its energy to nearby P molecule, before emitting a photon. This process is called "direct nonradiative energy transfer" (DET). This mechanism is know as dipolar Förster interaction in photophysics terminology. During film formation from latex particles, P and N labeled chains interdiffuse and can form a Förster domain where excited N molecule can transfer its energy to P molecule with 90% probability. (Experimentally determined from the completely mixed, chloroform cast films). Then the excited P molecule release a photon which then travels by satisfying equation 1 and emitted from the film surface according to equation 3. After the DET, the number of P photons emitted from the front surface of the film, is presented by N_{PNR}. The total number of P photons, emitted from the front surface of the film is then given by $N_P = N_{PR} + N_{PNR}$.

Interdiffusion by Fluorescence Method. In bulk state the interdiffusion of polymer chains can be explain by several theories (17,18) that are based on the reptation model of de Gennes (19). In the bulk state, polymer chains have a Gaussian distribution of segments. When chains are confined to the space adjacent to the interface they have distorted conformations. Interdiffusion across the interface leads to conformational randomization and recovery of a Gaussian distribution of segments (20,21). Entanglements prevent macromolecules larger than a certain (critical) length from undergoing large amplitude sideways diffusion. These macromolecules are pictured as being confined to a tube, and their diffusion occurs by coherent back and forth motion along the center line of the tube with a curvilinear diffusion coefficient, keeping the arc length of the chain constant. This worm-like motion is referred to as "reptation". The reptation time, T_r, is the time necessary for a polymer to diffuse a sufficient distance so that all memory of the initial tube is lost. The diffusion rate of this reptation across a polymer-polymer interface should be sensitive to the location of the chain ends. Since there is more free volume at the interface than in the bulk, an enrichment of chain ends at the interface is expected.

Tirrell et al (22) studied interdiffusion in terms of the steady state fluorescence of an acceptor emission intensity due to DET from a donor, in our case DET process can be represented by:

$$N^* + P \rightarrow N + P^*. \tag{7}$$

Here N^* and P^* are the excited naphthalene and pyrene molecules respectively. When the donor is excited exclusively and can transfer its energy to nearby acceptor groups the fluorescence intensity of the acceptor, $I(t)$, increases with time:

$$I(t) - I(0) = I_0 \alpha \int_{-\infty}^{+\infty} C(x, t) [C_0 - C(x, t)] \, dx \tag{8}$$

where $I(0)$ is $I(t)$ at time zero, and the x-axis is taken as normal to the polymer-polymer interface. Here I_0 represent the incident light intensity, α is a constant, $C(x,t)$ is the concentration profile of an acceptor at time t and C_0 is its interface concentration. At times longer than T_r, concentration profiles can be obtained from Fick's law of diffusion (23)

$$\frac{\partial C(r, t)}{\partial t} = D \nabla^2 C(r, t). \tag{9}$$

Then equation 8 can be rewritten as (22):

$$I(t) - I(0) = 0.165 \alpha C_0^2 (Dt)^{1/2} \tag{10}$$

where D is the center-of mass diffusion coefficient of the polymer chain.

Result and Discussion

Transparency and Photon Diffusion in Latex Film.   Emission (I_{op}) and scattered (I_{sc}) spectra of a latex film annealed at 180°C in elevated time intervals and excited at 345 nm, are shown in (Figure 1). Upon annealing P intensity (I_{op}) increased for 15 min interval then decreased, by increasing time intervals however I_{sc} decreased continuously from the beginning by increasing annealing time. The behavior of (I_{op}) and (I_{sc}) versus annealing time are plotted in (Figuare 2a) and (b) respectively.

The behavior of (I_{op}) and (I_{sc}) can be interpreted by results of Monte Carlo simulations. N_{op} and N_{sc} are plotted versus square of mean free path ($<r>^2$) of a photon in (Figuare 3a) and (b) respectively. Here it is assumed that the relation $<r> = \sqrt{Dt}$ is obeyed, where t is the annealing time and D is given in equation 9. In (Figuare 3a), N_{op} first increases suddenly then decreases by increasing $<r>^2$. These indicate that for very short $<r>^2$ values s is short but as $<r>^2$ values increase the optical path, s of a photon becomes longest and the probability of encountering a pyrene in the film becames highest, as a result N_{op} reaches a maximum. However, for longer $<r>^2$ values s becomes shorter again and photons can easily escape from the back surface of the film and as a result both N_{op} and N_{sc} decrease continuously by increasing $<r>^2$ values. Cartoon representation of film formation from high-T latex particles and its relation with the mean free and optical paths ($<r>$ and s) are presented in (Figuare 4). Early stage of film formation is shown in (Figuare 4a), where heptane evaporates and close packed particles form a powder film which includes many voids. This film yields low I_{op} and high I_{sc} values due to very short $<r>^2$ and short s values. (Figuare 4b) presents a film where, due to annealing, particle boundaries start to heal and disappear which gives rise to short $<r>^2$ and the longest s values. In such a film one can observe high I_{op} and low I_{sc} intensities. Finally

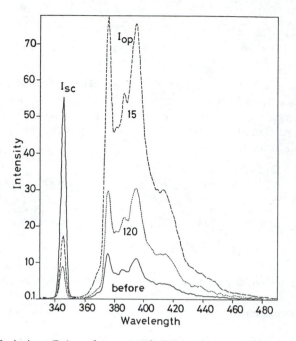

Figure 1. Emission (I_{op}) and, scattered (I_{sc}) spectra of a latex film before annealing and annealed at 180°C in 15 and 120 min time intervals. Film is formed from N and P particles and excited at 345 nm.

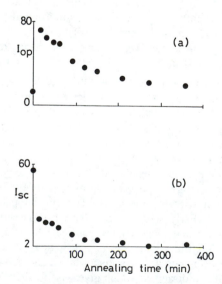

Figure 2. Variation of a- (I_{op}) and b- (I_{sc}), intensities versus annealing time. Film is excited at 345 nm.

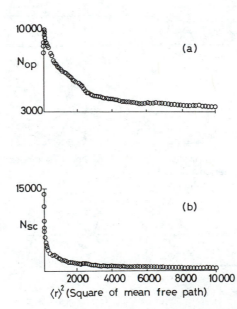

Figure 3. Variation of a- (N_{op}) and b- (N_{sc}), versus , square of mean free path of a photon. Results are obtained monte Carlo simulations.

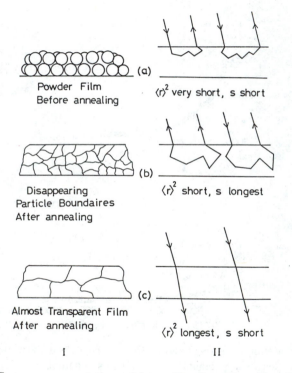

Figure 4. Cartoon representation of latex film formation from high-T latex particles, a- before annealing, powder film, b- annealed film, beginning of disappearance of particle boundaries, c- highly annealed, Almost transparent film.

(Figuare 4c) shows a almost transparent film with the longest $<r>^2$ but smaller s values. This film naturally presents both low I_{op} and I_{sc} intensities.

DET and Monte Carlo Simulations . As the interdiffusion process of polymer chains proceeds, particle boundaries disappear and DET between N and P molecules takes place. This picture can be quantitatively modelled by the results of Monte Carlo simulations. The total number of P and N photons (N_P and N_N) emitted from the front surface of the film are plotted versus $<r>^2$ of a photon in (Figuare 5a) and (b) respectively, where N_P decrease continuously however N_N increase at the beginning and then decrease by increasing $<r>^2$. As annealing time, t increases, $<r>^2$ also increases due to disappearance of interparticle voids and interfaces, then one should expect an increase in P and decrease in N intensities due to energy transfer from N to P molecules. Since, N_P contains photons from both radiative and non-radiative processes, above contradiction can be resolved using Monte Carlo simulations. (Figuare 6a) and (b) present the plot of N_{PR} and N_{PNR} versus , where one can observe continuous increase in N_{PNR} photons due to non radiative energy transfer process from N to P molecules. Number of photons, due to radiative energy transfer (N_{PR}) process present very high values at the beginning of film formation, then decrease imidiatly due to increase in values. Now (Figuare 5b) and (6b) look consistent by showing decrease in N photons and increase in P photons respectively but still this correspondence is not one to one. In other words this picture still does not fit to Tirrels Model in equation 7 and 8. Here, emission from naphthalene due to variation in film quality (N_{op}) dominates the total N_N emission. This contribution can be rationalized by comparing (Figuare 5b) and (Figuare 3a) where the quality of the film (or $<r>$) effects the number of emitted photons in both case. In order to eliminate the photons (N_{op})created due to the variation in the quality of the film and to separate the contributions solely from the radiative and non-radiative energy transfer processes we normalize the total number of photons emitted from the front surface of the film as follows

$$N_P' + N_N' = 1000 \qquad (11)$$

where, N_P' and N_N' are the normalized total number of P and N photons emitted from the front surface of the film respectively. N_P' and N_N' are plotted versus of a photon in (Figuare 7a),and (b) respectively in which one to one correspondence between P and N emissions now can be seen. When (Figuare 6b) is compared with (Figuare 7a) it is seen that N_P' is created by non-radiative energy transfer process(N_{PNR}) except at very early N_{PR} contribution. The normalized photon emission from naphthalene N_N' now decrease due to non-radiative energy transfer from N to P as the mean free path $<r>$ of a photon increase, except at the early stage, radiative energy transfer contribution.

Diffusion Coefficient. The emission spectra of a mixture of N and P labeled latex films before annealing and after annealed at 180°C for various time intervals are measured, when the film is excited at 286 nm, where I_P and I_N both increase suddenly by annealing the film sample for 15 min, then decrease continuously by increasing

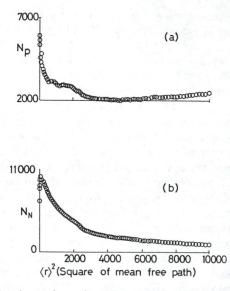

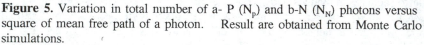

Figure 5. Variation in total number of a- P (N_p) and b-N (N_N) photons versus square of mean free path of a photon.   Result are obtained from Monte Carlo simulations.

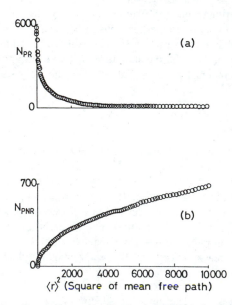

Figure 6. Variation in number of a- P photons (N_{PR}) due to radiative transfer, b- P photons (N_{PNR}) due to non-radiative transfer from N to P molecules, emitted from the front surface of the film.

annealing time intervals. The variation in I_N and I_P intensities versus annealing time are shown in (Figuare 8a). From the point of view of DET, the decrease in I_N is expected but the decrease in I_P is quite surprising. One would naturally expect an increase in I_P due to DET processes as I_N decrease. In previous section Monte Carlo simulations have predicted that this anomalous behavior in I_P is produced due to the variation in the quality of the film, which can be resolved by using equation 11. The emission due to the variation in optical quality of the film can be eliminated by normalizing the total P and N intensities to 1000. This can be done by adjusting the total area under N-P spectra $(I_N + I_P)$ to 1000 at each annealing step. Then the area of the pure P spectra (I_P') is calculated and supracted from 1000 to produce the N spectra (I_N'). The variation in normalized P and N intensities $(I_P'$ and $I_N')$ versus annealing time are presented in (Figuare 8b). As expected from Monte Carlo simulations, now as I_P' increase I_N' decrease by increasing annealing time. This behavior in I_P' and I_N' intensities show one to one correspondence which indicates that the energy transfer processes between N and P are purely radiative and non-radiative. In order to quantify the above results equation 10 can be used for I_P' intensity as follows

$$\frac{I_P'(t) - I_P'(0)}{I_P'(\infty) - I_P'(0)} = \frac{(Dt)^{1/2}}{R}. \tag{12}$$

Here, R is the radius of the particle and $I_P'(\infty)$ is the P intensity at $t=\infty$. Equation 12 is fitted to the data in (Figuare 8b) and chain diffusion coefficient, D is produced from the slope of this fit in (Figuare 9a). The linear relation in (Figuare 9a) supports the Tirrel's model, given in the theoretical section where Fickian type of diffusion is considered for times longer then the reptation time T_r of polymer chain. The diffusion coefficient, that was observed in this work is found to be as $D=5.9 \times 10^{-13}\ cm^2/sec$, which is order of magnitude faster than it was observed in the similar system at 170°C by using transient fluorescence technique (6).

The linear curve in (Figuare 9a) does not go through the origin, which means that there is a delay in chain interdiffusion which may corresponds to healing time(τ_H) (*12*). Presumable during first 15 min chains move halfway across the interface surface and particle boundaries start to disappear. This picture can be visualized by the result of Monte Carlo simulation. When the normalized total number of photons N_P' are plotted against mean free path of a photon $<r>$, which now corresponds to $(time)^{1/2}$, it is seen that a certain delay is required for non-radiative energy transfer to start. (Figuare 9b) presents the plot of N_P' versus $<r>$ where at the beginning, for very short $<r>$ values number of P photons due to radiative energy transfer (N_{PR})dominates to N_P', then decrease by increasing $<r>$. As $<r>$ values continue to increase, which is equivalent to saying that interdiffusion increase in time, the number of P photons due to DET (N_{PNR}) increase.

In conclusion, this work has presented simple steady state fluorescence method to measure backbone diffusion coefficient, D during film formation from high-T latex particles, where the experiments are easy to perform and spectrometer is inexpensive to obtain. Here, we have performed Monte Carlo simulations to introduce a novel correction technique for the energy transfer measurements for the steady-state fluorescence method which is very handy for measuring D value.

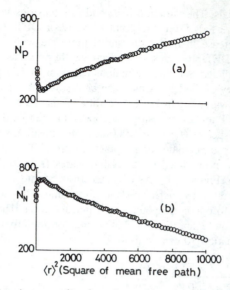

Figure 7. Variation in normalized total number of a- P (N_P') and b- N(N_N') photons versus , square of mean free path of a photon. Normalization is done according to Eq(11).

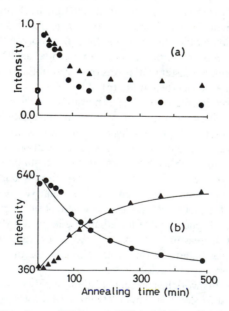

Figure 8. Variation in a- P(I_P) and N(I_N) b-Normalized P(I_P') and P(I_N') intensities versus annealing time. Film is excited at 286 nm.

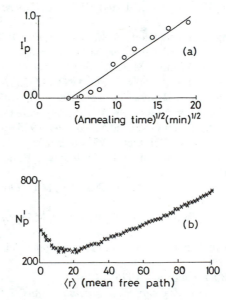

Figure 9. a-Plot of the data in Fig 8a versus square root of annealing time. Slope of the linear plot produced diffusion coefficient as D=5.9 x 10-13 cm²/sec. b-Plot of the data in Fig 7a versus <r>, mean free path of a photon.

Acknowledgement.   We thank to Turkish Academy of Sciences (TÜBA) for their financial support and Professor M.A. Winnik for providing us with the latex material.

Literature Cited

1. Wang, Y.; Kats, A.; Juhue, D.; Winnik, M.A.; Shivers, R. R.; an Dinsdale, C.J. *Langmuir* **1992**, *8*, 1435.
2. Roulstone, B.J.; Wilkinson, M.C.; Hearn, J. and Wilson, A.J. *Polym. Int.* **1991**, *24*, 87.
3. Linne, M.A.; Klein, A.; Miller, G.A.; Sperling, L.H. and Wignall, G.D. *Macromol. Sci. Phys. B2* **1988**, *(2-3)*, 217.
4. Yoo, J.N.; Sperling, L.H.; Glinka,C.J. and Klein, A. *Macromolecules* **1990**, *23*, 3962.
5. Kim, K.D.; Sperling, L.H. and Klein, A *Macromolecules* **1993**, *26*, 4624.
6. Pekcan, Ö.; Winnik, M.A. and Croucher, M.D. *Macromolecules* **1990**, *23*, 2673.
7. Wang, Y.; Zhao, C. and Winnik, M.A. *J.Chem.Phys.* **1991**, *95*, 2143.
8. Wang, Y. and Winnik, M.A. *J. Phys. Chem.* **1993**, *97*, 2507.
9. Wang, Y. and Winnik, M.A. *Macromolecules* **1993**, *26*, 3147.
10. Pekcan, Ö.; Canpolat, M. and Göçmen, A. *Eur.Polym. J.* **1993**, *29*, 115.
11. Pekcan, Ö.; Canpolat, M. and Göçmen, A. *Polymer* **1993**, *34*, 3319.
12. Canpolat M. and Pekcan, Ö. Polymer **1995**, *36*, 2025.
13. Winnik, M.A.; Hua, M.H.; Hongham, B.; Williamson B. and Croucher, M.D. *Macromolecules* **1984**, *17*, 262.
14. Pekcan,Ö.; Egan, L.S.; Winnik M.A. and Croucher, M.D. *Phys.Rev.Lett.* **1988**, *61*, 641.
15. Pekcan, Ö.; Egan, L.S.; Winnik M.T. and Croucher, M.D. *Macromolecules* **1990**, *23*, 2210.
16. *Optical Properties of Polymers;* Meeten, G.H., Elsevier Sci. Publishers Ltd, New York, USA., 1989
17. de Gennes, P.G. *Macromolecules* **1976**, *11*, 587.
18. Doi, M. and Edward, S.F. *J. Chem. Soc. Faraday Trans. 2* **1978**, *74*, 1802.
19. de Gennes, P.G. *Chem. Phys.* **1971**, *75*, 5194.
20. Prager, S. and Tirrell, M. *Chem. Phys.* **1981**, *75*, 5194.
21. Wool, R.P. and O' Connor, K.M. *J. Appl. Phys.* **1981**, *52*, 5953.
22. Tirrell, M.; Adolf, D. and Prager, S. *Springer Lecture Notes Appl. Math.* **1984**, *37*, 1063.
23. Crank, J. *the Mathematics of Diffusion* Clarendon Press, New York, 1975

Chapter 6

Small-Angle Neutron Scattering Studies of Polymer Interdiffusion During Latex Film Formation

Ming-Da Eu[1] and Robert Ullman

Macromolecular Research Center and Department of Nuclear Engineering, University of Michigan, Ann Arbor, MI 48109

An equation relating the growth of the radii of gyration of labeled latex particles, commonly observed in small-angle neutron scattering (SANS) experiments on interdiffusing latex films, to the interdiffusion coefficient of the polymers was derived from Fick's law of diffusion. SANS experiments were conducted on interdiffusing latex films consisting of isotopic latex particles with large mismatch in molecular weight. The interdiffusion coefficient, the interpenetration depth, and the extent of intermixing were extracted from the scattering data. The interdiffusion process can be described by Fick's law of diffusion with a time dependent interdiffusion coefficient. The measured interdiffusion coefficients decreased with increasing annealing time, suggesting that faster moving species control the interdiffusion process at short times. An analysis of the scattering data from completely interdiffused latex films based on de Gennes' scattering function for polymer blends reveals that molecular-level intermixing has been achieved in these latex films.

Over the past years, the interdiffusion of polymer chain molecules during the final stage of latex film formation has been the subject of much theoretical and experimental attention (*1*). It is obvious that the extent of interdiffusion across the latex boundaries has a dramatic effect on the properties of latex films. In practical applications, increasing amounts of latex films are used in paper coating, paints, textile sizing, and adhesives, due to the worldwide efforts to reduce the use of volatile organic solvents in these applications. All of these applications require the latex to be film-forming, at least to a degree. Therefore, an understanding of polymer-polymer interdiffusion process during latex film formation is essential to the successful development of new latex films.

Earlier studies on latex film formation process postulated that either polymer-air surface tension (*2*), or "autoadhesion" (*3*), i.e., the interdiffusion of polymer molecules across the particle-particle interfaces, is the driving force for the further coalescence of

[1]Current address: Abbott Laboratories, D–4E1, AP8A, 100 Abbott Park Road, Abbott Park, IL 60064–3500

dried latex films upon aging. Until recently, it was not possible to measure polymer interdiffusion during latex film formation. In the past several years, direct nonradiative energy transfer (DET) and small-angle neutron scattering (SANS) have been used to measure the interdiffusion coefficient and the molecular interpenetration depth during film formation directly (4-15). The DET technique is sensitive to the intermixing of fluorescent dye-labeled latex particles over a distance of a few nanometers, the interdiffusion process is followed by measurements of the changes in fluorescence decay curves. The interdiffusion coefficient can be extracted from the fluorescence decay data by assuming appropriate forms of the interpenetration depth or fitting the data with the solutions to the differential equation of Fick's law (4-10). SANS has been used to follow the growth of deuterated latex particles as interdiffusion progresses (11-15). The interdiffusion coefficient and the interpenetration depth were determined from the increase in particle size due to interdiffusion. In both DET and SANS studies, there is no consistent definition of the interpenetration depth, a quantitative measure of the extent of interdiffusion across particle-particle boundaries. Depending on the model assumed, various forms of the interpenetration depth have been proposed (5,8,13-15). In this paper, we will show that a simple equation relating the growth of the radii of gyration of labeled latex particles during interdiffusion to the interdiffusion coefficient can be derived directly from Fick's law of diffusion. The interpenetration depth defined through this equation also follows the scaling laws for the molecular weight and time dependence of the interpenetration depth. A generalized diffusion-scattering equation which reduces to the equation used by previous researchers will be introduced. SANS experiments were conducted on interdiffusing latex films composed of isotopic latex particles with large mismatch in molecular weight. The interdiffusion coefficient, the interpenetration depth, and the extent of intermixing will be obtained from the scattering data. The time dependence of the measured interdiffusion coefficients will be discussed. Finally, the scattering data from completely interdiffused latex films will be analyzed to study the miscibility of the polymers in the final films.

Theory

The Growth of Labeled Particles due to Interdiffusion. In this section, we derive the relation between the size of a labeled latex particle during interdiffusion and the interdiffusion coefficient. Consider a spherical latex particle labeled with either fluorescent dye or deuterium. As interdiffusion progresses, the boundary of the particle becomes diffused as shown schematically in Figure 1. The time evolution of the concentration profile for the particle under consideration can be described by Fick's law of diffusion

$$\frac{\partial c}{\partial t} = D\nabla^2 c \tag{1}$$

where c or c(r,t) is the concentration of the polymer in the particle. D is the interdiffusion coefficient. The origin of the coordinate system is the mass center of the particle. Because of spherical symmetry of the particle, equation 1 reduces to the following form:

$$\frac{\partial c}{\partial t} = D[\frac{1}{r^2}\frac{\partial}{\partial r}(r^2\frac{\partial c}{\partial r})] \tag{2}$$

where r is the radial distance from the origin. Multiplying both sides of equation 2 by r^4 and integrating over r, we obtain

$$\frac{\partial}{\partial t}\int_0^\infty dr \ r^4 c = D\int_0^\infty dr \ r^2 \frac{\partial}{\partial r}(r^2\frac{\partial c}{\partial r}) \tag{3}$$

The integral on the right hand side of equation 3 can be evaluated by integration by part as follows:

$$\int_0^\infty dr \ r^2 \frac{\partial}{\partial r}(r^2\frac{\partial c}{\partial r}) = r^4\frac{\partial c}{\partial r}\Big|_0^\infty - 2\int_0^\infty dr \ r^3\frac{\partial c}{\partial r}$$

$$= -2[r^3 c\Big|_0^\infty - 3\int_0^\infty dr \ r^2 c]$$

$$= 6\int_0^\infty dr \ r^2 c \tag{4}$$

where we have used the boundary conditions at $r = \infty$ where $\partial c/\partial r = 0$ and $c = 0$. Note that the last integral on the right hand side of equation 4 is equal to the total number of polymer molecules divided by 4π, and is therefore a constant. Inserting this result in equation 3, we obtain

$$\frac{\partial R_g^2(t)}{\partial t} = 6D \tag{5}$$

where

$$R_g^2(t) = \frac{\int dr \ r^4 c(r,t)}{\int dr \ r^2 c(r,t)} \tag{6}$$

It is obvious that $R_g(t)$ is the radius of gyration of the latex particle at time t. Integration of equation 5 yields the desired result

$$R_g^2(t) = R_g^2(0) + 6Dt \tag{7}$$

Equation 7 demonstrates that the relation between the growth of the radius of gyration of a latex particle due to interdiffusion and the interdiffusion coefficient D can be described analytically. Hahn et al. (*11,12*) studied the poly (butyl methacrylate) (PBMA) latex film formation process by SANS. They found an increase in the radii of gyration of the deuterated latex particles during the annealing process. To extract D from the measured $R_g(t)$, they obtained an analytical solution for c(r,t) in equation 2. The solution was then used in equation 6 to calculate $R_g(t)$ and to construct a master curve numerically for the time dependence of $R_g(t)$ as an universal function of $R_g(0)$ and Dt. The interdiffusion coefficients were then determined from the measured $R_g(t)$ and the master curve. It is obvious from above discussion that these data analysis procedures can be avoided since one can use equation 7 directly to obtain D from the measured $R_g(t)$. The same master curve can be constructed from equation 7 without the numerical computation step involved in obtaining $R_g(t)$ from equation 6.

The relation $R_g^2(t) = R_g^2(0) + 6Dt$ is perhaps not too surprised, considering the fact that the difference $R_g^2(t) - R_g^2(0)$ is the "mean" square displacement of the centers of mass of polymer molecules. Instead of an ensemble average of the square displacements of a particular molecule in time t, the mean square displacement here is an average displacement over all of the polymer molecules in the latex particle. Therefore $R_g^2(t) - R_g^2(0)$ must equal 6Dt. The interpenetration depth d(t) can then be defined as follows:

$$d(t) = [R_g^2(t) - R_g^2(0)]^{1/2} = (6Dt)^{1/2} \qquad (8)$$

For diffusion time greater than τ, the reptation time of the polymer molecule, $D \propto M^{-2}$, where M is the molecular weight of the polymer (16). Following equation 8, d(t) is found to scale as follows:

$$d(t) \propto M^{-1}t^{1/2} \qquad t > \tau \qquad (9)$$

This result agrees with the scaling law proposed by Kim and Wool (17) based on the reptation theory. At $t = \tau$, d(t) equals the radius gyration of the polymer R_{gi}. This is also consistent with equation 8 since $Rgi = (6D\tau)^{1/2}$. For $t < \tau$, the interdiffusion coefficient is ill defined since the center of mass of polymer molecule has moved over a short distance smaller than it's own size. The reptation theory predicts the following scaling law for d(t) in this time regime (16,17):

$$d(t) \propto M^{-1/4}t^{1/4} \qquad t < \tau \qquad (10)$$

To account for this scaling relation, the D in equation 8 becomes an apparent interdiffusion coefficient D_{app}, and D_{app} should have the following scaling law:

$$D_{app}(t) \propto M^{-1/2}t^{-1/2} \qquad t < \tau \qquad (11)$$

In other words, D_{app} decreases with increasing annealing time for $t < \tau$, and equation 1 becomes

$$\frac{\partial c}{\partial t} = D_{app}(t)\nabla^2 c \qquad t < \tau \qquad (12)$$

Equation 11 is essentially the same as the argument used by Wool and Whitlow (18) to explain the decrease in D values with increasing diffusion time observed in their secondary ion mass spectroscopy (SIMS) studies on polymer interdiffusion. This point will be discussed in detail later.

Diffusion-Scattering Theory. Summerfield and Ullman (19) derived an equation to obtain the self diffusion coefficient from SANS data on polymers composed of randomly mixed latex particles of protonated and deuterated polymers of the same molecular weight. A more generalized diffusion-scattering theory based on Summerfield and Ullman's equation was developed (20), the generalized diffusion-scattering equation extends the applicability of the equation to polymers with different molecular weight or chemical identities, and has the following form:

$$S(q,t) = S(q,0)\exp(-2q^2Dt) + S(q,\infty)[a(t) - a(0)\exp(-2q^2Dt)] \qquad (13)$$

where $S(q,t)$ is the scattering intensity from an interdiffused latex film at time t, q is the magnitude of wave factor q, and is equal to $4\pi/\lambda \sin(\theta/2)$, λ is the neutron wavelength , θ is the scattering angle. $S(q,0)$ and $S(q,\infty)$ are the scattering intensities of the initial and completely interdiffused latex films, respectively. Equation 13 assumes that the interdiffusion is Fickian, i. e., equation 1 is valid in describing the interdiffusion process. $a(t)$ is a space-time correlation function defined as follows:

$$a(t) = \frac{\bar{c} - <c^2(r,t)>}{\bar{c} - \bar{c}^2} \tag{14}$$

where $\bar{c}$ is the mean concentration of labeled molecules. The angular brackets represent an average over the space. If the labeled latex particles are well separated from the unlabeled latex particles initially, $a(0) = 0$. $a(t)$ increases from 0 to 1 as interdiffusion progresses. For a completely interdiffused latex film, $a(t) = 1$. In other words, $a(t)$ is a measure of the extent of intermixing during the interdiffusion process. This is expected from equation 14 since $<c^2(r,t)>$, the mean square concentration, is always greater than $\bar{c}^2$, the square of the mean concentration, for an interdiffusing latex film with concentration gradients throughout the space. Evidently, $<c^2(r,t)>$ approaches $\bar{c}^2$ during the interdiffusion process. The physical meaning of the diffusion-scattering equation can be explained qualitatively by considering a latex film composed of well segregated latex particles initially. This is shown schematically in Figure 2. Working at low q is equivalent to using a low power magnifying glass (21). We see large long wave length concentration fluctuations. The latex film is highly inhomogeneous. One should observe strong scattering at low q. As the polymer molecules start to interdiffuse across the latex boundaries, the long wave length concentration fluctuations begin to decrease; the latex film becomes more homogeneous; and the scattering intensity decreases as time increases (Figure 2a). This part of the scattering is represented by the first term of the diffusion-scattering equation $S(q,0) \exp(-2q^2Dt)$. At high q, on the other hand, the dimension we observe is less than the size of the latex particle since we are using the high power magnifying glass. The scattering from either a labeled particle or an unlabeled particle is almost zero since there is no concentration fluctuation at all in each particle initially. Thus we expect the scattering intensity to be very small in this high q region at $t = 0$. As interdiffusion progresses, however, each latex particle becomes more "inhomogeneous" since the labeled particles begin to mix with the unlabeled ones, we expect the scattering intensity to increase with time at high q. The physical picture at high q is shown schematically in Figure 2b. The above qualitative argument is consistent with the prediction from the diffusion-scattering equation. At high q, equation 13 reduces to $S(q,t) = S(q,0)a(t)$, since $a(t)$ is an increasing function of time. It follows that $S(q,t)$ increases with time at high q.

   Equation 13 can be used to extract the extent of intermixing and the interdiffusion coefficient from the scattering data. $S(q,0)$ is obtained from a sample containing originally segregated latex particles, $S(q,\infty)$ is obtained from a sample that is completely interdiffused, and $S(q,t)$ is measured at any intermediate time. Since for any particular time, there are data points for many values of q, $a(t)$ and D can be obtained from the scattering data.

   Notice that equation 13 is quite general for an interdiffusing latex film composed of randomly mixed labeled latex particles. It reduces to the equation used by both Hahn et al. (11,12) and Sperling et al. (13-15) in their SANS studies on interdiffusion during latex film formation. Since either the interpenetration depth or the interdiffusion coefficient was obtained from the measured radii of gyration of the labeled particles $(R_g(t))$ in their studies. Low concentrations of labeled latex particles were used to

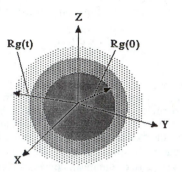

Figure 1. The growth of a labeled latex particle due to interdiffusion of polymer molecules across the particle boundary. The radius of gyration of the particle at t = 0 is $R_g(0)$. $R_g(0)$ increases to $R_g(t)$ after time t. This phenomenon has been observed in previous SANS experiments on interdiffusing latex films (11-15).

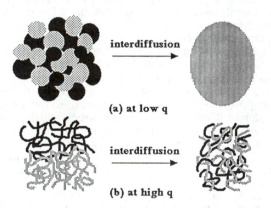

Figure 2. The physical meaning of the diffusion-scattering equation. (a) At low q, the concentration gradient decreases as interdiffusion progresses. The scattering intensity decreases with increasing annealing time. (b) At high q, on the contrary, the concentration gradient within a labeled or unlabeled latex particle increases since they start to mix with each other. The scattering intensity increases slowly with increasing annealing time at high q.

ensure that $R_g(t)$ can be determined from their scattering data. In this case, $S(q,\infty)$ is much smaller than $S(q,0)$, equation 13 reduces to

$$S(q,t) = S(q,0)\exp(-2q^2 Dt) \tag{15}$$

In the Guinier region, where $qR_g < 1$, $S(q,0) = S(0,0)\exp(-q^2 R_g^2(0)/3)$. $R_g(0)$ is, of course, the radius of gyration of the deuterated latex particle before interdiffusion. Equation 15 can be rewritten as

$$S(q,t) = S(0,0)\exp[-q^2(R_g^2(0) + 6Dt)/3] = S(0,0)\exp[-q^2 R_g^2(t)/3] \tag{16}$$

Equation 16, the so-called Guinier approximation, was utilized by Hahn et al in their SANS studies on polymer interdiffusion during PBMA latex film formation as mentioned earlier (*11,12*). More recently, Sperling and co-workers (*13-15*) also used equation 16 to study the correlation between the interpenetration depth and the tensile strength build-up of polystyrene (PS) latex films prepared by either emulsion polymerization or direct mini-emulsification. The interpenetration depth in their studies was defined as follows:

$$d(t) = [R_g(t) - R_g(0)](5/3)^{1/2} \tag{17}$$

Notice that the relation $R_g^2(t) = R_g^2(0) + 6Dt$ is also recovered in equation 16. This is again due to the fact that the growth of the labeled latex particle as a consequence of interdiffusion can be obtained directly from Fick's law of diffusion.

In fact, the use of low concentration deuterated latex particles in latex films to ensure that D and d(t) can be determined experimentally is not necessary. We can use high concentration deuterated latex particles and apply equation 13 to obtain D and a(t). The interpenetration depth can be obtained from the measured D according to equation 8. The use of high concentration deuterated latex particles offers the advantage of stronger scattering intensities from interdiffusing latex films. The extent of intermixing a(t) can also be determined from the high q portion of the scattering data. Finally, the scattering from a completely mixed sample $S(q,\infty)$ can be analyzed to obtain the Flory-Huggins interaction parameter and the correlation length of the polymers in the final film. These parameters give a quantitative measure of the compatibility of the components in the final film.

Experimental Methods

The polymers used in this study were monodisperse polystyrene (PS) (Pressure Chemical Co., Mw/Mn<1.1), and deuterated polystyrene (dPS) (Polymer Laboratories, Mw/Mn<1.1). Latex particles containing the two diffusing species dPS and PS were prepared by an ultrasonic emulsion method developed by Anderson and Jou (*22*). The emulsions were made as follows: an aqueous surfactant solution (0.1g sodium lauryl sulfate/2ml distilled water) was added to the dPS solution (1g of dPS/10ml benzene) followed immediately by ultrasonic agitation for 30 seconds. The dPS/benzene solution and the aqueous solution of the emulsifier were both kept at 65°C before mixing since sodium lauryl sulfate does not dissolve completely in water below 65°C. Ultrasonic agitation produces a stable emulsion containing dPS latex particles. The PS emulsion was prepared via the same procedures. dPS and PS emulsions were then mixed together, freeze-dried to remove solvent, and washed with hot distilled water to remove emulsifier. The resulting wet powder was dried under vacuum at 85°C until constant weight was obtained. The sizes of the latex particles range from

0.05 μm to 0.5 μm on the basis of SEM examination on the initial latex particles. Film samples for SANS experiments were prepared by pressing the dried polystyrene particles in a vacuum mold held in a hot press for 30 minutes at a temperature 110°C and mold pressure 7×10^6 Pa. These conditions are just sufficient to produce transparent films while minimizing chain interdiffusion between neighboring particles.

Thermal annealing was done in a vacuum oven. To prevent sample distortion during heating, the latex films were placed within a steel O-ring spacer sandwiched between two thin steel disks. Tubing clamps were used to hold the assembly together. Two thermocouples, one on the top steel disk, the other on the bottom disk, were used to monitor the sample temperature. For each experiment, 8 ~ 10 minutes were allowed for temperature equilibrium before the clock was started to measure the annealing time. As soon as the samples were taken out of the oven, they were quenched in an ice/water bath to inhibit any further diffusion. The annealing temperatures was 120°C. Completely interdiffused samples were prepared by annealing the films at 160°C for 24 hours followed by annealing at 120°C for an additional 24 hours to establish the required thermodynamic equilibrium at the annealing temperature.

The SANS measurements were performed with the SANS facilities at the Oak Ridge National Laboratory. The main results were obtained with a 14 m sample to detector distance, a neutron wavelength of 4.75Å, and source/sample slit diameters of 2.5 and 1.0 cm, respectively. A 8.0 cm beam stop was used. Neutrons were counted with a 64 cm × 64 cm two-dimensional position-sensitive detector. In this configuration useful SANS data were obtained over the range 4.6×10^{-3}Å^{-1} < q < 3.78×10^{-2}Å^{-1}.

Results and Discussion

Scattering Profiles and Fickian Interdiffusion. Figure 3 shows scattering profiles taken at different annealing times for latex films composed of equal amounts of dPS (Mw = 68,000) and PS (Mw = 600,000) latex particles initially. The number in the parentheses indicates the molecular weight of the polymer. Kratky plots ($S(q,t)q^2$ versus q) were used in order to expand the high q portion of theses curves. As expected from the diffusion-scattering theory, the scattering intensity decreases with increasing annealing time at low q. The $S(q,0)\exp(-2q^2Dt)$ term in the diffusion-scattering equation dominates. On the other hand, scattering at high q (q > 0.02Å^{-1}) is dominated by the $a(t)S(q,\infty)$ term and increases slowly with increasing annealing time. A "crossover" region is clearly seen in Figure 3. These results are consistent with the qualitative features of the diffusion-scattering equation discussed above. The extent of intermixing a(t) was obtained from the high q parts of the data where a(t) = S(q,t) / S(q,∞). To extract the interdiffusion coefficients from the scattering data, we define a function G(q,t) as follows:

$$G(q,t) = \frac{S(q,t) - a(t)S(q,\infty)}{S(q,0) - a(0)S(q,\infty)} = \exp(-2q^2Dt) \tag{18}$$

If the interdiffusion process can be described by Fick's law, then a semilog plot of G(q,t) versus q^2 at fixed annealing time should yield a straight line. Figure 4 shows the semilog plots of G(q,t) versus q^2. A series of lines with increasing slopes is clearly seen indicating that the interdiffusion is indeed Fickian. We present the first 14 data points only since we are interested in low q region (qR_{gi} < 1, R_{gi} is the radius of gyration of the polymer) where the transitional diffusion of centers of mass of polymer

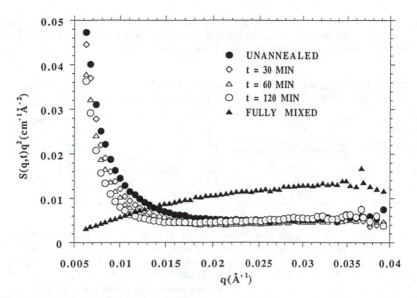

Figure 3. Kratky plots of latex films composed of deuterated polystyrene (Mw = 68,000) and polystyrene (Mw = 600,000) latex particles at different annealing times. At low q, the scattering intensity decreases with increasing annealing time. At high q, the scattering intensity increases slowly with increasing annealing time.

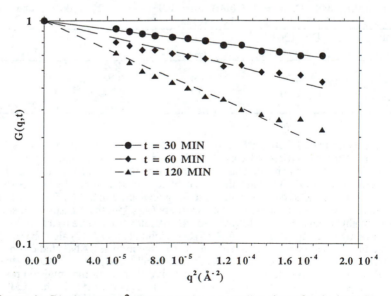

Figure 4. G(q,t) versus q^2 plots at various annealing times for the latex films in Figure 3. These straight lines indicate that the interdiffusion between deuterated polystyrene and polystyrene molecules is Fickian.

molecules takes place. The interdiffusion coefficients were determined from the slopes of these lines. Equation 8 was used to calculate the interpenetration depth d(t). These results as a function of annealing time are presented in Table I.

Table I. Summary of SANS Results of Polystyrene Latex Films containing dPS (Mw = 68,000) and PS (Mw = 600,000) Latex Particles

annealing time (min.)	D at 120°C $(10^{-17}$ cm^2/sec)	d (Å)	a (t)	$\exp(-2q^2Dt)$ at $q=0.03$ Å^{-1}
0			.414	1
30	5.99	80	.392	0.14
60	5.48	109	.365	0.029
120	5.11	149	.432	0.0012

The Kratky plots for latex films consisting of dPS (Mw = 129,000) and PS (Mw = 2750,000) latex particles at different annealing times are presented in Figure 5. Again, the scattering intensity decreases with increasing annealing time at low q. In high q region, the scattering increases with annealing time. Figure 6 shows the semilog plots of G(q,t) versus q^2 for these latex films. As seen from Figure 6, the interdiffusion processes is also Fickian at each annealing time for these latex films. The values of D, d(t), and a(t) determined from the scatting data for these latex films are summarized in Table II.

Table II. Summary of SANS Results of Polystyrene Latex Films containing dPS (Mw = 129,000) and PS (Mw = 2750,000) Latex Particles

annealing time (min.)	D at 120°C $(10^{-17}$ cm^2/sec)	d (Å)	a (t)	$\exp(-2q^2Dt)$ at $q=0.03$ Å^{-1}
0			.464	1
30	4.32	68	.399	0.25
60	2.91	79	.379	0.15
240	1.79	124	.389	0.0099

Time Dependence of Interdiffusion Coefficients. As seen from Tables I and II, the interdiffusion coefficient decreases slightly with increasing annealing time for the first set of latex films. Such a decrease with annealing time is more pronounced for the second set of latex films. The interdiffusion coefficients determined by both SANS and DET were often found to decrease with increasing annealing time (4-15). In the DET experiments performed by Winnik and co-workers (4,6,7,9), the measured D values for poly(methyl methacrylate) (PMMA) and PBMA latex films were found to decrease with annealing time. Sperling et al. (13-15) and Hahn et al. (11-12) also observed a decrease in D values with an increase in annealing time in their SANS studies on polystyrene and PBMA latex films, respectively. In all these previous studies, the decrease in D values was attributed to the large polydispersity of the emulsion polymers used, where short chains dominate the interdiffusion at short times. In this study, we intentionally used polystyrene latex films composed of two kinds of polystyrene molecules with mismatched molecular weight. It seems to us that the greater the difference in molecular weight, the more pronounced is the decrease in D values with increasing annealing time. Therefore, this study supports the proposition that faster

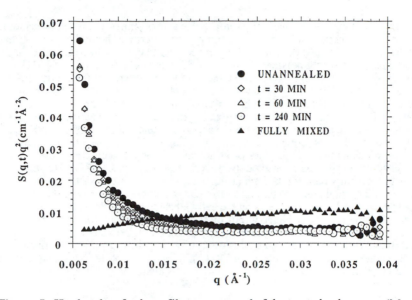

Figure 5. Kratky plots for latex films composed of deuterated polystyrene (Mw = 129,000) and polystyrene (Mw = 2750,000) latex particles at different annealing times. At low q, the scattering intensity decreases with increasing annealing time. At high q, the scattering intensity increases slowly with increasing annealing time.

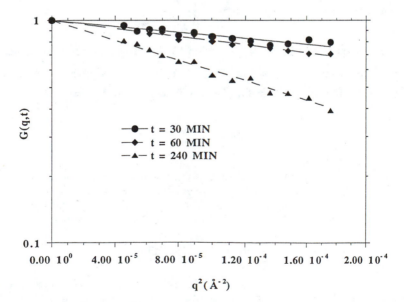

Figure 6. $G(q,t)$ versus q^2 plots at various annealing times for the latex films in Figure 5. These straight lines indicate that the interdiffusion between deuterated polystyrene and polystyrene molecules is Fickian.

moving species control the interdiffusion process at short times and the slower moving species become more significant at long times. However, we should be aware of the fact that the values of d(t) determined here are smaller than the radii gyration of the higher molecular weight polymers (211 Å for PS with Mw = 600,000 and 451 Å for PS with Mw = 2750,000) for all of the annealing times covered in this study. For the second set of latex films, d(t) is even smaller than the radius gyration of the smaller molecular weight dPS (98 Å) for annealing times less than one hour. Thus the measured D values can decrease with annealing time according to equation 11. In this time regime, D should be considered as an apparent interdiffusion coefficient D_{app}. It is also possible that conformational relaxation of high molecular weight polymers constrained within the latex particles provides an additional driving force for interdiffusion initially, as pointed out by several researchers (*4,18,23*). This effect becomes more important for the high molecular weight polymers whose chain dimensions are comparable to the sizes of the latex particles. To answer this question, we can conduct SANS experiments on partially labeled latex particles to determined the radius of gyration of the polymer within these particles. This approach has been successfully used by Cohen et al. (*24*) in their work on the polymer chain conformations within the microdomains of styrene-butadiene copolymers. In any event, we need to use a time dependent $D_{app}(t)$ in Fick's law of diffusion to account for the observed decrease in D with annealing time. As mentioned earlier, the diffusion equation becomes

$$\frac{\partial c}{\partial t} = D_{app}(t)\nabla^2 c \tag{12}$$

The solution of equation 12 in the wave vector space **q** takes the following form:

$$c(\mathbf{q},t) = c(\mathbf{q},0) \ \exp(-q^2 \int_0^t dt' \ D_{app}(t')) \tag{19}$$

Equation 19 shows that the term Dt in equation 13, the diffusion-scattering equation, should be replaced by the integral $\int_0^t dt' \ D_{app}(t')$. In other words, the measured D at annealing time t is an average interdiffusion coefficient D_t between t = 0 and t :

$$D_t = \frac{\int_0^t dt' \ D_{app}(t')}{t} \tag{20}$$

Our scattering data suggest that equation 12 is valid in describing the interdiffusion process for PS latex films since logG(q,t) shows q^2 dependence for all annealing times covered in this study. But the interdiffusion coefficient D_t defined in equation 20 should be used in interpreting the data. Similarly, d(t) has the following form:

$$d(t) = [6\int_0^t dt' \ D_{app}(t')]^{1/2} = (6D_t t)^{1/2} \tag{21}$$

Extent of Intermixing. Notice that the a(t) in Tables I and II was obtained from the high q portion of the scattering profiles where a(t) = S(q,t) / S(q,∞), independent of the data in low q region where the interdiffusion coefficient was determined. The relation a(t) = S(q,t) / S(q,∞) is valid provided that exp(-2q²Dt) << 1. As seen from Tables I and II, a(t) increases slowly with increasing annealing time except for short annealing times where $\exp(-2q^2 Dt)$ is not much smaller than one. The values of $a(t)$ obtained at short annealing times are the apparent values only. True values of $a(t)$ at these times should be less than those at longer annealing times because the initial states of the latex films were farther away from the equilibrium (completely mixed) state. It is interesting to comment on the apparent $a(0)$ values. We postulate that some degree of interdiffusion has been introduced during the sample preparation step (hot press). Since the $a(t)$ values increase slowly with annealing time as shown in Tables I and II, it is reasonable to say that $a(0)$ should be very close to the $a(t)$ value at 30 minutes annealing time. This is also supported by the Porod plots ($\log S(q,0)$ versus $\log q$) of the unannealed latex films at high q region ($q > 0.015$ Å) where the roughness of the boundaries between deuterated and protonated latex particles can be determined. If the boundaries are sharp and smooth, $S(q,0) \sim q^{-4}$ (25). As shown in Figure 7, the power-law exponents determined from the Porod plots for the unannealed films are smaller than 4, indicating that the initial boundaries between latex particles are not smooth. Some form of intermixing or coalescence between latex particles has been introduced in the unannealed films. As a result of this early interdiffusion, the time zero mentioned in the theory section previously should be shifted to some later time t_1. The shift of the origin of time to some unknown time t_1, however, does not alter the results obtained in this or previous studies with the same sample preparation method (11-15) .

Theoretically, equation 7 is valid between any two annealing times t_1 and t_2:

$$R_g^2(t_2) = R_g^2(t_1) + 6D(t_2 - t_1) \tag{22}$$

D can still be directly obtained from either equation 7 or 22 since the additional annealing time ($t_2 - t_1$) after sample preparation was determined in these experiments. Similarly, equation 13, the diffusion-scattering equation, takes the same form between annealing times t_1 and t_2 (replacing $\exp(-2q^2 Dt)$ with $\exp[-2q^2 D(t_2 - t_1)]$ in equation 13). Therefore, the early interdiffusion does not change the results obtained so far.

Another measure of the extent of intermixing, the volume fraction of mixing, $fm(t)$, has been obtained from DET measurements on interdiffusing latex films (4-10, 23). $fm(t)$ is defined in terms of the fractional growth in fluorescence energy transfer efficiency, and is the key parameter used to determine the interdiffusion coefficient from the fluorescence decay curves. The present results show that the extent of intermixing after 4 hours is about 0.4 for the PS latex films, at an annealing temperature about 20°C above the Tg of PS (103°C). This value appears to be higher than the $fm(t)$ value for PMMA latex films (~ 0.2) determined by DET under similar annealing conditions (6). In future studies, it will be worthwhile to compare the $a(t)$ value determined by SANS with the $fm(t)$ value measured by DET for the same polymer latex film under the same experimental conditions.

Miscibility of Polymers in the Fully Mixed Films. To assess the "final" extent of intermixing or more appropriately the miscibility of the polymers in the completely mixed films. We analyze the scattering data $S(q,\infty)$ for these films. For binary polymer blends in thermodynamic equilibrium in one phase region, using the so called "random-phase approximation" (RPA), de Gennes derived for $S(q,\infty)$ at small q the following scattering function (26):

$$S(q,\infty) = \frac{S(0,\infty)}{1 + q^2 \xi^2} \tag{23}$$

$$S(0,\infty) = K(\chi_s - \chi)^{-1} \tag{24}$$

$$\xi^2 = \frac{1}{6}(\frac{R_{g,A}^2}{z_A\phi_A} + \frac{R_{g,B}^2}{z_B\phi_B})(\chi_s - \chi)^{-1} \tag{25}$$

$$\chi_s = \frac{1}{2}(\frac{1}{z_A\phi_A} + \frac{1}{z_B\phi_B}) \tag{26}$$

ϕ_i is the volume fraction of component i. z_i and R_{gi} are the degree of polymerization and radius of gyration of component i, respectively. K is a contrast factor depending on the scattering lengths of the components. χ is the Flory-Huggins interaction parameter, χ_s is the value of χ at the spinodal point (phase separated). ξ is the correlation length which is a quantitative measure of the miscibility of the polymers in the final films. When χ approaches χ_s, the ξ value becomes a large number, and the scattering intensity increases dramatically. Equation 23 is valid when $qR_{g,i} < 1$. According to equation 23, a plot of $S(q,\infty)^{-1}$ versus q^2 (Ornstein-Zernike-Debye or OZD plot) at low q should yield a straight line. χ and ξ can be obtained from the intercept and slope of this line. Figures 8 and 9 show the OZD plots for the two sets of latex films used in this study. The values of ξ obtained are 84 Å and 163 Å for the low and high molecular weight latex films, respectively. These dimensions are either smaller than or comparable to the radii gyration of the PS or dPS molecules, indicating that molecular-level intermixing has been achieved in the fully mixed latex film. These results demonstrate that an analysis of the scattering data from completely mixed latex films by RPA provides detailed molecular-level information about the miscibility of the polymers in the final films. This analysis is especially useful if mixtures of latex particles composed of chemically different polymers are utilized to form the films.

The χ values for both latex films are within the range of 10^{-5} to 10^{-4}, as expected for these partially labeled isotopic mixtures. The χ values found here are too small to cause thermodynamic slowing down of the interdiffusion process due to deuteration of one of the diffusion partner (Eu, M.D.; Ullman R. to be published). However, for latex film composed of partially miscible polymer pairs, the effect of χ on the rate of interdiffusion can be significant (5). SANS measurements on these latex films can provide a quantitative measure of the magnitude of the thermodynamic force.

Summary

We showed that the growth of the radii of gyration of labeled latex particles, commonly observed in SANS experiments on interdiffusing latex films, is related to the interdiffusion coefficient of the polymers through a simple equation derived from Fick's law of diffusion. The time and molecular weight dependence of the interpenetration depth d(t) defined through this equation is consistent with the scaling law for d(t) based on the reptation theory. We used the generalized diffusion-scattering equation, which reduces to the equation used in previous SANS studies on latex film formation, to analyze the scattering data from SANS experiments on interdiffusing latex films consisting of isotopic latex particles with large mismatch in molecular weight. The interdiffusion coefficient, the interpenetration depth, and the extent of intermixing were obtained from the scattering data. Based on the observed q^2 dependence of the function $\log G(q,t)$ for all of the annealing times covered in this study, we concluded that Fick's law of diffusion is adequate to follow the interdiffusion process during

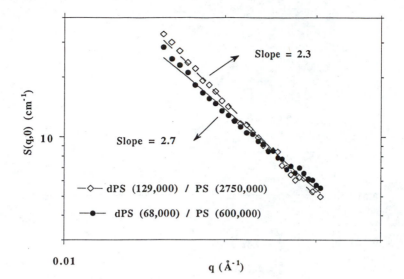

Figure 7. Power-law dependence of the scattering intensities from unannealed latex films (s(q,0)) at high q. The power exponents for both latex films are smaller than 4, suggesting that some degree of interdiffusion has been introduced during the sample preparation step.

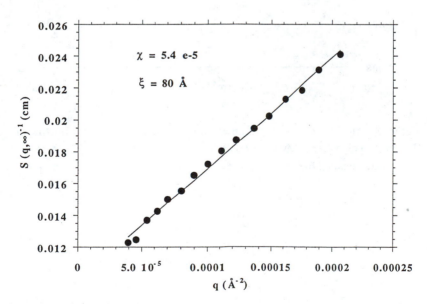

Figure 8. An OZD plot of the completely interdiffused latex film consisting of deuterated polystyrene (Mw = 68,000) and polystyrene (Mw = 600,000). χ and ξ were obtained from the intercept and slope of the line in the Figure.

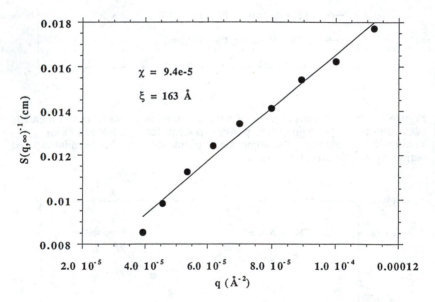

Figure 9. An OZD plot of the completely interdiffused latex film consisting of deuterated polystyrene (Mw = 129,000) and polystyrene (Mw = 2750,000). χ and ξ were obtained from the intercept and slope of the line in the Figure.

polystyrene latex film formation. We also found that the measured D values decrease with increasing annealing time. It seems to us that the greater the difference in molecular weight, the more pronounced is the decrease in D values with increasing annealing time. Therefore, this study supports the proposition that faster moving species control the interdiffusion process at short times and the slower moving species become more significant at long times. To account for such time dependence of the measured D, we were compelled to use a time dependent interdiffusion coefficient in Fick's law of diffusion. The measured D at annealing t is then an average interdiffusion coefficient D_t between t = 0 and t. The scattering data from completely interdiffused latex films were analyzed based on de Gennes' scattering function. Molecular-level intermixing has been achieved in the these latex films since the correlation lengths obtained from this analysis are either smaller than or comparable to the radii of gyration of the PS or dPS molecules. We believe that the use of the generalized diffusion-scattering equation and de Gennes' scattering function in SANS studies on interdiffusing latex films will allow us to gain deeper insight into the mechanism of polymer interdiffusion during latex film formation.

Literature Cited

1. Peckan, ö. *Trends in Polym .Sci.*, **1994**, 2, 236.
2. Vanderhoff, J. W. *Br. Polym. J.*, **1970**, 2, 161.
3. Voyutskii, S. S. *J. Polym. Sci.*, **1958**, 32, 528.
4. Zhao, C. L.; Wang Y.; Hruska, Z.; Winnik, M. A. *Macromolecules*, **1990**, 23, 4082.
5. Boczar, F. M.; Dionne, B. C.; Fu, Z. W.; Kirk, A. B.; Lesko, P. M.; Koller, A. D. *Macromolecules*, **1993**, 26, 5772.
6. Wang, Y.; Winnik, M. A. *Macromolecules*, **1993**, 26, 3147.
7. Kim, H. B.; Winnik, M. A. *Macromolecules*, **1994**, 27, 1007.
8. Juhué, D.; Lang J. *Macromolecules*, **1994**, 27, 695.
9. Kim, H. B.; Winnik, M. A. *Macromolecules*, **1995**, 28, 2033.
10. Juhué, D.; Lang J. *Macromolecules*, **1995**, 28, 1306.
11. Hahn, K.; Ley, G.; Schuller, H.; Oberthür, R. *Colloid Polym. Sci.* **1986**, 264, 1092.
12. Hahn, K.; Ley, G.; Oberthür, R. *Colloid Polym. Sci.* **1988**, 266, 631.
13. Yoon, J. N.; Sperling, L. H.; Glinka, C. J.; Klein, A. *Macromolecules*, **1990**, 23,3962.
14. Kim, K. D.; Sperling, L. H.; Klein, A. *Macromolecules*, **1993**, 26, 4624.
15. Kim, K. D.; Sperling, L. H.; Klein, A.; Hammouda B. *Macromolecules*, **1994**, 27,6841.
16. de Gennes, P. G. *J. Chem. Phys.*, **1971**, 55, 572.
17. Kim, Y. H.; Wool, R. P. *Macromolecules*, **1983**, 16, 1115.
18. Wool, R. P.; Whitlow, S. J. *Macromolecules* **1991**, 24, 5926.
19. Summerfield, G. C.; Ullman, R. *Macromolecules*, **1987**, 20, 401.
20. Eu, M. D. Ph.D. Dissertation, The University of Michigan, **1992**.
21. Higgins, J. S.; Benoit, H. C. *Polymers and Neutron Scattering*; Clarendon Press: Oxford, **1994**.; Chapter VI, pp 166-167.
22. Anderson, J. E.; Jou, J. H. *Macromolecules*, **1987**, 20, 1554.
23. Wang, Y. C.; Zhao, C. L.; Winnik, M. A. *J. Chem. Phys.*, **1991**, 95, 2143.
24. Bates, F. S.; Berney, C. V.; Cohen, R. E. *Polymer*, **1983**, 24, 519.
25. Debye, P.; Bueche, A. M. *J. Appl. Phys.* **1949**, 20, 518.
26. de Gennes, P. G. *Scaling Concepts in Polymer Physics*,: Cornell University Press, Ithaca, New York, **1979**; Chapter IV, pp 108-111.

Chapter 7

In Situ Sensor for Monitoring Molecular and Physical Property Changes During Film Formation

D. E. Kranbuehl, D. Hood, C. Kellam, and J. Yang

Departments of Chemistry and Applied Science, College of William and Mary, Williamsburg, VA 23187−8795

A frequency dependent electromagnetic sensing technique has been successfully used to continuously monitor, insitu, the cure process of commercially available epoxy and latex paints. The effects of environmental conditions such as temperature and humidity, as well as coating formulation differences due to pigments on the cure process is measured insitu during cure. The ability of the FDEMS sensor to monitor the extent to which a second coating can soften the initial coating is discussed. The ability of the FDEMS sensors to monitor the extent to which water can diffuse into the coating is also described. Overall, the FDEMS sensor shows considerable promise as an insitu on-line means of monitoring the buildup in durability of a coating during cure in a variety of processing environments including thermal, UV, and varying ambient conditions.

An insitu frequency dependent electrical measurement sensor (FDEMS) has been successfully used to monitor cure and the buildup in properties during film formnation. The planar micro-sensor is able to make continuous uninterrupted measurements of the film while it cures as a coating with only one side exposed. It is able to monitor reaction onset, cure rate, viscosity, buildup in hardness, cure completion and related processes such as latex coalescence and evolution of volatiles (1-5). Effects of storage, temperature, humidity, thickness, and variations in composition on the cure process can readily be detected. The sensor monitors the changes in the rate of translational motion of ions and rotational motion of dipoles through frequency dependent complex permitivity measurements.

In this report the ability of the sensor to monitor cure of epoxy-polyamide and latex films is discussed. The effects of the environmental conditions such as temperature and humidity, as well as coating formulation differences due to pigments,

0097−6156/96/0648−0096$15.00/0

on the cure process will be reported. The ability of the sensor technique to detect and monitor the three phases of latex film formation are shown. The ability of the FDEMS sensor to monitor the extent to which a second coating can soften the initial film is described. The FDEMS sensor is also used to monitor the extent to which water or solvents diffuse into the film. Overall, the FDEMS sensor shows considerable promise as an insitu on-line means of monitoring film formation as well as the buildup in durability of a film insitu in a variety of processing conditions including thermal, UV, and *varying ambient conditions.

Experimental

Measurements at frequencies from 5 to $10°$ Hz are taken continuously throughout the entire cure process at regular intervals and converted to the complex permitivity, ϵ^* = ϵ'' - $i\epsilon''$. Measurements are made with a commercially available geometry independent inter digitated DekDyne microsensor which was recognized with an IR-100 award in 1989. The sensor is planar, 1 inch by 1/2 inch in size and 3 mils thick. Resin is spread on the sensor as a film ~ 1 to 10 mils thick. The DekDyne sensor-bridge PC computer assembly is able to make continuous uninterrupted measurements of both ϵ'' and ϵ' over 10 decades in magnitude of ϵ'' and ϵ' at all frequencies. Measurements can be made remotely at distances of several hundred feet away from the computer-bridge system. Automated measurements can be made simultaneously on up to 6 sensors or more through multiplexing. Thereby samples can be compared side-by-side under identical ambient conditions of humidity, air flow, light flux, temperature etc. The sensor is inert and has been used at temperatures of $400°C$ and 1000 psi pressure. A more detailed description of the equipment and procedures used to monitor cure has been published (1,2). A schematic is shown in Figure 1.

Theory

Frequency dependent sensor measurements of the coatings' dielectric impedance as characterized by its equivalent capacitance, C, and conductance, G, are used to calculate the complex permitivity, $\epsilon^* = \epsilon' - i\epsilon''$, where $\omega = 2\pi f$, f is the measurement frequency and CO is the air replacement capacitance of the sensor.

$$\epsilon'(\omega) = \frac{C(\omega)\ material}{C_o} \tag{1}$$

$$\epsilon''(\omega) = \frac{G(\omega)\ material}{\omega C_o} \tag{2}$$

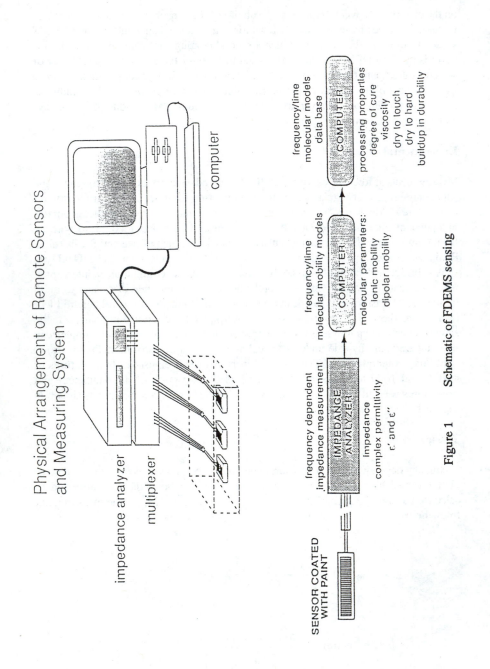

Figure 1   Schematic of FDEMS sensing

This calculation is possible when using the sensor whose geometry is invariant over all measurement conditions.

Both the permitivity ϵ' and loss factor ϵ'' can have an ionic and dipolar component.

$$\epsilon'' = \epsilon''_d + \epsilon''_I \qquad \text{and} \qquad \epsilon' = \epsilon'_d + \epsilon'_I \qquad (3)$$

The dipolar component ϵ_d arises from rotational diffusion of molecular dipole moments or bound charge. The ionic component ϵ_i arises from the translational diffusion of charge. The dipolar term is generally the major component of the dielectric signal at high frequencies and in highly viscous media. The ionic component dominates $\epsilon^*(\omega)$ at low frequencies, low viscosities and/or higher temperatures and generates a peak in ϵ'' when the measurement frequency $2\pi f = 1/\tau$, where τ is the mean rotational relaxation time.

One general form for the dipolar component's frequency dependence is the Havrilak-Negami equation

$$(4)$$

$$\epsilon'_d - i\epsilon''_d = \frac{(\epsilon_o - \epsilon_\infty)}{(1 + (i\omega\tau)^\alpha)^\beta} + \epsilon_\infty$$

where ϵ_0 and ϵ_∞ are the limiting low and high frequency values of ϵ_d. The rotational diffusion rate of the polar moments is characterized by a mean relaxation time τ. The parameters α and β characterize the distribution of the relaxation time.

Effects of ionic conduction in polymeric media subjected to an alternating electric field are a function of the mechanism for ion transport and as such reflect the viscosity of the medium, boundary conditions at the electrode surfaces, size of the ions involved, and heterogeneities in the system. In fluids, ions migrate to the electrodes and may form an electric double-layer at their surface causing electrode polarization. These ions may then diffuse through this layer and discharge at the electrode. As the viscosity increases, this motion is slowed, or sometimes blocked altogether.

Considerable research has focused on the effects of ionic conductivity on the frequency dependence of the dielectric constant ϵ' and the apparent conductivity, Gapp of the dielectric loss term, ϵ''. Friauf suggested that for both + and - ions blocked, $\log \epsilon'$ vs. $\log f$ would have a slope of -2.[6] However, if only one ion is blocked and the other free to translate, then the frequency dependence of ϵ' is proposed to vary as $f^{3/2}$ while G should be proportional to $f^{1/2}$.[7]

Johnson and Cole derived an empirical formula for the behavior of an ionizable material (formic acid) in liquid and solid states[9]:

$$\epsilon_{app} = \epsilon_d + C_o Z_o \sin\left(\frac{n\pi}{2}\right) \omega^{-(n+1)} \left(\frac{\sigma}{8.85x10^{-14}}\right)^2 \tag{5a}$$

$$\epsilon_{app}'' = \epsilon_d'' + \frac{\sigma}{8.85x10{-14}\omega} - C_o Z_o \cos\left(\frac{n\pi}{2}\right) \omega^{-(n+1)} \left(\frac{\sigma}{8.85x10^{-1}}\right) \tag{5b}$$

where C_o is the replaceable capacitance in Farads, Z is the electrode impedance induced by the ions, $\sigma(\text{ohm}^{-1} \text{ cm}^{-1})$ is the specific conductivity and n is between 0 and 1. In their experiments the value of n for the liquid state was 1 and that for the solid formic acid was 0.5. At particular values of n, Johnson and Cole's equations correspond to various theoretical representations. Then $n= 1$ corresponds to the case when electrode polarization effects become dominant. When $n=0.5$, equation (5) reduces to that for diffusion controlled ionic effects. The boundary condition $n=0$ is equivalent to pure d.c. conductance.

The d.c. ionic component of ϵ'' is the second term in eq (5b) where the specific conductivity $o(\text{ohm}^{-1} \text{ cm}^{-1})$ is an intensive variable, in contrast to conductance $G(\text{ohm}^{-1})$ which is dependent upon sample size σ and G reflects primarily the translational motion of ions through the resin medium. The value ϵ', by contrast, is affected by electrode impedance and charge layer effects and is observed to be less informative for monitoring coating cure.

Discussion

Plots of angular frequency times the loss factor $\omega\epsilon''(\omega)$ make it relatively easy to visually determine when the low frequency magnitude of ϵ'' is monitoring ionic translational motion and when dipolar rotational motion dominates. A detailed description of the frequency dependence of $\epsilon^*(\omega)$ due to ionic, dipolar and charge polarization effects has been previously described (1,2). If we test for and avoid charge polarization effects, which are usually small at frequencies above 10 Hz, the magnitude of the low frequency overlapping values of $\omega\epsilon''(\omega)$ can be used to monitor the time dependence of the ionic mobility. Peaks in a particular $\epsilon''\omega$ line (i.e. at a particular frequency) with time can be used to monitor dipolar mobility. Together the ionic and dipolar mobility can be used to monitor the occurrence of critical cure points such as dry-to-touch, dry-to-hard and the buildup in end-use durability.

Figure 2 shows a typical plot of log $(2\pi f\epsilon'')$ vs time monitoring the polymerization of a commercial Seaguard 151 marine epoxy-polyamide polymer coating during the first twelve hours after the application under normal conditions, i.e., 24°C and 45%RH. The cure is dominated by ionic diffusion for the first 100 minutes at the lower frequencies as indicated by the overlapping $2\pi f\epsilon''$ lines. The rapid decrease in the signal monitors the decrease in the mobility of the ions reflecting

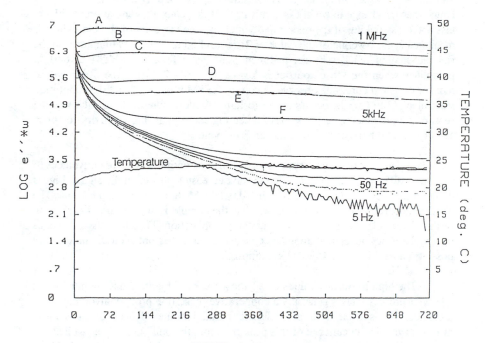

Figure 2   Log ($\epsilon''\omega$) versus time, 5Hz to 1 MHz, for a Seaguard marine epoxy polyamide coating curing at 24°C, 45% relative humidity.

a rapid buildup in viscosity. In general, the overlapping $2\pi f\epsilon''$ lines are proportioned to the inverse of the macroscopic viscosity where, $2\pi f\epsilon'' = A(\dfrac{1}{\eta})^x$. At the time of application the coating is in the 'wet' stage. In this stage the viscosity is in very low and solvent evaporation occurs rapidly. Solvent loss from the coating was measured by thermogravimetric analysis (TGA) in a pan of approximately the same film thickness. The 'wet' stage of solvent loss is seen in the rapid weight loss in the first 144 minutes as seen in Figure 3. The evaporation slows down markedly after six hours when the change in weight is barely noticeable. Thus a combination of solvent loss and cross-linking of the resin rapidly increase the viscosity of the film which is monitored in the lowering of the value of $2\pi f\epsilon''$ in Figure 2. At 100 minutes into the run the coating reaches its set-to-touch point as defined by ASTM D1640-83 guidelines when the sensor output is $\log (2\pi f\epsilon'') = 4.2$. Using the 50Hz line to monitor the initial cure, one observes again by ASTM inspection that the calibration value of $\log (2\pi f\epsilon'') = 2.8$ occurs at dry-to-hard. Achievement of these values by the sensor can be used as an insitu means of detecting and monitoring when dry-to-touch and dry-to-hard have occurred in any environment.

To monitor the long term properties of the coating it is more convenient to examine $2\pi f\epsilon'''$ on a non-log scale due to the decreasing change in the signal. Figure 4 monitors the sample during the 3rd day, 44-56 hours after application. The continual decrease in the signal shows that the sample is still curing. There is no further detectable change in ϵ'' 68 hours after application. Therefore the sample, as seen by frequency dependent impedance measurements, has obtained the fullest cure possible under 24°C and 45%RH conditions.

The high frequency values of ϵ'' show peaks in Figure 2 and monitor the α relaxation process. This is generally a cooperative relaxation process involving many molecules. It is seen both in dielectric relaxation and dynamical mechanical measurements. The occurrence of this peak monitors the buildup in Tg as well as the corresponding final use properties of the curing coating. The peaks in the 5 kHz to 1 MHZ $2\pi f\epsilon''$ lines, points A,B,C,D,E and F, indicate the time when the characteristic relaxation time for dipolar relaxation $\tau = 1/2\pi f$ has occurred. Thus, point A, 15 minutes, marks the cure time when $\tau = (2\pi x10^6)^{-1}$ sec. and the point F, 400 minutes, marks the time when the rotational relaxation time has slowed to $\tau = 1/(2\pi*5x10^3)$ sec. The relationship between the value of τ and the value of Tg or any other use property can be quantitatively determined by correlating measurements of τ with measurements of Tg or the use property of interest. Either using a correlation plot or a mathematical WLF type fit where $\ln \tau \propto A/(B-(T-Tg))$, the FDEMS output can be used to continuously monitor the buildup in an end-use property such as Tg insitu under the actual cure conditions.

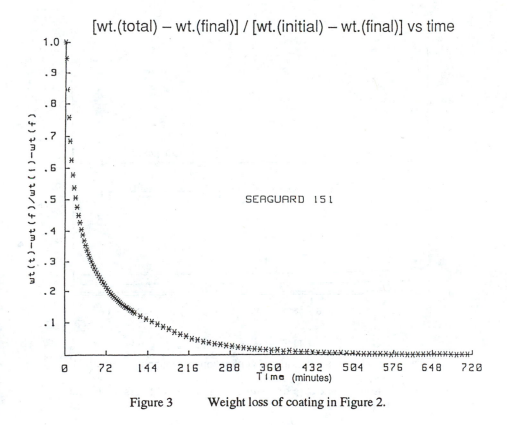

Figure 3   Weight loss of coating in Figure 2.

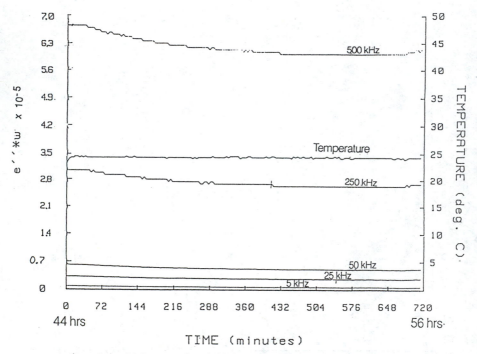

time 0 is 44 hours after initial application of coating in Figure 3

Figure 4   Longer time cure of coating in Figure 2, 44 hours to 56 hours
after application.

FDEMS output similar to Figure 2 and 4 can be used to monitor the variation in cure rate with temperature, humidity, airflow, pigment loading, catalyst concentration, thickness, age, batch, etc.

For example, it was thought epoxy-polyamide coatings would cure at temperatures as low as $10°C(50°F)$, even though the time required to reach a given stage is at least double that required at $22°C$[4]. To examine the effects of decreased temperature the polymerization process was monitored in an environmental chamber, held at $11°C$ and 30%RH, for 3 weeks. Figures 5 and 6 show plots of $\log(e''*2\pi f)$ measured with the FDEMS sensor. The time to set-to-touch was 420 minutes. Dry-to-hard occurred at 695 minutes. The value of $\log(\epsilon°*2\pi f)$ at set-to-touch was 4.1 and at dry-to-hard was 2.9. The drop in $e''*2\pi f$ is much slower than at $24°C$. Figure 6 shows further curing of the sample during the 6th day. The sample continued to cure as seen by continual drop in e'' even during the third week since application. After 500 hours the 50kHz line still has not reached the degree of polymerization that the same coating does in 48 hours at $24°C$ and the rate of cure as monitored by the change in ϵ'' is relatively flat.

These FDEMS results clearly show that below $10°C$, the curing is greatly retarded and full cure is not achieved. More important this epoxy-polyamide will not reach full until the temperature rises. Although the partly-cured film may feel dry, poor resistance to abrasion, moisture and chemicals result.

The effect of humidity on the cure rate is shown in Figure 7. A 75% relative humidity increased the time to dry-to-touch about 100 minutes to 300 minutes as monitored by the time for $\log(\epsilon''\omega)$ where $\omega=2\pi f$ to drop to 4.2. Thus the presence of moisture appears to significantly reduce the rate of cure as monitored by the FDEMS sensor.

The ability of the FDEMS sensor to monitor cure in a Glidden-ICI Paints proprietary latex coating is shown in Figure 8. Initially there is a rather rapid decay in the $\log(\epsilon''\omega)$ overlapping ionic lines form 10 to 8.5 at 30 minutes. This is followed by a rapid drop of over 2 decades from 8.5 to 6.2 in several minutes. We ascribe the initial 30 minutes to phase I packing of the latex spheres. There is a loss of volatiles to the point where the latex spheres touch. After this point water is no longer the continuous medium, the latex spheres rather than the suspending fluid are the conducting medium. This transition generated the large drop in $\epsilon''\omega$. This rapid drop is followed by a gradual decrease in $\log(\epsilon''\omega)$. There appears to be another shift in the slope around 72 minutes. We believe this shift in the rate of the drop of $\log(\epsilon''\omega)$ at 72 minutes is due to the transitions from what is described as a predominately phase II deformation and packing cure process to a phase III autohesion and diffusion cure process.

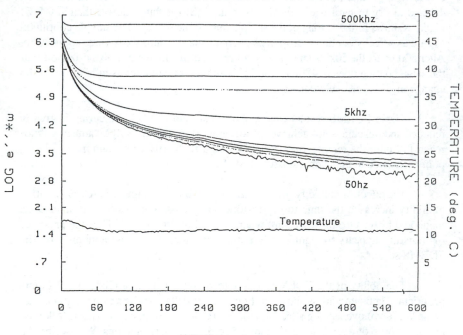

SEAGUARD 152: 11C

Figure 5   Same epoxy polyamide curing at 11°C, 30% relative humidity.

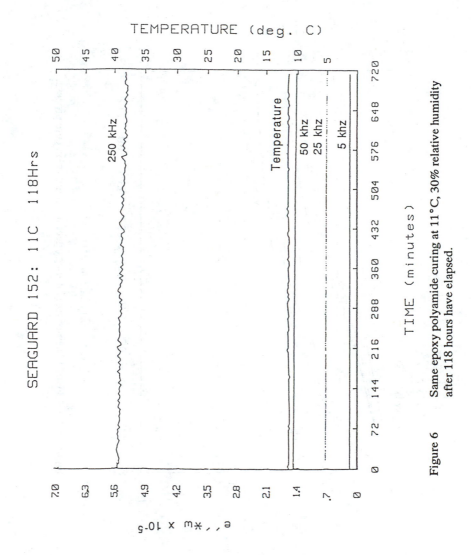

Figure 6   Same epoxy polyamide curing at 11°C, 30% relative humidity after 118 hours have elapsed.

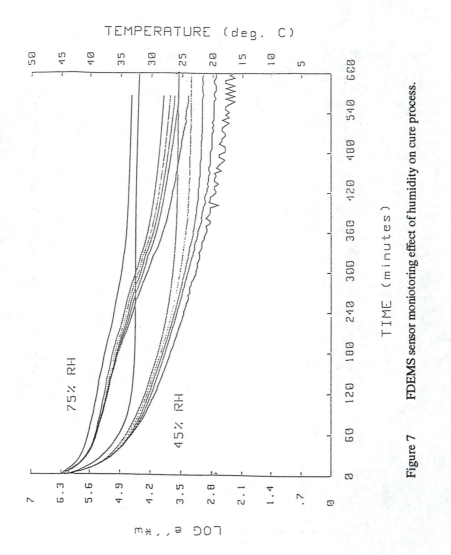

Figure 7   FDEMS sensor moniotoring effect of humidity on cure process.

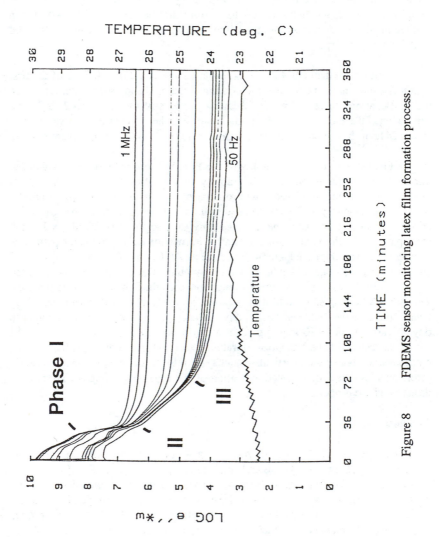

Figure 8   FDEMS sensor monitoring latex film formation process.

It is also interesting to demonstrate that the sensor can monitor the extent to which the first coating of the Seaguard epoxy polyamide is softened by the second for varying elapsed times since the initial coats application. Figure 9 displays the softening of the first coating by the second when the second coating has been applied 144 minutes after the first for the epoxy-polyamide system shown in Figures 2-6. The initial coating's values of log ($\epsilon''\omega$) have only dropped from 5.5 to 4.5. Upon application of the second coating the values rise to 6.0 over 60 minutes as the 2nd coat resoftens the 1st layer.

Next we use the FDEMS sensor to examine the extent of softening of the first coat if 24 hours elapse before the 2nd coating. Figure 10 shows that after 24 hours, the first coat has decreased its low frequency values of log ($\epsilon''\omega$) to 2 to 3. The application of the 2nd coat on the next day resoftens the 1st coat back to values 5.5. Hence 24 hours has no effect on the resoftening of the 1st coat for this paint system.

Finally, we examine the ability of the FDEMS sensor to monitor buildup in durability. Customarily buildup in durability of a coating for marine systems is evaluated over a period of weeks and months by immersing the coating in an 80° water bath and monitoring the time to blistering as well as the number and size of the blisters. Figures 11 and 12 monitor the coating on a series of standard 8 x 10 inch steel panels over the initial 45 minutes after immersion in the 80° water bath. The coating which was immersed after 4 days cure at 24°C, 45% RH shows a rapid and large change in the values of log ($\epsilon''\omega$) with 10 minutes. The coating which was cured 14 days under these conditions shows a much more gradual change over 45 minutes. The 4th day coating showed blistering based on medium ASTM blister frequency failure criteria on the 38th day. The 14 day cured coating showed failure on the 58th day. Clearly the FDEMS sensor gives a much more rapid, 1 hour versus 6 to 12 weeks indication of the buildup in durability and end-use properties. The sensor output once calibrated for a particular paint system to the industries standard criteria offers an immediate instrumented indication of the time the coating has attained its desired end use properties.

Conclusions

It has been shown FDEMS sensing provides a sensitive method for monitoring the cure of coatings, both latex systems and epoxy polyamide coatings. FDEMS can track the varying cure rates which result from changing environmental cure conditions and the sensor output provides an instrumental means for monitoring buildup in durability under varying environments such as temperature and/or humidity. Exceptional sensitivity can be attained with these sensor measurement techniques. The FDEMS sensors can detect and monitor buildup in cure for up to 30+ days, up to and beyond the time of attainment of acceptable service life properties. The FDEMS sensor output monitors the buildup in durability and service life properties and the output can be correlated with ASTM service life tests such as time to blister failure.

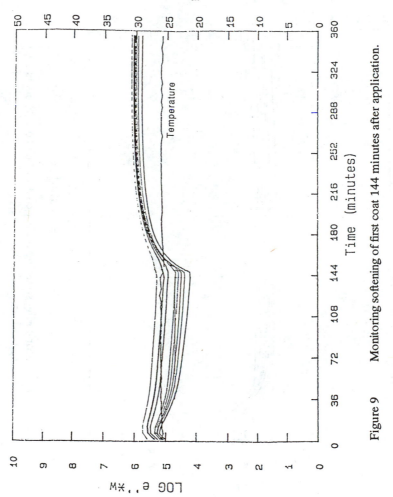

Figure 9   Monitoring softening of first coat 144 minutes after application.

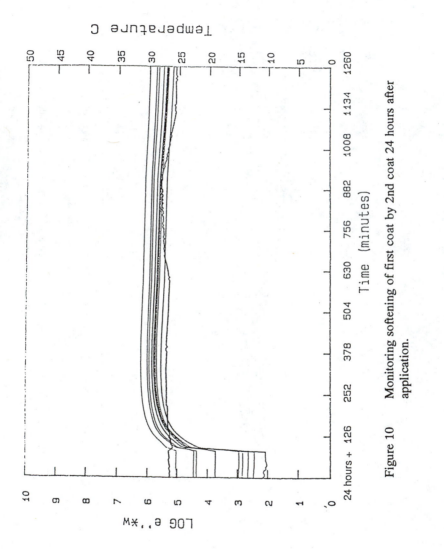

Figure 10   Monitoring softening of first coat by 2nd coat 24 hours after application.

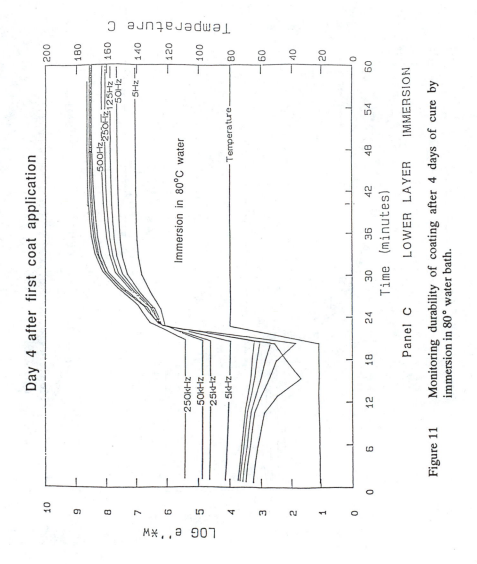

Figure 11   Monitoring durability of coating after 4 days of cure by immersion in 80° water bath.

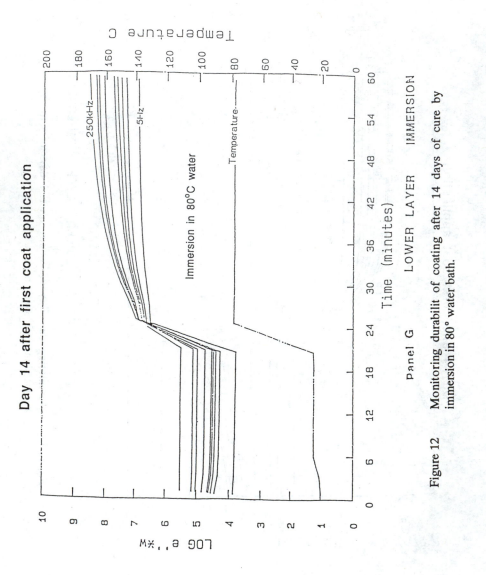

Figure 12   Monitoring durabilit of coating after 14 days of cure by immersion in 80° water bath.

Further FDEMS can monitor durability. As such, FDEMS should be able to be developed as an effective sensitive means for studying degradation or knockdown in service life properties or durability with time under use conditions such as heat, uv light, moisture, etc. It is an ideal technique due to its sensitivity for accelerated aging studies.

Acknowledgment

Partial support from the NSF Center of Excellence in Polymeric Adhesives and Composites NSF #MR912004 at Virginia Tech, Blacksburg, Virginia, is gratefully acknowledged.

Inquiries regarding the sensor and measurement equipment should be directed to DEK.

References

1.   D. Kranbuehl, "Cure Monitoring," *Int. Encyclopedia of Composites*, ed. S. Lee, 531-543 (1990).
2.   D. Kranbuehl, P. Kingsley, S. Hart, G. Hasko, B. Dexter, A. C. Loos, *Polymer Composites*, 15(4) 299-305 (1994).
3.   Y. Wang, M. Argiriadi, W. Limburg, S. Mahoney, D. D. Kranbuehl, D. E. Kranbuehl, *Polym. Mat. Sci. and Eng.* 70, pp. 279-280 (1994).
4.   D. Kranbuehl, *ACS Division of Polymeric Materials: Science and Engineering*, **63**, 90-93 (1990).
5.   S. Hart, D. Kranbuehl, D. Hood, A. Loos, J. Koury, J. Harvey, Int. SAMPE Symp. Proc. 39(1), pp. 1641-51 (1994).
6.   Friauf, R. J. *J. Chem. Phys.*, 1954, **22**, 1329.
7.   MacDonald, J. R., *J. Chem. Phys.* 1971, **54**, 2026.
8.   Johnson, J. F. and Cole, R. H. *J. Am. Chem.* Soc. 1951, **73**, 4536.

MECHANICAL PROPERTIES

Chapter 8

Dynamic Viscoelastic Properties of Polymer Latex Films

Role of Core Chains and Particle–Particle Interfaces

J. Richard

Centre de Recherches Rhône-Poulenc, 52 rue de la Haie Coq, 93308 Aubervilliers Cedex, France

The linear viscoelastic behavior of latex films is studied using Dynamic Micromechanical Analysis (DMA). Dried latex films are considered as cellular structures. The core of the cells consists of the hydrophobic particle core, whereas the array of interfacial membranes arise from the hydrophilic shells of the particles which provide colloidal stability in the initial dispersions. Both cases of electrostatically- and sterically-stabilized particles are considered. For the former latex particles which bear copolymerized carboxylic species at their surface, the influence of the cross-linking density within the particle cores is investigated, together with the modification of interfacial membranes induced by H bonding or ionic interactions. The viscoelastic behavior is correlated with the structure of the films which is determined by Small Angle Neutron Scattering (SANS) experiments and Transmission Electron Microscopy (TEM). Segment diffusion coefficients can be deduced from DMA spectra and make it possible to predict membrane preservation or further coalescence. The viscoelastic properties of films obtained from the latter latex particles which bear a steric stabilizer partly grafted in their shell, are also studied. The effect of thermal annealing of the films is more particularly investigated. A specific rebuilding effect of the membranes upon annealing is evidenced for this kind of film and the influence of various parameters (particle size, volume fractions, features of polymer shell) is semi-quantitatively investigated using an equivalent mechanical model and the Kerner equation.

Latexes consist of water dispersed polymer particles currently prepared by the well controlled emulsion polymerization technique. They are mainly used as binders in paints, paper coatings and adhesives, in which they play one of the most important roles by achieving the final cohesive and adhesive properties. The polymer particles are kept in stable aqueous dispersions, provided that repulsive interactions have been created between the particles. The repulsions originate from the surface layers of the

0097–6156/96/0648–0118$19.00/0

particles which are referred to as the membranes (*1*). These membranes very often consist of hydrophilic polymer chains copolymerized or grafted on the particle core. Colloidal stability of the dispersed system may originate from either electrostatic repulsions when the polymer is charged or ionogenic in a given pH range (*2*), or a steric mechanism when the hydrophilic shell polymer is non ionic but water-soluble (*3*). Upon dehydration of the dispersions, continuous polymer films are formed which exhibit a cellular structure ; the deformed hydrophobic particle cores form polyhedral cells, while the array of particle-particle interfaces, i.e. the cell walls, consist of the hydrophilic interfacial membranes (*1*).

This chapter presents a review of the work we have carried on to characterize the effects of core and membrane features on the linear viscoelastic behavior of latex films, in close correlation with structure modifications. Investigations are mainly based on dynamic viscoelastic measurements, while structural information is gained from Small Angle Neutron Scattering (SANS) and Transmission Electron Microscopy (TEM) experiments (*1, 4*). Both cases of electrostatically- and sterically-stabilized particles are considered. The model system chosen for electrostatically-stabilized latexes consists of partly cross-linked styrene-butadiene copolymer particles, whose colloidal stability arises from carboxylic species copolymerized at their surface. The influence of the cross-linking density within the cores and the modifications of interfacial membranes induced by H bonding or ionic interactions, are investigated. The viscoelastic behavior is correlated with the structure of the films and more precisely to its ripening through the further coalescence process. The model system chosen for sterically-stabilized latexes consists of poly(vinyl acetate) particles, on which poly(vinyl alcohol) chains are anchored either by chemical grafting or physical adsorption. The effect of thermal annealing on the viscoelastic behavior of these films is more particularly investigated and correlated with phase rearrangement and strong modifications in the morphology of the interfacial membranes' network .

Experimental Methods

The latex preparation method for both model systems has been previously described elsewhere (*5, 6*). For styrene-butadiene (SB) copolymers, the resulting latexes have a solid content of 50 % by weight and a pH of 4.5. Particle size distribution, which is determined using a classical turbidimetric method and TEM, is found to have a narrow width and an average value of ~ 0.18 µm for all the samples. The average composition of the SB copolymers is kept constant, namely 69 wt % of S and 27 wt% of B, for the whole series of samples. Classical chain transfer agents such as an alkylmercaptan C1 and a polyhalogenated organic compound C2 are used to vary the cross-linking density of the SB copolymers. The corresponding gel fraction values G are measured in chloroform using the extraction method previously described (*7*). The colloidal stability of the dispersions is obtained by copolymerizing an acrylic -type comonomer, so that the interfacial membranes in the films consist of highly carboxylated polymer chains. The content of this carboxylic comonomer is varied to study the effect of this parameter on the viscoelastic properties of the films. As a part of the acid comonomer is expected to be burried within the particle cores, the amount of carboxylic acid groups actually located at the surface of the particles has been determined by a conductimetric titration method. It has been also checked that the amount of water-soluble low molecular weight carboxylated species is very limited in our samples. Two reference latexes without any copolymerized carboxylic group have also been prepared and stabilized with sodium dodecylsulphate (SDS) surfactant molecules. Finally, in order to study the effect of membrane modifications on viscoelastic behavior of the films, the neutralization conditions are varied for one of the latexes, which exhibit a very low gel fraction value (15 wt%), i.e. a very low cross-linking density of the core. The pH of the latex has been adjusted to 9 using various bases such as : ammonia, sodium hydroxide, a diamino-ended polyoxyethylene (DAPOE) of low molecular weight (600 gmol^{-1}), and an aminoalcohol (AMA). The main data concerning the series of model electrostatically-stabilized SB copolymer particles are given in Table I.

Table I. Main features of the series of model SB copolymer latex particles

Latex	CTA type	CTA concn (wt %)	Gel fraction (wt %)	Total/Surface carboxylic group contents (μmol/cm^{-3} pol.)	Neutralization conditions (pH/Base)
SB1	C1	0	89.5	608/200	4.5/Ammonia
SB2	C1	0.4	76	608/230	" "
SB3	C1	0.8	69.5	608/258	" "
SB4	C1	1.25	43.5	608/252	" "
SB5	C1	2	15	608/241	" "
SB6	C1	"	"	" "	9/Ammonia
SB7	C1	"	"	" "	9/NaOH
SB8	C1	"	"	" "	9/AMA
SB9	C1	"	"	" "	9/DAPOE
SB10	C1	2	15	304/133	4.5/Ammonia
SB11*	C1	2.4	10	0/0	" "
SB12*	C1	0.6	72	0/0	" "
SB13	C2	0.4	81.5	608/208	" "
SB14	C2	0.8	71.5	608/220	" "
SB15	C2	1.6	40	608/238	" "
SB16	C2	3.2	17	608/200	" "
SB17	C2	3	20	470/174	" "
SB18	C2	3	20	780/188	" "
SB19	C2	3	20	1090/461	" "

* Latexes stabilized with SDS surfactant molecules instead of copolymerized carboxylic groups.

For poly(vinyl acetate) (PVAc) homopolymers, a part of the PVA steric stabilizer (Rhodoviol 25/140 from Rhône-Poulenc) has been introduced in the initial charge of the reactor at various volume fractions (betwen 3.15 and 6.5 %). Depending on this initial concentration, the resulting latexes have a rather broad particle size distribution for the highest initial content and a rather narrow for the lower ones. The average particle size is found to range between 0.55 and 2 μm. In order to study the effect of PVA volume fraction and features on the viscoelastic properties of the films, various amounts of PVA with viscosity grade 25 (namely the viscosity of a 4 wt % aqueous solution of PVA at 20 °C) and different hydrolysis level (HL) ranging between 80 and 99 mole %, have been introduced in the dispersions after the polymerization (final addition) (7). Thus, the samples exhibit total PVA volume fractions ranging from 3.15 % to 13.5 %. The main data concerning the series of PVAc homopolymer particles sterically-stabilized by PVA are given in Table II.

Table II. Main features of the series of the PVAc latex particles sterically-stabilized by PVA

Latex	PVA volume fraction (%)		PVAc average particle size (d)	PVAL hydrolysis level* (HL)
	initial charge (ϕ_i)	final addition (ϕ_a)	(μm)	(mole %)
A1	3.15	0	0.7	-
A2	5.2	0	1.2	-
A3	5.2	3.6	1.2	88
A4	5.3	4	2	80
A5	5.3	4	0.55	80
A6	5.3	5.7	0.55	80
A7	5.3	5.7	0.55	99
A8	6.5	0	1.2	-
A9	6.5	3.6	1.2	88
A10	6.5	7	1.2	88

* This is the hydrolysis level of PVA added at the end of polymerization (final addition).

Homogeneous transparent polymer films have been prepared from both series of products by casting the latexes in silicone molds and evaporating water in an oven for 6 hours at 50 °C, which is a temperature higher than their minimum film formation temperature (MFT). Samples with a constant thickness ranging between 0.7 and 1.6 mm are then cut to dimensions of ~ 1 x 5 cm^2 and stored under vacuum in a dessicating vessel prior to use. DMA experiments are performed on these dried latex films using a Rheometrics RDS-LA viscoelastometer which works in the linear simple extension deformation mode (6-9). It is equipped with an attachment for temperature control in the range - 50 °C / + 150 °C. Thus, the isochronal temperature dependence of the storage modulus (E'), loss modulus (E") and loss tangent (tan δ) can be measured over a wide range of temperature at a frequeny of 1 Hz under a nitrogen atmosphere. The strain amplitude is fixed at 10^{-3} which is checked to lie in the linear viscoelastic regime. For the SB copolymer latexes, the isothermal frequency dependence is also deduced by sweeping the frequency of the sine deformation between 10^{-1} and 500 rad/s, at different temperatures close to the glass transition temperatures of the polymers. Recorded spectra have been superimposed by shifting them along the frequency axis until a reliable superposition could be obtained for the three viscoelastic functions (E', E", and tan δ) with the same WLF shift factor. Criteria of applicability of this method of reduced variables are checked to be fulfilled (8). For PVAc homopolymer latexes, the effect of annealing has been studied by running the samples once from 25 °C to 150 °C, then annealing them at 150 °C for half an hour under a nitrogen atmosphere, and finally running them again in the same temperature range. It is found that longer annealing periods do not induce any additional alteration of the viscoelastic behavior of the films (6). Finally, it is worthwhile noticing that the annealing temperature is chosen to be much lower than the decomposition temperature of PVA under nitrogen atmosphere, namely 200 °C (10).

Structure of the films obtained from these latexes are fully investigated using SANS and TEM techniques with appropriate selective labelling of the hydrophilic particle-particle interfaces (4,6,9).TEM experiments are peformed on ultramicrotomed samples. Contrast is enhanced by selectively staining the carboxylic or PVA interfacial membranes with uranylacetate (9) or OsO_4 (11), respectively. As for SANS experiments which are only performed on the model SB latexes, high contrast is obtained by rehydrating the dried films with D_2O vapor, which selectively labels the hydrophilic carboxylated interfacial membranes (4,12). The diffraction patterns are radially averaged to yield the classical scattering curves where intensity I is plotted against the magnitude of the scattering factor Q. In this work, only the presence and intensity of diffraction peaks in the logI - logQ plots will be discussed. The vanishing of these peaks is considered as an evidence of the destruction of the network of ordered interfacial membranes and hence interdiffusion of polymer chains.

Results and Discussion

Electrostatically-Stabilized Particles

Effect of Cross-linking Density within the Particle Cores. Figure 1 shows the isothermal master curves recorded for the storage moduli E' of films from the series SB1-SB5 in which CTA C1 is used to restrict the cross-linking density. All the curves exhibit the classical strong relaxation from a glass-like behavior in the high frequency region to a rubber-like one in the lower frequency zone. The effect of CTA content on the viscoelastic properties of the films appears to be twofold : (i) First, it is responsible for a large shift in the location of the transition zone on the frequency scale over ~ 4 frequency decades. As proposed by Ferry (13), this

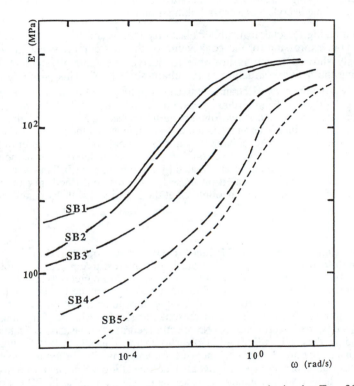

Figure 1. Log-log plot of the isothermal master curves obtained at $T_0 = 301$ K for the storage modulus E' of the series of model latexes SB1-SB5 synthesized with various contents of CTA C1.

displacement reflects strong variations in the relaxation times and local molecular motions at the chain segment level. (ii) Second, it controls the level of the storage modulus approached in the low frequency region, which represents the modulus of the rubber-like network. A high CTA content even leads to an almost complete collapse of the modulus in this region. The same kind of behavior is obtained for films from the series SB13-SB16 synthesized with CTA C2. However, as evidenced from Figure 2, the shift of the relaxation frequency and the variations of the rubbery modulus are found to be less pronounced. Interestingly, the above effects have been also observed recently by Zosel et al. (14) for poly(n butyl acrylate) (PBA) latexes whose cross-linking density has been controlled by adding a mercaptan as a CTA and methallyl methacrylate as a cross-linking monomer.

Combining the classical rubber elasticity theory and the phenomenological Marvin-Oser expression for the peak value of the loss compliance J''_m (15), we previously showed that it is possible to derive the main physical parameter describing the network structure at the chain scale (7,8). This is the density of elastically effective network strands between coupling loci v_c. In our system, these coupling loci arise from either covalent multifunctional links or physical intermolecular forces, such as hydrogen bonding between carboxylic groups or only topological constraints. This analysis led us to conclude that the nature of CTA used for the synthesis of the SB copolymers is of major importance : as a matter of fact, in addition to the reduction of gel fraction of the polymer, it determines the density of non covalent molecular interactions created by the free chains (7). Transverse proton relaxation experiments performed on these latex films confirmed this analysis by probing the state of constraints encountered at the semi-local scale by chain segments embedded in the network (7).

Relationships between Molecular Dynamics and Further
Coalescence. It seems to be of particular interest to connect the molecular dynamics of the chains with the structure of the films. More precisely, the occurence of the further coalescence process, which induces fragmentation of the membranes and fusion of the cores, is expected to strongly depend on the mobility of polymer chains within the particles. For this reason, the structure of the films has been studied using SANS expriments. SANS spectra recorded for the series of latex films SB1-SB5 synthesized with CTA C1 are shown in Figure 3. In this Figure, only relative intensities are significant, since spectra are vertically shifted for the sake of clarity. Inspection of these spectra reveals that the CTA content clearly controls the structure of the latex films and the coalescence process. First, for the SB copolymer particles which does not contain any chain transfer agent (sample SB1), the diffusion pattern exhibits two strong intensity peaks, which originate from the highly ordered array of non fully coalesced latex particles. This spectrum reflects the high degree of organisation existing in the network of hydrophilic membranes, which is obviously preserved in the film. Secondly, the amplitude of the intensity maxima appears to gradually decrease as the CTA concentration increases in the particles, i.e. as the cross-linking density decreases within the particle cores. These peaks finally vanish almost completely for the sample SB5 with the lowest gel fraction value of the series. These variations in film structure with CTA content were confirmed by TEM experiments (4). The same SANS and TEM experiments have been performed on the series SB13-SB16. It turns out that even for the sample (SB16) with the lowest gel fraction value of the series (17 wt%), further coalescence does not occur when CTA C2 is used to restrict cross-linking density of the particles, as evidenced from the TEM photomicrograph shown in Figure 4. This

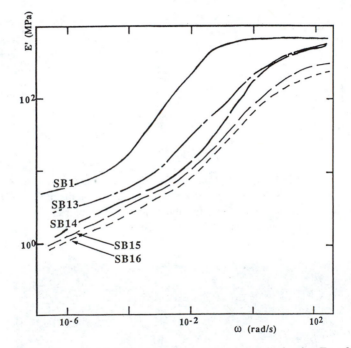

Figure 2. Log-log plot of the isothermal master curves obtained at $T_0 = 301$ K for the storage modulus E' of the series of model latexes SB13-SB16 synthesized with with various contents of CTA C2. The corresponding curve for sample SB1 is included in the diagram for comparison.

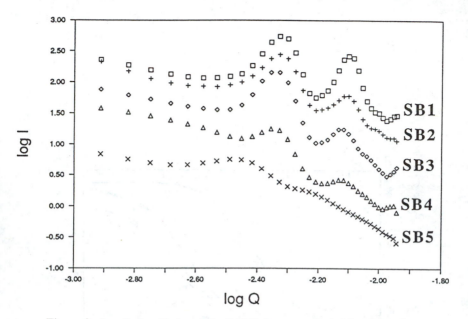

Figure 3 . Log I - log Q plot of the SANS spectra recorded for the series
of model latexes SB1-SB5 synthesized with various contents of CTA C1.
Vanishing of the network of interfacial membranes and interdiffusion
of polymer chains is evidenced for sample SB5.

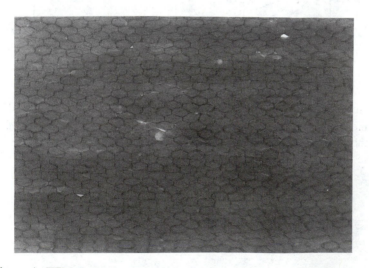

Figure 4 . TEM photomicrograph of stained ultramicrotomed sample SB16
(magnification X 44 000).

outcome rules out the gel fraction value as the key parameter controlling the occurence of further coalescence. Furthermore, it points out that molecular interactions between free chains play a major part in the dynamics and diffusion properties of these systems.

Then, in order to perform a more quantitative analysis of the above observations, we have derived the diffusion coefficients of chain segments and correlated their values and variations with the occurence of further coalescence. For this purpose, we propose to describe the chain segment dynamics at the local scale, using the monomeric friction coefficient ξ_0, which can be derived from the DMA spectra. From its definition, this parameter reflects the local segment mobility and hence can be related to the translatory diffusion coefficient of chain segments D_S in a simple way $(14, 16, 17)$:

$$D_S = kT \, \mu_S = kT / \xi_0 \qquad (1),$$

where T is the temperature, μ_S the chain segment mobility and k the Boltzmann's constant. At sufficiently long, but finite, diffusion times ($t > t_g$ where t_g is the characteristic time for glass-rubber relaxation(16)), the over-all diffusion of the chains which allows fusion of particle cores is expected to become Fickian (18) and can be consistently described using the curvilinear diffusion coefficient D_C. The over-all diffusion of a chain trapped in the fixed curvilinear tube formed by its neighbouring molecules is related to the chain segment mobility μ_S by the expression $(16, 17)$:

$$D_C = kT \, \mu_S / N = kT / N \, \xi_0 = D_S / N \qquad (2),$$

where N is the number of monomer units in the chain. Equation 2 assumes that monomeric frictions are additive as stated for the free Rouse chain. It does not remain valid for long entangled chains whose diffusion coefficient is expected to follow a larger molecular weight dependence $(18, 19)$:

$$D_C = D_S \, N_e / N^2 \qquad (3),$$

where N_e is the average number of monomer units between entanglement points. Hence, following this frame, the values of ξ_0 and D_S determined from DMA spectra can be correlated to the diffusion of polymer chains through particle-particle interfaces, whatever the molecular weight dependence of the diffusion coefficient. Then, according to Ferry(13), the average monomeric friction coefficient ξ_0 can be derived from the low frequency end of the relaxation zone appearing in Figure 1 and Figure 2. As previously shown, the most convenient method to derive ξ_0 is to use the frequency dependence of the complex compliance J* which is the inverse of the complex modulus E* $(4, 6, 7)$. Then, the values of ξ_0 can be determined experimentally from the low frequency end of the transition zone, following the work by Stratton et al. (20) for a series of rubber vulcanizates cross-linked with various agents. For the frequency dependence of J' and J" in this region, the following equation applies :

$$\log J' = \log J" = - 1/2 \log \omega - \log (a\rho N_0/4M_0) - 1/2 \log (4 \, \xi_0 kT/3) \qquad (4),$$

where M_0 is the monomer molecular weight, ρ the density of the polymer, N_0 Avogadro's number, and a the root-mean-square end-to-end length per square root of monomer units (*13*). In the present work, this latter is taken to be the average value calculated from the large number of values reported by Ferry for this parameter in various vinyl, diene and acrylate copolymers (*13*), namely a = 7 Å. Then, from the above isothermal DMA master curves, the corresponding values of ξ_0 and D_S are calculated using equation 4 and equation 1 respectively, and considering that the monomer molecular weight M_0 is equivalent to the weight of a statistical SB copolymer unit. The values of ξ_0 and D_{s0} which are obtained at temperature T_0 are shown in Table III. T_0 is close to the glass transition temperature T_g of the films and is used as a reference temperature to obtain the DMA spectra. It is chosen to be equal to 301 K for the whole series of samples.

Table III. Monomeric friction coefficients ξ_0 and segment diffusion coefficients D_{s0} of a series of SB copolymer latex films at reference temperature T_0

Latex	ξ_0 (Ns/m)	D_{s0} (cm^2/s)
SB1	11×10^2	$3,8 \times 10^{-20}$
SB2	87	4.8×10^{-19}
SB3	86×10^{-1}	4.8×10^{-18}
SB4	71×10^{-2}	5.8×10^{-17}
SB5	31×10^{-2}	1.3×10^{-16}
SB11	3×10^{-2}	1.4×10^{-15}
SB12	59×10^{-2}	7.1×10^{-17}
SB13	47	8.8×10^{-19}
SB14	6	6.9×10^{-18}
SB15	35×10^{-1}	1.2×10^{-17}
SB16	29×10^{-1}	1.4×10^{-17}

Then, a clear correlation between the value of the diffusion coefficient D_{s0} deduced from the DMA spectrum and the occurence of further coalescence can be established. Inspection of Figure 3 on one hand and Table III on the other hand brings out strong arguments in favor of this correlation. The large increase of the diffusion coefficient (over ~ 4 orders of magnitude) as CTA concentration in the SB copolymer increases, is clearly correlated to the decrease of the amplitude of the diffused intensity in the SANS spectra indicating less ordered structures. Moreover, it appears that further coalescence leading to the fusion of particle cores only occurs in a massive way when D_{s0} exhibits a value higher than 10^{-16} cm^2/s. This is the case for samples SB5 and SB11 (*4*). On the opposite, several samples (e.g. samples SB4 and SB12) which exhibit D_{S0} values smaller than but still very close to this threshold value, does not undergo further coalescence under identical film formation and storage conditions, even if the interfacial membranes are expected to be highly mobile as in the case of surfactant double layers (sample SB11)(*4*). This result clearly demonstrates that mobility of the core chains determines the occurence of the further coalescence process. Although D_{S0} is only related to local segment dynamics, these observations suggest that there exists a minimum threshold value for the diffusion coefficent of chain segments which allows further coalescence to occur and fusion of particle cores to take place. Indeed, the factor which controls local mobility is also found to affect macroscopic diffusion, as expected from equations 2 and 3. In the present system, the threshold value for D_{S0} ranges ~ 10^{-16} cm^2/s at $T_0 = 301$ K.

Constraints Induced by Free Chains. Inspection of Table I and Table III leads to the conclusion that D_{S0} is not correlated to the gel fraction value G of the SB copolymer. For instance, in sample SB16 which exhibits almost the same low gel fraction value as sample SB5, the segment diffusion coefficient D_{S0} is found to be one order of magnitude lower than in SB5, and further coalescence does not occur precisely in SB16. This behaviour can be more readily observed in Figure 5, in which the segment diffusion coefficient D_{S0} at 301 K is plotted against the gel fraction value G for the series of samples SB1-SB5 and SB13-SB16. It points out that not only covalent cross-links but also physical intermolecular interaction loci exert constraints on chain segments and thus restrict local mobility. Moreover, for the same fraction of covalently cross-linked chains, additional constraints are generated by free chains in SB copolymers containing CTA C2. These constraints lead to a strong restriction of the chain segment diffusion coefficient and limit coalescence. CTA C2 seems to act as if while decreasing the constraints arising from covalent knots in the polymer, it induces an increase of the density of physical interaction loci, thus decreasing local segmental mobility and preventing further coalescence from occuring. Hence, this work emphasizes the key part played by non covalent physical intermolecular interactions in polymer chain diffusion and restriction of further coalescence in latex films.

Effect of Modifications of the Interfacial Membranes. Films obtained from SB copolymer particles with the lowest gel fraction values (i.e. G ~ 15-20 wt%) and stabilized by copolymerized carboxylic surface goups, are investigated. The hydrophilic membranes of the cellular domains are modified either by varying the content of the carboxylic groups located at the surface of the particles or by neutralizing them with various bases.

Figure 6 and Figure 7 show the isothermal master curves recorded for the storage moduli of films from latexes with various amounts of copolymerized carboxylic groups synthesized with CTA C2 and CTA C1, respectively. The principle of time-temperature superposition (or method of reduced variables) is checked to be valid for the whole series of samples with the classical WLF temperature dependence for the shift factors. The main feature of these isothermal master curves is the classical relaxation phenomenon from a glasslike behaviour in the high frequency region to a rubberlike one in the lower range. The value of the rubbery storage modulus in the low frequency region is found to be strongly dependent on the content of carboxylic groups in the polymer. This behaviour brings out clear evidence of the specific interactions between carboxylic groups which contribute to increase the density of

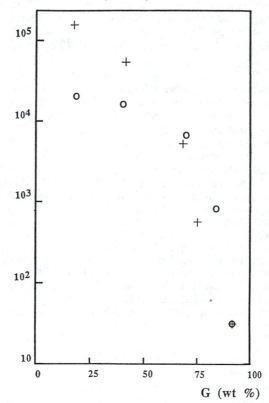

Figure 5 . Plot of the variations of the segment diffusion coefficient D_{s0} (x 10^{21} cm^2/s) derived from the DMA spectra at $T_0 = 301$ K against the gel fraction value G (wt%) for the two series of samples SB1-SB5 (+) and SB 13-SB16 (O).

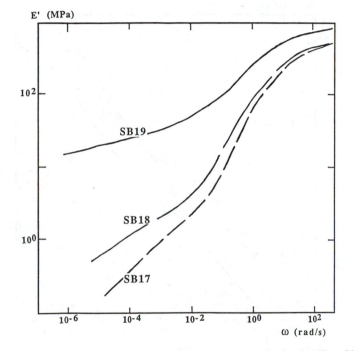

Figure 6 . Log-log plot of the isothermal master curves obtained at $T_0 = 301$ K for the storage modulus E' of the series of model latexes SB17-SB19 synthesized with CTA C2 and containing different amounts of copolymerized carboxylic groups.

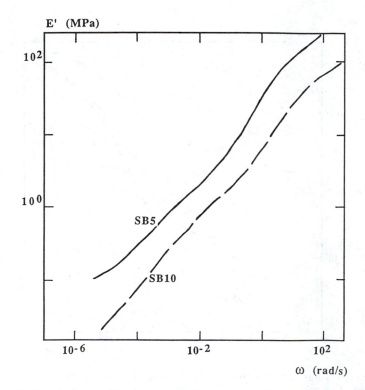

Figure 7 . Log-log plot of the isothermal master curves obtained at $T_0 = 301$ K for the storage modulus E' of the model latexes SB5 and SB10 synthesized with CTA C1 and containing different amounts of copolymerized carboxylic groups.

coupling loci between the chains of the polymer network. These interactions likely find their origin in the hydrogen bonding between carboxylic groups, leading to the formation of dimers (*21*). A semi-quantitative analysis has been recently performed to specify the part played in the viscoelastic behavior of the films by the carboxylic groups which are located at the surface of the particles and those which are not (*9*). The observed effect of rubbery modulus enhancement upon copolymerization of carboxylic monomers was also reported by Zosel et al. (*22,23*) for model latexes of n-butylacrylate homopolymers which contain acrylic acid as the stabilizing species.

The question arises whether a second continuous phase consisting of an acid-rich copolymer with a high glass transition temperature is formed in the films. In the present case, even when the content of copolymerized carboxylic groups is high (latex SB19), the DMA spectrum exhibits only a single relaxation phenomenon from glasslike to rubberlike behaviour. This observation brings out strong arguments against the formation of a continuous segregated carboxylic phase within the films. Otherwise, a second transition related to this acid-rich phase whould have appeared in our DMA spectra. From data recently published by Zosel (*24*), we infer that the appearence of this second phase only occurs for very high acrylic acid (AA) contents, i.e. higher than 15 wt%. For such a high AA content, low molecular weight water-soluble polyacrylic acid chains are expected to be formed in the aqueous phase and be responsible for the formation of the second phase in the films. This behavior is not observed in our samples. Only clustering of carboxylic groups into separated microphases is suspected to occur in the membranes, when carboxylic group content becomes higher than 900 μmol. cm^{-3} in the films (*9*). TEM experiments performed on stained samples of latex films SB17-SB19 confirmed that no segregated carboxylic phase exist in the films : only thin interfacial membranes located at the boundary between particles can be observed in the photomicrographs of stained ultramicrotomed samples (*9*).

A second qualitative comment arises from inspection of Figure 6. The position of the relaxation frequency does not significantly shift on the frequency scale, when the content of copolymerized carboxylic groups is varied. It strongly suggests that the copolymerization of carboxylic groups does not affect the local monomer dynamics, at least as far as concentration of this species remains lower than 1090 μmoles per polymer volume unit (cm^3). Hence, the creation of a new coupling loci through H bonding interactions does not sufficiently increase the tightness of the polymer network for the local segment dynamics to be perturbed. However, the DMA method only gives average information over the whole sample volume and the monomer dynamics must obviously be affected in the interfacial membranes. It is not detected in the DMA spectra, likely because only a small volume fraction of the films is concerned. For latexes synthesized with CTA C1 (Figure 7), a noticeable shift of the relaxation frequency is observed. SANS experiments performed on the films clearly show that interfacial membranes have vanished and interdiffusion of polymer chains taken place. For this reason, acid groups are likely to be more uniformly distributed in these films and the probability of acid dimer formation through H bonding is expected to depend on the carboxylic acid content. Thus, the observed effect on the local chain segment dynamics is consistent with the more uniform distribution of acid groups within the samples.

Figure 8 shows the isothermal master curves recorded for the storage moduli of films from the series of model latexes SB5-SB9. It is worthwhile noticing that these samples consist of the same carboxylated copolymer of low gel fraction, whose neutralization conditions (pH and base) have been varied. Basically,

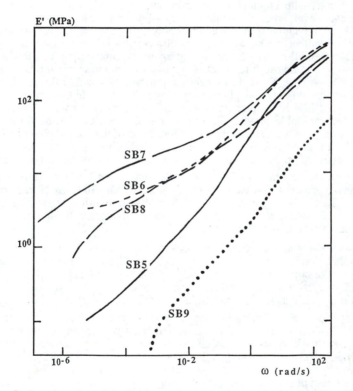

Figure 8 . Log-log plot of the isothermal master curves obtained at $T_0 = 301$ K for the storage moduli of films from the series of model latexes SB5-SB9 which are prepared under different neutralization conditions (pH, base).

neutralization of surface carboxylic groups of the particles (from pH 4.5 to pH 9) is found to produce two different effects, depending on the nature of the base used :

i) - When the base is chosen among ammonia, NaOH or AMA, neutralization induces a large increase of the rubbery storage modulus in the low frequency range, which is the signature of a large increase in the cross-linking density of the polymer. This effect is attributed to the formation of physical ionic interactions in the membranes, which act as cross-linking knots. It is reminiscent of the well-known ionomer effect observed in bulk ion-containing polymers. In these systems, due to the low polarizability of the surrounding matrix, ionic species tend to form dipoles, or multiplets or even to cluster into larger domains, which link the chains together (*21,25-27*) and yields a large increase of the tensile strength of the materials (*28*). Very interestingly, the above changes in the viscoelastic behavior of the latex films upon neutralization are correlated with structure modifications. SANS experiments performed on the above samples (Figure 9 and Figure 10) clearly show that neutralization induces the preservation of the network of interfacial membranes. As a matter of fact, the SANS spectra of films from latexes neutralized to pH 9 exhibit 2 intensity peaks characteristic of a highly ordered structure, while the spectrum of the low pH sample does not exhibit any peaks. For these reasons, it can be inferred that in our latex films, ionic dipoles, multiplets or clusters which should arise from ion pair interactions, remain confined in the polar interfacial membranes separating the SB particle cores of low polarizability. It is worthwhile noticing that this ionomer effect is found to be very sensitive to water and tends to disappear in the presence of moisture. This result is quite consistent with data recently reported by Kim et al. for bulk poly(styrene-co-sodium methacrylate) ionomers, in which water has been shown to act as a polar plasticizer (*29*).

ii) - on the opposite, when the base used for neutralization is DAPOE, two main alterations of the DMA spectrum are noticed. The position of the relaxation zone on the frequency scale is shifted towards the higher frequency region and the rubbery behavior in the low frequency region tends to completely disappear. It is replaced by a rapid decrease of the storage modulus towards values typically reported for liquid-like polymers. These two alterations are the signature of some kind of plasticization of the latex film induced by DAPOE, which is actually a totally opposite effect compared to the previous ionomer one. It could be related the size and the polymeric nature of the neutralizing agent, which can give rise to particular non ionic interactions with the chains belonging to the latex particles. SANS experiments performed on this sample show that neutralization with DAPOE still induces preservation of the array of hydrophilic interfacial membranes, accompanied by the presence of large hydrophilic aggregates (Figure 10) (*12*). This leads us to conclude that physical ionic cross-linking still takes place in the membranes, but the presence of DAPOE is likely to strongly reduce the intermolecular forces among carboxylic moieties. The same effect has been observed in bulk carboxylic ionomers based on styrene-methacrylic acid copolymers, plasticized by alkyl phthalates (*27,30*).

From a further inspection of Figure 8, it can also be inferred that the nature of the counter-ion used for neutralization plays a major part in the ionic cross-linking effect of the membranes. For instance, the value of the rubbery modulus in the low frequency region appears to be much higher when NaOH is used, compared to the case of ammonia or AMA. It means that Na^+ counter-ions form a large number of very stable cross-links and according to Navratil et al. (*27*), this behavior could be attributed to the small size of the counter-ion, which is in favor of a stability of the

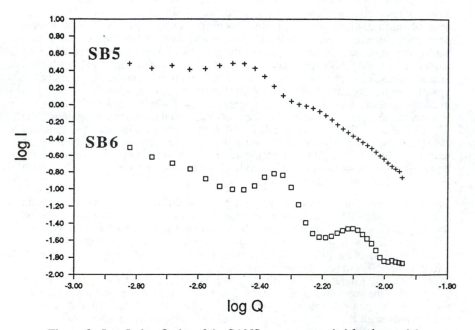

Figure 9 . Log I - log Q plot of the SANS spectra recorded for the model latexes SB5 and SB6 whose pH is 4.5 and 9 respectively. Only relative intensities are significant. Neutralization of the carboxylic groups at pH 9 leads to the preservation of a network of interfacial membranes.

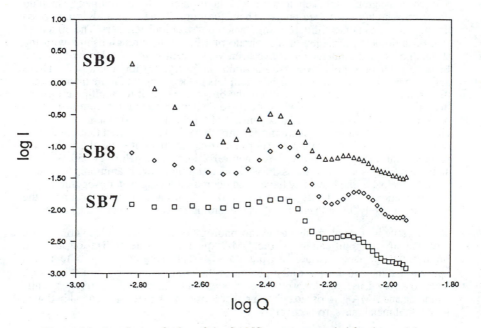

Figure 10 . Log I - log Q plot of the SANS spectra recorded for the model latexes SB7, SB8 and SB9 which are neutralized to pH 9 with NaOH, AMA and DAPOE respectively. Only relative intensities are significant.

ionic multiplets. The isochronal temperature dependence of E' and tanδ recorded for the samples SB5-SB8 makes it possible to perform a more refined analysis of the ionic cross-linking effect (Figures 11 to 13). First, the strong ionomer effect resulting in an overwhelming increase of the rubbery modulus in the high temperature region is confirmed, with very stable cross-links over a wide range of temperature for Na$^+$ counter-ions. Secondly, latex films SB7 and SB8 neutralized with NaOH and AMA respectively, clearly exhibit a second transition located around 55 °C which can be easily observed on both E' and tanδ spectra (Figure 12 and Figure 13). The appearence of this transition is the signature of the formation of highly connected phase of ionic clusters located in the network of interfacial membranes. When the second transition does not appear, as in the case of sample SB6, the ionic groups are likely to remain in the form of isolated multiplets which do not form a highly connected phase. Then, they only behave as additional cross-linking knots and not as a different segregated phase of ionic clusters. The effects of size of multiplets, formation of a separate phase and their relationships with the strength of the ionic interactions have been very recently investigated by Kim and Eisenberg for bulk poly(styrene-co-sodium methacrylate) ionomers (29). These authors have correlated the size of the multiplets to the strength of ionic interactions : the stronger the electrostatic interaction, the bigger the size of the multiplets and the higher the T$_g$ of the clusters. Following these authors, we can infer from inspection of Figures 12 and 13 that in our latex films, AMA is likely to lead to a more effective clustering effect. Moreover, Kim and Eisenberg have also specified that the degree of neutralization of the carboxylic groups controls the effect of ionic clustering, this phenomenon being not observed for neutralization level below ca. 50 %. This statement suggests that in our sample SB6 neutralized with ammonia which is volatile, the pH of the latex may have been decreased during water evaporation and film formation so that the degree of neutralization of the carboxylic groups in the dried film is likely lower than 50%, thus preventing the formation of ionic clusters.

Detailed information about the morphology of the films in close correlation with the strong modifications of the DMA spectra upon neutralization are also brought out by inspection of TEM photomicrographs (Figures 14, 15 and 16). It appears that the counter-ion which generates the more stable cross-links, i.e. Na$^+$, leads to a very dense, regular, well-ordered network of interfacial membranes, while ammonia and AMA give rise to either a rather loose, incomplete or even a thin dotted network of membranes, respectively.

Finally, we would like to emphasize that the effect of ionic cross-linking in latex films was earlier reported by Zosel for carboxylated poly(n-butyl acrylate) (PBA) particles whose surface acrylic acid groups were cross-linked with a multivalent system, namely Zn(NO3)2 (31).

Sterically-Stabilized Particles

Specific Effect of the Membranes upon Annealing. Films obtained from the sterically-stabilized model particles of Table II have also been studied as for their viscoelastic behavior. More particularly, the effect of thermal annealing has been investigated. Figure 17 shows the typical isochronal temperature dependence of the storage modulus E' and loss tangent tanδ recorded before and after annealing at 150 °C for sample A10 in which the PVA volume fraction is 13.5 % (Table II).

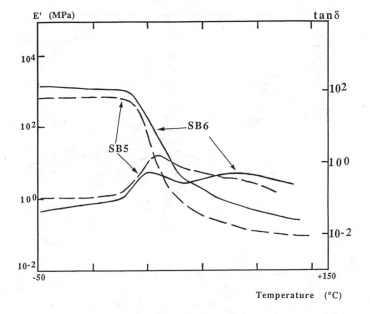

Figure 11 . Log plot of the isochronal temperature dependence of E' and tanδ recorded for samples SB6 which is neutralized to pH 9 with ammonia. The corresponding spectra for the non neutralized sample SB 5 (pH 4.5) are also shown for comparison.

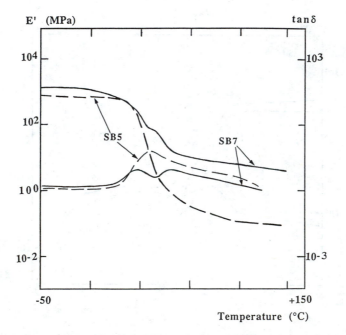

Figure 12 . Log plot of the isochronal temperature dependence of E' and tanδ recorded for samples SB7 which is neutralized to pH 9 with NaOH. The corresponding spectra for the non neutralized sample SB5 (pH 4.5) are also shown for comparison.

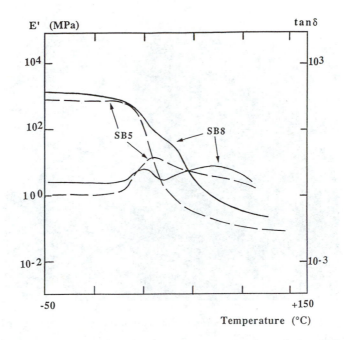

Figure 13 . Log plot of the isochronal temperature dependence of E' and tanδ recorded for samples SB8 which is neutralized to pH 9 with AMA. The corresponding spectra for the non neutralized sample SB5 (pH 4.5) are also shown for comparison.

Figure 14 . TEM photomicrograph of stained ultramicrotomed sample SB6 (magnification X 44 000).

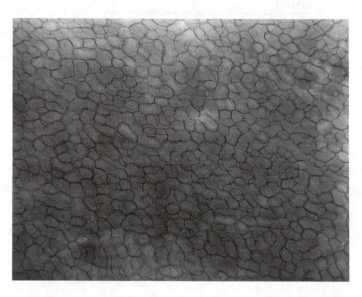

Figure 15 . TEM photomicrograph of stained ultramicrotomed sample SB7 (magnification X 44 000).

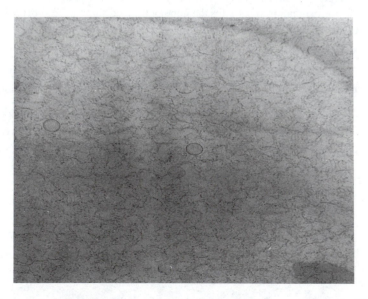

Figure 16 . TEM photomicrograph of stained ultramicrotomed sample SB8 (magnification X 44 000).

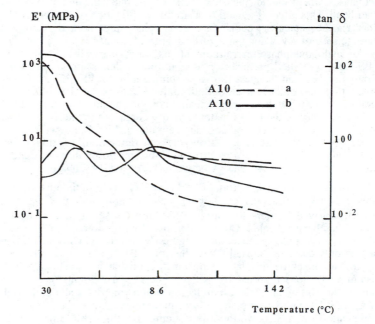

Figure 17 . Log plot of the isochronal (f = 1 Hz) temperature dependence of the storage modulus E' and loss tangent tanδ recorded before (a) and after (b) annealing at 150 °C for sample A10 in which the PVA volume fraction is 13.5 %

Annealing appears to result in dramatic changes in the viscoelastic behavior of the films. Apart from the main transition observed at ~ 35 °C before annealing, a second relaxation phenomenon clearly stands out at ~ 75 °C after annealing. The former transition is associated with the glass transition of the PVAc cores, whereas the latter is related to the PVA phase. The appearence of this transition is accompanied by an overwhelming increase of the storage modulus E' in the temperature range between the two transitions. In this zone, E' only exhibits a smooth variation with temperature. In addition, after annealing, tanδ exhibits two well-identified loss peaks, which are the signature of a phase-separated blend of polymers, since each component appears to experience its own glass transition (32). All these observations strongly suggest that after annealing the film contains two clearly distinct phases, which should consist of PVAc on the one hand and a PVA phase on the other hand. This effect is related to a strong modification of the morphology of the films. The mechanical detection of the second phase after thermal treatment arises from an increase of PVA domain size or degree of connectivity within the PVA phase. An analogous phase rearrangement has been observed for films obtained from PS-PBA core-shell particles or homopolymer blends, upon annealing at a temperature higher than the glass transition temperature for PS (32,33). It has been attributed to a more or less extended coalescence of the PS phase at high temperature. In the present case, phase rearrangement should stem from a thermally induced coalescence of the PVA clusters initially contained in the films. Molecular mobility of PVA chains likely becomes high enough at 150 °C, i.e. well above their glass transition temperature, to allow the rearrangement of this phase via macromolecular diffusion. The resulting equilibrium morphology is expected to minimize the interfacial surface energy between PVAc and PVA phases.

Using Transmission Electron Microscopy (TEM) associated to a selective staining of the PVA phase with OsO$_4$ (11), the exact morphology of the films, and more particularly the extent to which PVAc and PVA domains coalesce after annealing, has been revealed. Comparison of TEM photomicrographs taken on sample A10 before and after annealing (Figure 18) clearly evidences that, while the initial film only contains isolated PVA clusters, a PVA lattice or matrix in which PVAc particles are embedded, has been formed upon annealing. This result is fully consistent with those reported by Kast (11), who also found out such a film structure in the case of an ethylene-vinylacetate copolymer latex sterically-stabilized by PVA, and related it to the presence of two main relaxations in the DMA spectra of the composite system. It is worthwhile noticing that this rebuilding effect of the network of interfacial membranes is completely opposite to that generally observed for films from electrostatically-stabilized particles (e.g. the above carboxylated particles), in which annealing induces fragmentation and expulsion of the membranes into large hydrophilic lumps dispersed in the matrix consisting of the coalesced cores of the particles (1, 4).

Semi-quantitative Analysis of the Rebuilding Effect of the Membranes. The rebuilding effect of the hydrophilic interfacial membranes upon annealing has been studied as a function of PVA volume fraction (Figure 19), average PVAc particle size (Figure 20) and PVA hydrolysis level (Figure 21).

Figure 19 shows the typical isochronal temperature dependence of the storage modulus E' recorded after thermal annealing for latex films A1, A3, A6 which contain various PVA volume fractions ranging from 3.15 to 11 %. Spectra recorded for samples A2 and A8, whose PVA volume fractions are 5.2 % and 6.5 %

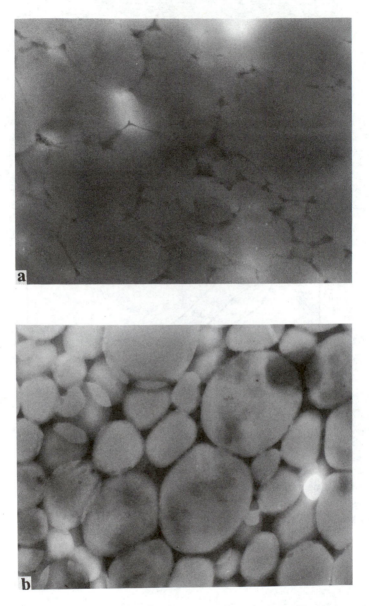

Figure 18 . TEM photomicrographs of stained ultramicrotomed sample A10 before (a) and after (b) thermal annealing. The PVA phase is selectively stained with OsO4. Comparison of the pictures gives evidence of the rebuilding of the PVA interfacial membranes upon annealing (magnification X 70 000).

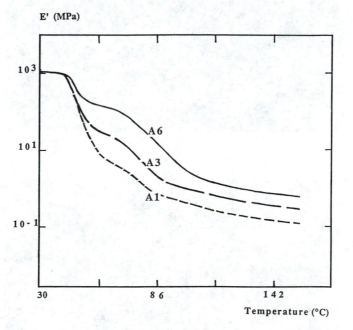

Figure 19 . Log plot of the isochronal (f = 1 Hz) temperature dependence of the
storage modulus E' recorded after thermal annealing for samples A1, A3,
A6 whose PVA volume fractions are 3.15, 8.8 and 11 %, respectively.

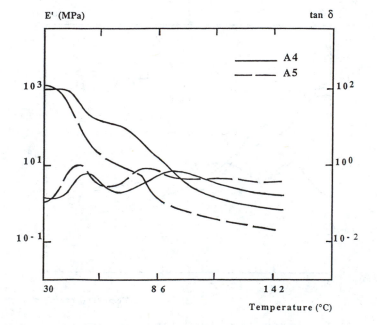

Figure 20 . Log plot of the isochronal (f = 1 Hz) temperature dependence of E' and tanδ recorded after thermal annealing for samples A4 and A5, whose average PVAc particle size is 2 μm and 0.55 μm, respectively.

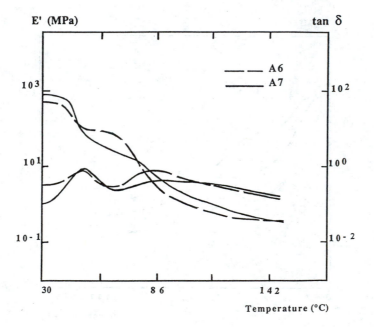

Figure 21 . Log plot of the isochronal (f = 1 Hz) temperature dependence of E' and tanδ recorded after thermal annealing for samples A6 and A7, for which the HL value of PVA added at the end of polymerization is 80 and 99 mol%, respectively.

respectively, are not included in Figure 18, because they are quite similar to that obtained for film A1 and they do not actually show any modification upon annealing. Inspection of Figure 19 reveals that the value of E' in the temperature zone located between the two main transitions strongly depends on the PVA volume fraction. Moreover, it qualitatively shows that there exists a threshold value for the PVA volume fraction ϕ_t, above which the system undergoes a profound phase reorganization upon annealing. For PVA content lower than ϕ_t, some changes may occur in the distribution of PVA in the films, but they do not significantly alter their thermomechanical properties. In the present system, it turns out that ϕ_t ranges between 6.5 % and 8.8 %. A semi-quantitative analysis of these data will show that the value of E' measured in the region located between the two transitions can be related to the mechanically effective volume fraction of PVA involved in the segregated continuous phase formed upon annealing (*34,35*).

Figure 20 and Figure 21 show that both PVAc particle size and PVA hydrolysis level have a marked influence on the viscoelastic behavior of the annealed films. From this result, it can be inferred that these two parameters determine the connectivity enhancement of the PVA phase upon annealing. They likely control the distribution of the PVA chains around the PVAc cores in both the dispersions and the final films (*6*).

From these experimental data, we have performed a quantitative analysis of thermally-induced phase rearrangement, and derived the values of the mechanically effective volume fractions of PVA involved in the segregated continuous phase after thermal annealing (v_m). This corresponds to PVA fraction which gives rise to the enhancement of the storage modulus E' of the composite in the temperature range located between the two transitions. For this purpose, the results have been interpreted on the basis of a mechanical composite model which makes it possible to derive the modulus of the multiphase system from the modulus of each constituent phase, provided that some microstructural prerequisites are fulfilled. Other work has suggested that a satisfactory description of the elastic behavior of composite latex films can be provided using the Kerner equation (*34-36*). Under certain simplifiying assumptions previously discussed (*6*), the Kerner equation for the storage modulus of the composite films (soft inclusions/hard matrix type) can be written in the form :

$$E'/E'_m = \frac{(1 - v) E'_m + (\alpha + v) E'_i}{(1 + \alpha v) E'_m + \alpha (1 - v) E'_i} \qquad (5),$$

where E'_m is the storage modulus of the matrix, E'_i the storage modulus of the inclusions, v the volume fraction of the inclusions. The parameter α is a function of the Poisson ratio of the matrix v_m, and can be expressed as :

$$\alpha = 2 (4 - 5 v_m) / (7 - 5 v_m) \qquad (6).$$

Equation 5 offers the opportunity to derive the actual volume fraction of PVA involved in the continuous matrix after annealing. For a composite system, the temperature dependence of the storage modulus in the range located between the two transitions, is expected to parallel that of the material forming the continuous matrix. This behavior is indeed observed experimentally ; in this temperature zone, the storage modulus of the inclusions E'_i is much lower than that of the matrix E'_m, since the former have already experienced their glass transition, whereas the latter is still in a glassy state. Then, in this particular temperature range, where E' exhibits almost a plateau, equation 5 can be simplified to :

$$E'/E'_m = \frac{(1 - v) E'_m}{(1 + \alpha\, v) E'_m} \qquad (7),$$

This expression can be inverted to give :

$$v = \frac{1 - E'/E'_m}{1 + \alpha\, E'/E'_m} \qquad (8)$$

Finally, v_m can be expressed as :

$$\begin{aligned} v_m &= 1 - v \\ &= \frac{(1 + \alpha)\, E'/E'_m}{1 + \alpha\, E'/E'_m} \end{aligned} \qquad (9)$$

Then, following Dickie (*34*) and Jourdan et al. (*35*), we have taken a constant value of 0.35 for v_m, and hence a constant value of 0.86 for α. The values of v_m deduced from equation 9 for the samples which exhibit the profound mofification of their viscoelastic profile upon annealing are given in Table IV.

Table IV. Values of the volume fraction of the mechanically effective PVA (v_m), derived from the Kerner equation

Latex	PVA volume fraction (%)		PVAc average particle size (d) (µm)	v_m* (%)
	initial charge (ϕ_i)	final addition (ϕ_a)		
A3	5.2	3.6	1.2	5
A4	5.3	4	2	9
A5	5.3	4	0.55	3
A6	5.3	5.7	0.55	11
A7	5.3	5.7	0.55	4
A9	6.5	3.6	1.2	3
A10	6.5	7	1.2	9

* The accuracy of the values of v_m is estimated to be within 15 % in relative value.

Because of the slow relaxation sometimes observed between the two glass transition temperatures, the accuracy of the determination is not very high (within 15 % in relative value). However, it allows a reasonable discussion of the results. Inspection of Table IV reveals that the whole fraction of PVA introduced in the films is not necessarily involved in the continuous matrix after annealing. Depending mainly on the size of the PVAc particles and the hydrolysis level of PVA added at the end of polymerization, the value of v_m is often found to be only a part of the total PVA volume fraction in the sample; in this case, it ranges about the volume fraction of PVA added at the end of polymerization. This behavior has not yet been elucidated. However, in many respects, these semi-quantitative results confirm the above qualitative analysis and emphasize the pronounced effect of PVAc particle size and PVA hydrolysis level on dynamic viscoelastic properties of our composite films.

Conclusion

This review points out the role played by core chains and interfacial membranes in the dynamic viscoelastic properties of latex films. The first conclusion concerns the core chain dynamics. As expected, cross-linking the particle cores has been found to enhance the rubbery modulus of the films. In addition, it has also been quantitatively shown to affect the local chain segment dynamics and restrict the mobility of the core chains so that the further coalescence process cannot occur, even if the membranes are very mobile. Then, polymer chain interdiffusion and further coalescence are restricted not only by covalent cross-links, but also by non covalent physical intermolecular interactions which strongly enhance the level of constraints in the system. Basically, molecular mobility within the particle cores which can be derived from viscoelastic spectra appears to be the key parameter to predict the occurrence of further coalescence.

The second conclusion concerns the interfacial membranes. Based on both thorough investigations of the viscoelastic properties and structural observations, this work also brings out strong evidence of the major part played by the interfacial membranes in latex films. Furthermore, this statement has been proved to be valid for both kinds of particles with either electrostatic or steric stabilization.

For films from particles electrostatically-stabilized by copolymerized carboxylic species, the interfacial cross-linking effect has been shown to greatly enhance the rubbery behavior of the films. Modifications of the membranes which induce this cross-linking originate from either H bonding between carboxylic acid species or ionic dipolar interactions within multiplets or clusters. For instance, increasing the content of carboxylic monomers results in a strong enhancement of the rubber-like behavior, although these groups are not found to form a second separate, macrosopic phase in the films. Upon neutralization of the surface carboxylic groups, physical ionic cross-linking has been shown to occur in the membranes and additional effects have been evidenced depending on the nature of the counter-ion. A second glass transition temperature appears in the isochronal temperature dependence of the viscoelastic features, when the counter-ion of the carboxylate anions is sodium or originates from an aminoalcohol. This is attributed to a segregation of the ionic species into connected ionic multiplets or clusters within the membranes, due to the low polarizability of the particle cores. Then, based on structural observations, the enhancement of the rubber-like modulus has been clearly related to the order and density of the interfacial network of membranes : preservation of a highly dense, ordered array of membranes has been observed when these membranes contain highly connected ionic clusters which behave as both a second more rigid phase and very stable cross-linkers. However, it has also been shown that neutralization of the surface carboxylic groups sometimes lead to an unexpected plasticization of the film, for instance when the base used for neutralization contains POE chains.

Besides, latex films from sterically-stabilized particles have been shown to exhibit an unexpected specific behavior upon annealing. For this kind of film, thermal annealing does not induce fragmentation and expulsion of the interfacial membranes; on the contrary, it gives rise to an enhancement of connectvity within the array of polymeric particle shells, which consist of the steric stabilizer. This strong phase rearrangement which leads to a rebuilding of the network of interfacial membranes throughout the films, is responsible for a profound modification of their viscoelastic properties, with the appearence of a second transition. The effect has been shown to depend on the steric stabilizer volume fraction and features, and also on the particle size. A semi-quantitative analysis of the enhancement of connectivity of the PVA phase has been performed. Based on a classical equivalent model for polymer-polymer composite and the correlative application of the Kerner equation, it leads to rather consistent results.

Finally, from a more general viewpoint, it is worthwhile noticing that this work gives many guidelines to properly design and manipulate the interfacial membranes, so as to adjust and modify the viscoelastic properties of latex films in a given temperature and frequency range.

Acknowledgements

The author wishes to acknowledge K. Wong for his contribution to the SANS investigation of the latex films, J. Maquet for TEM experiments on these samples, and C. Mignaud for the viscoelastic measurements. This work used the neutron beams of Institut Laue-Langevin (ILL) in Grenoble and Laboratoire Léon Brillouin (LLB) in Saclay. The staff of LLB and ILL are acknowledged for providing us guidance and support during neutron scattering experiments.

Literature Cited

1.   Joanicot, M. ; Wong, K. ; Richard, J. ; Maquet, J. ; Cabane, B.
      Macromolecules **1993**, *26*, 3168.

2.   Overbeek J.Th.G. *J. Colloid Interface Sci.* **1977**, *58*, 408.

3.   Napper D.H. *J. Colloid Interface Sci.* **1977**, *58*, 390.

4.   Richard, J. ; Wong, K. *J. Polym. Sci. Part B : Polym. Phys.* **1995**, *33*,
      1395.

5.   Charmot, D. ; Guillot, J. *Polymer* **1992**, *33*, 352.

6.   Richard, J. *Polymer* **1993**, *34* , 3823.

7.   Cohen-Addad, J.P. ; Desbat, F. ; Richard, J. *Macromolecules* **1994**, *27*
      2111.

8.   Richard, J. *Polymer* **1992**, *33* , 562.

9.   Richard, J. ; Maquet, J. *Polymer* **1992**, *33* , 4164.

10. Marten, F.L. in *Encyclopedia of Polymer Science & Engineering* ; Mark, F.H.; Bikales, N.M. ; Overberger, C.G. ; Menges, G. ; Wiley : New York, 1987, Vol. 17; pp 168-198.

11. Kast, H. *Makromol. Chem. Suppl.* **1985**, *10/11*, 447.

12. Richard, J. ; Mignaud, C. ; Wong, K. *Polymer International* **1993**, *30* , 431.

13. Ferry, J.D. *Viscoelastic Properties of Polymers* ; Wiley : New York, 1980 ; pp 224-436.

14. Zosel, A. ; Ley G. *Macromolecules* **1993**, *26*, 2222.

15. Marvin, R.S. ; Oser H.J. *J. Res. Nat. Bur. Stand. (B)* **1962**, *66*, 171.

16. de Gennes, P.G. *J. Chem. Phys* **1971**, *55*, 572.

17. de Gennes, P.G. *Macromolecules* **1976**, *9*, 587.

18. Wang, Y. ; Winnick, M.A. *J. Phys Chem.* **1993**, *97*, 2507.

19. de Gennes, P.G. *Macromolecules* **1986**, *19*, 1245.

20. Stratton, R.A., Ferry, J.D. *J. Phys. Chem.* **1963**, *67*, 2781.

21. Marx, C.L. ; Caufield, D.F. ; Cooper, S.L. *Macromolecules* **1973**, *6*, 344.

22. Zosel, A. ; Heckmann, W. ; Ley, G. ; Mächtle, W. *Colloid Polym. Sci.* **1987**, *265*, 113.

23. Zosel, A. ; Heckmann, W. ; Ley, G. ; Mächtle, W. *Makromol. Chem., Makromol. Symp.* **1990**, *35/36* , 423.

24. Zosel, A. *Polym. Adv. Technol.* **1995**, *6*, 263.

25. Jalal, N. ; Duplessix R. *J. Phys France* **1988**, *49*, 1775.

26. Eisenberg, A. *Macromolecules* **1970**, *3*, 147.

27. Navratil, M. ; Eisenberg, A. *Macromolecules* **1974**, *7*, 84.

28. Hara, M. ; Sauer, J.A. *J. Macromol. Sci. - Rev. Macromol. Chem. Phys.* **1994**, *C34*, 325.

29. Kim, J. ; Eisenberg, A. *J. Polym. Sci. Part B : Polym. Phys.* **1995**, *33*, 197.

30. Chapoy, L.L. ; Tobolsky, A.V. *Chem. Scr.* **1972**, *2*, 44.

31. Zosel, A. *Double Liaison* **1987**, *34*, 19.

32. O'Connor, K.M. ; Tsaur, S.L. *J. Appl. Polym. Sci.* **1987**, *33*, 2007.

33. Cavaillé, J.Y. ; Vassoille, R. ; Thollet, G. ; Rios, L. ; Pichot, C. *Colloid Polym. Sci.* **1991**, *269*, 248.

34. Dickie, R.A. *J. Appl. Polym. Sci.* **1973**, *17*, 45.

35. Jourdan, C. ; Cavaillé, J.Y. ; Perez., J. *Polym. Eng. Sci.* **1988**, *28*, 1318.

36. Dickie, R.A. ; Cheung, M.F. *J. Appl. Polym. Sci.* **1973**, *17*, 79.

Chapter 9

Role of Interdiffusion in Film Formation of Polymer Latices

Albrecht Zosel and Gregor Ley

Polymer Research Laboratory, BASF AG, 67056 Ludwigshafen, Germany

The mechanical strength of films from polymer latices develops by the interdiffusion of chain segments and the formation of entanglements across particle boundaries during film formation.

Measurements of the fracture energy of poly-n-butylmethacrylate films, tempered above the glass transition temperature for different times, show a transition from brittle to tough (yielding) behaviour with a strong increase of the fracture energy at short temper times, which is followed by an increase proportional to the square root of the temper time until a constant fracture energy is reached after long times. The film strength is correlated to the interdiffusion length determined by small angle neutron scattering of deuterated films.

The formation of interparticular entanglements is hindered in latices with crosslinked particles, as has been studied with a series of latices, crosslinked with various amounts of a bifunctional monomer.

In most applications polymer latices are transformed into coherent, transparent films which generally have a considerable mechanical strength comparable to that of films from polymer solutions or melts. The film formation of polymer latices has accordingly found wide-spread interest, as a great part of the end-use properties of emulsion polymers develops during this process. The film formation of latices can be regarded as a three stage process consisting of
1. the concentration and packing of the latex particles by evaporation of water,
2. the deformation of particles, and
3. the coalescence of particles by interdiffusion of chain ends and segments across particle boundaries and the formation of entanglements.

0097–6156/96/0648–0154$15.00/0
© 1996 American Chemical Society

It is supposed that the mechanical strength of latex films develops during this third step, which will be the subject of this paper.

The research on diffusion accross polymer polymer interfaces started with the pioneering work of Voyutskii [1]. In the last two decades, the investigation of the roleof interdiffusion in the healing and welding of polymer polymer interfaces has experienced a tremendous progress, mainly connected with the work of de Gennes [2], Kausch [3], Tirrell [4] and Wool [5]. The interdiffusion in latex films has also been studied during the last years, mainly by SANS [6-8] and by steady state fluorescence decay measurements [9]. The work presented here deals with the development of mechanical strength during this third stage of film formation. Other studies of the mechanical behaviour during film formation have been published by Klein, Sperling and coworkers [8] and by Eckersley and Rudin [10].

The development of film strength by interdiffusion and the formation of interparticular entanglements will be increasingly hindered in latices with crosslinked particles, especially when the mean molecular mass between crosslinks, M_c, becomes smaller than the mean molecular mass between entanglements, M_e. This has been studied with films formed from latices with crosslinked particles [11,12].

A crucial point in studying interdiffusion during film formation is the unambiguous definition of the beginning of the interdiffusion phase. As interdiffusion in particles above the glass transition temperature starts immediately upon particle contact, the time "zero" for interdiffusion may be different at different parts of the particle surface. Furthermore, it may be different in different regions of the drying film, as it is well known that often a drying front propagates through a film in the compaction stage.

One way to overcome this difficulty is to carry out the first two steps of the film forming process below the glass transition temperature and subsequently to heat up the film to an annealing temperature above T_g, so that the starting time of the inter-diffusion process is well defined. This strategy has been applied in this study.

Experimental Part

Samples. For this purpose we used model emulsion polymers of n-butyl-methacrylate (BMA). The latices were prepared by batchwise emulsion polymerization with sodium laurylsulphate as emulsifier, using a standard recipe at 80 °C. They had typically a solid content of about 30% and a mean particle diameter of about 60 nm. More details on the latex synthesis are given in a previous publication [6]. PBMA latices are well suited for studies of interdiffusion as PBMA has a glass transition temperature of 29 °C, determined by dynamic mechanical analysis with a frequency of 1 Hz, but nevertheless forms coherent, brittle films at 23 °C. No significant interdiffusion takes place at room temperature owing to the low segmental mobility of the polymer, as has been shown earlier by SANS studies of the same samples [6]. Interdiffusion thus starts only on annealing the films at temperatures above T_g.

The particles of the BMA homopolymer are uncrosslinked. In order to study the effect of crosslinking on interdiffusion, copolymers of BMA and the bifunctional monomer methallyl-methacrylate (MAMA) with molar concentrations of MAMA between 0 and 2% were prepared. The BMA/MAMA copolymer latices have crosslinked particles with a crosslink density increasing with the concentration of the bifunctional monomer.

Experimental Methods. The latex films were characterized by measurements of the dynamic shear modulus as a function of temperature and frequency, using a dynamic mechanical analyzer with a parallel plate geometry [12]. The glass transition temperature of PBMA, the mean molecular mass between crosslinks, M_c, and the entanglement length M_e are determined from the dynamic modulus which consists of the storage modulus G' and the loss modulus G".

The mechanical strength of the films is characterized by tensile tests at 23 °C which measure the tensile strength σ_B, i. e. the tensile stress at break, and the strain at break, ε_B. The energy per unit volume, W_B, which is necessary to break the sample, is calculated by integration of the stress strain curve:

$$W_B = \int \sigma \, d\varepsilon$$

A crosshead speed of 1.67 mm/s and a sample length of 24 mm were employed in the tensile tests.

Film Formation of Uncrosslinked Latices

Films were formed from the PBMA latex without bifunctional monomer at 23 °C, and were tempered above T_g, i. e. at 90 °C, for different times after film formation. Specimens for the tensile tests were cut before the samples cooled down to room temperature. As the unannealed films are very brittle they had to be heated to about 50 °C for a short time of less than 2 min in order to cut specimens.

Some examples for stress strain plots are given in the left part of Figure 1 for the uncrosslinked PBMA. The untempered film shows a rapid increase of σ with increasing tensile strain ε and brittle fracture at a low strain of about $4 \cdot 10^{-2}$. After an annealing time of 5 min already, this fracture behaviour changes completely. A steep stress peak is observed at low strains, too, but it is followed by yielding which gives rise to a high ultimate strain of about 3. In the case of yielding, the area under the stress strain plot is much higher than for brittle fracture. That means that the fracture energy increases drastically. At an annealing time of 5 min or less, thus, a transition from brittle fracture to tough or yielding behaviour is observed. Longer annealing times only gradually change the shape of the stress strain characteristics.

In Figure 2 the fracture energy W_B is plotted versus the temper time at 90 °C for the uncrosslinked PBMA showing a strong increase by more than two orders of magnitude at annealing times below 5 min in correspondence with the transition from brittle to tough behaviour. This transition is followed by a further gradual increase of W_B which is not finished in the time interval of 360 min, shown in Figure 2. Annealing at a lower temperature, e. g. 50 °C, results in a slower increase of the fracture energy with the tempering time.

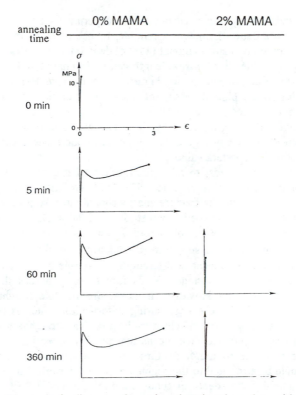

Figure 1: Stress-strain diagrams for poly-n-butylmethacrylate with 0 and 2% methallylmethacrylate after different annealing times

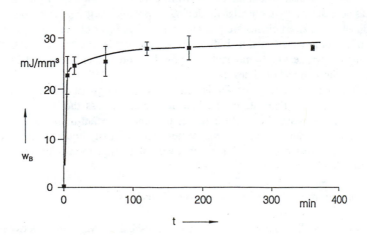

Figure 2: Fracture energy W_B per unit volume for uncrosslinked PBMA as a function of the annealing time at 90 °C

Hahn, Ley and coworkers carried out SANS experiments on films from similar latices at the Institute Laue - Langevin, Grenoble, France [6,7]. From measurements of the radius of gyration of a small amount (1%) of deuterated particles in dependence on the annealing time, the interdiffusion length can be evaluated and correlated with the increase of the fracture energy. In Figure 3 the fracture energy W_B and the interdiffusion depth d are plotted versus the square root of the annealing time t at 90 °C. The penetration depth is proportional to $t^{1/2}$, as predicted by the reptation theory of de Gennes [13]. W_B increases by about two orders of magnitude within the first 5 min, as already shown in Figure 2, then gradually increases linear with $t^{1/2}$ and seems to level off with a constant value.

At present, there is no theory to relate the mechanical strength of the interface to the arrangement of chains in the interface. The models of de Gennes, Prager and Tirrell, and Kausch agree that the fracture energy should increase proportional to $t^{1/2}$, too, what is verified by various experimental studies and the work, presented here.

If we assume that no considerable interdiffusion occurs during film formation below T_g, the untempered film is supposed to consist of a packing of the hard latex spheres, which has a certain mechanical strength, i.e. tensile stress at break, due to intermolecular forces such as van der Waals forces acting across the boundaries between the packed particles. However, it lacks a measurable toughness, i.e. the ability to store and to dissipate energy during deformation which is related to the slippage and the disentanglement of chain molecules. This toughness requires the interdiffusion of chain segments. That means that the first steep increase of W_B and the transition from brittle to tough fracture cannot be caused by a further "dry" sintering but should be attributed to the beginning of the interdiffusion process, which gives rise to an interdiffusion depth of about 2 nm after 5 min at 90°C, as estimated from the SANS data in Figure 3. This compares to about 3 nm for the length of an extended chain of the molecular mass between entanglements, M_e. A value of $2.1 \cdot 10^4$ g/mole has been evaluated for M_e from dynamic mechanical measurements.

It has been shown by molecular modelling of polymer aggregates in solution [14] that chain ends and short chains are present in a surface layer of the aggregates with a higher concentration than in the "bulk" of the aggregates. As this material diffuses faster than the average weight molecules, it possibly contributes to the fast rise of film toughness at short annealing times, too.

The further slow increase of W_B is caused by the progressive interdiffusion and the formation of entanglements of the long molecules across the boundaries of the former particles. At least, the fracture energy becomes constant at an interpenetration depth of about 40 nm, which has the same order of magnitude as the radius of gyration of the PBMA molecules for which a weight average molecular mass of about $5 \cdot 10^5$ g/mole has been found.

Film Formation of Latices with Crosslinked Particles

The mean molecular mass between two crosslinks, M_c, and between two entanglements, M_e, can be calculated from the storage modulus G' according to the theory of rubber elasticity [15]. M_c was additionally determined by swelling measurements in tetrahydrofurane. It follows from Figure 4 which shows M_c as a function of the molar concentration of MAMA   that M_c equals M_e at a molar concentration of about 1.5%.

Stress strain measurements were carried out with films of the PBMA latex with 2% MAMA, i. e. above this "critical" concentration, following the same procedure as for the uncrosslinked films. In the right part of Figure 1, stress strain plots are shown for films, annealed 60 and 360 min, resp. It follows that the PBMA with 2% MAMA exhibits brittle fracture up to the longest annealing times without any significant change of the very low fracture energy. This conclusion can also be drawn from Figure 5 showing the fracture energy of the uncrosslinked and the crosslinked PBMA in dependence on the annealing time. For the films crosslinked with 2% MAMA, the fracture energy is by more than two orders of magnitude lower than in the case of the uncrosslinked PBMA up to an annealing time of 360 min. These films remain extremly brittle upon annealing. The highly crosslinked particles cannot perform any significant interdiffusion, they are only deformed by Van der Waals attractions.

Figure 6 shows the fracture energy as a function of the temper time at 90°C for films with MAMA concentrations of 1 and 1.5% resp. together with the results for 0 and 2% MAMA. The PBMA with 1% MAMA exhibits exclusively brittle fracture until 15 min, both fracture types at 30 min, and yielding at 60 min and longer times. For the samples with 1.5% MAMA brittle as well as tough fracture are observed between annealing times of 60 and 360 min. In the Figure the data points are plotted for tough behaviour only. Films with the MAMA concentration near the critical value, thus, show some kind of transition between both fracture types.

It follows, thus, that interdiffusion leading to entanglements is essential for the toughness of latex films. Measurements on uncrosslinked films from a latex with a mean molecular mass smaller than the entanglement length fit into this scheme. This can be concluded from Table 1 which gives a summary of the measurements and shows the low mechanical strength of films with a $M_w < M_e$ which is due to brittle fracture of this polymer.

Conclusion

As a conclusion we propose the following model for the development of mechanical strength of the uncrosslinked latex film:
1. Very quick formation of a continuous, though very brittle film by sintering of the particles, increasing the area of close contact between the particles.
2. Rapid transition from brittle to tough fracture by fast interdiffusion of chain ends and small chains in the interparticular boundaries, and the formation of the first entanglements.
3. Slow development of the final mechanical strength by interdiffusion and entanglements of the long chain molecules.

These processes are more and more slowed down and hindered with increasing MAMA concentration. Stage 2 and 3 of this model are absent in the formation of films from crosslinked particles with $M_c < M_e$.

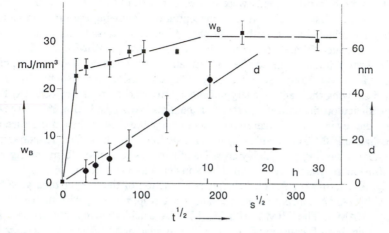

Figure 3: Fracture energy W_B and interdiffusion length d for uncrosslinked PBMA, plotted against the square root of the annealing time at 90 °C

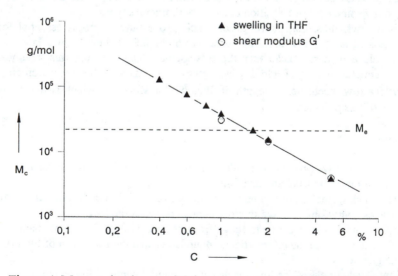

Figure 4: Mean molecular mass M_c between crosslinks for BMA/MAMA copolymers as a function of the MAMA concentration

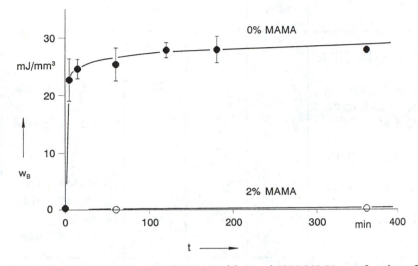

Figure 5: Fracture energy W_B of PBMA with 0 and 2% MAMA as a function of the annealing time at 90 °C

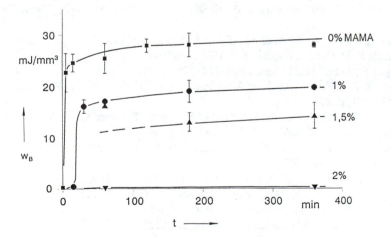

Figure 6: Fracture energy W_B of PBMA with various concentrations of MAMA versus the annealing time at 90 °C

Table 1: Mechanical strength of films from PBMA latices

Sample	σ_B MPa	ε_B	W_B mJ/mm^{-3}	Characterization of sample
Uncrosslinked PBMA				
$M_W \gg M_e$, untemp.	9	0.04	0.25	$M_W = 5 \cdot 10^5$ g/mole
$M_W \gg M_e$ tempered	12	3.3	31	
$M_W < M_e$, "	4	0.07	0.2	$M_W = 2 \cdot 10^4$ g/mole
Crosslinked PBMA				
$M_c > M_e$, tempered	13	1.9	20	1% MAMA, $M_c = 3 \cdot 10^4$ g/mole
$M_c < M_e$, "	10	0.03	0.15	2% MAMA, $M_c = 1.5 \cdot 10^4$ g/mole

$M_e = 2.1 \cdot 10^4$ g/mole

Literature Cited

1 Voyutskii, S. S. Autoadhesion and Adhesion of High Polymers; Wiley: New York, 1963.
2 de Gennes, P. G. C. R. Acad.Sc. (Paris) **1980,** B291, 219.
3 Kausch, H. H. Colloid Polymer Sci. **1981,** 259, 917.
4 Prager, S.; Tirrell, M. J.Chem.Phys. **1981,** 75, 5194.
5 Wool, R. P. In Fundamentals of Adhesion; Lee, L.-H., Ed.; Plenum Press: New York, 1991, p. 207.
6 Hahn, K.; Ley, G.; Schuller, H.; Oberthür,R. Colloid Polymer Sci. **1986,** 264, 1092.
7 Hahn, K.; Ley, G.; Oberthür, R. Colloid Polymer Sci. **1988,** 266, 631.
8 Yoo, J. N.; Sperling, L. H.; Glinka, C. J.; Klein, A. Macromolecules **1990,** 23, 3962 and **1991,** 24, 2868.
9 Winnik, M. A.; Wang, Y.; Haley, F. J. Coatings Technol. **1992,** 64(811), 5.
10 Eckersley, S. T.; Rudin, A. J. Appl. Polymer Sci. **1994,** 53, 1139.
11 Zosel, A.; Ley, G. Polymer Bulletin **1992,** 27, 459.
12 Zosel, A.; Ley, G. Macromolecules **1993,** 26, 2222.
13 de Gennes, P. G. J. Chem. Phys. **1971, 55,** 572.
14 Evers, O. A.; Ley, G.; Hädicke, E. Macromolecules **1993,** 26, 2885.
15 Ferry, J. D. Viscoelastic Properties of Polymers, 3rd ed., J. Wiley: New York, 1980.

Chapter 10

Film Formation and Mechanical Behavior of Polymer Latices

C. Gauthier[1], A. Guyot[2], J. Perez[1], and O. Sindt[1]

[1]Groupe d'Etudes de Metallurgie Physique et de Physique de la Matihre, Institut National des Sciences Appliquées, 20 avenue Albert Einstein, F—69621 Villeurbanne, France
[2]Laboratoire de Chimie des Procedes de Polymerisation, CPE Lyon, B.P. 2077, F—69616 Villeurbanne, France

The aim of this work is to obtain relevant information on the film forming process in order to control the development of mechanical properties of the latex film. As appropriate mechanical behavior cannot be reached if coalescence is not achieved, we first investigate the coalescence of the core-shell latex by performing weight loss measurements at different temperatures and humidities. After a brief survey of the literature, we propose a model of coalescence with a new description of the deformation of latex beads, based on the diffusion of structural units under the polymer / water interfacial forces. Calculations based on this model suggests that particle deformation and water evaporation may display different kinetics. Therefore, the end of coalescence may be related to the slowest process. At the end of coalescence stage, mechanical behavior of the film can be investigated. Then, we perform tensile tests for different time of annealing during the autohesion phenomenon. During this film forming stage, the behavior of the film changes drastically from brittle to ductile. We observe two distinct evolution for strain at break and stress at rupture. The former is attributed to the formation of entanglements through the particles boundaries whereas the latter might concern water diffusion throughout the film.

The process consisting of driving a latex from its colloidal form to a continuous film needs a good understanding since it is of great importance for industry. Film formation is a critical aspect of all applications that involve coating a surface or forming a layer with good cohesive properties. Consequently, great efforts have been devoted to the study of this phenomenon.

Several stages during film formation have been observed experimentally and a phenomenological description of the process, divided into three parts, is generally accepted (1) :

0097—6156/96/0648—0163$15.00/0
© 1996 American Chemical Society

Stage 1- Water evaporation and colloid concentration. As water evaporates, a uniform shrinkage of the inter-particle distance occurs and the voids are gradually filled by particle sliding, until a dense packing of spheres is obtained.

Stage 2- Particle deformation and evaporation of the bounded water (i.e. water perturbed by the high concentration of surfactants at the particle surface). This stage, also called coalescence, results in a honeycomb like structure of deformed particles in dodecahedra.

Stage 3- Interdiffusion of macromolecules. In this stage the mechanical strength increases and the water permeability of the film decreases. Under certain conditions, the polymer chains can diffuse through the particle boundaries. The honeycomb like structure disappears and an homogeneous, continuous film is formed.

These three different stages have been investigated both theoretically and experimentally. Much of the early work on latex film formation focused on the early stages of the process: water evaporation accompanied by particle concentration followed by the particle coalescence and deformation into dodecahedra. Various models have been proposed to describe the process of film formation. They will be briefly reviewed in the first part of this paper. Considering some of the main arguments of the theories which were previously established, a physical model that describes the coalescence process is advanced. This model will be confronted to our experimental data obtained from weight loss measurements at various temperature and humidity. Then, the evolution of mechanical behavior of latex films during stage 3 will be studied.

Mechanisms and Theories for Film Forming Process : a survey

From the standpoint of thermodynamic, the process of coalescence of polymer droplets into a film is favourable because of the decrease in free energy achievable with minimization of total surface. All contiguous particles would flow into deformed spheres representing a minimization of surface and gravitational energy. During stage 1 of film formation, as water evaporates from the surface, the particles centers approach each other but particles remain separate until they are forced into contact by spatial limitations. The second stage begins when the particles can no longer slide each other into new positions. Then, particles are brought into close proximity so that on can assume that their stabilizing layers may collapse, resulting in polymer-polymer contact.

Coalescence Stage. In a first attempt to propose a coalescence criterion, Dillon et al. *(2)* stated that particle coalescence resulted from the viscous flow of polymer induced by the surface tension between water and polymer particles. Nevertheless their theory could hardly describe the film forming process of crosslinked particles (stated on viscous flow only). A few years later, Brown *(3)* noted the influence of the water evaporation rate and concluded on its driving role. He proposed to make a force balance and, by neglecting some parts of his equation, concluded that capillary forces were driving coalescence (taking into account the water/air surface tension). This model has been completed by Mason *(4)* and Lamprecht *(5)* who improved the mathematical description.

Vanderhoff et al. *(6)* argued that capillary forces calculated by Brown were too low to insure coalescence of particles bigger than 200 nm in diameter. They proposed to take into account the driving forces arising from the polymer water interfacial tension. They used the mechanism stated by Dillon et al. *(2)* and concluded that capillary forces were driving coalescence at the beginning, but after a certain threshold, the intra-particle pressure caused by the polymer/water surface tension provoke coalescence. This concept of capillary forces and interfacial forces acting in tandem to promote latex coalescence has been developed by Eckersley and Rudin *(7)* . They compared the values of deformation of the latex beads related to the radius of the circle of contact of the spheres in both theories. In order to fit the experimental observations, they proposed to add the contributions of both forces.

Sheetz *(8)* described the process as follows : under water/air surface tension, the beginning of the film formation is caused by capillary forces. These forces are not only perpendicular to the film's surface, but a component parallel to the surface exists. The perpendicular part of the force can be described by Brown's equations. But, under the parallel force action, a rapid close of the capillary takes place at the surface of the film, resulting in a lower water evaporation rate. Since the remaining water has to diffuse through the surface of the polymer film, one should consider that the free energy of water evaporation is transformed into mechanical compression work that ensures the coalescence.

Most of these analyses describe the deformation of the latex particles using the Hertz contact solution for the deformation of two elastic spheres that were pressed into contact. Since the polymer is in fact viscoelastic, the error may be compensate by replacing the elastic shear modulus G by a time dependant shear modulus $G(t)$. Lamprecht *(5)* and Eckersley and Rudin *(7)* have proposed to introduce the mathematical development generating the equivalent viscoelastic solution to the Hertz contact problem.

Recently, Sperry et al. *(9)* proposed comparative examinations of the compacting stage in the absence as well as in the presence of water. Int is generally admitted that particle deformation occurs only above the minimum film formation temperature (MFT) which often lies close to the glass transition temperature (Tg) of the latex polymer. MFT is defined to be the minimum temperature at which a latex cast films becomes clear . Below this temperature , the dry latex is opaque owing to interparticle voids. These authors have observed that a film pre-dried well below the MFT displays a transition from turbid to clear film as it is heated. They were convinced of the lack of a special role for water in this process and proposed to decompose it into separate water evaporation and film compacting events.

The Third Stage of Film Forming Process. Once the second stage has correctly occurred, the so-called stage 3 can begin. In this final stage, interdiffusion of polymer chains of adjacent particles across particle boundaries causes further coalescence. Several studies have been published on direct non radiative energy transfer (DET) using fluorescence decay measurements *(10-11)*. These measurements required to label particles with appropriate donor and acceptor dyes. Small angle neutron scattering (SANS) is also a technique which provides an effective way to study the third stage. It measures the growth in particle size of deuterated latex particles as self-diffusion progresses. For the first time in 1986, Hahn and co-workers *(12)* reported work on coalescence study by SANS on butyl methacrylate (BMA) polymers. Linne et al *(13)* published SANS measurements on the actual interdiffusion depth in polystyrene. From these data, it results that the molecules originally constrained to ~ 160 Å in the latex expand to ~ 660 Å as the polymer is annealed above glass temperature to form a polymer film. They highlight that the molecular weight and the location of chain ends are important parameters of polymer interdiffusion process.

To obtain quantitative information, one should fit experimental data to a model describing the polymer diffusion. Two different approaches are found in literature. One assumes that polymer diffusion can be evaluated by using models based on the Fick's second law. One the other hand, different theories based on De Gennes' reptation diffusion model were developped. They describe the development of material strength, due to the formation of entanglements between polymer chains across the interface. Along this second line, four theoretical models have been proposed *(14-17)* ; they are reviewed and compared to experimental SANS data in *(18)*. These models agree that fracture stress should increase with healing time to the one-fourth power whereas a one-half power law is deduced from Fick's second law. Different studies *(17-18)* present analysis based on these models. Nevertheless, considering the whole literature, no evidence is driven from experimental results to invalidate one or the other approach.

Besides SANS experiments, Yoo et al. *(18)* measured the tensile strength of polystyrene films and reached the conclusion that penetration depth for full mechanical strength may be partially controlled by the spatial distribution of chains ends in the interface and the dimensionless ratio of polymer chains' radii of gyration to the latex particle's radius. From their experimental results, the order of magnitude for the minimum penetration depth, that gives a cohesive film, is around a few nanometers. The development of film strength has been also studied by Zosel et al. *(19)* with poly (butyl methacrylate) (PBMA) . Tensile tests were performed after increasing annealing time at 90°C. The tensile behavior changed completely from brittle to ductile within the first five minutes and the corresponding penetration depth measured by SANS lies once again in the range of a few nanometers. Finally, fracture energy seems to rise to a constant value when the interdiffusion depth has the same order of magnitude as the radius of gyration of the PMBA molecules.

In conclusion, effective correlation can be obtained from comparison between SANS measurements and mechanical tests in the third stage of film formation process. Nevertheless, direct studies of the third stage requires heavy equipments and specific modifications of the latex samples. In this study, we are looking for relevant information on the film forming process in order to control the development of mechanical properties of the latex film. Our investigations were first devoted to stage 1 and 2 since appropriate mechanical behavior can not be reached if coalescence is not achieved. In the second part of this work, the third stage of film forming process is analysed through evolution of tensile behaviour of the latex film with increasing time.

Experimental

Numerous parameters should be considered in order to understand the film forming process. It is generally accepted that these parameters can be divided into two parts. The first one is related to latex properties such as particle diameter, glass transition and viscoelastic properties of the polymer , the nature of surfactants, etc. . The second set concerns the physical parameters of the film forming process and it includes temperature, humidity, the interfacial energy (either water/air or water/polymer). In this study, the parameters of the latex are kept constant and we focused our attention on physical ones.

Latex Preparation. Structured core-shell lattices are prepared by seeded emulsion polymerization. Homopolymerization of polystyrene seed is first performed in batch using Sodium Dodecyl Sulfate surfactant (SDS). The size of the seed particles is 100 nm in diameter. In a second stage, polymerization of styrene and butyl acrylate is achieved by a semi-continuous process. The shell composition is 50 weight % Styrene - 49 weight % Butyl acrylate with a low content (1 weight %) of acrylic acid. The final size of particles is 200 nm. The feed rates of monomer and the initiator concentration are adjusted in order to avoid any composition drift during the shell polymerization. Moreover, a parallel continuous addition of SDS water solution is performed in order to maintain the stability of the particles during and after polymerization. We check that no nucleation of new particle population occurs, and that the number of particles is constant and equal to the number of seed particles.

Film formation is obtained after dehydration of the latex and maturation in a regulated oven (temperature and humidity) . During film formation, the core-shell particles deform into a film with a soft matrix including hard spherical inclusions of poly(Styrene) (Tg $\approx$ 100°C). Glass transition temperature of the soft matrix has been determined by DSC measurements and stand around 20°C (heating rate : 10°/min).

Weight Loss Measurements. The film forming process is studied by weight loss measurements. A Teflon mould is filled with the latex and weighted on a balance in a regulated oven. Data are recorded on line by a computer . The weight loss curves versus time are recorded for different temperature and humidity conditions. On these curves, the three stages of film forming process can be determined (Figure 1) :

Stage 1- Linear decrease of the mass with time; the bulk water evaporates resulting in latex concentration at a rate depending only on temperature, hygrometry and electrolyte concentration in the water phase.

Stage 2- Evaporation continues and particles are emerging at the surface of the film. The rate of evaporation decreases; coalescence occurs and the honeycomb like structure is formed.

Stage 3- No important variation of mass with time; only polymer diffusion can occur.

The times for stage 1 and stage 2 are measured as well as the water evaporation rate in the stage 1 (from the slope of the curve). Experiments are carried maintaining all parameters except of temperature and humidity constant (air velocity, mould size, latex weight)

Tensile tests. The mechanical properties of latex films at large strains and their fracture behaviors have been studied in uniaxial elongation by means of a commercially available tensile tester (Instron 8561) . Tensile tests are performed at 23 °C after various times during stage 3 in order to obtain nominal stress σ = F/So versus elongation λ = l/lo. Temperature of testing is carefully controlled with maximum deviation of +/- 0.2°C. The dimensions of the specimen (H3 type) are 50x4x1 mm^3 and the cross head speed is 200 mm/min. A minimum of five samples are tested for each experiment. True stress - strain curves are derived from conventional curves using the iso - volume assumption. A series of tensile curves is obtained when film formation is achieved at T = 32°C

Results

Water Evaporation. Considering the weight loss measurements, all the curves first display a linear dependence versus time. Although this first stage does probably not determine the final properties of the film, we believe, considering Sperry's study *(9)* , that the rate of water evaporation event should be estimated independently of film forming process. Before the beginning of stage 2, when all the mold surface is free, water evaporation is a function of water pressure, humidity ,air velocity and evaporation surface. The evolution of water evaporation rates, respectively, with temperature and humidity are reported in Table I and Table II. At a fixed temperature, water evaporation rate decreases linearly when humidity increases. On the opposite, it increases with temperature when humidity is constant.

Table I. Water evaporation rate versus Temperature (Relative Humidity = 75%)

T (K)	283	293	297	301	305	309	313
Evaporation rate (mg/min)	4.6	6.2	8.2	9	9.2	9.9	10.1

Table II. Water evaporation rate versus Relative Humidity (Temperature = 20°C)

R. H. (%)	52	55	65	75	85	90	98
Evaporation rate (mg/min)	9.8	9.2	7.7	6.2	5.2	4.8	3.7

Water evaporation is constant since the end of stage 1. In the second stage, the process is slowed down. As described in literature *(20)*, during coalescence, water evaporation is not homogeneous on the whole surface of the mould. Nevertheless, we propose to estimate roughly the total time $(t_1 + t_2)$ which is necessary to evaporate the amount of water (M_{water}) present in the sample. We decided to calculate two limiting values from the value of water evaporation rate obtained from the slope of the linear decrease of weight during stage 1. In a first assumption, we fixed a water evaporation rate equal in stage 1 and stage 2.

$$t_1 + t_2 = \frac{M_{water}}{rate_{evaporation}}$$

It seems quite evident that this calculated value overestimates the real one since it neglects the variation of the surface of evaporation. Hence, a second limit is calculated taking into account the variation of water evaporation rate due to the decrease of free surface. We consider that at the end of stage 1, polymer particles with a defined a radius are emerging and that water evaporation occurs only in the interstitial regions between spheres in close packed array (Scheme 1).

$$\frac{S_{free}}{S_{total}} = \frac{S_{hexagon} - S_{particle}}{S_{hexagon}}$$

Calculations have been made for each experiment varying temperature and humidity. For example, at T= 40°C and R.H. = 75% , the time for water evaporation during 2nd stage is 60 min as calculated with assumption 1 whereas experimental time for coalescence is about 150 min. Assumption 2 leads to t_2 = 800 min. For any temperature and hygrometry, experimental data are in the range of the two calculated limits. The results related to temperature are reported on Figure 2.

However, one could be surprised by the fact that the measured times are always nearer the values obtained with the first assumption, which seems in a first view, quite unrealistic. Thus, since time to evaporate water is well approximate with the first assumption, there should be a mechanism occurring during stage 2 which maintains a high level of free surface for water evaporation, higher than the only interstitial surface. We consider that this mechanism consists in particle deformation, which is assumed to control the coalescence.

Physical Model for Coalescence Mechanism. The calculation we proposed first considers that coalescence process is governed by the value of compacity. At the beginning of stage 2, compacity of the latex is 0.74. which corresponds to a close-packed array of spheres. At this point, latex particles are emerging at the surface. In fact, the whole coalescence stage will consist in an increase of compacity from 0.74 (close packed array) to 1

In order to obtain a theoretical model for coalescence, one should first define the driving forces acting on latex particles. From our point of view, the two mechanisms , called by Rudin capillary force or polymer water interfacial forces, are the expression of the same physical principle. Practically, both forces are calculated from Young and Laplace law, one considering the water / air interface (capillary force) and the second the polymer / water interfaces. Capillary forces lead to restrict the mechanism to the surface of the mould which can not be entirely true in macroscopic samples, as already discussed by Sheetz. So, our description will take

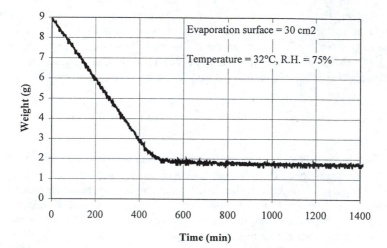

Figure 1 : Weight loss curve at 32°C and 75% Relative Humidity.

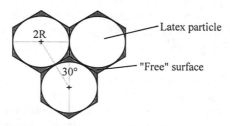

Scheme 1 : Surface evaporation.

into account the driving force derived from the Young- Laplace law using the polymer / water interfacial energy.

Thus, the assumptions of our model are that the Young and Laplace law describes the driving forces for coalescence which are opposed by viscoelastic deformation of the polymer. In adequate conditions (at temperatures above the glass transition temperature) polymer diffusion is involved. In this model, the calculations are made in terms of polymer diffusion whose description is based on the formal statement proposed by Nabarro-Herring in order to describe the high temperature creep of polycrystals *(21)*.

The system is presented on Scheme 2. We have to consider two distinct points: A in the zone between two adjacent particles (called coalescence zone) and B somewhere in the particle. Due to the difference of curvature between the coalescence zone and the rest of the particle, a difference of pressure exists between A and B which could be minimized if the curvature radius of the coalescence zone (called r) decreases. This induces polymer diffusion from the particle to the coalescence zone. In the absence of other particles, the system tends to form one bigger particle. In the close-packed latex, particles will be transformed into dodecahedra. Calculations are made only on two particles considering the symmetrical aspect of the problem. In fact, some refinement should be obtained taking into account gravitational force which introduces some anisotropy.

The pressures in points A and B are derived from Young and Laplace equation. Pressures are function of the polymer / water interfacial energy and are of opposite signs.

$$P_A = -\frac{2\sigma}{r} \qquad\qquad P_B = \frac{2\sigma}{R}$$

Also, some geometrical considerations enable us to establish a relation between r, R and a (the half of the height of the coalescence zone) :

$$(a+r)^2 + R^2 = (R+r)^2 \qquad \rightarrow \quad r = \frac{a^2}{2(R-a)} \approx \frac{a^2}{2R}$$

The driving force related to the difference of pressure can be expressed by the following relation where μ is the chemical potentiel of a structural unit of the macromolecule and v_0 the volume of a structural unit.

$$F = -\text{grad}\mu$$

In point A and B : $\qquad \mu_A = \mu_0 + P_A \cdot v_0 \qquad\qquad \mu_B = \mu_0 + P_B \cdot v_0$

Then , $\qquad\qquad F = -\text{grad}\mu = -\frac{\Delta P \cdot v_0}{a}$

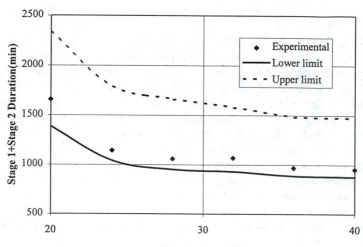

Figure 2 : Duration of Stage 1 and Stage 2 versus Temperature at 75% Relative Humidity.

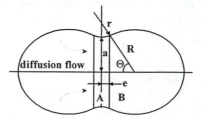

Scheme 2 : Coalescence mechanism : description of the system.

Using the formal statement of Nabarro-Herring *(21)* , we can relate this force with the diffusion of polymer structural units via the rate of diffusion ($V_{dif.}$) where D = diffusion parameter , k = Boltzmann constant :

$$V_{dif.} = \frac{DF}{kT}$$

From the rate of diffusion, we can deduce the diffusion flow (J). By definition, it is the number of structural units which cross a unitary surface per second and corresponds to the rate of diffusion divided by the volume :

$$J = \frac{V_{dif.}}{v_o} = -\frac{DF}{v_o kT} = \frac{D}{kT} \cdot \frac{(P_A - P_B)}{a}$$

$$= \frac{D}{kT} \cdot \frac{2\sigma}{a} \cdot \left\{ \frac{1}{r} + \frac{1}{R} \right\} = \frac{D}{kT} \cdot \frac{2\sigma}{a} \cdot \left\{ \frac{2R}{a^2} + \frac{1}{R} \right\}$$

The diffusion flow leads to the growth of the volume v_{coal} of the coalescence zone by diffusion of structural units through its surface πa^2 :

$$\frac{dv_{coal}}{dt} = J v_o \pi a^2 = \frac{Dv_o \pi \sigma}{kT} \cdot \left\{ \frac{2a}{R} + \frac{4R}{a} \right\}$$

v_{coal} can be calculated once again from geometrical considerations :

$$v_{coal} = \pi a^2 . e$$

$$e = r \cos\theta = \frac{rR}{r+R} \approx \frac{a^2}{2R}$$

$$\frac{dv_{coal}}{da} = \frac{2\pi a^3}{R}$$

Therefore the derivative expression of a on time:

$$\frac{da}{dt} = \frac{2\pi D\sigma v_o}{kT} \left(\frac{2R}{a} + \frac{a}{R} \right) \cdot \frac{R}{2\pi a^3}$$

Then, we can express the variation of a(t), height of the coalescence zone :

$$a(t) = \int_{t_0}^{t_2} \frac{Dv_o \sigma}{kT} \left[\frac{1}{a^2} + \frac{2R^2}{a^4} \right] dt$$

The expression of a(t) takes into account the following parameters : R (particle radius) σ (water/polymer interfacial energy $\approx$ 35 mJ/m^2) and vo (volume of the structural unit). As the length of a structural unit of the polymer chain λ is about 5 Å, the volume of the structural unit (vo $\approx \lambda^3$) can be estimated as 125 Å^3 .

In this description, the diffusion coefficient D can be related to the time necessary to move a structural unit over a distance of its own length (called τ_{mol}) . The determination of this time τ_{mol} is explained elsewhere *(22)* .

$$D\,(T) = \frac{\lambda^2}{\tau_{mol}(T)}$$

The coalescence time (stage 2) is obtained when the spherical particle is deformed into a decahedron. That corresponds to a value of a(t) at the end of stage 2 which is about 55 nm whereas our particles have a radius of 100 nm (see Scheme 1).

Results and Discussion. All results are reported in Figure 3. Calculated times for coalescence (line) are compared to experimental data obtained from the weight loss measurements (dots). We also plot the calculated times for water evaporation as calculated in the first assumption. Two distinct zones are observed. Above 300 K, the calculated times for coalescence are lower than measured values. In this region, particle compaction is very easy due to the low value of polymer elastic modulus meaning a rapid diffusion of polymer units. Although particle deformation can be very fast, complete coalescence is not reached since films still contain lot of water. Moreover, in this range of temperature, we find that experimental data are close to calculated times for water evaporation (obtained from experimental water evaporation rates). Therefore, we can consider that water evaporation is the governing phenomenon in this range of temperature. As proposed by Sheetz *(8)* , there is probably a rapid close of the capillary at the surface of the film, resulting in a lower water evaporation rate because water has to diffuse through the surface of the polymer film to evaporate.

On the contrary, below 300 K, the prediction of particle deformation event is longer than measured data from weight loss curves. In this range, particle deformation becomes more difficult as temperature moves closer to Tg. Moreover, one could argue that, since no more water is present in the film, calculations should be revised after t_2, taking into account the polymer / air interfacial energy . In fact, this may only lead to increase even more of the calculated times. Below 300 K, we observe that experimental data are not any more close to calculated values for water evaporation. To our point of view, this means that, even if the calculated curve does not fit exactly the experimental data, we can consider that deformation is the phenomenon which controls the kinetic of film formation. In this range of temperature, the films produced can contain nano voids, corresponding to the uncomplete filling of interstitial domains.

In conclusion, in agreement with Sperry's results, we think that deformation of particles and evaporation events have different kinetics. Consequently, the time for coalescence is related to the slowest process. For our samples, at T>300 K, the slowest event is water evaporation. Below this temperature, the deformation of particles determines the end of coalescence. From a practical point of view, the choice of film forming conditions should take into account this aspect. So, we decide to adopt a film forming temperature so that both phenomena display comparable times. Further studies are performed with a film forming temperature of 32°C.

Development of mechanical strength during film formation. Since no information can be obtained about the third stage from weight loss curvs, mechanical properties were investigated. We decide to study the influence of annealing time during autohesion on the tensile properties of the films. At 32°C, the time for evaporation and compaction processes are similar. In this case, we can without any problem define the initial time for stage 3. The results of tensile tests which were performed at 23°C after different annealing times at 32°C are plotted on Figure 4.

During the third stage, the behavior of the material changes drastically with increasing times. At low annealing times, the film displays a brittle behavior with a strain at break (ε_r) lower than 0.1. The behavior changes from brittle to very ductile for long annealing times and a plastic consolidation domain can be observed. The strains at break become higher than 5. In fact, this drastic change results from two different types of evolution of the mechanical properties. For short annealing times, the stress at break increases first, and then (one decade later) the strain at break goes up quickly to a plateau value. For times longer than 500 min, the strain at break stabilizes (Figure 5). One the contrary, for the same time of scale, the plastic stress still increases slowly (Figure 6).

The rapid change of tensile behavior from brittle to ductile has been observed by Zosel et al. *(19)*. The fast increase of the film toughness has been interpreted in terms of progressive interdiffusion and formation of entanglements of the macromolecules. In our opinion, (i) the increase of stress shows formation of Van der Waals bounds between particles, resulting in the cohesion of the whole material and (ii) the increase of the strain at break is due to the diffusion of macromolecules through particles boundaries which increases the number of entanglements between two particles. Moreover, the slow increase of plastic stress for longer annealing times can be related to a redistribution of the macromolecules in the film which increases intermolecular Van der Waal's interactions. The main cause of such a phenomenon should be the diffusion of water molecules through out the film. This assumption has been checked by more precise water loss measurements and tensile tests on wet and dry films. Due to the annealing temperature chosen by Zosel (T=90°C), the plastic stress evolution due to water diffusion could not have been observed in his experiments.

Some confirmations of these results have been obtained by forming the film at another temperature. At 40°C, we observe once again both types of evolution. The time necessary to reach the steady state strain at break is reduced, as the diffusion becomes easier at higher temperature. Moreover, a plateau for plastic flow stress is also reached for annealing time of about 3000 min. In a future work, we propose to establish a correlation between the kinetics of the evolution of the mechanical properties and that of the diffusion of polymer units or of water through the film.

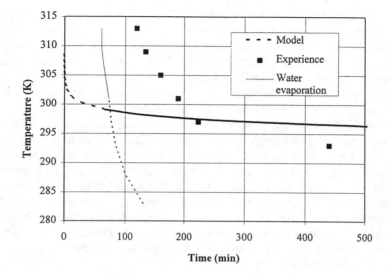

Figure 3 : Comparison of calculated coalescence times with experimental data.

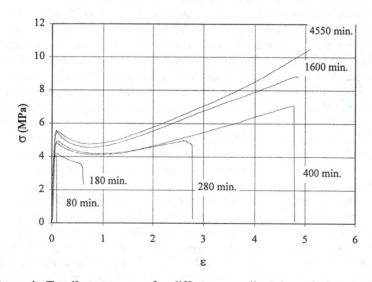

Figure 4 : Tensile test curves after different annealing times during stage 3.

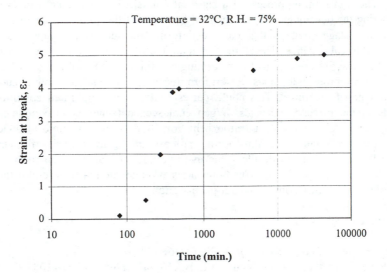

Figure 5 : Strain at break after different annealing times during stage 3.

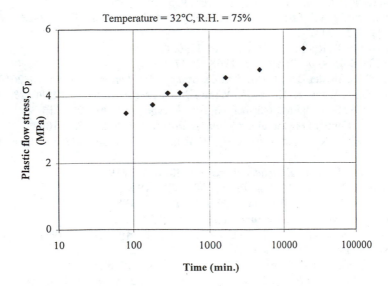

Figure 6 : Plastic flow stress after different times during stage 3.

Conclusion

The film forming process of a classical core-shell latex has been investigated considering the development of mechanical behavior of the film. The analysis of the coalescence stage suggests that particle deformation and water evaporation may display different kinetics. Therefore, the end of coalescence may be related to the slowest process. The selection of film forming conditions should take into account this aspect and we propose to choose a film forming temperature so that both phenomena display comparable times. This could correspond to some optimum conditions in regards to the presence of voids. When coalescence (stage 2) is complete, film displays a cohesion but with a brittle behavior. With increasing time of annealing during the third stage of the film forming process, the behavior of the film changes drastically from brittle to ductile. This evolution is attributed to the formation of entanglements through the particles boundaries whereas the plastic stress of the film may reflect water diffusion throughout the material.

Literature Cited

1. Rios Guerrero., *Macromol. Symp.* **1990**, *35/36*, **389**.
2. Dillon R.E., Matheson L.A., Bradford E. B., *J. Colloid Sci.* **1951**, *6*, 108.
3. Brown G. L., *J. Polym. Sci.***1956**, *22*, 423.
4. Mason G., *Br. Polym. J.* , **1973**, *5*, 101.
5. Lamprecht J., *Colloid & Polym. Sci.* **1965**, *9*, 3759.
6. Vanderhoof J.W., Tarkowski H.L., Jenkins M.C., Bradford E.B., *J. Macromol. Chem.* **1966**, *1*, 131.
7. Eckersley S.T., Rudin A., *J. Coat. Tec.*, **1990**, *62*, 89.
8.Sheetz D. P. ,*J. Appl. Polym. Sci.*, **1965** , *9*, 3759.
9. SperryP.R., Snyder B.S., O'Dowd M.L., Lesko P.M., *Langmuir*, **1994**, *10*, 2619.
10. Zhao C.L. , Wang Y., Hruska Z., Winnik M. A., *Macromolecules*, **1990**, *23*, 4082.
11. Peckan O., Canpolat M., Göçmen A., *Eur. Polym. J.* , **1990**, 29,*1*, 115.
12. Hahn K., Shuller G., Oberthür, *Colloîd & Polym. Sci.*, **1986** , *264*, 1092.
13. Linne M. A., Klew A., Sperling H., Wingall G.D., *J. Macromol. Sci. Phys.* **1988**, *B27*, 181.
14. De Gennes P. G., *C. R. Acad. Sci. Paris* **1980**, *B291*, 219
15. Prager S., Tirrel M., *J. Chem. Phys.*, **1981** , *75*, 5194
16. Jud K., Kausch H.H., Williams J.G., *J. Mater. Sci.*, **1981** , *16*, 204
17. Kim Y.H., Wool R.P., *Macromolecules*, **1983** , *16*, 1115
18. Yoo J.N., Sperling L.H., Glinka C.J., Klein A., *Macromolecules*, **1991** *24*, 2868
19. Zosel A., Ley G., *Macromolecules*, **1993** , *26, 9*, 2221
20. Feng J., Winnick M.A., proc. ACS PMSE Chicago, **1995**,90
21. Poirier J.P., in *"Plasticité à haute température des solides cristallins"* Eyrolles Ed., Paris 1976 p.152
22. Perez J. in *"Physique et mécanique des polymères amorphes"* Lavoisier Ed., Paris 1992.

Chapter 11

Healing and Fracture Studies in Incompletely Annealed Latex Films and Related Materials

M. Sambasivam[1,2,4,6], A. Klein[1–3,5], and L. H. Sperling[1,2,4,5,7]

[1]Polymer Interfaces Center, [2]Center for Polymer Science and Engineering, [3]Emulsion Polymers Institute, [4]Materials Research Center, [5]Department of Chemical Engineering, and [6]Department of Chemistry, Lehigh University, Bethlehem, PA 18015–3194

The molecular aspects of film formation and subsequent fracture in latex films have been reviewed. The film formation studies are based on SANS and DET techniques, and the fracture studies are based on tensile tests and more recently on a grinding fracture technique using a custom-built Dental Burr Grinding Instrument (DBGI). Strength development in latex films occurs by an interdiffusion process at temperatures above the glass transition temperature, and full strength is achieved at an interdiffusion depth equal to one radius of gyration of the chains. Molecular aspects of fracture in these films are considered in terms of chain scission and pullout processes. Under fully annealed conditions, for polystyrene at 32,000 g/mol, the chain pullout energy contribution is substantially 100% of the total fracture energy. At higher molecular weights 151,000 g/mol, and 420,000 g/mol, the chain pullout contributions decrease to about 60% and 10%, respectively. Correspondingly, the chain scission and pullout contributions for PMMA (485,000 g/mol) are 75% and 25%, respectively. Using the molecular weight dependence of reptation time, τ, it is possible to predict the fractional chain scission energy contribution during fracture in PS for a given molecular weight. Chain pullout energies calculated from theoretical equations agree fairly well with the experimental values. Frictional coefficients calculated from the chain pullout energy for PS and PMMA allowed the estimation of the temperature of chain pullout, which is about 242-368°C and about 220°C, respectively, well above the glass transition temperature. The thickness around the crack tip to which this temperature is confined was estimated to be about 1.5-2.3 nm using fracture mechanics calculations.

Water-based polymer latexes are gaining more attention in the coatings and adhesives industries over conventional solvent-based systems, mainly due to restrictions imposed

[7]Corresponding author

0097–6156/96/0648–0179$15.25/0
© 1996 American Chemical Society

by environmental requirements. A better understanding of the properties of latex films became very critical to suit the requirements of their growing applications. Similar to bulk polymers, the mechanical properties of latex films are dependent on the molecular weight and its distribution (1-3), and are sensitive to the presence of low molecular weight additives, such as surfactants (4). In addition, the strength of these films, for a given molecular weight, depends on the annealing time and annealing temperature (1,5-7).

The process of film formation from latexes takes place via three different steps: i. water evaporation; ii. particle deformation due to capillary forces and evaporation of interstitial water, and; iii. interdiffusion of chains across the particle-particle interface. Randomization of the chains takes place by interdiffusion across the particle-particle interfaces, which determines the mechanical properties of the final film.

Latex films provide a good model material for studying the healing and fracture at polymer-polymer interfaces (1,8,9). The chain interdiffusion in latex films is analogous to the healing process in bulk polymers, where the interface between the two surfaces is bridged by chains reptating and subsequently forming entanglements. Fracture in the latex films physically re-creates the interface. Important questions that arise are: how far must the chains interdiffuse into the neighboring particle for good strength; how does molecular weight affect the interdiffusion rate and the final strength? The healing and the fracture processes can be reversible up to a certain depth of interdiffusion and at certain molecular weights (8).

The strength development at the interface is followed by measuring the chain interpenetration depth, number of chains crossing the interface, fracture energy, or tensile strength, among other properties. Several workers have studied the film formation process using different techniques such as small-angle neutron scattering (SANS) (4,9-13) and direct nonradiative energy transfer (DET) (5,6,15-17).

The SANS technique primarily measures the radius of gyration of deuterated probe particles, which increases with annealing, from which the diffusion coefficient is obtained. Figuratively, the model resembles an exploding star.

The DET technique involves labelling of the polymer chains with donor and acceptor fluorescent groups by chemical reaction. The intensity of energy transfer is followed as a result on interdiffusion which yields information on the extent of mixing (f_m) and diffusion coefficients. There are several advantages and disadvantages between the two techniques. However, they provide valuable information at the molecular level. Winnik et al. (15) compared the diffusion coefficients of poly(butyl methacrylate) (PBMA) obtained from DET and SANS techniques, and inferred that the values were nearly identical within experimental error, suggesting that both techniques yield the same results despite the different approaches.

Combining the film formation results from SANS or DET with the mechanical tests, such as tensile, provides a better level of understanding the film formation process in these films (8). In this review, the emphasis is mainly on the mechanical properties of incompletely annealed latex films of polystyrene (PS) and poly(methyl methacrylate) (PMMA). Also, the film formation studies are briefly reviewed.

Theory of Healing and Fracture at Polymer Interfaces

The chain motions in a bulk amorphous polymer are characterized by the de Gennes reptation theory (*18*), which was further developed by Doi and Edwards (*19*). This model has also been used to describe chain motions at polymer-polymer interfaces by de Gennes (*20*), Prager and Tirrell (*21*), Wool et al. (*22*), and Jud et al. (*23*). The conformations of the chains at the interface are not in equilibrium when the interfaces are brought into contact, such as-molded latex films. When diffusion of the species, on either side of the interface, is thermodynamically favorable, interdiffusion of the chains takes place at the interface. Consequently, the chains begin to disengage from their tubes. The portion of the chain that disengaged from the original tube at time t, is termed a minor chain (*22*).

Based on the Wool's minor chain reptation theory (*22*), there are two different ways of defining the reptation time, τ: i. the time at which only 1/e fraction of the chain still remains in the initial tube, and; ii. the time at which the chains at the interface have diffused through a distance equal to one radius of gyration, R_g. The reptation time is given by,

$$\tau = R^2/(3\pi^2 \, D) \tag{1}$$

$$D = <l>^2.(\pi/2t) \tag{2}$$

$$\tau \sim M^3 \tag{3}$$

where R is the end-to-end distance, D is the self-diffusion coefficient, M is the molecular weight, and $<l>$ is the diffused distance at time t. For a healing interface, the mechanical properties are expected to increase up to the reptation time, after which the system is considered equilibrated. However, experimentally, properties change asymptotically for two or three times the reptation time (*2,3*).

Molecular weight and annealing time are important variables in determining the strength of the interface. According to the scaling laws, the fracture toughness (G_{IC}) of the interface scales with annealing time, t, as,

$$G_{IC} \propto t^{1/2} \quad (t < \tau) \tag{4}$$

Some other important scaling relationships for both incompletely and fully annealed systems are listed in Table I. Even though the different models predict slightly different exponents, they all describe the same physical phenomenon, which is interdiffusion resulting in the formation of entanglements on both sides of the healing interface. The exponent is also dependent on the real time, t, with respect to the reptation time, τ.

Another important molecular parameter is the number of chains crossing the interface bound by physical entanglements on both sides, N, which scales as:

$$N \propto t^{1/4} \, M^{-5/4} \quad (t<\tau) \tag{5}$$

$$N \propto M^0 \quad (t>\tau) \tag{6}$$

Table I. Time and Molecular Weight Dependence of Fracture Energy (G_{IC})

Author	Time (t) and molecular weight (M) dependence	
	$t < \tau$	$t > \tau$
de Gennes (20)	$G_{IC} \sim t^{1/2}M^{-3/2}$	$\sim M^{0.0}$
Prager and Tirrell (21)	$G_{IC} \sim t^{1/2}M^{-3/2}$	$\sim M^{0.0}$
Kim and Wool (22)	$G_{IC} \sim t^{1/2}M^{-1/2}$	$\sim M^{1.0}$
Jud et al. (23)	$G_{IC} \sim t^{1/2}M^{-1.0}$	$\sim M^{1/2}$

τ is the reptation time

In the authors' work, the number of chain scissions are approximated to be the number of chains crossing the interface, N.

Latex Film Formation Studies

A number of people have investigated the self diffusion coefficients of polystyrene, Table II (8), and poly(methyl methacrylate), Table III (8). All data were converted to 135°C and 150,000 g/mol for better comparison. Although each instrument used has a somewhat different sensitivity, and the samples were prepared by a variety of methods, the diffusion coefficients for the polystyrene agree within a factor of ten. The major exception is the data of Yoo, et al. (12b), which is two orders of magnitude lower than the others. It is of special interest to compare this data with that of Kim, et al.[14], since the data were generated in the same laboratory. Although the molecular weight of the Kim, et al. sample was higher, it had hydrogen end groups, compared with the liberal supply of sulfate end groups on the Yoo, et al. samples. Also, the latter had a much broader molecular weight distribution, which may have contributed somewhat to a "smearing effect," i.e., the lower molecular weight species may have diffused almost completely across the latex particle interfaces before the SANS experiments were initiated. However, since the tensile strength of the Kim, et al. sample increased with time much more rapidly than the Yoo, et al. samples, it is tentatively concluded that the sulfate end groups actually slowed the interdiffusion down. Possible reasons include the ionic nature of the sulfate end group, contributing to a positive χ (1) effect, ionomer segregation, and/or the large size of the sulfate group. The critically important point here is that almost all commercial latexes are prepared via ordinary emulsion polymerization, while most of the polymer physics experiments are based on anionic polymerizations and subsequent emulsification or miniemulsification. If these diffusion coefficients actually differ by a factor of 100, then let the technologist beware!

In Table III, there is a variation of a factor of 50 in the converted D* data for PBMA, ranging from 1.1×10^{-11} to 1.8×10^{-13}. In addition, it might be expected that PAMA should have a slightly larger diffusion coefficient than PBMA, since its glass transition temperature is lower. The poly(methyl methacrylate) data yields the best comparison to the polystyrene data, since they have nearly the same glass transition temperatures, 100°C for the polystyrene, and 106°C for the poly(methyl methacrylate). After suitable corrections for temperature and molecular weight, the data do, in fact, almost agree.

Table II. Comparison of Self-Diffusion Coefficients of Polystyrene Measured by SANS with Other Techniques

M_n (g/mol)	Temp. (°C)	D (cm²/sec)	D* (cm²/sec)	Method	First Author
233,000	120	5.5×10^{-18}	7.2×10^{-16}	FRS	Green(a)
68,000	130	1.8×10^{-15}	1.2×10^{-15}	SANS	Anderson(b)
150,000	135	2.4×10^{-16}	2.4×10^{-16}	SANS	Kim(c)
199,000	136	1.7×10^{-15}	3.0×10^{-15}	IMS	Whitlow(d)
400,000	140	9.5×10^{-16}	2.3×10^{-15}	SANS	Stamm(e)
185,000	144	4.6×10^{-16}	1.2×10^{-16}	SANS	Kim(f)
69,000	144	6.6×10^{-17}	2.3×10^{-18}	SANS	Yoo(g)
115,000	160	8.1×10^{-14}	6.6×10^{-16}	SANS	Brautmeier(h)
110,000	212	2.5×10^{-11}	2.0×10^{-15}	SANS	Stamm(e)
96,000	212	1.6×10^{-11}	8.2×10^{-16}	Rayleigh Scattering	Barlow(i)

* Converted to the same annealing condition (M_n=150,000 g/mol, T=135°C) by using $D \alpha\ M^{-2}$ and WLF equation

(a) P. F. Green, E. J. Kramer, J. Mat. Res., **1**, 201 (1986).
(b) J. E. Anderson, J. H. Jou, Macromolecules, **20**, 1544 (1987).
(c) K. D. Kim, B. Hammouda, L. H. Sperling, A. Klein, Macromolecules, **27**, 6841 (1994).
(d) J. S. Whitlow, R. P. Wool, Macromolecules, **24**, 5926 (1991).
(e) M. Stamm, J. Appl. Cryst., **24**, 651 (1991).
(f) K. D. Kim, G. D. Wignall, L. H. Sperling, and A. Klein, Macromolecules, **26**, 4624 (1993).
(g) J. N. Yoo, L. H. Sperling, C. J. Glinka, A. Klein, Macromolecules, **23**, 3962 (1990).
(h) D. Brautmeier, M. Stamm, P.linder, J. Appl. Cryst., **24**, 665 (1991).
(i) A. J. Barlow, A. Erginsay, J. Lamb, Proc. R. Soc., **A298**, 481 (1967).

Table III. Selected Diffusion Coefficients via Direct Energy Transfer Technique (DET)

Polymer	M_n (g/mol)	Temp. (°C)	D (cm^2/sec)	D* (cm^2/sec)	Reference
PBMA	76,000	70	1.3×10^{-15}	1.1×10^{-11}	(a,b)
PBMA	590,000	90	3.0×10^{-16}	1.6×10^{-12}	(c)
PBMA	500,000	90	3.5×10^{-16}**	-	(c)
PBMA	240,000	90	2.0×10^{-16}	1.8×10^{-13}	(c)
PBMA	300,000	90	2.3×10^{-16}**	-	(c)
PBMA	360,000	80	5.0×10^{-15}	6.7×10^{-11}	(d)
PMMA	140,000	149	3.0×10^{-15}	6.9×10^{-17}	(e)
PAMA	180,000	80	4.0×10^{-16}	2.0×10^{-13}	(d)

* Converted to the same annealing condition (M_n=150,000 g/mol, T=135 °C) by using $D \alpha M^{-2}$ and WLF equation
** SANS results for comparison.
PBMA- Poly(butyl methacrylate); PMMA- Poly(methyl methacrylate); PAMA-Poly(amyl methacrylate)

(a) Y. Wang, C. L. Zhao, and M. A. Winnik, J. Chem. Phys., **95**, 2143 (1991).
(b) M. A. Winnik, Y. Wang, and C. L. Zhao, in "Photochemical Processes in organized Molecular Systems," K. Honda, Ed., Elsevier Science Pub., Amsterdam, 1991.
(c) K. Hahn, G. Ley, H. Schuller, and R. Oberthur, Colloid Polym. Sci., **264**, 1029 (1986); **266,** 631 (1988).
(d) E. M. Boczar, B. C. Dionne, Z. Fu, A. B. Kirk, P. M. Lesko, and A. D. Koller, Macromolecules, **26**, 5772 (1993).
(e) Y. Wang and M. A. Winnik, Macromolecules, **26**, 3147 (1993).

Source: Reproduced with permission from reference 8. Copyright 1994 John Wiley.

Hahn, et al. (*10*) were the first investigators to use SANS to study the latex film formation process in PBMA. The effect of molecular weight during the coalescence of latex particles was studied. Also, they studied the effect of crosslinking on interdiffusion. Their results indicated that the particle-particle coalescence is due to a center-of-mass diffusion. In crosslinked matrix, this interdiffusion is highly restricted.

Winnik, et al. (5) considered the film formation process in PBMA latexes using DET measurements. The interdiffusion in the latex films was studied as a function of molecular weight and annealing temperature, and in the presence of plasticizers.

Winnik, et al. (17) also studied the interdiffusion in melt pressed PMMA films. Diffusion coefficients were calculated as a function of temperature, about 10^{-15} cm^2/s at 149°C and 10^{-18} cm^2/s at 120°C, see Table III, which yielded an activation energy of 100 kcal/mol.

Boczar, et al. (6) carried out interdiffusion studies during latex film formation in PBMA and poly(amyl methacrylate) (PAMA) using DET technique. They investigated the effect of particle size, temperature, polymer compatibility etc. on interdiffusion. According to their spherical diffusion model, they observed that a change in the particle size affected the intermixing rate proportional to the particle surface area to volume ratio; however, the diffusion coefficient was not altered. Figure 1 shows that the intermixing rate in the 100-nm sized particles is faster than that in the 300-nm sized particles. Also, their results support the fact that the temperature dependence of the diffusion coefficient follows both WLF and Arrhenius equations in a narrow temperature range of T_g to T_g + 60°C. Activation energies of 33-37 kcal/mol and 36-39 kcal/mol for PBMA and PAMA, respectively were obtained from the two equations. In the case of blends containing the two polymers, the extent of mixing was low due to the lack of miscibility between the two components.

Studies on latex film formation at Lehigh University focussed on two aspects: i. interdiffusion using SANS, and; ii. fracture using tensile and grinding tests. Linne, et al. (11) studied the interdiffusion in compression molded polystyrene latex films using SANS. Molded latex films were annealed at 40°C above T_g (=104°C) for different times. Also, the tensile strength of the films was studied as a function of the annealing time. Since the latex particle size was small (about 38 nm) compared to the dimensions of the chain (R_g= 79 nm, where R_g is the radius of gyration), the chains were highly constrained. On annealing, the entropic forces dominated the center-of-mass diffusion process, leading to a faster interdiffusion and a rapid increase in the tensile strength.

Following this work, Yoo, et al. (12) investigated the correlation between SANS interdiffusion depth and tensile strength in polystyrene latex films of M_w = 2.59x10^5 g/mol (intermediate M), and 1.8x10^6 g/mol (high M) (Figure 2), prepared via conventional emulsion polymerization. In this work, sub-micron sized latex particles were prepared, but large enough to avoid entropic constraint of the chains. As Figure 2 indicates, the tensile strength increased with interdiffusion depth in the intermediate molecular weight sample and reached a plateau value at a depth corresponding to the 0.8R_g of the chains, as predicted by Zhang and Wool (24). The high M, however, showed a rapid increase in the tensile strength with depth initially before reaching a plateau value at a depth much below 0.8R_g. This was attributed to the location of chain ends, which in the latter case was probably close to the particle surface, thereby leading to a faster interdiffusion rate.

Anderson and Jou (13), prepared latexes of anionically synthesized polystyrene via a direct emulsification process. Polystyrene dissolved in benzene was emulsified

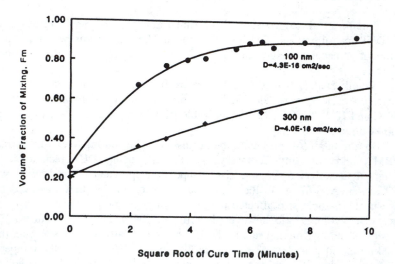

Figure 1. Plots of f_m versus cure time for PBMA latexes of two different particle sizes. Although the diffusion coefficients are similar, interparticle mixing is faster in the 100-nm latex, in accord with predictions from the model. (Reproduced from reference 6. Copyright 1993 ACS.)

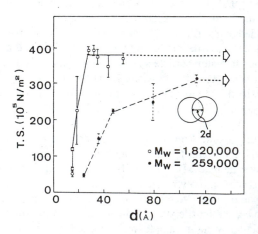

Figure 2. Plot showing that the tensile strength for the high molecular weight polystyrene yielded a higher ultimate tensile strength than the medium molecular weight (*12*). Unexpectedly, the tensile strength of the higher molecular weight also increased faster with interdiffusion. (Reproduced from reference 12. Copyright 1990 ACS.)

in water using surfactants. Then, the solvent and surfactant were removed. However, the latex particle size distribution in their study was very broad. At Lehigh University, Mohammadi, et al. (*25*) developed a direct miniemulsification technique that yielded a narrow size distribution latex particles from anionically synthesized polystyrene. Table IV gives the direct miniemulsification recipe developed by Mohammadi et al[25].

Table IV. Recipe for Direct Miniemulsification

Ingredient	weight (g)
	Aqueous phase
DDI water	100.0
sodium lauryl sulfate	0.435
cetyl alcohol	0.255
stearyl alcohol	0.109
	Oil phase
Polystyrene	2.0
Cyclohexane	20.0
cetyl alcohol	0.255
stearyl alcohol	0.109

Source: Reproduced from reference 14. Copyright 1993 ACS.

Using this direct miniemulsification technique, Kim et al. (*14*) studied the film formation process in PS with medium molecular weights (185,000 g/mol). Tensile strength was measured and correlated with the interdiffusion depth data from SANS (Figure 3). Similar to Yoo, et al.'s (*12*) results for medium molecular weight PS, Kim et al. observed a maximum tensile strength at an interdiffusion depth of about $0.8R_g$. This is also in agreement with the reptation theory, according to which the chains are completely randomized at an interdiffusion depth equal to one R_g. As discussed above, Kim, et al. (*14*) found that the diffusion coefficients obtained were much higher than those obtained by Yoo et al. for emulsion polymerized latex films.

Kim, et al. (*4*) also studied the effect of annealing temperature and residual surfactant on the interdiffusion kinetics in polystyrene (M=151,000 g/mol). The interdiffusion was faster at higher annealing temperatures and in the presence of surfactant, which acts as a plasticizer. Apparent activation energies of 55 and 48 kcal/mol for molecular weights of 185,000 g/mol and 151,000 g/mol, respectively, in the temperature range of 125-155°C were calculated, which compared well with that from dynamic mechanical measurements (*26*), about 56-80 kcal/mol.

Molecular Basis of Fracture in Latex Films

Mohammadi et al. (*1*) designed and built a Dental Burr Grinding Instrument (DBGI), which uses a fine dental burr to grind the latex films at a depth of 500 nm per pass.

This experiment allows the measurement of two quantities independently: i. by measuring the torque at the burr, the energy required to grind a unit volume of the film, in other words, the fracture energy per unit volume or (by measuring the size of the ground particles and computing their surface areas) per unit area. ii. molecular weight measurements made before (initial sample) and after (ground powder) grinding using gel permeation chromatography (GPC). The last allows the determination of the number of chain scissions and chain pullouts per unit volume or per unit area. The size of the ground powder, measured using dynamic light scattering, allows the estimation of the surface areas generated.

Mohammadi et al. (*1*) studied the film formation in high molecular weight polystyrene latex films (420,000 g/mol) as a function of the annealing time and frequency of the rotating burr in the DBGI. The fracture energy increased with annealing time, and exhibited a peak at the reptation time (96 min.), as shown in Figure 4 (*9*). Also, the number of chain scissions, and the tensile strength exhibited a peak around the reptation time. At short annealing times, the fracture surface showed the characteristics of the individual latex particles, like a basket of eggs. At long annealing times, a mirror-like fracture surface was obtained. Thus, the peak in the mechanical properties was attributed to the change in the crack path from the particle-particle interface to through the particles.

The total fracture energy was divided into three contributions: chain stretching, scission, and pullout. The chain scission energy and the chain stretching energy were calculated from the number of chain scissions using the Lake and Thomas theory (*27*), according to which all the bonds between the entanglement points are stretched to the maximum limit before one breaks. The basic concepts of the Lake and Thomas theory received partial confirmation from studies on the deformation characteristics of aromatic polyamides (*27a,b*). In that case, elongation of the rod-shaped chains under stress was attributed to the changes of the bond angles and bond lengths up and down the chain. The point is that the whole chain was involved, not just one bond.

From the total fracture energy, the energy for all of the chain scissions was known. Then, the remainder portion was assumed to be the chain pullout energy. For fully annealed films, of the 420,000 g/mol material, the contribution from chain scission energy was about 90% and about 10% from chain pullout. A minor contribution (less than 1%) was delegated to the uncoiling process.

Sambasivam et al. (*2,3*) investigated the fracture behavior of a series of polystyrene latex films of narrow molecular weight distribution (M=151,000 g/mol, 32,000 g/mol, 600,000 g/mol) prepared via the direct miniemulsification process, using the DBGI. In addition, latex films of commercial polystyrene (Styron 6069) (M_n = 180,000 g/mol; PDI=1.5) were studied. In the latex form, the molecular weight slightly lowered to 142,000 g/mol due to processing. An annealing temperature of 144°C was chosen for direct comparison with Mohammadi et al.'s data. Results from these studies are also included in Figure 4. It is clear that the fracture behavior of these latex films strongly depends on the molecular weight. The medium molecular weights (151,000 g/mol and 142,000 g/mol) show considerable increase in fracture energy on annealing. The blend system contained a 50:50 mixture of low molecular weight PS (34,000 g/mol) and high molecular weight PS (600,000 g/mol) (*3*). Due to the presence of the low molecular weight, which is right at the critical limit for entanglements, the fracture energy is much lower than the medium molecular weight

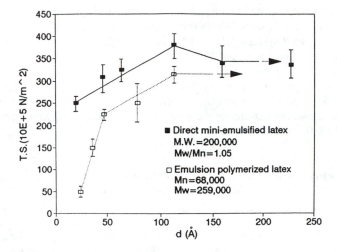

Figure 3. Tensile strength of polystyrene films first increases, then goes through a maximum, and finally levels off (*4*). The appearance or non-appearance of a maximum in both tensile strength measurements and dental burr fracture experiments (below) is both a function of molecular weight and reptation time relative to the time frame of the experiment. (Reproduced from reference 4. Copyright 1993 ACS.)

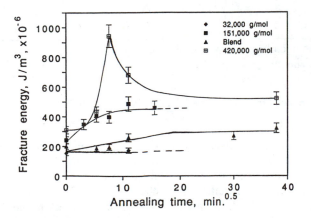

Figure 4. Plot of fracture energy versus square root of annealing time for different narrow molecular weight distribution PS latex films. (Adapted from reference 3.)

films. These results are similar to those obtained by Yang et al. (28) for the bulk PS blend system. At the critical entanglement molecular weight, 32,000 g/mol, there is no change in the fracture energy with annealing time. The chains fail to form effective physical entanglements on interdiffusion. The peak in the fracture energy, observed in Mohammadi et al.'s work (420,000 g/mol) (1), was not seen for lower molecular weights or the blend systems. This is because the high annealing temperature caused significant interdiffusion at short times, and the reptation times in all these samples were less than 10 minutes.

Table V gives the energy contribution results for medium molecular weight PS (151,000 g/mol) (2) as a function of the annealing time. It can be seen that the chain scission energy contribution increases with time before reaching a plateau value of about 40% beyond 60 minutes.

Table V. Energy Contributions for Medium Molecular Weight (151,000 g/mol; PDI=1.02) PS Latex Films

Annealing time, t, min.	No. of scission m^{-3} $\times 10^{-23}$	Contributions		Total Fracture energy, J/m^3 $\times 10^{-6}$
		%E_S [*]	%E_P	
0	4.5	28.0	72.0	242
10	3.2	14.0	86.0	349
30	5.9	21.0	79.0	406
60	12.0	44.0	56.0	399
120	14.0	42.0	58.0	486
240	12.0	39.0	61.0	459

[*] Includes the uncoiling energy contribution (<1%)
E_S= Scission energy
E_p = Pullout energy

Source: Reproduced from reference 2. Copyright 1995 ACS.

It is evident that the chain pullout or scission depends on the molecular weight at room temperature. Also, the chain pullout process in the present model (2) is limited to chain ends for glassy polymers like PS. In other words, the extent of chain scission is dependent on the effective number of physical entanglements, and for a given molecular weight, M, the portion of a chain forming entanglements is M - 32,000 g/mol in PS (the portion of the chain sticking out beyond the last physical entanglement on either side is taken as 16,000 g/mol (29)).

The maximum relaxation time, τ, which according to equation 3 scales as M^3, can be used to gain some insight as to the fracture relationships among the various molecular weights. The fraction of the energy contribution from chain scission, F, scales as (3)

$$F = [(M-32,000)/M]^3 \tag{7}$$

It should be noted that this assumes the chains are in a reptative motion under fracture conditions in the glassy state. It shall be shown below that the chains involved in scission and pullout are actually well above their glass transition temperatures, and hence able to reptate.

The values of F based on experiments and equation 7 for different molecular weights of PS are shown in Table VI (*3*). It is readily seen that there is very good agreement between the theoretical and experimental values. The infinite molecular weight corresponds to acrylic acid anhydride crosslinked PS (5.0 mol% crosslinker).

Table VI. Comparison of Chain Scission Energy Contribution as a Function of Molecular Weight in Polystyrene

M (g/mol)	F^* (theory)	F (expt.)
32,000	0.0	0.0
151,000	0.48	0.40
180,000	0.55	0.67
420,000	0.79	0.90
Infinite	1.0	0.99[**]

[*] From equation 7
[**] Acrylic acid anhydride (AAA) crosslinked PS (5.0 mol% crosslinker)

Source: Reproduced with permission from reference 3. Copyright 1995 John Wiley.

Sambasivam et al. (*7*) studied the effect of annealing temperature on the fracture behavior of emulsion polymerized PMMA latex films (M_n =485,000 g/mol) (Figure 5). Two different annealing temperatures, 140 and 180°C were chosen. When the films were annealed at 140°C, a peak in the fracture energy was observed, similar to Mohammadi et al.'s results (*1*). But when the annealing temperature was increased to 180°C, no peak was observed, and the fracture energy increased monolithically. Similar to the PS latex films, this behavior was attributed to the reptation time, which at 180°C is about 82 seconds compared to a reptation time of 24 hours at 140°C. When the reptation time is short compared to the experimental annealing time, the peak is not observed.

A similar maximum in the fracture toughness (K_{IC}) of molded PMMA was observed by Danusso et al. (*30*) When the molding of the pellets was carried out at lower temperatures, the fracture of the sheets occurred between the pellets (intergranular), and at higher molding temperatures, the fracture occurred through the pellets (transgranular). Also, the K_{IC} maximum occurred around the transition in the crack path from intergranular to transgranular fracture.

Zosel and Ley (*31*) studied the effect of crosslinking on the mechanical properties of PBMA latex films using SANS and tensile measurements. By using a bifunctional monomer, methallyl methacrylate (MAMA), they made a series of latexes ranging from uncrosslinked to highly crosslinked polymers. Figure 6 shows the fracture energy results for the latex films with different levels of crosslinker as a

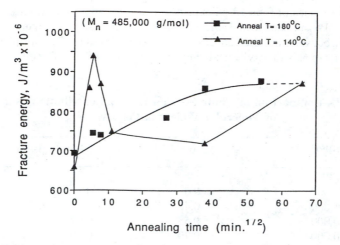

Figure 5. Fracture energy per unit volume versus annealing time for PMMA latex films (M_n = 485,000 g/mol; PDI=1.85) at two different annealing temperatures. Average error: ± 10%. (Reproduced with permission from reference 7. Copyright 1995 John Wiley.)

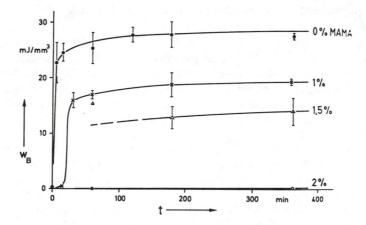

Figure 6. Fracture energy vs annealing time at 90°C for PBMA with various concentrations of MAMA, as indicated. (Reproduced from reference 31. Copyright 1993 ACS.)

function of annealing time. It can be seen that the tensile strength increases with annealing time in the uncrosslinked latex films, whereas in the crosslinked films, the strength does not appreciably increase on annealing. Also, it was observed that in the case of highly crosslinked films, where the molecular weight between chemical crosslinks (M_C) were larger than the entanglement length (M_e), the films remained brittle even after annealing. They concluded that the interdiffusion step is hindered in crosslinked films.

Sambasivam et al. (*32*) studied the fracture behavior in bulk crosslinked PS and PMMA films, prepared via photopolymerization, as a function of mol % crosslinker. Reversible crosslinkers, acrylic anhydride (AAA) and methacrylic anhydride (MAA) were used separately. Figure 7 shows the results of this study. A maximum is observed around 1.5-5.0 mol% crosslinker for both crosslinked PS and PMMA. The maximum in Sambasivam et al.'s data (*32*) was explained in terms of the apparent change in the number of physical entanglements with crosslinking.

However, Zosel and Ley's (*31*) data for the t > 300 minutes, which approximates a fully healed system (or bulk state), i.e. t >> τ, does not go through a maximum; the fracture energy of PBMA decreases with increasing crosslinker (MAMA). The main difference between the two sets of data is that in the Zosel and Ley materials, each latex particle was a separate network. In the film form, the micro-networks were not attached to each other by primary bonds. In the Sambasivam, et al. studies, the material was a single macroscopic network.

Fracture per Unit Volume Versus per Unit Surface Area. Most of the fracture results from the authors' work have been expressed in terms of per unit volume instead of the conventional per unit area. Evidence from the scanning electron microscopy (SEM) of the ground surface revealed that there is considerable sub-surface damage (*8*). Also, the fracture surface generated with the dental burr is not smooth and planar. Hence, the assumption of a spherical model (for fractured latex films) to describe a jagged surface might create substantial error. These problems could be minimized by expressing the results on a per unit volume of fractured material. However, for better comparison of results with literature, measuring the average size of the ground powder allowed the determination of the surface area generated due to grinding and allowed the expression of the results on a per unit area basis (Table VII) (*7*). Although the fracture energy in terms of per unit area from the grinding experiment, G_{BG} (where BG represents Burr Grinding), probably contains mixed modes of fracture, the values are comparable to reported G_{IC} values (*33;34*).

Theoretical Chain Pullout Energy. Two different theories were used to calculate the chain pullout energy in PS and PMMA latex films. The first one is based on Evans[35] equation,

$$E_P = kTN_e^2 \qquad (8)$$

where E_P is the chain pullout energy, k is Boltzmann constant, T is the temperature, and N_e is the number of mers between entanglement points. The second theory was based on Mark's approach (*36*). Here, the force necessary to scission a chain is assumed to be that force necessary to stretch a C-C bond from 1.54 Å to 2.54 Å. If

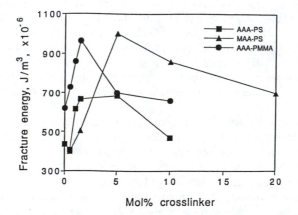

Figure 7. Plot of fracture energy versus mol% crosslinker for XPS and XPMMA-(AAA-acrylic acid anhydride; MAA-methacrylic acid anhydride). (Reproduced with permission from reference 32. Copyright 1996 John Wiley.)

the force required to pullout a chain is, on average, about half the force required for scission (this assumes all values from zero to actual fracture are equally likely), the pullout energy can be calculated. The results of these calculations are given in Table VIII for both PS and PMMA. The two theories yield values differing nearly by a factor of ten. It must be noted that the present experimental values agree much more closely with the Mark approach values.

Table VII. Comparison of Fracture Data in Per Unit Volume Versus Per Unit Area

Annealing time (min.)	Clump diameter µm	No. of scissions m^{-3} $\times 10^{24}$	m^{-2} $\times 10^{17}$	Fracture energy E_T J/m^3 $\times 10^6$	G_{BG} J/m^2
Polystyrene M=151,000 g/mol (12)					
0	1.4	0.4	1.0	242	55
240	3.0	1.2	7.0	459	230
M=32,000 g/mol (13)					
0	0.6	-	-	154	15
120	0.6	-	-	174	17
Poly(methyl methacrylate) M=485,000 g/mol (14)					
0	2.8	3.8	0.9	695	150
2880	3.5	7.1	2.5	880	511

G_{BG} - Fracture energy from dental burr grinding experiment
Source: Reproduced with permission from reference 7. Copyright 1995 John Wiley.

Frictional Coefficients

Under conditions favoring chain pullout, according to Prentice (*37*), the fracture energy, γ, is related to the molecular frictional coefficient, μ_0, as,

$$\gamma = (\mu_0 \cdot v \cdot n \cdot L^2)/2 \qquad (9)$$

where v is the velocity of chain pullout, n is the number of chains, and L is the length of the chain segment being pulled out.

Table VIII. Theoretical and Experimental Chain Pullout Energies for PS and PMMA

Basis	Chain pullout energy (E_P) $J/m^3 \times 10^{-6}$	
	PS	PMMA
Evans equation (7)	49 (N_e = 150 mers)	27 (N_e = 95 mers)
Mark approach	311 (150 mers)	197 (95 mers)
Experimental*	260 (151,000 g/mol)	211 (485,000 g/mol)

* Fully annealed latex films

The following assumptions were made to obtain molecular frictional coefficient of chain pullout. The chain pullout energy was considered to be γ; two values were assumed for the velocity in equation 9: i. velocity of the burr (8.3×10^{-3} m/sec, lower bound value) and; ii. crack propagation velocity in glassy polymers under impact conditions (620 m/sec, upper bound value). The quantity n is obtained from the difference between the total number of chains and the number of chain scissions. the quantity L was assumed to be equal to the length of the chain end, which is roughly $1/2 \ M_e$, where M_e is the molecular weight between physical entanglements.

Substituting these values in equation 9, the molecular frictional coefficients were obtained for PS and PMMA. From the molecular frictional coefficients, the mer frictional coefficient was obtained by multiplying equation 9 by the length of one mer (2.54 Å/mer). Usually, the mer frictional coefficients are obtained from dynamic mechanical measurements in the melt state. The mer frictional coefficient, ζ_0, was calculated to be about 1.5×10^{-5} and 2.0×10^{-10} dyn.s/cm for medium molecular weight PS, and about 1.2×10^{-4} and 1.8×10^{-9} dyn.s/cm for PMMA, for the lower and upper bound velocities, respectively. Comparison of these values with that from literature[26] yielded an estimate of the actual temperature of the pullout process to be about 150-250°C for PS (2) and about 220°C for PMMA (7). These temperatures are all above the glass transition temperatures of the individual polymers, suggesting that at the actual instant of pullout, a chain is activated to a high energy state. Being above T_g, it is able to undergo reptation motions during pullout.

In order to verify the temperature at the crack tip in the grinding experiments (using DBGI), some other calculations were made. It must be noted that the grinding experiment was carried out under water-cooled conditions. The first calculation was based on the total fracture energy, and the second one was based on the energy associated with water evaporation from the surface of the cooling water drop. For the former, the total fracture energy for the fully annealed medium molecular weight (151,000 g/mol) PS (=460x10^6 J/m^3) was used, and for the latter, the standard enthalpy and entropy values associated with water evaporation at room temperature were used[38]. Table IX lists the temperature values obtained from each source. It can be readily seen that the temperature values from the three different sources range from 150 to 368 °C. The lowest value, 150°C, based on the dental burr surface velocity, seems to be too low.

Table IX. Polystyrene Fracture Temperature Calculations

Source	Energy, J/m^3	Temperature, °C
1. Experimental[a] fracture energy (grinding experiments)	460x10^6	368
2. Frictional coefficient[a]	260x10^6 (pullout energy)	150-250
3. Water evaporation @ room temp.	1000x10^6	242

[a] M. Sambasivam, A. Klein, and L. H. Sperling, *Macromolecules*, **28**, 152 (1995). Source: Reproduced with permission from reference 9.

This temperature rise is localized around the crack tip, with the bulk of the sample at ambient conditions. In order to estimate the size of this localized zone, it was assumed that it is equal to the plastic zone size at the crack tip. The plastic zone size was calculated to be about 1.5 nm from the data given in reference 39 under plane-strain conditions (for PS with a molecular weight of 218,000 g/mol). By assuming that the pullout in polystyrene is limited to chain ends, this length is about 7,500 g/mol. The radius of gyration of such a chain end segment is about 2.3 nm, which is in reasonable agreement with the plastic zone size. In other words, the local heating at a crack tip is confined to a very small length, of the order of a few nanometers.

Summary and Conclusions

The film formation and subsequent fracture studies in PS and PMMA latex films have been reviewed. The interdiffusion process in the latex films follows Wool's minor chain reptation theory. Tensile strength in these films increases due to interdiffusion and subsequent formation of physical entanglements. Full strength can be achieved at an interdiffusion depth equal to one radius of gyration of the chains, as predicted by the minor chain reptation theory. The annealing temperature has a significant effect on the interdiffusion rate, following the Arrhenius and WLF equations within a given range. The activation energies for the diffusion process are about 33 kcal/mol for poly(butyl methacrylate), about 55 kcal/mol for PS, and about 100 kcal/mol for PMMA. In crosslinked systems, the interdiffusion is hindered due to the presence of permanent chemical crosslinks.

The molecular basis of fracture in plastics has been reviewed. Fracture energy and the number of scissions increase with annealing time. This is due to the interdiffusion and subsequent formation of physical entanglements. For PS, at the critical molecular weight for entanglements, about 32,000 g/mol, the entanglements barely resist the crack growth, resulting in substantially 100% chain pullout. With increasing molecular weights, however, the chain scission energy contribution increases at the expense of the pullout process from 0% for M=32,000 g/mol to about 90% for 420,000 g/mol. At the medium molecular weight, about 151,000 g/mol, the scission and pullout contributions are roughly equal. From the molecular weight dependence of reptation motion, a simple equation has been used to predict the extent of chain scission energy contribution during fracture in PS. The values predicted by this equation agree with the experimental values.

The theoretical chain pullout energies calculated from the Evans equation and the Mark approach were compared with the experimental values, for both PS and PMMA. The Mark approach yields better agreement.   Frictional coefficients calculated from the chain pullout energy indicate that the actual temperature of pullout is about 242-368°C for PS and about 220°C for PMMA. In the case of polystyrene, this temperature seems to be confined to a thickness of 1.5-2.3 nm.

Acknowledgments

The authors would like to thank National Science Foundation for support through Grant No. ECD-9117064, and the companies that make up the Polymer Interfaces Center at Lehigh University for their collective support. The authors would also like to thank Dow Chemical Company for Styron 6069, Shell Development Company for medium molecular weight PS, and Rohm & Haas Company for PMMA latex.

Literature Cited

1. Mohammadi, N.; Klein, A.; Sperling, L. H., *Macromolecules*, **1993**, *26*, 1019.
2. Sambasivam, M.; Klein, A.; Sperling, L. H., *Macromolecules*, **1995**, *28*, 152.
3. Sambasivam, M.; Klein, A.; Sperling, L. H., *J. Appl. Polym. Sci.*, **1995**, *58*, 357.
4. Kim, K. D.; Sperling, L. H.; Klein, A.; Wignall, G. D., *Macromolecules*, **1993**, *26*, 4624.
5. Wang, Y.; Winnik, M. A., *J. Phys. Chem.*, **1993**, *97*, 2507.
6. Boczar, E. M.; Dionne, B. C.; Fu, Z.; Kirk, A. B.; Lesko, P.; Koller, A.D., *Macromolecules*, **1993**, *26*, 5772.
7. Sambasivam, M.; Klein, A.; Sperling, L. H., accepted, *Polymers for Advanced Technologies*, 1995.
8. Sperling, L. H.; Klein, A.; Sambasivam, M.; Kim, K. D., *Polym. Adv. Technol.*, **1994**, *5*, 453.
9. Sperling, L. H.; Klein, A.; Sambasivam, M., submitted, *J. Polym. Mater.*, 1995.
10. Hahn, K.; Ley, G.; Schuller, H.; Oberthur, R., *Coll. Polym. Sci.*, **1986**, *264*, 1092.

11. Linne, M. A.; Klein, A.; Sperling, L. H.; Wignall, G. D., *J. Macromol. Sci. Phys.*, **1988**, *B27 (2&3)*, 217.
12. (a) Yoo, J. N.; Sperling, L. H.; Klein, A.; Glinka, C. J., *Macromolecules*, **1991**, *24*, 2868 (b) Yoo, J.N.; Sperling, L. H.; Glinka, C. J.; and Klein, A., Macromolecules, **1990**, *23*, 3962.
13. Anderson, J. E.; Jou, J. H., *Macromolecules*, **1987**, *20*, 1544.
14. Kim, K. D.; Sperling, L. H.; Klein, A.; Wignall, G. D., *Macromolecules*, **1993**, *26*, 4624.
15. Wang, Y.; Zhao, C. -L.; Winnik, M. A., *J. Chem. Phys.*, **1991**, *95(3)*, 1.
16. Winnik, M. A.; Wang, Y.; Haley, F., *J. Coat. Technol.*, **1994**, *64*, 51.
17. Wang, Y.; Winnik, M. A.; *Macromolecules*, **1993**, *26*, 3147.
18. de Gennes, P. G., *J. Chem. Phys.*, **1971**, *55*, 572.
19. Doi, M.; Edwards, S. F.; *Faraday Trans.*, **1978**, *2*, 1789.
20. de Gennes, P. G., *C. R. Acad. Sci.*, **1981**, *292*, 1505.
21. Prager, S.; Tirrell, M., *J. Chem. Phys.*, **1981**, *75*, 5194.
22. Kim, Y. H.; Wool, R. P., *Macromolecules*, **1983**, *16*, 1115.
23. Jud, K.; Kausch, H. -H.; Williams, J. G., *J. Mater. Sci.*, **1981**, *16*, 204.
24. Zhang, H.; Wool, R. P., *Macromolecules*, **1989**, *22*, 3018.
25. Mohammadi, N.; Kim, K. D.; Klein, A.; Sperling, L. H., *J. Coll. Int. Sci.*, **1993**, *157*, 124.
26. Ferry, J. D., *Viscoelastic Properties of Polymers*, 3rd ed., John Wiley & Sons, New York, 1980.
27. Lake, G. J.; Thomas, A. G., *Proc. R. Soc. Lond. Ser. A*, **1967**, *A300*, 108.
27a. Tashiro, K.; Kobayashi, M.; and Tadokoro, H.; *Macromolecules*, **1977**, *10*, 413.
27b. Tadokoro, H., *Structure of Crystalline Polymers*, Wiley Interscience, New York, **1979**, P. 400.
28. Yang, A. C. -M.; Lee, C. K.; Ferline, S. L., *J. Polym. Sci. Polym. Phys. Ed.*, **1992**, *30*, 1123.
29. Wool, R. P., *Macromolecules*, **1993**, *26*, 1564.
30a. Danusso, F.; Tiegi, G.; Lestingi, A., *Polym. Commun.*, **1986**, *27*, 56.
30b. Danusso, F.; Tiegi, G.; Lestingi, A., *Polym. Commun.*, **1985**, *26*, 221.
31. Zosel, A.; Ley, G.; *Macromolecules*, **1993**, *26*, 2222.
32. Sambasivam, M.; Klein, A.; Sperling, L. H., submitted, *J. Appl. Polym. Sci.*, 1995.
33. Kinloch, A. J.; Young, R. J., *Fracture Behaviour of Polymers*, 2nd ed., Applied Science Publishers, London and New York, 1983.
34. Kausch, H.-H., *'Polymer Fracture'*, 2nd ed., Springer-Verlag, Berlin, 1987.
35. Evans, K. E., *J. Polym. Sci. Polym. Phys. Ed.*, **1987**, *25*, 353.
36. Mark, H., *"Cellulose and Cellulose Derivatives"*, *Vol. IV*, Emil Ott, editor, Interscience Publishers Inc., New York, 1943.
37. Prentice, P., *Polymer*, **1983**, *24*, 344.
38. *Handbook of Chemistry and Physics*, 60th edition, Weast, R. C., ed., CRC Press Inc., Boca Raton, Florida, 1980.
39. Wool, R. P., *Polymer Interfaces- Structure and Strength*, Hanser Publishers, New York, 1995.

Chapter 12

Development of Properties During Cure of Epoxy and Acrylate Coating Materials

Jakob Lange[1], Anders Hult[1], and Jan-Anders E. Månson[2]

[1]Department of Polymer Technology, Royal Institute of Technology,
S−100 44 Stockholm, Sweden
[2]Laboratoire de Technologie des Composites et Polymères, Ecole
Polytechnique Fédérale de Lausanne, CH−1015 Lausanne, Switzerland

The development of time-dependent properties with conversion during cure of step-wise reacting epoxy and chain-wise reacting acrylate coatings has been investigated. Using a torsional dynamic mechanical analyser, dynamic shear modulus and change in sample thickness was monitored simultaneously, thus giving information on both the physical properties and the progress of the reaction in one experiment. Isothermal cure below and above the ultimate glass transition temperature, i.e. with and without vitrification at the cure temperature, was compared. The epoxy and acrylates were found to behave similarly on cure above the ultimate glass transition temperature but to exhibit differences during vitrification on cure below this temperature. From the shrinkage and storage moduli, approximate values of the relaxation time and relaxation modulus as a function chemical conversion were calculated.

The solidification on cure plays a key role in determining the properties of a thermosetting coating material. Understanding the events and changes in the material during cure is essential if its full potential is to be employed. Such information is valuable both for designing cure cycles and optimising processing parameters, as well as for examining the mechanical behaviour, e.g. residual stress build-up, during the cure process [1].

In general thermoset materials cure by either a step-wise or a chain-wise mechanism [2]. They can furthermore be cured either at a temperature above the ultimate glass transition temperature ($T_{g\infty}$), experiencing only gelation, or at a temperature below $T_{g\infty}$, encountering both gelation and vitrification at the cure temperature [3]. Thoroughly investigating the cure process of thermoset coatings therefore involves studying all these cases. Furthermore, both the chemical and the physical changes in the material should be followed. Unfortunately, the techniques

0097−6156/96/0648−0200$15.00/0

normally employed to measure the progress of the chemical reaction are unable to detect the physical changes on cure, whereas the methods commonly used to follow changes in mechanical properties do not provide any direct information as to the state of cure of the sample. To get the full picture the two kinds of analysis have to be run separately and the results correlated afterwards, which often is difficult. A method permitting both the chemical and physical changes to be detected therefore offers major advantages.

The curing and change in properties with cure of thermosets is well documented in literature for step-wise as well as chain-wise reacting systems [3, 4, 5]. In general, chain-wise reacting materials exhibit gelation at lower degrees of conversion and yield materials with higher crosslink densities than do step-wise reacting systems. Regarding the development of viscoelastic properties with cure, little has been reported on chain-wise reacting systems, whereas step-wise systems have been well investigated. Plazek and Chay [6] measured the development of the creep compliance and the retardation spectrum, and Adolf and Martin [7] and Hodgson and Amis [8] measured dynamic moduli at different frequencies near the gel point and showed that viscoelastic functions could be superposed at different extents of reaction.

In this chapter the evolution of properties during cure is studied using simultaneous dynamic mechanical analysis and dilatometry [1]. Three different model systems, one epoxy and two acrylates, were used to achieve step-wise and chain-wise cure above and below $T_{g\infty}$. The shear modulus, at different frequencies, and the sample contraction were measured as a function of cure time in a torsional parallel-plate dynamic mechanical analyser. Using this data the development of the relaxation time and relaxation modulus as a function of cure in the different materials is described [1, 9].

Experimental

Materials The diglycidyl ether of bisphenol F (PY 306), **1**, 2,2-di(4-aminocyclo-hexane)propane (HY 2954), **2**, were received from Ciba, Switzerland. N-methyldiethanolamine, 98%, was obtained from Fluka, Switzerland. Tripropyleneglycol diacrylate, **3**, and benzopinacole, 99% was purchased from Aldrich, Germany. Di-ethoxylated bisphenol A dimethacrylate (Diacryl 101), **4**, 84% was obtained from Akzo Chemicals BV, the Netherlands. All chemicals were used without further purification. The monomers are presented in Figure 1.

Methods Experiments were performed on one epoxy-amine mixture and two acrylates. The epoxy-amine mixture was **1+2** in stoichiometric amounts. The two acrylates, **3** and **4**, where cured separately using 2 mole% benzopinacole and 0.3 mole% diethanolamine as initiator.

Samples were cured in a rotational parallel plate rheometer (Rheometrics RDA 2 dynamic mechanical analyser). Aluminium plates of 8 mm diameter were used. The dynamic shear modulus was measured, at regular intervals, at five frequencies between 0.001 and 10 Hz and at zero normal force, while simultaneously monitoring the change in plate distance. The strain was varied automatically by the instrument and was between 0.05 and 0.001. In a typical experiment the liquid monomer was applied between the plates of the rheometer and the distance set to 1 mm. The sample was then heated to the cure temperature, 115°C for both acrylates and 100°C or 140°C for the epoxy, and the dynamic shear modulus and the change in plate distance measured as a function of cure time. During one run measurements were performed at three frequencies simultaneously by superposition of the strain input signal and decomposition of the stress output signal. A complete set of five frequencies thus required two runs.

Results and discussion

Stiffness and shrinkage versus cure time In Figure 2 typical evolutions of shear modulus and sample thickness contraction for isothermal cure *above* $T_{g\infty}$ of the epoxy and the first acrylate (**3**) are presented. The thickness contraction, c, is proportional to the free linear shrinkage, s, of the polymer according to

$$dc = \frac{1+v_p}{1-v_p} ds, \tag{1}$$

where v_p is the Poissons's ratio of the polymer. This relationship will be further discussed below. It should be noted that only the contraction after gelation was measured; prior to gelation the liquid sample does not exert any force on the plates and accommodates to the shrinkage by transversal flow. At first, during the heating-up and reaction until gelation, the modulus is neglectable. At gelation the modulus rises rapidly, and the shrinkage becomes detectable. The modulus and shrinkage then gradually level off as the reaction goes to completion.

Figure 2 points to some differences between the systems. The acrylate exhibits a much higher rubbery modulus and a much larger shrinkage after gelation than the epoxy. This is due to the combined influence of the structure of the monomers and the reaction mechanism. The ring-opening polymerisation of the epoxide group and the step-wise reaction mechanism lead to a low overall shrinkage and late gelation for the epoxy system, whereas the vinyl polymerisation and chain-wise mechanism of the acrylate yield a high overall shrinkage and early gelation in the acrylate system. Furthermore, chain-wise cure of low molar mass monomers (e.g. the present diacrylate) is known to produce more densely crosslinked networks than step-wise cure of similar structures (i.e. the di-epoxy/di-amine mixture). Since the modulus of a crosslinked material above T_g is proportional to the crosslink density, this explains why the acrylate has a higher rubbery modulus than the epoxy. The modulus curves in Figure 2 were obtained at a frequency of 1 Hz. It is worth noting that varying the measurement frequency between 10 and 0.01 Hz did not significantly change the results (data not shown here), thus indicating a low frequency dependence during the whole reaction. This is consistent with the low time-dependence of the properties expected in a material far above its T_g.

The cure *below* $T_{g\infty}$ of the epoxy and the second acrylate (**4**) is illustrated in Figure 3. The modulus was measured at five different frequencies (three shown here). Comparing Figures 3 and 2 shows the effect of vitrification during cure. The shrinkage is hardly affected, whereas the modulus exhibits some differences. Regarding the

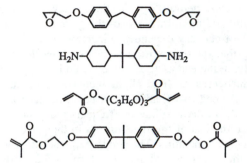

Diglycidyl ether
of bispenol F, 1

2,2 di(4-aminocyclo-
hexane)propane, 2

Tripropylene glycol
diacrylate, 3

Di-ethoxylated bisphenol
A dimethacrylate, 4

Figure 1. Epoxy, amine and acrylate monomers.

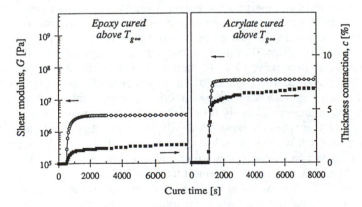

Figure 2. Modulus (measured at 1 Hz) and thickness contraction on cure above $T_{g\infty}$.

epoxy, vitrification can be identified as a separate event following gelation, as indicated by the tendency towards a knee in the modulus curves before the rise to the glassy value. In the case of the acrylate the modulus rises evenly from gelation towards the final value, which suggests that vitrification begins immediately after gelation. The separation of gelation and vitrification during cure below $T_{g\infty}$ in stepwise reacting systems but not in chain-wise reacting systems has been observed by Babayevsky [4].

As opposed to cure above $T_{g\infty}$, both the epoxy and the acrylate exhibit strong frequency dependencies during cure below $T_{g\infty}$. In the epoxy the frequency dependence is limited in the liquid and the initial part of the gelled state, but as the sample approaches vitrification the frequency dependence increases dramatically. The rise in modulus accompanying the glass transition is detected earlier at higher measurement frequencies. Finally, at the end of the reaction deep in the glassy state, the frequency dependence again decreases. The slight increase in modulus with increasing frequency observed at the end is typical of all viscoelastic materials in the vicinity of the glass transition [10]. In the acrylate the frequency-dependent region extends from just after gelation to the end of the reaction. As in the epoxy, a higher measurement frequency detects a higher modulus. The strong frequency dependence at the end of the reaction shows that the material is very close to, i.e. still in, the transition zone. In all, this shows that in the acrylate the glass transition is wide, and that vitrification commences just after gelation and lasts until the end of the reaction.

Frequency - cure time superposition Figure 3 shows that the measurements at low frequencies could not be carried out at the beginning of the reaction. Here the properties of the system changed quickly, making measurements lasting more than about 10 s unrealistic. Since the time of measurement is proportional to the inverse of the measurement frequency, $t \geq 1/f$, measurements below 0.1 Hz will thus be difficult. One way of addressing this problem would be to extrapolate data obtained at higher frequencies into the desired range. Work on natural rubber and polyurethanes has pointed to an interchangeability between crosslink density and time [11, 12]. The results indicate that when the crosslink density changes, it is mainly the relaxation time that is affected, whereas the other aspects of the viscoelastic behaviour remain unchanged. Studies on reacting systems (epoxies and silanes) also indicate that there is an interchangeability between measurement frequency and crosslink density (cure time in this case) [7, 8]. It is worth noting that these studies all concerned materials far away from their glass transition, where the relaxation time *decreases* with increasing crosslink density, whereas the properties of the materials in the present work are dominated by the glassy state where the relaxation time instead *increases* with increased degree of crosslinking [13]. The basic time-dependence of the properties is believed to be the same, however.

Studying the sets of curves at different frequencies for the two materials in Figure 3 a certain resemblance in shape within each set is visible. By postulating that the gelation occurs at one single time (t_{gel}), independent of frequency [7], and assuming the time to rise in modulus (t_{rise}) to be a characteristic of the behaviour at each frequency, two characteristic values on the cure time scale for each modulus curve are obtained. If the time is set to zero at gelation and the time-scale for each modulus curve then divided by the time to rise (defined as the time to reach 3/4 of the final modulus at that frequency) a scaling, or normalisation, along the time axis is obtained. In other words the time is normalized according to $t_{norm} = (t - t_{gel}) / t_{rise}$, where $t_{rise} = t$ at $G = 0.75G_{final}$. The result of such a scaling of time, together with a normalization of the modulus with respect to final value; $G_{norm} = G / G_{final}$, applied to the five experimental frequencies is shown in Figure 4. As can be seen the scaling works well, the curves superimpose nicely. The choice of 3/4 of the final value as point of reference is of course arbitrary, what is required is a characteristic time for the rise in modulus. The epoxy was not sensitive to the choice of the second reference time, all tested values between 0.2 and 0.8 produced a good superposition. The acrylate was more sensitive, however, and 0.75 gave the best fit.

One single curve thus describes all the modulus-cure time data obtained over five decades of measurement frequencies. As a consequence, once the relationship between modulus and cure time is established, it suffices to know the characteristic time and final modulus value at any frequency to reconstruct the full curve at that particular frequency. Since these values are possible to obtain also at low measurement frequencies, in spite of the difficulties in observing the early stages of the reaction, it is possible to reconstruct the initial part of the curve.

Relaxation time The viscoelastic behaviour of a polymeric material may be characterised by its relaxation time, τ. This parameter roughly describes the time required for e.g. a residual stress state in the material to decay. A measure of the development of the relaxation time with cure can be obtained from the modulus-cure time data at different frequencies. Each frequency, f, can be taken to correspond to a certain relaxation time, $\tau \approx 1/f$. A criterion determining when the modulus at a particular frequency becomes high and enters into the glassy state can then be defined. At the cure time when this criterion is fulfilled for a certain frequency, the system is assumed to have the relaxation time corresponding to that frequency. In this way each measurement frequency will provide a point on the curve of relaxation time versus cure time. It is of course important to chose a proper criterion for when the modulus is taken to enter the glassy region. In the previous section a time to rise was defined as the time when the modulus reached 3/4 of its final value. This criterion appears to be a characteristic of each modulus curve, as indicated by the fact that division by this time made the curves superposable. The point where the modulus arrives at 3/4 of its end value was therefore chosen. Relaxation time versus cure time for the two systems cured below $T_{g\infty}$ is shown in Figure 5. As can be seen the development can be taken to be linear as a first approximation, and the slope is roughly the same for both materials.

The degree of conversion can be calculated from the thickness contraction, if the conversion at the gel point is known, and by assuming that c is zero until gelation and then directly proportional to the conversion. The change in Poisson's ratio with cure is thus neglected. The gel-point conversion can be calculated theoretically for step-wise reacting systems using the classical theory of gelation [14], but has to be measured or estimated for chain-wise systems. For the epoxy the gel-point conversion was calculated to be 0.6, whereas it was taken to be 0.05 for the acrylate [5]. The degree of conversion, x, was calculated from c, the gel point conversion, x_{gel}, and the total thickness contraction after gelation, c_c, according to the following expression:

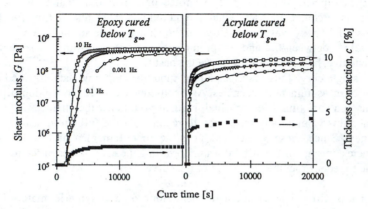

Figure 3. Modulus at three different frequencies (same legend in both graphs) and thickness contraction for cure below $T_{g\infty}$.

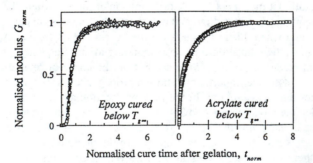

Figure 4. Normalised dynamic shear modulus($G_{norm} = G / G_{final}$) at 10 ($\square$), 1 ($\Diamond$), 0.1 (∇), 0.01 (Δ) and 0.001 (o) Hz versus normalized cure time after gelation ($t_{norm} = (t - t_{gel}) / t_{rise}$) for all systems.

$$x(t) = \frac{c(t)}{c_c}(1 - x_{gel}) + x_{gel} \qquad (2)$$

It should be noted that the degree of cure as calculated here is relative to full cure (end of reaction) at the cure temperature, thus not taking any residual reactivity into account. Shown in Figure 5 is the development of relaxation time as a function of degree of conversion. It is interesting to note that the relaxation time as a function of conversion is very similar for the two systems, and that the relationship appears to be linear for a large part of the relaxation times.

Relaxation modulus A more detailed description of the relaxation behaviour of a viscoelastic material is given by the relaxation modulus. It can be obtained experimentally by subjecting a sample to a step strain and then monitoring the decay in stress with time. However, this kind of experiment is difficult to perform on reacting systems where the properties change with time. The relaxation modulus can instead be approximated by the storage modulus. This is done by substituting the relaxation modulus at a certain time with the storage modulus measured at a frequency corresponding to the inverse of that time: $G_{rel}(t) \approx G'(1/f)$ where G_{rel} is the relaxation and G' the storage modulus [10]. In this way data of the kind presented in Figure 3 can be converted into estimates of the evolution of the relaxation modulus with cure.

Curves of relaxation modulus versus conversion calculated using the data in Figure 3 and Equation 2 are shown in Figure 6. It may be noted that some points (indicated by asterisks) at low frequencies and low conversions have been obtained using the reconstruction technique described in the previous section. The behaviour of the epoxy system agrees well with the traditional description of the glass transition in a viscoelastic solid [10]. Early in the reaction, just after gelation, the modulus is at its relaxed or equilibrium value, which is independent of time. The level of the relaxed modulus then increases as the sample cures as indicated by the vertical change with conversion. As the material vitrifies, the modulus rises to the glassy value, beginning at the short time end. Finally, the behaviour over the whole experimental window is governed by the unrelaxed, glassy modulus, which is largely independent of time as well as of conversion [3, 10].

As can be seen in Figure 6 the acrylate exhibits a different behaviour. Instead of the three distinct regions present in the epoxies a set of parallel curves is observed. This is a reflection of the characteristics observed in Figure 3, i.e. the immediate onset of vitrification after gelation, the wide transition and the fact that the material remains in the transition zone at the end of the reaction. The rise in level of the relaxation modulus with increasing conversion in the acrylate is related to the change in the equilibrium modulus with cure. In conclusion, whereas three distinct regions are observed in the epoxy, i.e. gel state, transition region and glassy state, the acrylate exhibits essentially the central part of the relaxation curve, i.e. the transition region, together with a rise in equilibrium modulus.

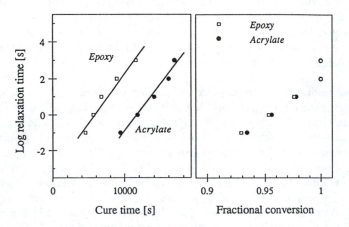

Figure 5. Relaxation time of the epoxy and acrylate cured below $T_{g\infty}$ estimated from dynamic mechanical data at different frequencies.

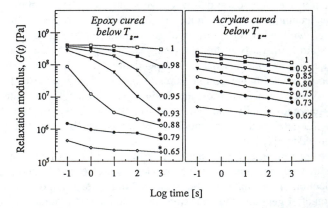

Figure 6. Relaxation modulus calculated from dynamic shear modulus ($t=1/f$) at different degrees of conversion for the acrylate systems. Points obtained using the reconstruction technique are indicated by asterisks.

Conclusions

A torsional dynamic mechanical analyser can be used to collect simultaneous information on the stiffness and the progress of the chemical cure reaction during cure. From the data of sample contraction and modulus at different frequencies provided by this method, the evolution of the time-dependent mechanical properties, e.g. relaxation time and relaxation modulus, with conversion can be described.

Comparing a step-wise curing epoxy and chain-wise curing acrylates points to differences in the vitrification process. On cure above the ultimate glass transition temperature and during the initial part of cure below the ultimate glass transition temperature, i.e. until gelation, the same behaviour is observed, irrespective of reaction mechanism. Vitrification, however, differs between the systems. In the epoxy it is a distinct event, occurring separately from gelation and ending with the end of the cure reaction. In the acrylate vitrification commences immediately after gelation, the two events being indistinguishable, and lasts until the end of the reaction, leaving the sample still in the transition zone.

References

1. Lange, J., Toll, S., Hult, A. and Månson, J.-A. E., *Polymer* 1995, **35**, 3135.
2. Odian, G., "Principles of Polymerisation", John Wiley & Sons, New York, 1981.
3. Enns, J. B. and Gillham, J. K., *J. Appl. Polym. Sci.* 1983, **28**, 2567.
4. Babayevsky, P. G., *Progr. Colloid & Polym. Sci.* 1992, **90**, 57.
5. Kloosterboer, J. G. and Lijten, G. F. C. M., *Polym. Commun.* 1987, **28**, 2.
6. Plazek, D. J. and Chay, I.-C., *J. Polym. Sci.: Part B: Polym. Phys.* 1991, **29**, 17.
7. Adolf, D. and Martin, J. E., *Macromolecules* 1990, **23**, 3700.
8. Hodgson, D. F. and Amis, E. J., *Macromolecules* 1990, **23**, 2512.
9. Lange, J., Hult, A. and Månson, J.-A. E., submitted to *Polymer*.
10. Ferry, J. D., "Viscoelastic Properties of Polymers", John Wiley & Sons, New York, 1980.
11. Plazek, D. J., *J. Polym. Sci.: Part A-2* 1966, **4**, 745.
12. Chan, Y.-W. and Aklonis, J. J., *J. Appl. Phys.* 1983, **54**, 6690.
13. Mangion, M. B. and Johari, G. P., *J. Polym. Sci.: Part B: Polym. Phys.*, 1991, **29**, 1127.
14. Billmeyer, F. W., "Textbook of Polymer Science", John Wiley & Sons, New York, 1984.

Chapter 13

Film Formation and Physical Aging in Organic Coatings

D. Y. Perera, P. Schutyser, C. de Lame, and D. Vanden Eynde

Coatings Research Institute, avenue Pierre Holoffe, B–1342 Limelette, Belgium

Physical aging, a phenomenon occurring in all polymeric materials stored at a temperature below their glass transition temperature, affects practically all coating characteristics such as mechanical, thermal and electrical. Although, on molecular scale is not completely understood, it is associated with coating densification, a fact which can influence the film formation process. DSC and measurement of stress were used to study the evolution of physical aging in three types of coatings: a thermoplastic acrylic (latex) and two thermosetting systems (a water-borne polyester/melamine and a polyester/triglycidyl isocyanurate powder coating). The influence of coalescents, cross-link density, and presence of pigments is discussed. This work also shows that the film formation and physical aging are interdependent and therefore that the study of physical aging is useful in understanding the film formation process.

Film formation and physical aging are important processes which determine the properties of a coating including its durability.

Physical aging occurs in all polymeric materials stored at a temperature (T) below their glass transition temperature (T_g). Briefly, it can be explained by considering that cooling from a temperature (T) above the glass transition temperature (T_g) to one below it, brings a polymer to a non equilibrium state since the values of volume, enthalpy and entropy are higher than in equilibrium state (see Figure 1) (1-3). In the polymer approach towards the equilibrium, these thermodynamic quantities decrease, inducing important changes into material properties such as mechanical, thermal and dielectric. This process has been known in the literature under many terms, related mainly to the property investigated (e.g., volume relaxation, enthalpy relaxation, structural relaxation) until Struik referred to it as physical aging (2).

In contrast with chemical aging (e.g., irreversible composition changes induced in

0097–6156/96/0648–0210$15.00/0

a material by photo-oxidation, temperature and moisture), physical aging is reversible by heating the material at a $T > T_g$.

A great number of publications were dedicated to various aspects of this process (e.g., 1-13). Although it can be described by certain phenomenological models (e.g., Kohlrausch - Williams - Watts) on a macroscopic scale, on a molecular scale this process is not well understood. Nevertheless, it is considered that physical aging is associated with conformational arrangements, increased molecular packing and densification, suggesting that the film formation, a process affecting practically all properties of a coating, and physical aging might be mutually dependent.

In this study thermal stress and enthalpy relaxation were used to evaluate physical aging. While enthalpy relaxation has already been used for a long time to investigate physical aging (3-5,9-11) this is not the case for thermal stress.

For physically aged coating applied on a substrate, the stress dependence on temperature is characterized by a "U" shaped curve (trough) in the T_g region (Figure 2 ; the numbers 1,2 and 3 represent the glassy state, the glass transition and the rubbery state, respectively). This dependence was explained by the effects of physical aging on the properties directly determining the stress magnitude, i.e., thermal expansion coefficient (α^T_F), elastic modulus (E) and Poisson's ratio (see reference 14 for a detailed discussion). With increasing aging, the magnitude of this trough increases as does the maximum peak temperature (T_p). The first effect is due to the fact that physical aging induces not only higher E but, at higher temperatures, also higher α^T_F. The second effect (the increase of T_p with aging), is a consequence of the fact that with aging the coating continues to densify and the erasure of physical aging necessitates more energy. This work investigates the physical aging process of three coatings: a thermoplastic acrylic (latex) and two thermosetting systems (a waterborne polyester/melamine and a polyester/triglycidyl isocyanurate powder coating). The influence of coalescents, cross-link density and the presence of a pigment is discussed.

Experimental

Coatings Characteristics and Instrumental Methods

Thermal stress was evaluated with CoRI-Stressmeter (Braive Instruments, Liège, Belgium) under dry conditions (RH $\approx$ 0 %) at a temperature scan rate of 0.2 K/min. Details on this apparatus and the mathematical equation used to calculate the stress are described in reference 15. The deflection of the uncoated substrate was taken as zero stress.

Enthalpy relaxation (ΔH) was obtained with a Mettler TA-4000 DSC-30, controlled by a TC-11 microprocessor and the Graphware TA-72 software. The measurements were performed on about 10 mg samples at a heating rate of 20 K/min under dry nitrogen purge.

Relative storage modulus (E'_{rel}) and loss tangent (tan δ) were determined with the Polymer Laboratories Dynamic Mechanical Analyzer in double cantilever bending mode under dry nitrogen purge. The frequency was 3 Hz and the temperature scanned at a rate of 2 K/min.

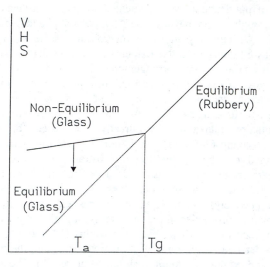

Figure 1. Schematic representation of the dependence of volume (V), enthalpy (H) and entropy (S) on temperature. T_a = aging temperature; T_g = glass transition temperature.

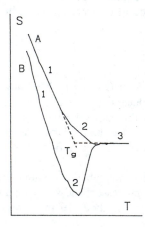

Figure 2. Schematic representation of the stress (S) dependence on temperature (T) for physically aged (B) and nonaged coatings (A).

T$_g$-TMA (Softening Point) was determined on coated tin plates with the Mettler TA-4000 Thermo-Mechanical Analyzer (TMA-40) under Helium purge; heating rate = 10 K/min.

Coalescent level in the film was determined by gas chromatography.

Materials

The materials investigated were:

1) a high T$_g$ - acrylic latex (T$_g$-midpoint ≈ 57°C as determined by DSC at 20 K/min.) containing 2,2,4-trimethyl 1,3-pentanediol monoisobutyrate (TPM)(varnish L1) or butylglycol (BG)(varnish L2); film formation: 50°C for 16 hours;

2) two polyesters crosslinked with hexamethoxy methylmelamine cured under such conditions that their T$_g$-midpoint were similar (≈40°C by DSC at 20 K/min); and

3) a pigmented (TiO$_2$; PVC ≈ 10 %) and a non- pigmented powder coating (carboxyl-functional polyester crosslinked with triglycidyl isocyanurate) cured at 200°C for 15 minutes (T$_g$-midpoint ≈ 75°C by DSC at 20 K/min).

The coatings were applied on precalibrated stainless steel strips, tin plates and glass microfibre filters for stress, TMA, and DMA measurements, respectively. When necessary, free films were prepared by applying the coatings on Teflon or other appropriate substrates.

Physical aging

The experimental conditions for physical aging were as follows:

latex varnishes: 21 ± 1.5 (°C) ; 50 ± 1.5 (% RH)

polyester/melamine: 21 ± 1.5 (°C) ; 0 (% RH)

powder coating: 65 ± 0.5 (°C) ; 0 (% RH).

Results

The results obtained are presented in figures 3 to 13. They show how physical aging is affected by the type of coalescent, the cross-link density and the presence of a pigment.

Coalescent

Figures 3 and 4 show that the stress dependence on temperature for an (acrylic) latex varnish is influenced by the coalescent used.

During the period investigated :

1. L1 is only slightly physically aging (Figure 3) ;
2. L2 starts to physically age as soon as the film is formed followed by a slow down of the aging process (see the decrease of the maximum trough height with time in Figure 4) ;

3. for L2, the trough displacement to higher temperatures is important (Figure 4)

We relate the above observations to the characteristics of the coalescent used (evaporation rate, partition coefficient and plasticizing effectiveness). Indeed, while BG is preferentially located at the surface of the particles, TPM is probably uniformly distributed in the particles (16-18). Moreover, BG evaporates much more rapidly from the film than TPM. With respect to butylacetate (100), the evaporation rate of BG and TPM are 6 and ≪ 1, respectively (19).

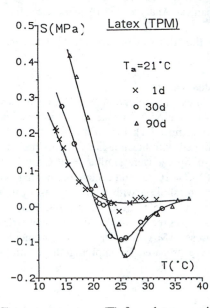

Figure 3. Stress (S) vs. temperature (T) for a latex varnish containing TPM physically aged at 21°C and 50% RH for different periods of time (d, days).

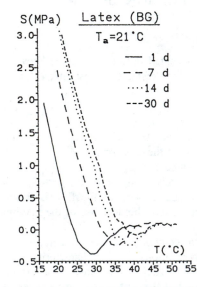

Figure 4. Latex varnish containing BG (L2)(idem Fig. 3).

The insignificant physical aging of L1 (Point 1) can be explained as resulting from the presence of a relatively large amount of TPM present in the film which, combined with its plasticizing effect and its low evaporation rate, induces a T_g close to the aging temperature ($T_a = 21°C$) during a long period of time. It must be added that the real T_g-values at 50% RH are, most likely, even lower than those shown in Table I. As previously discussed, when $T_a \geq T_g$ physical aging is non existent or negligible.

The decrease of the maximum trough height with physical aging (Figure 4)(Point 2), a fact contrary to expectation, can be explained by assuming that the decrease of free volume occurring during physical aging is partially compensated by the creation of free spaces resulting from the coalescent evaporation. Once the film is free of coalescent the maximum trough height should increase with physical aging, a fact confirmed in Figure 5 and previously discussed in references (14).

The significant trough displacement to higher temperatures for L2 (Figure 4) is due to the large T_g increase during aging (see Table I), a direct consequence of the relatively fast evaporation of BG from the film.

Table I : Evolution of the level of coalescents TPM and BG in the film (% by weight of varnish solids) and T_g-TMA (°C) with time (d, days) for L1 and L2 physically aged at 21°C and 50% RH

		wet varnish	*1d*	*7d*	*30d*	*90d*	*without coalescent*
L1	TPM	15	12	11.9	11,5	11.2	≈ 0
	Tg-TMA	-	≈ 24	25	≈ 28	≈ 29	≈ 57
L2	BG	40	5.5	3.8	2.4	1.7	≈ 0
	Tg-TMA		≈ 36.5	≈ 43	≈ 51.5	≈ 53.5	≈ 57

Cross-link density

This section tackles the question whether or not and how the cross-link density, an essential film property of thermosetting coatings, affects the process of physical aging.

With this aim in view, we used an experimental design procedure (20) to determine the cure conditions (bake time and bake temperature) for two polyester/melamine coatings (A and B), containing the same melamine, which produce T_g- midpoint values as similar as possible. Since the kinetics of physical aging is determined by the difference T_g-T_a, the similarity in T_g-values of the two coatings enabled us to carry out the aging experiments at the same temperature ($T_a = 21°C$).

The results of DSC and stress measurements are presented in figures 6 to 11. Figures 6 and 9 illustrate the dependence of heat flow and stress on temperature. They are typical of results obtained with physically aged coatings. Figures 10 and 11 show, respectively, the dependence of enthalpy relaxation and maximum trough height (maximum compressive stress,ΔS) as a function of aging time (t_a).

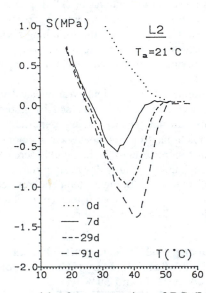

Figure 5. Latex varnish after evaporation of BG (L2) (idem Fig. 3)

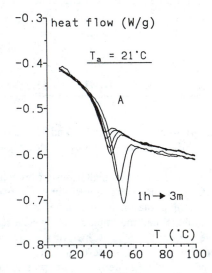

Figure 6. Heat flow vs. temperature (T) for coating A physically aged at 21°C for different periods of time (1 hour = the smallest peak; 3 month = the largest peak).

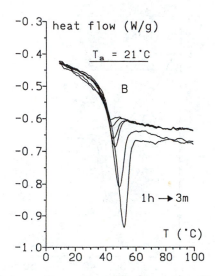

Figure 7. Heat flow vs. temperature (T) for coating B physically aged at 21°C for different periods of time (1 hour = the smallest peak; 3 month = the largest peak).

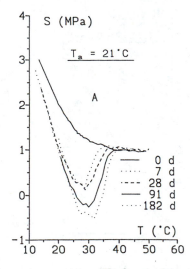

Figure 8. Stress (S) vs. temperature (T) for coating A physically aged at 21(°C) for different periods of time (d, days).

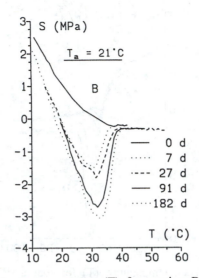

Figure 9. Stress (S) vs. temperature (T) for coating B physically aged at 21(°C) for different periods of time (d, days).

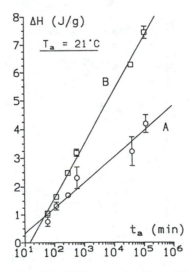

Figure 10. Enthalpy relaxation (ΔH) vs. aging time (t_a) for coatings A and B physically aged at 21°C.

These figures show that for both properties investigated (enthalpy relaxation and stress) the coating B is physically aging more rapidly (about twice) than coating A. To understand this behavior, DMA measurements were also performed. They clearly show (see Figure 12) that coating A is more crosslinked than coating B as indicated by the difference in storage modulus (E') in the plateau region above and below T_g, and by the peak maximum of the loss tangent [tan δ(max)]. The larger the tan δ (max) and the difference in E' the smaller the degree of crosslinking (21-24).

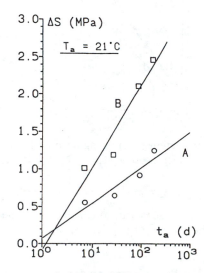

Figure 11. Maximum trough height (ΔS) vs. aging time (t_a)(d, days) for coatings A and B.

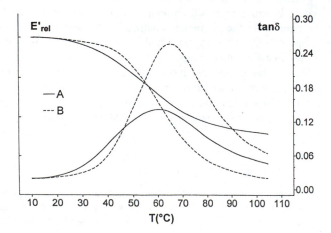

Figure 12. Relative elastic storage modulus (E'$_{rel}$) and tan δ vs. temperature (T) for coatings A and B physically non-aged.

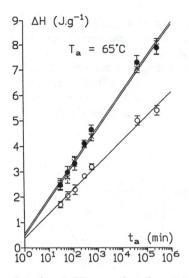

Figure 13. Enthalpy relaxation (ΔH) vs. aging time (t_a) for a non-pigmented (X) and pigmented powder coating ($\bigcirc$); ($\bullet$)-ΔH-values as calculated by amount of binder.

The analysis of all the results allows us to conclude that the cross-link density is a significant factor in the process of physical aging.

Presence of a pigment

In order to approach the reality of commercial paints, the effect of a pigment in a binder was also investigated. Figure 13 shows the enthalpy relaxation dependence on time (t_a) for a non-pigmented and a pigmented powder coating. It is interesting to note that the enthalpy increases approximately linearly with log t_a and that the pigmented coating is aging more slowly than the non-pigmented one. However, if the enthalpy relaxation for the pigmented coating is calculated by weight of binder, the resulting slope is, within the experimental error, the same as that of the binder. This fact indicates that the "slower aging rate" of the pigmented coating is a result of the binder dilution due to pigment incorporation. We do not exclude the possibility that, at higher pigment concentrations or for other systems a different behavior could be observed. This might be the case if the mobility of polymer segments participating in the physical aging process is affected by the pigment/binder interaction.

Conclusion

For high T_g-coatings (i.e. for any coating having its T_g higher than the environmental temperature) physical aging studies help in the understanding of the causes and processes of film formation. This study showed that physical aging is affected by the type of coalescent agent. Under certain circumstances the effect of free volume decrease occurring during physical aging can partially or totally be eliminated by the creation of free spaces resulting from the coalescent evaporation. Also, the study showed that cross-link density might be a determining factor in the physical aging process.The higher the cross-link density the lower the rate of physical aging. The effect of the addition of a pigment on the aging process, is most likely due to binder dilution.

Literature cited

1. Kovacs, A.J. *J. Fortsch. Hochpolym. Forsch.* **1963**, *3*, 394 .
2. Struik, L.C.E. *Physical Aging in Amorphous Polymers and Other Materials;* Elsevier: Amsterdam, Holland, 1978.
3. Goldstein, M. (p. 13); Johari, G.P. (p. 17); Hodge, J. H. (p. 65); Rendell, R.W. and Ngai, K.L. (p. 309) In *Relaxations in Complex Systems*; Ngai, K.L. and Wright, G.B., Eds; Naval Research Laboratory: Washington, DC, 1984.
4. Bauwens-Crowet; Bauwens, J-C. *Polymer*, **1986**, *27*, 709; **1987**, *28*, 1863 ; **1990**, *31*, 646.
5. ten Brinke, G.; Grooten, R. *Colloïd & Polymer Sci.* **1989**, *267*, 992.
6. Mangion, M.B.M.; Johari, G.P. *J. Polymer Sci. Part B. Polym. Physics* **1990**, *28*, 71.
7. Lee, A.; Mc Kenna, G.B. *Polymer* **1990**, *31*, 423.
8. Mijovic, J.; Devine, S.T.; Ho, T. *J. Appl. Polym.Sci.* **1990,** *39*, 1133.
9. Celli, A.; Scandola, M. *Polymer*, **1992**, *33*, 2699.
10. Montserrat, S. *Progr. Colloïd Polym.Sci.*, **1992,** *87*, 78-82.
11. Greidanus, P.J. *Proc. 19th FATIPEC Congress, Aachen*, Germany, **1988**, *1*, 485.
12. Scherer, G.W. *Relaxation in glass and composites*; Krieger Publishing Co.: Malabar, Fl, 1992.
13. McKenna, G.B. In *Comprehensive Polymer Sci., vol. 2, Polymer Properties*; Booth, C.; Price, C., Eds.; Pergamon: Oxford, 1990, 311.
14. Perera, D.Y.; Schutyser, P. *ACS (PMSE)* **1992**, *67*, 222; *Prog. Org. Coat.* **1994**, *24*, 299.
15. Perera, D.Y.; Vanden Eynde, D. *J. Coat. Technol.* **1987**, *59, n° 748*, 55.
16. Hoy, K.L. *J. Coat. Technol.* **1973**, *45, n° 579*, 51.
17. Winnik, M.A.; Wang, Y.; Haley, F. *J. Coat. Technol.* **1992**, *64, n° 811*.
18. Juhué, D.; Lang, J. *Double-Liaison* **1994**, *n° 464-465*, III.
19. *Industrial Solvents Handbook*; Flick, E.W., Ed.;nds, NJ,U.S.A., 1985.
20. CoRI, *Annual Research Report*, 1985.
21. Grentzer, T.H.; Holsworth, R.M.; Provder, T. *ACS (Org. Coat. Plastics)* **1981**, *44*, 515.
22. Skrovanek, D.J.; Schoff, C.K. *Prog. Org. Coat.* **1988**, *16*, 135.
23. Skrovanek, D.J. *Prog. Org. Coat.* **1990**, *18*, 89.
24. Hill, L.W. *J.Coat. Technol.* **1992**, *64, n° 808*, 29.

Chapter 14

Correlation Between Network Mechanical Properties and Physical Properties in Polyester–Urethane Coatings

James C. Scanlan, Dean C. Webster, and Allen L. Crain

Research Laboratories, Eastman Chemical Company, P.O. Box 1972, Kingsport, TN 37662–5150

A series of polyester polyols were prepared and evaluated in polyurethane coatings in order to study composition-property as well as structure-mechanical property relations. The polyesters were prepared using three diacids, namely isophthalic acid, 1,4-cyclohexanedicarboxylic acid, and adipic acid. The effects of polyester functionality and molecular weight also were evaluated. The polyesters were formulated into clear coatings and cured with a polyfunctional isocyanate. Dynamic mechanical thermal analysis was used to characterize the network structure. The crosslink density (XLD), calculated from the measured rubber modulus, compares favorably to the value predicted by Miller-Macosko theory. The glass transition temperature (T_g) is modeled in terms of composition and crosslink density to ± 5 °C. Hardness, as reflected by the room temperature modulus, is a function of composition and T_g. Impact strength and flexibility are functions of both T_g and XLD. The combination of hardness and flexibility can be optimized by combining low XLD with a high-T_g-contributing monomer.

Two-component (2K) polyurethane coatings are used in applications such as automotive refinish, aircraft coatings, industrial maintenance, and product finishes. Urethanes generally possess good flexibility, adhesion, and durability. While isocyanates will react with a wide variety of active hydrogen-containing materials, the most common co-reactants used in polyurethane coatings contain hydroxyl groups. The curing chemistry typically involves the reaction of a polyfunctional isocyanate with a polyfunctional polyol (1). Acrylic polyols are used when exterior durability is desired, while polyester polyols are used when higher solids and solvent resistance are needed (2).

The polyol typically contributes 40-70 % by weight to the binder, and hence requires major consideration during the polyurethane coating formulation. A number of approaches are available for resin design and formulation. Traditional resin formulation often begins with a "starting point" formulation that has a set of known properties. Modifications are made to the molecular weight, hydroxyl functionality, or monomer composition to achieve the balance of properties required for the intended

0097–6156/96/0648–0222$15.00/0

application. Here the formulator relies on experience to predict how changes in composition will affect performance, since quantitative relationships have not been established between polyester structure and coatings properties. A number of qualitative relationships are available to guide the formulator, however. For example, various diacid intermediates are classified as "hard" or "soft" depending on their effect on coating hardness. A drawback of this approach is that a large number of iterations in the cycle of resin synthesis and coating evaluation may be required to optimize composition. Interactions between the variables may be difficult to discern as well.

Another approach that has achieved some popularity is to employ statistical experimental design. Here, a set of variables and their ranges are chosen for study that are expected to have an effect on properties. Design tools are used to select a subset of combinations of parameters which span the design space. A polynomial model is used to describe the results and can be used to optimize the composition to achieve the desired combination of properties.

Statistical experimental design can be a very powerful technique since a large number of variables can be evaluated with a minimal number of experiments. Interactions between the variables can be identified. However, the results are usually expressed as composition-property relationships. The coatings formulation process can be aided further by generalizing the results in terms of composition-structure and structure-property relationships. Once structures that produce desirable properties are identified, new compositions may be suggested.

A key technique for structure determination of coatings networks is dynamic mechanical thermal analysis or DMTA (*3*). The elastic modulus, glass transition temperature, and crosslink density are determined quickly and accurately. Hill and Kozlowski (*4*) used DMTA to indirectly measure the crosslink density (XLD) of acrylic and polyester polyols crosslinked with melamine-formaldehyde (MF) resins. The measured XLD compared favorably with the theoretically predicted value. Details of the crosslinking reactions, such as self-condensation of the MF resin, were also discerned. Bauer and Dickie (*5*) related coatings properties to the theoretical XLD from the experimentally-determined extent-of-reaction between acrylic polyols and MF resins.

We seek to understand how polyester formulation variables, such as monomer composition, molecular weight, and hydroxyl functionality, affect network properties and the resulting coating properties. We find that DMTA characterization of the network provides an important link between polyester formulation variables and the physical properties of the coatings.

Experimental

Experimental Design and Analysis. A first designed experiment was constructed with 5 variables, namely, average hydroxyl functionality (nominal), f_{OH}, which spanned the range 2.5-3.5, number average molecular weight (nominal), M_n, which spanned the range 500-1250, and molar concentration of three acid-functional monomers, adipic acid, or ADA, isophthalic acid, or IPA, and 1,4-cyclohexanedicarboxylic acid, or CHDA. The polyols NPG® glycol (diol) and trimethylolpropane (triol) were used to adjust the functionality. Polyester functionality and molecular weight were not measured, but were assumed to be equal to the value expected after the resin synthesis. Isocyanate-to-hydroxyl stoichiometry, $r = [NCO]/[OH]$, was fixed at a value of 1.1. A D-Optimal designed experiment with these five variables required the synthesis of 22 polyester resins. The data were fit by least-squares regression to Scheffè polynomials (*6*) for purposes of visualizing the fractional-mixture-design results. Effects that were resolvable included linear and non-linear blending (of the polyester acids), effects of M_n and f_{OH} on linear blending, and the interaction between M_n and f_{OH} on linear blending. If any one of the three terms (for each diacid) of any of the five effects was statistically significant at the 95% confidence level, then all three of the terms were included in the model because of collinearity between the mixture components.

In a second designed experiment, polyester-acid type and stoichiometry were studied. Three polyester resins, each containing a single acid of either ADA, IPA or CHDA, were prepared to the same nominal molecular weight and hydroxyl functionality (1250 and 3.5, respectively). Each resin was then formulated at isocyanate-to-hydroxyl stoichiometry, of 0.50, 0.67, or 1.1.

Polyester Resin Synthesis. The polyester resins were synthesized in a one-liter two-piece glass reactor equipped with an overhead mechanical stirrer, thermocouple, inert gas inlet, and a steam-jacketed partial condenser. Raw materials were charged into the reactor and the temperature raised to form a homogeneous melt. The temperature was then raised until water evolution began, then raised ten degrees every 30 minutes until a maximum temperature of 230°C was reached. The resin was held at 230° until the acid value reached 10+/- 2. The resin was cooled and poured into a clean unlined metal can.

Urethane Coatings. The clear coatings formulations were prepared at 65% solids in a 1:1 solvent blend of MAK:xylene. Cure was catalyzed by dibutyltin dilaurate (0.005 wt. %). BYK-300 (0.01 wt %) was used for flow and leveling control. Coatings were crosslinked with Desmodur-N 3390 (Bayer Corp., Pittsburgh, PA), which is the triisocyanurate of 1,6-hexanediisocyanate with higher oligomers. It has a nominal average functionality of 3 and molecular weight of 585 g mol^{-1}. Clear coatings were applied by using a wire-wrapped draw-down bar to either iron phosphate pretreated (Bonderite 1000) panels or to panels coated with an automotive primer-surfacer. Coatings were force-dried at 80 °C for 45 minutes, and then aged at ambient conditions for one week before evaluation. Clear coatings for DMTA were formed on glass microscope slides and were readily peeled off.

Physical Properties. The glass transition temperature was determined by DSC using a DuPont 2100 at a heating rate of 20 °C/min. T_g was taken at the midpoint of the inflection during the second scan. The storage and loss moduli, E' and E", and the loss tangent, tanδ, were determined by DMTA on a Rheometrics RSAII. Films were evaluated in tensile deformation at 16 Hz and a heating rate of 20 °C/min. DMTA transition temperatures, the sub-T_g β-transition, T_β, and the glass transition, T_g, were identified from the local maxima of tanδ. The T_g by DMTA is approximately 14 °C higher than by DSC. The crosslink density, XLD, was calculated from the modulus above T_g (7) as E'/3RT (in units of mol cm^{-3}), where R is the ideal gas constant. DMTA replications were performed on a few samples to determine precision. The standard errors were found to be 5 °C for T_β, 1.5 °C for T_g, 10% for the modulus below T_g, and 5% for the modulus above T_g. The test error is larger below T_g because of difficulty in loading a specimen under uniform tension.

Coatings physical property testing was done using ASTM methods. Hardness was obtained by using a Koenig pendulum hardness (KPH) tester (ASTM D4366). Impact resistance was determined by using a Gardner Heavy Duty Variable Impact Tester (ASTM D2794). Reverse impact test results are reported. The conical mandrel bend test was used as a measure of coating flexibility (ASTM D522).

Results and Discussion

DMTA. DMTA, DSC, and physical properties are tabulated in Tables I and II. Representative DMTA spectra are presented in Figure 1 (low temperature tanδ data were smoothed for clarity). Sub-T_g β-transitions are evident in the tanδ spectra. The

Table I. Summary data for the first designed experiment. The stoichiometry is r = 1.1.

ADA	CHDA	IPA	fOH	M_n	Tg by DSC °C	XLD by DMTA $\times10^3$ (mol-cm³)	log E' (Pa) @ 22 °C	tanδ @ 22 °C	$\nu_e \times10^3$ (mol-cm³)	KPH	Reverse Impact (in-lb)
1	0	0	2.5	1250	-8.5	0.62	6.72	0.182	1.29	27	160
1	0	0	2.5	500	29.5	1.72	9.00	0.160	2.23	64	10
1/3	1/3	1/3	3	875	49.6	1.33	9.19	0.097	1.95	111	0
0	0	1	3	500	73.2	1.67	9.35	0.026	2.58	100	0
0	1	0	3.5	1250	59.1	1.15	9.26	0.026	1.73	88	10
0	0	1	3.5	1250	58.5	1.06	9.36	0.030	1.73	61	30
1/3	1/3	1/3	3	875	55.1	1.24	9.28	0.023	1.95	99	0
1/2	0	1/2	3.5	875	58.7	1.56	9.30	0.016	2.14	110	0
1	0	0	3.5	1250	5.8	1.63	7.76	1.016	1.73	17	80
0	1/2	1/2	3	1250	60.5	1.01	9.30	0.025	1.56	88	20
1	0	0	3.5	500	47.3	2.56	9.23	0.035	2.78	134	0
0	0	1	3.5	500	85	1.54	9.35	0.020	2.78	126	0
1/3	1/3	1/3	3	875	54.2	1.58	9.26	0.025	1.95	95	10
0	1	0	2.5	1250	44.9	0.91	9.11	0.054	1.29	81	40
0	1/2	1/2	3.5	875	73.2	1.45	9.32	0.027	2.14	95	0
0	1/2	1/2	2.5	500	69.1	1.29	9.31	0.026	2.23	107	0
1/2	0	1/2	2.5	1250	33.7	0.85	9.18	0.112	1.29	55	160
0	1	0	2.5	500	63.5	1.32	9.30	0.020	2.23	117	0
1/2	1/2	0	3.5	1250	36.8	1.45	9.16	0.052	1.73	70	20
1/2	1/2	0	2.5	875	29.3	1.06	9.05	0.187	1.64	46	110
0	1	0	3.5	500	75.8	1.40	9.19	0.059	2.78	103	0

Table II. Summary data for the second designed experiment. Fixed were the polyester f = 3.5 and M_n = 1250.

ADA	CHDA	IPA	r	Tg by DSC °C	XLD by DMTA $\times 10^3$ (mol-cm³)	log E' (Pa) @ 22 °C	tanδ @ 22 °C	$\nu_e \times 10^3$ (mol-cm³)	KPH	Reverse Impact (in-lb)
0	1	0	0.5	30.8	0.31	9.12	0.033	0.62	49	160
0	1	0	1.1	42.1	1.10	9.20	0.024	1.73	155	0
0	1	0	0.66	42	0.51	9.10	0.031	1.05	114	130
0	0	1	0.5	35.2	0.38	9.36	0.043	0.62	53	140
0	0	1	1.1	58.5	1.20	9.35	0.041	1.73	111	0
0	0	1	0.66	36.5	0.57	9.28	0.034	1.05	65	80
1	0	0	0.5	-13.7	0.50	6.61	0.203	0.62	31	140
1	0	0	1.1	10.1	1.50	7.63	0.993	1.73	24	120
1	0	0	0.66	-6.2	1.00	6.97	0.347	1.05	43	130

tanδ local maximum, tanδ$_\beta$, occurs at -48 °C for CHDA-based coatings, and at -77 °C for ADA-based coatings. IPA-based coatings may have a β-transition outside of the experimental window, that is, below -100 °C. For the representative samples depicted in Figure 1, the modulus decreases through the β-transition (-100 °C to -30 °C) from 2.8 to 2.2 GPa for IPA- and from 3.2 to 2.0 GPa for the CHDA- and ADA-based coatings. IPA has a slightly higher room temperature modulus than CHDA (2 vs. 1.5 GPa). T$_\beta$ and tanδ$_\beta$ for the blends-based coatings are described by slightly non-linear blending rules.

The coating T$_g$ was restricted to the ceiling value of 80 °C imposed by the curing conditions. It is known that the curing temperature limits the coating T$_g$ in practice (8). Post-cure during DMTA is evidenced by an E'/3RT function that is an increasing function of temperature above T$_g$. For samples demonstrating incomplete cure, the minimum in the E'/3RT function above T$_g$ was used to calculate the XLD. Some samples with T$_g$'s well below the ceiling temperature demonstrated post-cure, which indicates that the reaction conditions did not always bring about complete reaction. Crosslink densities of some samples increased by as much as 10% during the DMTA scan.

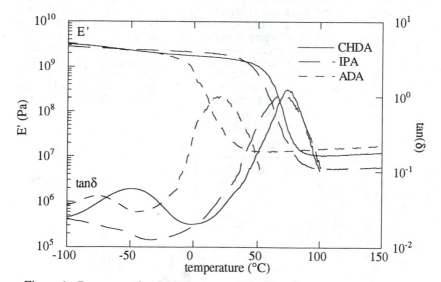

Figure 1. Representative DMTA spectra, E' and tanδ, of polyester-urethane coatings made from the three polyester-acids, as indicated by the legend.

Relationship Between Composition and Structure. Various theories have been proposed to predict thermoset polymer structure, that is, the crosslink density, concentration and mass distribution of sol phase, etc., from a knowledge of the structure of the precursor materials. The theory of Miller and Macosko (9) has achieved some popularity (10). Rubber elasticity theory provides a relation between the rubber modulus and the crosslink density (7). Literature values of the ratio of measured E'/3RT (above T$_g$) to the theoretically-calculated elastically-effective-crosslink-density, v_e, obtained from the Miller-Macosko theory, fall between 0.5 and 1.3 when the equivalent molecular weight of the precursor is less than 10,000 g mol^{-1} (11). Here, we calculated v_e from the code published by Bauer (12). The theoretical XLD calculation was made assuming complete reaction of the polyester hydroxyl groups and no self-condensation of the isocyanate. With our data, the ratio of experimental to theoretical crosslink density falls between 0.5 and 0.95. The correlation of the two measures is shown in Figure 2. We consider this acceptable agreement between experiment and theory considering the uncertainties in (1) the rubber elasticity and Miller-Macosko theories, (2) the precursor functionality and molecular weight, (3) the extent of reaction, and (4) the possibility of self-condensation of the isocyanate. CHDA and IPA resins generally have lower experimental XLD values than ADA resins at equivalent theoretical v_e, which may be due to our uncertainty of precursor molecular weight, functionality, and extent of reaction.

The Miller-Macosko theory predicts that the crosslink density depends on both M_n and f_{OH} but not just the ratio, or equivalent molecular weight, $M_e = M_{nOH}/f_{OH}$. The experimental crosslink density, $E'/3RT$, was fit with r, M_{nOH}, and f_{OH}, with an R^2 of 0.771 ($R^2_{max} = 0.777$, calculated as 1 minus the ratio of the sum of the squares of pure error to the total sum of the squares of the model) and by r and M_e with an R^2 of 0.746 ($R^2_{max} = 0.777$). We infer that in the range covered here, M_n and f_{OH} may be grouped into the equivalent molecular weight without significant loss of fit. The fit is improved by expanding the number of model terms, which only serves to describe experimental biases due that result from uncertainties of precursor molecular weight and functionality. We conclude that the XLD from DMTA is an easier and more precise measure of network structure than that provided by calculation of theoretical XLD obtained after determination of extent of reaction and precursor composition.

The T_g of thermosets is known to be a strong function of XLD. Several theoretical predictions have been made (13), and there have been several recent experimental attempts at establishing the relationship (14-16). T_g and XLD exhibit a linear relationship with our data in the range of $3.1 \times 10^{-4} \leq \text{XLD} \leq 2.6 \times 10^{-3}$ mol cm^{-3}, so we chose to fit out data to

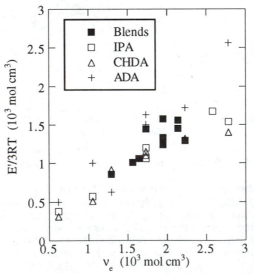

Figure 2. Correlation between the measured XLD, $E'/3RT$, and the theoretical elastically-effective XLD, v_e, calculated from the Miller-Macosko theory. The "blends" represent coatings made from polyols containing two or three of the diacids.

$$\frac{T_g - T_{g0}}{T_{g0}} = K \frac{E'}{3RT},$$

where E'/3RT is the XLD obtained by DMTA, K is a constant, and T_{g0} is the glass transition temperature of the uncrosslinked polymer having the same copolymer composition as the crosslinked polymer. We evaluated the copolymer effect by incorporating a non-linear blending rule for T_{g0}

$$T_{g0} = \sum_{i=1}^{4} \gamma_i x_i + \sum_{i < j}^{3} \gamma_{ij} x_i x_j,$$

where the x_i are mass fractions of the components, and the γ_i represent the linear blending effects of the polyester and isocyanate while the γ_{ij} represent the non-linear blending of the polyester acids (the experiment was not designed to determine the effects of non-linear blending of the isocyanate). Fitting our data, we find K to be 12 cm³ mol⁻¹, and the γ_i to be -55 °C for ADA, 12 °C for CHDA, 18 °C for IPA, and 100 °C for the isocyanate. These values compare reasonably to resin T_{g0}'s by DSC of -43 °C, 8 °C, and 46 °C for polyester resins of ADA, CHDA, and IPA respectively. However, the γ_i for the isocyanate is much higher than the experimental T_{g0} of a n-butanol-Desmodur N3300 adduct of -47 °C, which reflects a significant difference between monomeric and polymeric urethane. The root-mean-square of the residuals of the model fit is 5 °C, and the residuals are independent of the prediction values. The non-linear blending terms are small but statistically significant. We consider this to be a good fit of the data, since the experimental T_g's range over 100 °C. Without the non-linear blending terms, which would not be known a priori, the root-mean-square of the residuals is 9 °C. Hence, we conclude that the precursor T_{g0} provides a means for predicting the coating T_g.

Relationship Between Structure and Mechanical Properties. The crosslink density is an increasing function of stoichiometry for r < 1, and a decreasing function for r > 1. We chose not to explore the region of stoichiometry [NCO]/[OH] > 1 because of the possibility of self-condensation of isocyanate. A consequence of this experimental plan is that the crosslink density is confounded with the mass fraction of polyester in the coating (the correlation between ν_e and mass fraction of polyester is R = 0.988). Thus, we cannot distinguish between effects of the mass fraction of polyester from effects of crosslink density, although our plan provides us with a greater range of both.

Hardness is closely related to the modulus. Scratch and indentation empirical coating tests can be large deformation tests and so are mixed measures of yield strength, tensile strength, and viscoelastic recovery (17). The Koenig pendulum hardness (KPH) has been found (18) to be a measure of damping rather than hardness. We confirm this finding. A scatterplot of KPH (on primed panels) and tanδ⁻¹ is shown in Figure 3. Since KPH is a measure of damping, we rely on our measurement of modulus for indications of hardness. Figure 4 shows the room temperature storage modulus plotted as a function of T_g (by DSC). E' @ 22 °C increases as T_g increases through room temperature because the glass transition is a broad transition that occurs over a range of approximately 50 °C for the materials studied here. It is apparent that IPA-based coatings have a slightly higher modulus than CHDA- or ADA-based coatings at equivalent T_g, which is likely to be a consequence of the differences in sub-T_g behavior.

Impact resistance was tested on coatings applied to both primed and non-primed panels. Reverse impact properties of the coatings applied to non-primed panels all

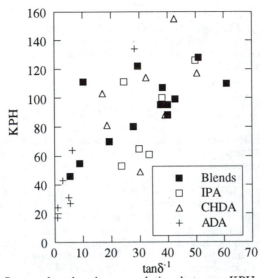

Figure 3. Scatterplot showing correlation between KPH and the room temperature tanδ.

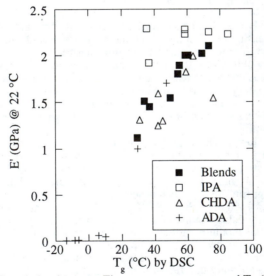

Figure 4. Correlation between E' at room temperature and T_g (by DSC).

exceeded the maximum of the impact tester (160 in-lb). Values for coatings on the primed panels ranged from less than the minimum (10 in-lb) to greater than the maximum. Failure on the primed panels occurred by delamination at the primer-panel interface, rather than at the primer-top coat interface. We fit the reverse impact data by multiple-linear-regression using T_g and XLD as factors, since the two factors are only mildly collinear. An R^2 of 0.77 and an root-mean-square-error (RMSE) of 31 in-lb results, and both effect's significances are shown as leverage plots *(19)* in Figure 5. (A leverage plot presents a graphical display of an effect's significance test, and is created such that the distance from the point to the sloped line is the residual and the distance from the point to the horizontal line of the mean shows what the residual error would be without the effect in the model.) Both factors are statistically significant at the 95% confidence level. The goodness-of-fit is improved by restraining the fitted values to the range of 0 to 160. The restrained-model fit provides $R^2 = 0.90$ and RMSE = 20 in-lb, and the residuals are independent of the fitted equation as well as the stoichiometry, M_n and f_{OH}.

The conical mandrel bend (elongation) results generally correlate with the reverse impact results (R = 0.67), however, several samples with good elongation had poor impact. Thus we have relied on reverse impact resistance as a more severe test of flexibility than the conical mandrel test.

The three variables, diacid composition, T_g, and XLD, are interrelated and only two may be specified. Thus, when examining compositional effects on properties, either T_g or XLD must be specified. The usual formulator's practice is to compare properties at constant molecular weight and functionality (i.e., constant XLD). This view of the modulus and impact resistance is presented in Figure 6a at iso-XLD of 1.22×10^{-3} mol cm^{-3} (the median value of our dataset). In the figure, the T_g (by DSC) varies from 5 °C for ADA to 63 °C for IPA. The alternative view is shown in Figure 6b at an iso-T_g of 40 °C. The XLD ranges in Figure 6b from 0.4×10^{-3} mol cm^{-3} for IPA-based coatings to 2.1×10^{-3} mol cm^{-3} for ADA-based coatings. While the iso-XLD view may be simpler to prepare from an experimental standpoint, the iso-T_g view may have more significance to the formulator since other coating properties such as weatherability and chemical resistance are depressed if T_g approaches or becomes less than the use temperature *(20)*. In the traditional (iso-XLD) view, CHDA and IPA considered to be hard but weak, while in the iso-T_g view all three acids are considered as hard, but ADA is considerably less tough than CHDA or IPA. Some sub-T_g transitions are thought to improve impact strength *(3, 21)*. Here, the impact strength of these coatings are largely described by effects of T_g and XLD, that is, effects of sub-T_g behavior are too small to distinguish these monomers.

Conclusion

We have studied the relationship between polyester precursor composition and polyurethane network structure and the relationship between network structure and mechanical properties. The crosslink density determined by DMTA is in general agreement with the crosslink density predicted by Miller-Macosko theory. We can predict the T_g of the coating from a knowledge of the precursor composition and the crosslink density of the coating. Coating hardness depends on distance from T_g (that is, T_g-T) and the monomer composition of the polyester. The impact resistance is a function of both the coating T_g and the crosslink density. An especially interesting conclusion is that an optimum balance of hardness and flexibility might be achieved by combining a low crosslink density with a high T_g-precursor composition.

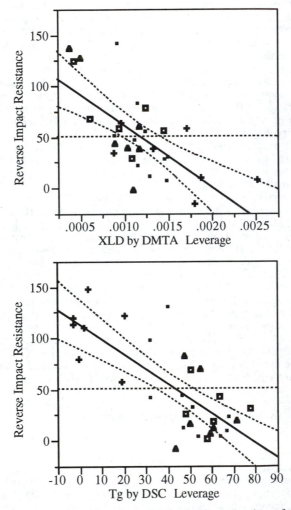

Figure 5. Leverage plots for the multiple-linear regression of the reverse impact resistance of primed coatings with regressors T_g (by DSC) and XLD (by DMTA). The solid line represents the regression, the dashed hyperbolic lines represent the 95% confidence intervals of the slope of the regression, and the dashed horizontal line represents the mean of the dataset.

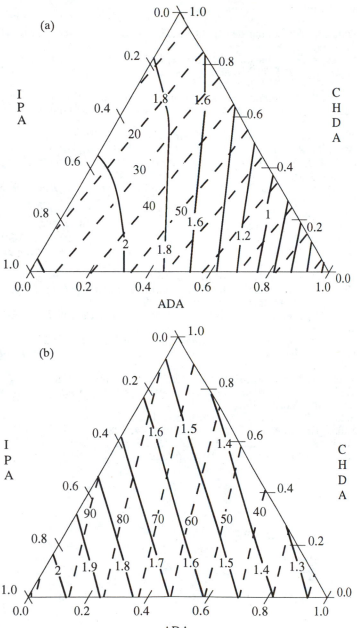

Figures 6a-b. Contour plots depicting the composition dependence of the fitted E' GPa at 22 °C (solid lines) and reverse impact strength (dotted lines) at **(a)** (top) constant XLD of 0.00122 mol cm^{-3} and at **(b)** (bottom) constant T_g of 40 °C.

Literature Cited

1. Potter, T. A.; Schmelzer, H. G.; Baker, R. D. *Prog. Org. Coat.* **1984**, *12*, 321-339.
2. Wicks, Z. W.; Jones, F. N.; Pappas, S. P. *Organic Coatings: Science and Technology. 1. Film Formation, Components, and Appearance*; John Wiley & Sons: New York, 1992.
3. Hill, L. W. *Mechanical properties of coatings;* Federation Series on Coatings Technology; Federation of Societies for Coatings Technology: Philadelphia, 1987.
4. Hill, L. W.; Kozlowski, K. *J. Coat. Tech.* **1987**, *59*, 63-71.
5. Bauer, D. R.; Dickie, R. A. *J. Coat. Tech.* **1982**, *54*, 57-63.
6. Cornell, J. A. *Experiments with mixtures: designs, model and analysis of mixture data*; John Wiley & Sons: New York, 1990.
7. Heinrich, G.; Straube, E.; Helmis, G. In *Advances in Polymer Science*; Springer-Verlag: New York, 1988; Vol. 85; pp 33-88.
8. Gillham, J. K.; Enns, J. B. *Trends Polym. Sci.* **1994**, *2*, 406.
9. Miller, D. R.; Macosko, C. W. *Macromolecules* **1976**, *9*, 206.
10. Eichinger, B. E.; Akgiray, O. *Comput. Simul. Polym.* **1994**, 263-302.
11. Queslel, J. P.; Mark, J. E. In *Advances in Polymer Science*; Springer-Verlag: New York, 1984; Vol. 65; pp 135-176.
12. Bauer, D. *J. Coat. Tech.* **1988**, *60*, 53-66.
13. Nielsen, L. E.; Landel, R. F. *Mechanical properties of polymers and composites*; Marcel Dekker: New York, 1994.
14. Stutz, H.; Illers, K. H.; Mertes, J. *J. Polym. Sci. Polym. Phys. Ed.* **1990**, *28*, 1483-1498.
15. Shefer, A.; Gottlieb, M. *Macromolecules* **1992**, *25*, 4036-42.
16. Hale, A.; Macosko, C. W.; Bair, H. E. *Macromolecules* **1991**, *24*, 2610-2621.
17. Wicks, Z. W.; Jones, F. N.; Pappas, S. P. *Organic Coatings: Science and Technology. II. Applications, Properties, and Performance*; John Wiley & Sons: New York, 1992.
18. Sato, K. *Prog. Org. Coat.* **1980**, *8*, 1-18.
19. Sall, J. *The American Statistician* **1990**, *44*, 308.
20. Wicks, Z. W. *J. Coat. Tech.* **1986**, *58*, 22-34.
21. Hill, L. W. *J. Coat. Tech.* **1992**, *64*, 29.

Chapter 15

Effect of Amine Solubilizer Structure on Cured Film Properties of Water-Reducible Thermoset Systems

L. W. Hill, P. E. Ferrell, and J. J. Gummeson

Specialty Resins Division, Monsanto Company, 730 Worcester Street, Springfield, MA 01151

A primary amino-alcohol (2-amino-2-methyl-1-propanol, AMP) and a tertiary amino-alcohol (N,N-dimethylaminoethanol, DMAE) differ greatly in their effect on extent of cure of thermoset films prepared from water-reducible, hydroxy and carboxy functional polyester (PE) polymers crosslinked with etherified melamine formaldehyde (MF) resins. Use of AMP as the neutralizer for -COOH groups on PE causes little or no inhibition of the acid-catalyzed curing reactions whereas use of DMAE causes strong inhibition. Results indicate that the primary amine group of AMP reacts with formaldehyde, evolved from MF during cure, to produce products of reduced base strength. Glass transition temperatures and crosslink densities of cured films were determined by dynamic mechanical analysis (DMA). Functional group conversion during cure was estimated by multiple internal reflectance, Fourier transform IR (MIR-FTIR).

The term "water-reducible" (WR) is used to describe hydrophobic coatings resins that have been modified to contain solubilizing groups (*1,2*). These resins are often synthesized in a solvent that eventually becomes the cosolvent in a waterborne formulation. Polyester polyols and acrylic polyols are widely used examples. These polyols are synthesized in alcohol-ether solvents and are modified to include carboxylic acid groups (carboxyl equivalent weight of approximately 1000 g/eq.; acid number ≈ 56 mg KOH/g) (*1*). Since -OH and -COOH groups, at the levels used, are not sufficiently hydrophilic to carry these resins into water, amino alcohols are added to neutralize a portion of the carboxylic acid groups. DMAE (N,N-dimethyl aminoethanol) is used frequently and AMP (2-amino-2-methyl-1-propanol) is used occasionally. The amine salts improve solubility or dispersibility in water. WR binders usually do not contain any low molecular weight surfactants (*2*), or stated in another way, part of the binder is converted to a crosslinkable surfactant. Many of the problems of other waterborne systems are avoided by eliminating low molecular weight surfactants. In many cases, WR resins are used in thermoset coatings.

0097–6156/96/0648–0235$15.00/0

Molecular weight build-up and crosslinking are achieved during bake by inclusion of etherified melamine formaldehyde resins (MF), which react with -OH groups and possibly with some of the -COOH groups of the polyol (*1*). These reactions between polyol and MF are called "co-condensation". In addition, the various groups present on MF resins can react with each other in a process called "self-condensation". For solvent-borne formulations, the usual bake conditions result in complete conversion of -OH groups on the polyol whereas the extent of self-condensation is moderate and variable (*3*). Since both co- and self-condensation are acid catalyzed, retention of amines during bake of WR films is expected to inhibit the curing reactions (*4*). Differences in amine content at the surface and in the bulk are reported to cause wrinkling due to differences in extent of cure (*4,5*). In MF-free acrylic polyol solutions with AMP as neutralizer, amidation and esterification reactions occur (*5*). In this work, we focus on reactions of amines with MF or with cure volatiles produced from MF. The effects of amine type (primary vs. tertiary) and MF type (high methylol vs. low methylol) on extent of cure are considered.

Experimental

A commercial water-reducible polyester (PE) polyol resin (72-7289, McWhorter) with approximate -COOH eq. wt. of 1020 g/eq and approximate -OH eq. wt. of 342 g/eq was formulated at 75/25 by wt. with MF1 or MF2. MF1 is a hexamethoxymethylmelamine (HMMM) type (Resimene 745, Monsanto). MF2 is a high methylol type (Resimene 735, Monsanto). The structure of HMMM, and an example of a structure likely to be present in MF2 is shown below.

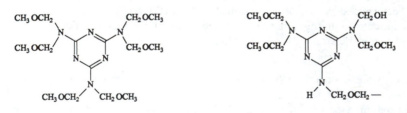

Hexamethoxymethylmelamine   A Structure Present in MF2

For WR films, AMP or DMAE was added at a calculated extent neutralization of 80%. The cosolvent was propyleneglycol-monopropylether (PP), and water was added to reach application viscosity. For solvent borne films, no neutralizer was used, and PP was added to reach application viscosity Cure catalyst, p-toluenesulfonic acid (pTSA), was used at 0.5 phr on resin solids. A flow and leveling agent designed for waterborne systems (Modaflow AQ 3000, Monsanto) was used at 1.0 phr on resin solids. Films were prepared by drawdown on tinplated steel, 3" x 6" panels (Q-Panel Co.) and baked 30 minutes at 130°C. Bars with various gaps, 40 to 80 μm, were used to obtain films having a range of dry film thickness (DFT). Cured films were removed from the substrate by the amalgamation method (*6*), and thickness was determined by micrometer.

Dynamic mechanical analysis (DMA) was used to determine extent of cure. DMA was carried out on free films at 11 Hz oscillating frequency with a temperature scan from 10° to 190°C at 2°/min. on an Autovibron instrument (Imass, Inc.). This DMA method and interpretation of the resulting DMA plots have been described in detail (6). Multiple internal reflectance, Fourier transform infra-red spectroscopy (MIR-FTIR) was carried out on the top and bottom surfaces of free films using a model IR/44 FTIR (Nicolet) with an out-of-compartment contact sampler having a ZnSe 45° horizontal, flat crystal (Spectra-Tech., Inc.).

Results and Discussion

Amine Properties. Properties of DMAE and AMP are given in Table I. These amines were selected for study because they are similar in many respects but differ in terms of primary vs. tertiary structure. AMP is a slightly stronger base than DMAE as indicated by pK_a values. AMP is slightly less volatile than DMAE as indicted by boiling points. Based on both base strength and volatility, we expect to lose less AMP than DMAE during cure. The experimental amine loss result is consistent with this expectation, but the authors (5) conclude that amine loss occurs by reaction with the polyol as well as by volatilization.

Table I. Properties of DMAE and AMP

	M. Wt.	Type	B. Pt.	pKa	Amine Loss[1]
DMAE	89.1	Tertiary	134°	9.3	70 %
AMP	89.1	Primary	165°	9.9	58 %

[1]From a water/PP solution of an acrylic copolymer containing -COOH groups (500 eq. wt.). Amine was initially present at 100% E.N. Amine loss was determined by titration after heating 20 minutes at 150°C (5).

pK_a = - $\log_{10}$ (dissociation constant of conjugate acid)

Determination of Extent of Cure. The effect of retained amine on extent of cure was determined by DMA (Figure 1) and by MIR-FTIR (Figure 2). As extent of cure increases, the glass transition temperature, T_g, increases; the height of the loss tangent peak for the glass transition, Tan δ(max), decreases; and the minimum in storage modulus in the rubbery plateau, E'(min), increases. The value of E'(min) is reported to be directly proportional to crosslink density for unpigmented films (6). The rise in E' at high temperatures (Figure 1) is attributed in part to additional curing during the DMA scan.

In the MIR-FTIR spectrum (Figure 2), the band at 729 cm⁻¹ is from the aromatic ring of the isophthalic acid used in preparation of the PE. The band at 815 cm⁻¹ is from the triazine ring breathing mode. There are many ether groups in MF cured films, but luckily the C-O stretching mode of the methoxymethyl ether at 912 cm⁻¹ is distinguishable from other ether bands. Since methoxymethyl (MM) groups are consumed during cure while triazine (T) rings are not, the peak area ratio MM/T (912 cm⁻¹/815 cm⁻¹) is an indication of extent of cure. MM/T decreases as extent of cure increases. Top and bottom surfaces of free films were analyzed in the film thickness study, but only the top surfaces were analyzed in the solvent borne versus

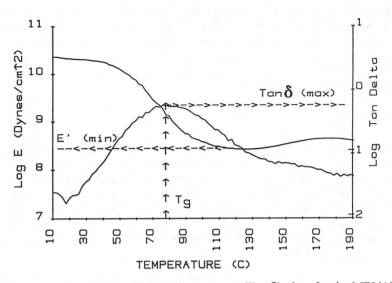

Figure 1. Storage modulus (E') and loss tangent (Tan δ) plots for the MF2/AMP WR film. Measures of extent of cure are indicated. The temperature scan was at 2°C/min. The film was cured 30 minutes at 130°C.

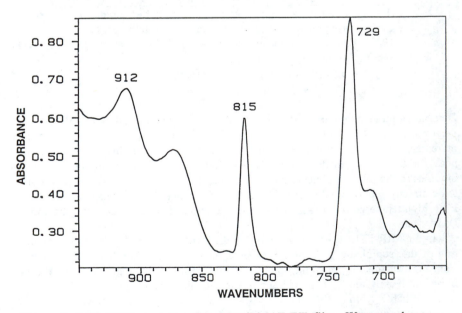

Figure 2. MIR-FTIR spectrum of the MF2/DMAE WR film. Wave numbers are indicated for an aromatic ring, the triazine ring and the methoxymethyl of MF2.

WR film comparison. Use of free films facilitates good contact between the MIR crystal and the sample so that strong spectra are obtained. FTIR results are considered to be semi-quantitative, not quantitative.

Comparison of SB and WR Films. DMA and FTIR results are summarized in Table II. Data in the first three rows are for films cured with MF1. For solvent-borne (SB) films, the only amine present was the small amount (0.26 wt. % on resin solids) used as a blocking agent for the pTSA catalyst. Other data are for WR films with amine present at 80% EN (5.6 wt.% on resin solids). The effects of DMAE and AMP are very different when SB and WR films are compared. Since MF cure is acid catalyzed, we expect cure inhibition by amines in WR formulations. The expected inhibition is observed for the MF1/DMAE film, but not for the MF1/AMP film. With DMAE, T_g decreases by 12°C and crosslink density is reduced by half whereas with AMP, T_g increases by 15°C and crosslink density increases by a factor of 1.2. Thus, when the crosslinker is MF1, AMP actually accentuates cure instead of inhibiting cure based on DMA results. In contrast, FTIR results indicate that both amines reduce the extent of conversion of methoxymethyl groups with DMAE having the greater effect.

Table II. DMA and FTIR of SB and W-R Films

MF	Film Type	Amine	T_g	Tan δ (max)	$10^{-8} \times E'$ (min)	[1]FTIR MM/T
			°C		(Dyn/cm^2)	
MF1	SB	[2]None	84	0.61	2.8	0.8
MF1	WR	DMAE	72	0.90	1.4	1.2
MF1	WR	AMP	99	0.57	3.4	1.0
MF2	SB	[2]None	90	0.55	3.6	0.5
MF2	WR	DMAE	74	0.63	1.3	0.9
MF2	WR	AMP	76	0.59	2.1	0.6

[1]MM/T = methoxymethyl/triazine, 912 cm^{-1}/815 cm^{-1}
[2]DMAE was used to block the catalyst

Auxiliary crosslinking by AMP is not fully understood. It may involve reactions between AMP and the PE to form amide or ester crosslinks as proposed by Wicks and Chen (5). However, reactions between -OH and -NH$_2$ groups of AMP with the MF resin are also likely (9). The primary amine group of AMP is likely involved because we do not observe enhanced cure when the amine neutralizer has a tertiary amine group as in DMAE.

With MF2 as crosslinker (Table II, last three rows), both DMAE and AMP cause cure inhibition in WR films relative to the SB film. In this case, the inhibition is indicated by both DMA and FTIR. Inhibition by DMAE is much stronger than by AMP based on E'(min) and MM/T values. It is proposed that AMP undergoes reaction with formaldehyde that is generated from MF2 during cure. The electron

$$
\begin{array}{ccc}
CH_3 & & CH_3 \\
| & & | \\
CH_3\text{-}C\text{-}CH_2\text{-}OH + H_2C{=}O & \rightarrow & CH_3\text{-}C\text{-}CH_2\text{-}OH \\
| & & | \\
NH_2 & & NH(CH_2OH)
\end{array}
$$

withdrawing effect of the methylol group is believed to reduce base strength relative
to that of unreacted AMP. The weakness of inhibition by AMP is attributed to this
decrease in base strength.. The tertiary amine group of DMAE cannot react with
formaldehyde in this way, which accounts for its stronger inhibition of cure.

The occurrence of the proposed AMP plus formaldehyde reaction has considerable
support in previous work. AMP has been used as a formaldehyde scavenger (7). This
reaction is the first step in synthesis of 4,4-dimethyloxazolidine from AMP (8).
Products consistent with formation of an AMP/formaldehyde adduct were identified
by gas chromatography/mass spectroscopy following heating of mixtures of MF and
AMP (9).

Effect of MF Structure. With MF1 as crosslinker, auxiliary crosslinking reactions by
AMP are proposed, but with MF2 as crosslinker reaction of AMP with evolved
formaldehyde is proposed. Reasons for this difference in emphasis are elucidated in
terms of the pattern of substitution on MF shown in Table III. This MF analysis
is based on size exclusion chromatography and 13C NMR using the peak assignments
of Tomito and Ono (10). The number of significant figures is very limited because
it has been assumed that oligomers are formed by methylene-ether bridging and that

Table III. Functional Groups Per Triazine in MF Resins

MF	1	2	3	4	Total
MF1	5.1	0.2	0.3	0.4	6.0
MF2	3.2	0.6	1.4	0.8	6.0

1. Methoxymethyl, $-CH_2OCH_3$
2. Hydrogen, -H
3. Methylol, $-CH_2OH$
4. Methylene ether bridge, $-CH_2OCH_2-$

no methylene bridges are present. Values have been normalized to total six,
corresponding to the six positions available for substitution on melamine.

It is evident that MF1 is much closer to the structure of HMMM than is MF2.
The higher content of $-CH_2OCH_3$ in MF1 is apparent, even after curing, in terms of
the MM/T ratios of Table II. MF2 is much richer than MF1 in $-CH_2OH$, -H and
$-CH_2OCH_2-$ bridges. Since methylol groups, $-CH_2OH$, on MF are the direct precursors
of formaldehyde (11), we expect MF2 to generate more formaldehyde than MF1
during cure. This is the rationale for emphasizing the AMP plus formaldehyde reaction
when the crosslinker is MF2. When the crosslinker is MF1, we expect less
formaldehyde evolution and we observe accentuated crosslinking. Thus, it is
suggested that auxiliary crosslinking by AMP occurs to a greater extent when AMP
is not diverted by reaction with formaldehyde.

It is well established that crosslinkers such as MF2 are more active than MF1
types, at least with regard to self-condensation (3,11). The difference in activity is
evident in DMA results for SB films of Table II. In general, inhibition by retained
amines is less severe than most coatings chemists would expect, and the reason
proposed here is that reaction with evolved formaldehyde reduces the overall system
basicity. Both inherent activity, based on the substitution pattern (Table III), and

potential for reduced system basicity, based on reaction with formaldehyde, favor MF2 over MF1. Thus, without proposing some type of auxiliary crosslinking in the AMP/MF1 system, it would be very difficult to explain the results of Table II.

Effect of Film Thickness. The effects of variation in dry film thickness (DFT) on T_g are shown in Table IV. For the MF1/DMAE and MF2/DMAE films, T_g decreases with increasing thickness. The effect is very large for the thickest cases, reaching a 30°C change in going form 23 to 79 μm for MF1/DMAE. In remarkable contrast, the films containing AMP show little (MF2/AMP) or no (MF1/AMP) decrease in T_g with increasing thickness. The effect of thickness on T_g is attributed to variations in extent of cure caused by retained amino-alcohol. Since AMP and DMAE have high boiling points (Table I), we do not expect complete loss by volatilization during cure. In thicker films, the diffusion path of the amine from bulk to surface is longer, and greater retention is believed to result. We propose that both AMP and DMAE are retained to a greater extent in thicker films and that the higher content of DMAE, but not AMP, inhibits cure more strongly as thickness increases. The weakness of inhibition by AMP, noted in previous discussion, is supported by the film thickness results shown in Table IV.

Table IV. Effect of Thickness on Tg and FTIR Peak Ratios

MF/Amine	DFT	T_g	MM/T Top	MM/T Bottom
	μm	°C		
MF1/DMAE	23	80	1.0	1.2
	31	76	1.1	1.4
	79	50	1.0	1.8
MF1/AMP	22	94	0.8	0.9
	31	94	0.8	0.9
	45	94	0.8	0.9
MF2/DMAE	27	70	0.6	0.8
	36	66	0.7	0.9
	49	57	0.8	1.1
MF2/AMP	20	82	0.4	0.7
	32	80	0.4	0.7
	41	76	0.4	0.7

The effects of variation in DFT on methoxymethyl (MM) content remaining after cure are also shown in Table IV. For the bottom surfaces of MF1/DMAE and MF2/DMAE films, MM/T increases with increasing thickness. The increase in content of unconverted MM groups is attributed to greater inhibition caused by greater retention of DMAE in thicker films. In contrast, MM/T does not change with thickness in the bottom surfaces of films prepared with AMP. We believe that AMP would not cause much inhibition even if more of it is retained in thicker films. Within experimental uncertainty, variation in DFT does not cause any change in MM/T for top surfaces of any of the films. This finding suggests that concentration of amino-alcohol at the shallow depth (2 to 4 μm) sampled by our MIR-FTIR method does not depend strongly on thickness. Presumably amine concentration very near the surface is nearly in a steady state established by surface escape and replenishment by diffusion from the bulk. The quasi-steady-state surface concentration is expected to be lower than the bulk concentration, which is consistent with a lower MM/T value for the top surface versus bottom surface in every case. Of course, MF2 gives lower MM/T than MF1 in corresponding cases because there is less MM in MF2 to start (Table III).

Summary

Retention of amino-alochol neutralizers in WR coatings does not inhibit cure as extensively as one would expect based on the need for acid catalysis of the curing reactions.   Weakness of inhibition is attributed, in part, to reactions between formaldehyde evolved from MF resins during cure and the amino-alcohols.   These reactions are believed to reduce overall system basicity.   Formaldehyde can add directly to the primary amine group of AMP, but it cannot add to the tertiary amine group of DMAE.   AMP is found to inhibit cure less extensively than does DMAE.

When AMP is used as neutralizer in WR formulations, cured films have higher crosslink density and T_g than obtained from AMP-free solvent borne formulations of the same binder components.   It is proposed that AMP participates in auxiliary crosslinking reactions that are not yet fully understood.   Previous work with MF-free systems (5) indicates that AMP can partially crosslink an acrylic resin that contains -OH groups and -COOH groups.   Previous work with a polyol-free system (9) suggests that AMP reacts with MF resins. The auxiliary crosslinking by AMP occurs to a greater extent if the MF resin is an HMMM type (low methylol). Less auxiliary crosslinking is observed when the MF resin is a high methylol type.

Acknowledgements.   The authors gratefully acknowledge the support of Monsanto Company.   Samples and helpful discussions were provided by Richard Johnson of McWhorter Technologies.

Literature Cited.
1. Wicks, Jr., Z.W.; Jones, F.N.; Pappas, S.P.; *Organic Coatings Science and Technology:* Wiley-Interscience, New York, NY, 1992; Vol. 1, pp. 111-119.
2. Padget, J.C.; *J. Coat. Technol.* 1994, 66, p. 89.
3. Hill, L.W.; Kozlowski, K.; *J. Coat. Technol.* 1987, 59, p. 63.
4. Hill, L.W.; Wicks, Jr., Z.W.; *Prog. Org. Coat.* **1980**, 8, p. 161.
5. Wicks, Jr., Z.W.; Chen, G.-F.; *J. Coat. Technol.* **1978**, 50, p. 39.
6. Hill, L.W. In *Paint and Coating Testing Manual*; Koleske, J.V., Ed., 14th Edition, Am. Soc. for Testing and Materials, Philadelphia, PA, 1995; pp. 534-546.
7. Tanaka, H.; Murakami, Y.; Morilei, Y.; Jap. Kokai 75,33,289; 1975, to Dainippon.
8. Robinson, G.N.; Alderman, J.F.; Johnson, T.L.; *J. Coat. Technol.* 1993, 65, p. 51.
9. Ferrell, P.E; Gummeson, J.J.; Hill, L.W.; Truesdell-Snider, L.J.; *J. Coat. Technol.* **1996, 68**.
10. Tomita, B.; Ono, H.; *J. Poly. Sci.* 1979, 17, p. 3205.
11. Jones, F.N., Chu, G.; Samaraweera, U.; *Prog. Org. Coat.* **1994**, 24, p. 189.

MORPHOLOGY AND FILM STRUCTURE

Chapter 16

Small-Angle Neutron Scattering Studies of Composite Latex Film Structure

Y. Chevalier[1], M. Hidalgo[2,4], J.-Y. Cavaillé[2], and B. Cabane[3]

[1]Laboratoire des Matériaux Organiques à Propriétés Spécifiques, Centre National de la Recherche Scientifique, B.P. 24, 69390 Vernaison, France
[2]Centre de Recherches sur les Macromolécules Végétales, Centre National de la Recherche Scientifique, B.P. 53, 38041 Grenoble Cedex 09, France
[3]Service de Chimie Moléculaire and Unité Mixte Rhône-Poulenc, Centre National de la Recherche Scientifique, CE Saclay, 91191 Gif sur Yvette, France

The structure of composite films made of high T_g polystyrene (PS) nodules dispersed in a low T_g polybutylacrylate (PBuA) matrix was studied by means of small angle neutron scattering. For films cast from mixtures of PS and PBuA latexes, segregation of PS particles leads to dense clusters of PS particles in a PBuA continuous medium. This segregation has the main features of a phase separation. For films cast from two-stage (core-shell) particles, this segregation phenomenon is prevented, depending on the coverage of the PS core by the PBuA shell. Upon annealing the films above the T_g of PS, extensive coalescence of PS particles occurred when large contacts were already present in the dry film at room temperature, whereas coalescence could be prevented when PS particles were taken apart by the presence of a PBuA shell. The extent of coalescence had a strong influence on the films mechanical properties.

In waterborne coatings applications, polymer films can be obtained by drying polymer particles colloidal suspensions. During the drying process, particles have to approach each other, stick, and finally get deformed from spherical to polyhedral as water evaporates (*1-3*). Particles made of polymers of low glass transition temperature (T_g) are thus required in such a film formation process but the final (dry) films mechanical properties are poor. A mechanical properties reinforcement can be obtained when solid particles are mixed with the low T_g polymer ones. Those solid particles can be either inorganic or organic when a high T_g polymer is used. In this latter case, a further structural transformation of the dry films can be obtained by heating the films above the glass transition temperature of the high T_g polymer. The weak bonds between solid particles at their points of contact are then strengthened by polymer

[4]Current address: Elf-Atochem, 95 rue Danton, 92300 Levallois-Perret, France

interdiffusion (coalescence), which results in a further enhancement of elastic modulus (*4-6*).

A large variety of mechanical properties reinforcements was observed, depending on high T_g polymer particles radii, surface chemistry or volume fraction(*4-6*). Efficient reinforcement can be achieved when the high T_g polymer particles form a continuous percolating network into the low T_g polymer matrix. In this context, the structure of the high T_g polymer particles dispersion in the dry film is a key parameter that one wishes to be able to control. This is the motivation for performing systematic studies about the structure of dry composite films made of solid particles dispersed into a low T_g polymer matrix. In the present work, the structure of films made of glassy polystyrene (PS) particles in a soft polybutylacrylate (PBuA) matrix was studied by means of small angle neutron scattering (SANS).

Materials and Methods

Materials. The PS/PBuA films were obtained by drying aqueous latex dispersions. The PS and PBuA polymers were brought into the films either by mixing PS and PBuA homopolymer latex particles before drying, or by drying aqueous suspensions of composite PS/PBuA particles. PBuA coalescence occurs during the drying process when two PBuA domains contact each other because of the low T_g of PBuA (-50°C). The two film formation processes are summarized in Figure 1.

It has been shown that efficient mechanical reinforcement could be achieved with mixtures above a PS particle content of about 30% volume fraction (*4*). The structure of films having two different PS contents were then studied: 10% and 45% PS, below and above this percolation threshold respectively. The PS and PBuA particle diameters were 45 nm and 80 nm respectively, as measured by quasi-elastic light scattering. For composite particles the PS volume fraction was 40%, and two different soft shells were studied. The first one (sample A) was a pure PBuA shell which gave imperfect core-shell morphologies (half-moon like) and the second type of shell (sample B) was made of a copolymer of butyl acrylate and 10% methacrylic acid (MA) which gave a better coverage of the PS core by the soft shell (Figure 2). Those particles were prepared in two steps: PS emulsion polymerisation gave a PS seed of 120 nm diameter and the PBuA or PBuA-coMA shell was obtained in a second polymerisation step (*7*).

Small Angle Neutron Scattering on Latex Films. SANS experiments were carried out with the D11 diffractometer (*8*) at the high flux reactor of the Institut Laue-Langevin (ILL) at Grenoble. Data reduction was performed with standard procedures available as program packages at the ILL, giving after incoherent background subtraction the differential scattering cross-section $d\sigma/d\Omega = I(q)$ (*9*). The scattering vector domain observed was $10^{-3} \text{ Å}^{-1} < q < 0.1 \text{ Å}^{-1}$, which defined the length scale explored in real space as $10^3 \text{ Å} > d > 10 \text{ Å}$. The PS particle diameters are well in this range, so that both the PS particle structure (radius and surface area) and ordering (PS-PS correlations) could be studied by this way (*10*).

A good contrast between PS and PBuA could be obtained with natural isotopic compositions. No deuterated materials were necessary. The scattering length densities of different polymers are the sum of the scattering lengths of nuclei b_i divided by the molecular volume V, $\rho = (\Sigma b_i)/V$. The contrasts $\Delta\rho = |\rho - \rho_{PBuA}|$ of

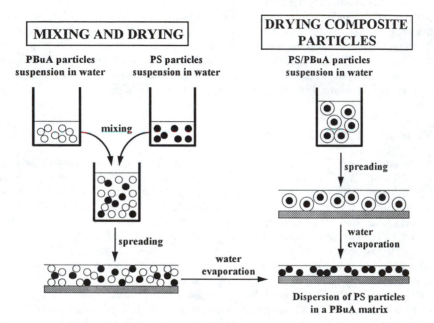

Figure 1.   Film formation processes either from mixtures of PS and PBuA particles or from core-shell composite PS/PBuA particles.

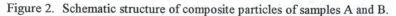

Figure 2.   Schematic structure of composite particles of samples A and B.

hydrogenated PS, PBuA and of the zwitterionic emulsifier used in emulsion polymerisation are given in Table I. The scattering length density of deuterated polystyrene is also given for comparison.

Table I. Scattering length densities of compounds found in dry films

Compound	Scattering length density ρ (cm^{-2})	Contrast $\Delta\rho$ (cm^{-2})
PBuA(^{1}H)	0.66 10^{10}	-
PS(^{1}H)	1.40 10^{10}	0.74 10^{10}
PS(^{2}H)	6.42 10^{10}	5.76 10^{10}
Emulsifier	0.08 10^{10}	0.58 10^{10}

It can be seen in Table I that the contrast between the emulsifier and the PBuA is of the same order of magnitude as that between PS and PBuA. Thus, the elimination of the emulsifier is very important, any emulsifier remaining after the latex washing procedures can give very strong scattering if it is segregated from the polymer into lumps of large size. An emulsifier which strongly adsorbs onto the polymer particles is necessary for obtaining monodisperse PS particles of small size (D < 100 nm), but its strong adsorption makes it difficult to completely wash out. The PBuA particles used in PS-PBuA mixtures were prepared using sodium dodecylsulfate (SDS) as an emulsifier because it is easy to remove. A larger particle diameter and size polydispersity is not a problem since PBuA particles coalesce into a continuous matrix during film formation at room temperature. It has already been observed that the zwitterionic emulsifier dodecyldimethylammoniopropylsulfonate ($C_{12}H_{25}{}^+N(CH_3)_2-(CH_2)_3-SO_3{}^-$) which was used for emulsion polymerisation (*11*) did give such a strong scattering at low angles in dry films of homogeneous poly(butylacrylate-co-styrene) (*2,3*).

Results and Discussion

Films Obtained at Room Temperature from Mixtures of PS and PBuA Particles. The SANS data obtained for dry films with 10% and 45% PS contents are shown in Figure 3 on a Log-Log scale. Going from the high-q region (high resolution) to the low-q one, one can observe a q^{-4} decay of scattered intensity, a broad peak around 0.015-0.02 Å^{-1}, and a strong rise of scattered intensity at low angles. Experimental data look similar for both PS contents although these compositions are located below and above the percolation threshold. This will be discussed from high to low q, that is, from the isolated particle structure to the large scale arrangement of PS domains.

At high q (from 0.02 to 0.1 Å^{-1}) the scattered intensity decays as the fourth power of q, following the asymptotic Porod's law, $I(q) = 2\pi (\Delta\rho)^2 A/V q^{-4}$, where A/V is the interfacial area (A) per unit sample volume (V). This behavior means that the PS-PBuA interface is a sharp one, their is no mixing region between PS and PBuA or, if any, it is thinner than the experimental resolution of 10 Å. This is quite an expected result since polystyrene and polybutylacrylate are non-miscible polymers.

The value of the prefactor, which is a measure of the interfacial area A, is 4.5 times larger for the 45% PS content sample than for the 10% one.

Oscillations around the Porod asymptotic behavior can be observed which allows the determination of the PS particles diameter as 42 nm, in good agreement (but slightly lower) with the quasi-elastic light scattering determination (Figure 4). Thus the initial PS particle diameter is found in the dry film and the correct interfacial area ratio $A(45\%)/A(10\%) = 4.5$ were measured, which show that no coalescence of PS particles had occurred at room temperature. This result was expected owing to the high T_g of polystyrene (100 °C).

A broad peak (hump) is observed for both the 10% and the 45% PS content films which is related to PS particle correlations. The position of this peak defines a mean interparticle distance in the films. According to the resolution of the experiment, this peak is too broad to be a Bragg peak. Moreover, no higher order diffraction peaks can be observed. Thus, the PS particles ordering is not periodic. The important observation is that the position of this peak is the same for the 10% and 45% PS content films. If the PS particles were distributed at random throughout the film, the mean distance between particles should be larger for the 10% PS content film and the peak position should be at a lower q value. Moreover, the peak position (0.0165 Å^{-1}) is at a higher q value (the mean interparticle distance is smaller) than predicted for a random dispersion of 42 nm diameter particles, even for the 45% volume fraction. Thus, *aggregation of PS particles has occurred*. The structure of the film is made of clusters of PS particles immersed in a continuous matrix of PBuA polymer. The position of the peak gives the volume fraction of PS particles inside the clusters. Since it is the same for the 10% and 45% PS content films, the PS volume fraction inside the clusters is the same for both samples. This aggregation phenomenon has thus the *features of a phase separation* between a phase containing a high density of PS particles and a phase of pure PBuA.

The low-q rise of scattered intensity could be related to the form factor of the clusters, giving their large scale structure (their size). The scattered intensity follows approximately a power law as $I(q) \propto q^{-a}$, the exponent a ranging between 3 and 4. The size of the clusters would then be very large (in the μm range) in order to account for the experimental neutron scattering and the films would also strongly scatter light. This is actually not the case, the films are transparent. On the contrary, the films become turbid after a thermal annealing which causes PS particles coalescence into large nodules as described in the following of this paper. This interpretation then seems in contradiction with the experimental data. The same power law behavior has already been observed for films made of homogeneous polymer particles containing residual emulsifier while the low-q intensity remained flat in the absence of emulsifier (3). This has been ascribed to large lumps of segregated emulsifier. It is then believed that the present low angle scattering is again due to the segregation of residual emulsifier. Indeed the emulsifier comes from the PS particles (PBuA has been polymerised with SDS), which accounts for the observed intensity ratio $I(45\%)/I(10\%)$ of 4.5 in the low-q domain. The data about PS particle correlations are then completely hidden by the strong scattering coming from the emulsifier in the q-range below 5×10^{-3} Å^{-1}, and the scattering from the emulsifier contributes significantly to the total scattering up to $q = 0.02$ Å^{-1}.

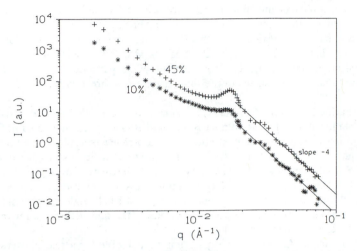

Figure 3. Scattered intensity of 10% and 45% PS content films dried at room temperature (Log-Log scale).

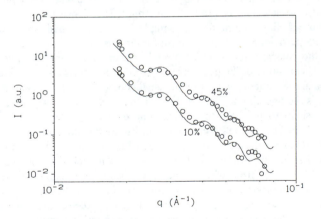

Figure 4. High-q region showing the oscillations around the Porod's asymptotic law. The full line is the theoretical scattered intensity for spheres of 42 nm diameter (smeared according to the experimental resolution).

The PS volume fraction inside the clusters can be estimated assuming the arrangement of PS particles inside these clusters is random. For a hard sphere dispersion of particles a PS volume fraction of 60% gives the right position of the peak. The complete theoretical scattering is then calculated as the sum of that of hard spheres at 60% volume fraction (*12*) for the clusters and a flat scattering for the remaining pure PBuA phase. As shown in Figure 5, the hard sphere model reproduces the peak position since the PS volume fraction inside the clusters was adjusted for that, but the width of the peak is also fairly reproduced without any additional fitting parameter. The arrangement of PS particles inside the clusters is thus close to random. A sketch of the films structure is shown in Figure 6.

Films Obtained at Room Temperature from Composite PS/PBuA Particles. The SANS data for films obtained from latexes A and B closely resemble data of the films cast from latex mixtures described above. The q-range investigated is however of lesser width (from 10^{-3} to $8 \cdot 10^{-3}$ Å^{-1}) but is well suited for this study because of the larger PS seed diameter. A broad peak and an intensity rise at small q can be observed. The q-range is not large enough towards high-q for observing the expected Porod's q^{-4} decay. The strong scattering at small angle is again ascribed to the surfactant used for emulsion polymerisation. The same analysis as for the above described blends can be performed.

For sample A, the phase separation model of PS particle clusters with a random distribution of particles inside it holds. The internal PS particle volume fraction was chosen so as to reproduce the position of the peak and this model was successful in predicting the full peak shape quite fairly (Figure 7). The PS particles volume fraction was found as 48% against 40% for the overall PS volume fraction. The PS particles clusters were thus found much less dense than for blends. The imperfect core-shell structure of the sample A particles had some efficiency in preventing PS particles aggregation during film formation.

On the contrary, this kind of model cannot fit to the experimental data of sample B. The position of the peak is at a lower q than for sample A, indicating that PS particles are separated by a larger mean distance. For a PS volume fraction of 40% which is the overall one, the position of the peak is correctly predicted but the peak is too sharp to be fitted to a hard sphere model. The origin of this is the protective coating around the PS core brought by the functionalized PBuA shell. The PBuA shell provides indeed a better coverage of the PS core than in sample A. This more ordered structure observed in the film can be modelled by introducing a shell around the PS spheres from where other PS particles are excluded. Thus the PS spheres have their correct diameter of 120 nm in the model but the excluded volume is larger than the particle, the hard sphere diameter exceeds the particle diameter. The thickness of the excluded shell is used as an adjustable parameter in looking for the best agreement between experiment and the theoretical model. The PS volume fraction of 40% has not to be considered as a free parameter as it is set at the overall value of 40% (Figure 7). The best excluded shell was found 4 nm thick, much less than in the original latex particles (22 nm).

The consequence of using composite particles with a core-shell structure is to prevent core aggregation. The best the coverage of the glassy core by the soft shell, the more efficient this aggregation inhibition. The model used for film B would mean

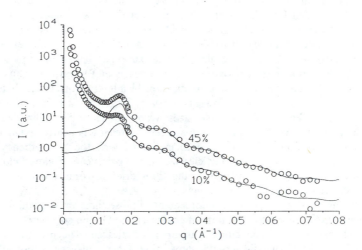

Figure 5. Scattering for the hard sphere model as explained in the text for films from PS and PBuA mixtures with 10% and 45% PS content.

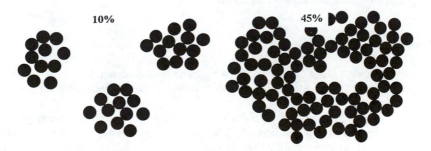

Figure 6. Sketch of the 10% and 45% PS content films structure.

complete inhibition of PS particles aggregation but we do not believe this is the actual situation. This is a phenomenological model, the small thickness of the excluded shell is an average value, so that the presence of some contacts between PS cores are allowed in the model. Such contacts actually should exist for PS particles coalescence to occur, but SANS does not give any experimental evidence for that.

Films Structure after Thermal Annealing above the PS T_g. The films described above were annealed for 3 hours at 140°C in order to allow coalescence of the PS particles. Dynamic mechanical analysis (DMA) has shown that this thermal treatment results in a considerable reinforcement of mechanical properties (increase of the elastic modulus) (4-6). For all the films studied, annealing led to a complete loss of the PS-PS correlation peak (Figure 8). Coalescence of PS particles thus occurs, giving larger PS nodules in the PBuA matrix. For blends, the Porod's law is followed in the whole observed q-range, showing that very large PS particles were formed. On the contrary, limited PS particle sizes were obtained with films cast from composite particles. With sample B, the scattered intensity is nearly constant at the lower observed q values, PS coalescence is limited with the best efficacy in this sample. It is clear that the PS particle size after coalescence follows closely the extent of aggregation of PS particles in the films dried at room temperature, or in other words the amount of PS-PS contacts.

The mechanical properties reinforcement of those films are in good qualitative agreement with the measured amounts of coalescence. On the sole basis of dynamic mechanical analysis, it was concluded that no PS coalescence occurred in sample B (5). SANS structural analysis actually shows that some coalescence took place but that its extent is limited. This is thus only an apparent contradiction between these two analyses coming from the very different scales DMA and SANS are sensitive to. This is a pedagogical example of the advantage of using several techniques of investigation in studying complex materials.

Conclusions

In composite films made of high T_g PS particles embedded in a low T_g PBuA matrix, the dispersion of PS particles is not random and depends drastically on the PS particles surface.

The aggregation of PS particles in films cast from blends have been shown, even for a low PS volume fraction of 10%. High density clusters are formed. This aggregation can be prevented by the use of composite particles with a correct core-shell morphology.

Strong mechanical reinforcement can be achieved by heating the dry films above the PS T_g if the PS particles coalescence is allowed in the as dried films structure: PS-PS particles contacts are required for rapid coalescence to occur during annealing.

The combined used of macroscopic (mechanical measurements) and local (SANS) investigation methods allows to progress in the understanding of the mechanisms of film formation.

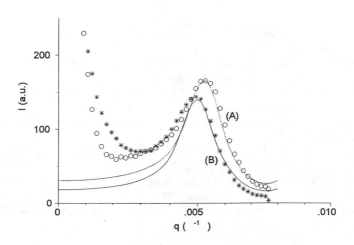

Figure 7. SANS data for films A and B compared with the predictions for the models as described in the text. Sample A (exptl: circles, calc: dashed line); sample B (exptl: stars, calc: solid line).

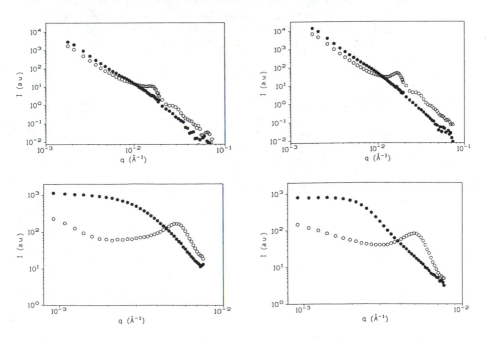

Figure 8. SANS data before (○) and after (●) annealing for 10% (top left) and 45% (top right) blends, and for samples A (bottom left) and B (bottom right).

References

1. Zosel, A.; Heckmann, W.; Ley, G.; Mächtle, W. *Makromol. Chem., Makromol. Symp.* **1990**, *35/36*, 423.
2. Joanicot, M.; Wong, K.; Maquet, J.; Chevalier, Y.;Pichot, C.; Graillat, C.; Lindner, P.; Rios, L.; Cabane, B. *Prog. Colloid Polym. Sci.* **1990**, *81*, 175.
3. Chevalier, Y.; Pichot, C.; Graillat, C.; Joanicot, M.; Wong, K.; Maquet, J.; Lindner, P.; Cabane, B. *Colloid Polym. Sci.* **1992**, *270*, 806.
4. Cavaillé, J.-Y.; Vassoille, R.; Thollet, G.; Rios, L.; Pichot, C. *Colloid Polym. Sci.* **1991**, *269*, 248.
5. Hidalgo, M.; Cavaillé, J.-Y.; Guillot, J.; Guyot, A.; Pichot, C.; Rios, L.; Vassoille, R. *Colloid Polym. Sci.* **1992**, *270*, 1208.
6. Hidalgo, M.; Cavaillé, J.-Y.; Cabane, B.; Chevalier, Y.; Guillot, J.; Rios, L.; Vassoille, R. *Polym. Adv. Technol.* **1995**, *6*, 296.
7. Rios, L.; Hidalgo, M.; Cavaillé, J.-Y.; Guillot, J.; Guyot, A.; Pichot, C. *Colloid Polym. Sci.* **1991**, *269*, 812.
8. Ibel, K. *J. Appl. Crystallogr.* **1976**, *9*, 296.
9. Cotton, J.-P. In *Neutron, X-ray and Light Scattering: Introduction to an Investigative Tool for Colloidal and Polymeric Systems*; Lindner, P.; Zemb, Th., Eds; North-Holland: Amsterdam, 1991; pp 19.
10. Cabane, B. In *Surfactants Solutions: New Methods of Investigation*; Zana, R., Ed.; Surfactant Science Series 22; Marcel Dekker: New York, 1987; pp 57.
11. Essadam, H.; Pichot, C.; Guyot, A. *Colloid Polym. Sci.* **1988**, *266*, 462.
12. Ashcroft, N. W.; Lekner, J. *Phys. Rev.* **1966**, *145*, 83.

Chapter 17

Structures of Stretched Latex Films

Y. Rharbi[1,4], F. Boué[1], M. Joanicot[2], and B. Cabane[3]

[1]Laboratoire Léon Brillouin, CE Saclay, 91191 Gif sur Yvette, France
[2]Rhône-Poulenc, 93308 Aubervilliers, France
[3]Equipe Mixte CEA-RP, Service de Chimie Moléculaire, CE Saclay, 91191 Gif sur Yvette, France

Small angle neutron scattering has been used to examine the structures of cellular latex films submitted to uniaxial deformation. Two modes of deformation have been found, depending on the relative strenths of the hydrophobic cell cores and hyhydrophilic cell walls. When the films were stretched in conditions where the cell walls were harder than the cell cores, it was found that the structures were stretched uniformly according to the macroscopic deformation of the sample. When the films were stretched in conditions where the cell walls were softer than the cell cores, correlations between cells were lost in the direction perpendicular to the stretching, indicating that the macroscopic stretching causes microscopic shear deformations.

Polymeric films can be made through evaporation of an aqueous dispersion of latex particles [1-4]. Each particle is made of a hydrophobic polymeric core that is copolymerized with a hydrophilic membrane. Upon evaporation, the dispersion passes through 3 stages:
a) Concentration: the particles are kept apart from each other by the repulsions between their membranes, and they organize into a crystalline like structures to minimize these repulsions [5, 11].
b) Contact: the removal of water forces each particle to come in physical contact with its 12 neighbours [6-8].
c) Compression: upon removal of the remaining water, capillary forces compress the particles into dodecahedral cells. The resulting film has a foam structure where the interior of cells are hydrophobic, and the cell walls are hydrophilic. [6, 8].

These cellular latex films are used to manufacture coatings. In such applications the important properties are the mechanical properties of the films and their permeability to water. The most difficult situations are those where mechanical stress and agression by water are combined, e.g. where the mechanical resistance in a wet environment is required.

[4]Current address: Department of Chemistry, University of Toronto, 80 St. George Street, Toronto, Ontario M5S 3H6, Canada

The behavior of a film under mechanical stress may be characterized by a map of the deformations. In this respect, it is important to realize that the films are composite materials, with a dispersed phase (the cell cores) surrounded by a continuous phase (the membranes). Depending on which phase is the softer, the macroscopic deformation of the film may be achieved through deformation of all cells, or through shear deformations of the membranes, causing relative displacements of cells. The relative importance of these 2 types of deformation must depend on the cohesion of the membranes. For instance, in a wet environment, the hydrophilic membranes may swell with water and lose their cohesive strength.

In order to measure the deformations and the relative displacements of the cells, we have used small angle neutron scattering (SANS). Contrast between cell cores and membranes was obtained through addition of a deuterated solvent that was located either in the cell cores or in the membranes. The addition of a selective solvent was also used to vary systematically the strength of membranes with respect to that of cell cores. Some films that were submitted to uniaxial stretching in the dry state where the membranes were the strongest component of the material; others were stretched when the membranes were weakened by addition of water. In this paper we present neutron scattering spectra of films stretched in these different conditions, and a discussion of the modes of deformation observed in these films.

Latex Dispersions and Film Formation.   Latex dispersions were synthesized through emulsion polymerization of styrene (S), butyl acrylate (BA) and acrylic acid (AA) in water [6]. The reaction produced polymeric particles that were 1200 Å in diameter. Each particle had a hydrophobic core made of PS and PBA sequences, surrounded by a thin layer of hydrophilic PAA and PBA sequences that were copolymerized with core polymers. A typical composition was S: 64%, BA: 32%, AA: 4%. In addition to the dispersed particles, the aqueous phase also contained soluble species (amphiphilic polymers, surfactant and salt). These were eliminated through dialysis and equilibrium with ion exchange resins.

Films were made through evaporation of concentrated dispersions in an oven. The drying temperature was kept slightly above the minimal film formation temperature T_f, which is slightly above the glass transition temperature T_g of the core polymer. The value of T_g was set according to the proportion of S and BA ($T_g=18C°$ for a 48:48 ratio, $T_g=55C°$ at 64:32). The rate of evaporation was controlled through the relative humidity of the oven.

We found that all films made from latex that had been washed and kept in the acid form took up very little water (on the order of 2%). however it could be rehydrated easily if the latex dispersion had been neutralized with NaOH. In this case the maximum water content after reswelling was 50%.

Small Angle Neutron Scattering

Technique.   Concentrated latex dispersions and the films made from them have long range order [5, 8]. Therefore they can diffract radiation of appropriate wavelength. The directions of diffraction and the relative intensities in these directions are related to repetitions of the unit cell in the lattice and to the structure of this unit cell. For crystalline systems that have periods of 0,1µm, small angle neutron scattering is appropriate.

The measurements were made on the instrument PAXY of LLB. The wavelength of incident neutrons was chosen between 3.5 and 15 Å, and the sample to detector distance was set between 2 and 7 m.The scattered neutrons were collected on a XY (64*64) multidetector. The map of collected intensities yields a 2-dimensional image of the reciprocal lattice of the sample along a plane normal to the beam. The range of the magnitude of scattering vector q is $2*10^{-3}$ $Å^{-1} < q < 2.5*10^{-2}$ $Å^{-1}$, corresponding to a range of real space distances 240Å $< 2\pi/q < 3000Å$.

In the case of latex films, it is desired to have contrast between cell cores and cell membranes, and also contrast for defects in the structure. A simple way of producingcontrast is to swell the films with deuterated solvents, e.g. D_2O, which is preferentially adsorbed by hydrophilic PAA polymers, or benzene, preferentially absorbed by PS.

Diffraction Patterns. When the latex particles are all of the same size, their packing in a film gives a periodic structure with face centered cubic symmetry (fcc) [5, 8]. Depending on the method of fabrication of the film, this structure may grow in many small crystallites that have random orientations, or in a large single crystal. In the former case the diffraction pattern is made of rings (Figure 1a); in the latter case, the diffraction pattern is made of spots corresponding to the points of the reciprocal lattice that are in the diffraction condition (Figure 1b).

The highest intensities in the diffraction pattern usually originate from diffraction by the densest planes of the structure [12]. In Figure 1a, the radius of the ring is at $q= 6.8*10^{-3}$ $Å^{-1}$; this corresponds to the distance between (111) planes of particles, which are the densest planes of the fcc structure. This is confirmed by the 6-fold symmetry of the single crystal pattern of Figure 1b, which matches the symmetry of the fcc structure in a (111) plane. The next step of spots corresponds to the 220 reflections

Mechanical Properties of Latex Films

In this section we describe the macroscopic deformations that occur in films when they are submitted to mechanical stress. Since the films are composite materials made of cell cores separated by membranes, their behavior under mechanical stress depends on the relative strengths of these 2 components.

Mechanical Description of the Cell Cores and Cell Membranes. The cell cores are made of a melt of PS-BA copolymers with a high molar mass $M_w > 5*10^5$. This material becomes fluid either at elevated temperatures or upon addition of a solvent that localizes in the cores and lowers their T_g.

The membranes are quite thin (10 to 20 Å) They are made of copolymers that contain a large proportion of PAA; in the dry state these polymers are held together by hydrogen bonds between their acid groups or by clusters of ion pairs originating from the neutralization of these groups. In the bulk, these polymers have a high glass transition temperature (T_g=140C°). This membranes may be softened by the introduction of a selective solvent: water, for films made of neutralized latex, or methanol, for all films containing AA and BA monomers in their membranes.

Effect of the Strength of the Membranes. Films made of neutralized latex were rehydrated through exposition to water vapor at various water contents, and then stretched at Tg+25C°, still in presence of water vapor; the results are presented in Figure 2. At low water contents (<5%) the modulus remains high, the films break at a rather low deformation ($\lambda r < 5$), and up to that point the deformation is recoverable upon releasing the applied force, as it is in dry films. At intermediate water contents (10 to 15%), the modulus drops and the maximum deformation reached before breaking increases considerably ($\lambda r > 17$). Only the moderate deformations ($\lambda < 3$) are recoverable upon releasing applied force; larger deformations are not recoverable, indicating that viscous flow (creep) took place in the sample. Finally, for the highest water contents (35%) the modulus is very low and the films break at low values of the deformation ($\lambda r < 5$).

Structures of Deformed Films

In this section we describe the microscopic deformations that occur in films when they are submitted to mechanical stress. The films were examined through neutron scattering after they had been stretched in conditions where the membranes were either the strongest or the weakest component of the composite.

Structures of Films with Strong Membranes. This situation is encountered in 2 cases: (i) dry films at temperature T>Tg of the cores, in which case the membranes are strongest; (ii) films swollen with an apolar solvent which weakens the cores in comparison with the membranes. In these conditions, the mechanical response of the films to a macroscopic deformation is mainly elastic .

Dry Films. These films were stretched in the dry state, at a temperature T = Tg + 25 °C, in a bath of siloxane oil. In these conditions the modulus of the films is low, and they can be stretched at same rate of deformation λ. The films were then cooled in air while the deformation was maintained. After cooling, the stress could be released without loosing the deformation, since room temperature was below Tg. Finally the films were rehydrated with D_2O (2% - 5%) for the observation through neutron scattering. We checked that this slight reswelling did not change the macroscopic deformation.

At low values of the deformation, the diffraction ring becomes an ellipse with the short axis in the stretching direction. The intensity is weakest in this direction, and strongest in the perpendicular direction (Figure 3a). As the deformation is increased, the length of the short axis decreases as λ^{-1}, while that of the long axis grows as $\lambda^{1/2}$ (Figure 4). These microscopic deformations match the macroscopic deformations, which are proportional to λ in the stretching direction and to $\lambda^{-1/2}$ in the perpendicular direction because the volume of the material is conserved during the deformation. Therefore each cell is deformed in the same way as the whole

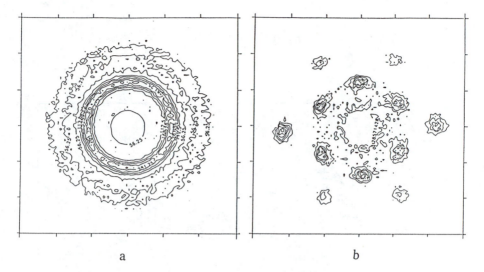

a b

Figure 1. (a) Diffraction pattern from a polycrystalline latex film. The figure is a contour map of the intensity levels obtained on the 2-dimensional detector. (b) Diffraction pattern from a single crystal latex film.

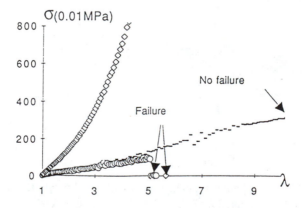

Figure 2. Apparent modulus of dry films stretched at a temperature $T_g+25C°$ and at a slow rate $S=(1/L)*(dL/dt)=0.0016s^{-1}$. Vertical axis: modulus, in 10^{-2} MPa. Horizontal axis: deformation $\lambda =L/L_0$. Diamonds: dry film made from latex neutralized at pH 10. Dashes: film swollen at 15% with water. Circles: film swollen at 35%.

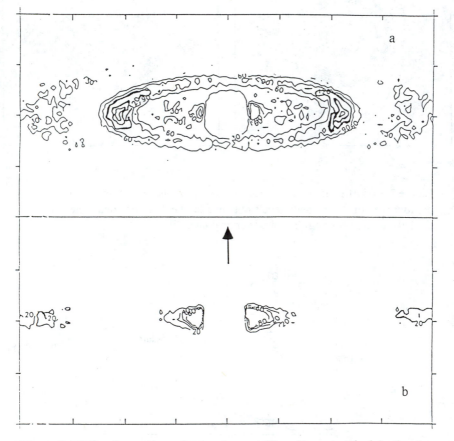

Figure 3. Diffraction patterns from polycrystalline films stretched in the dry state. (a) Final deformation λ =2,4. (b) Final deformation λ =7.

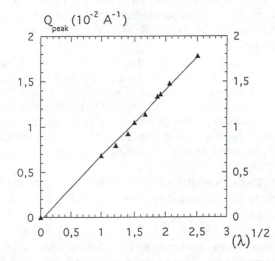

Figure 4. Deformation of the unit cell according to the diffraction patterns. Horizontal axis: square root of the macroscopic deformation λ. Vertical axis: relative variations of the large axis of the diffraction ellipse.

sample, indicating that the deformation field is affine. Moreover, the variations of intensity along the ellipse also match the scattering expected from an affine deformation of the sample [10]. The detailed analysis is presented in APPENDIX; it shows that the distribution of intensity on the ellipse can be reproduced if it is assumed that each crystallite is deformed and reoriented by an affine deformation field. These observations demonstrate that the deformation is affine at all scales, from the cell size all the way up to the sample dimensions.

At higher deformations the ellipse continues to stretch and strengthen in the direction perpendicular to the stretching, while it weakens and becomes nearly invisible in the stretching direction (Figure 4). In the end, the scattering is concentrated in 2 diffraction spots located at the extremities of the ellipse. This pattern indicates that most membranes have been stretched so much that they are nearly parallel to the stretching direction; the 2 spots correspond to diffraction by this lamellar structure. The distance between the spots measures the thickness of the cells in the perpendicular direction; it is still affine to the macroscopic deformation at least up to $\lambda = 7$ (Figure 3b).

In addition to this diffraction by the periodic lattice of membranes, the patterns also show diffuse scattering caused by defects. In films that have been quenched immediately after stretching, diffuse scattering is localized mainly next to the beam in a set of "whiskers" that are elongated in the direction perpendicular to the stretching. This scattering corresponds to defects that are elongated in the stretching direction; they may be air bubles or water pools that preexisted the stretching and were stretched so much that their thickness caused scattering in the range of accessible q vectors.

Those observations are reproducible for all latex films, in acid or neutralized state. The difference is the maximum deformation before breaking: (acid state $\lambda > 10$; neutralized state $\lambda < 5$).

Films Swollen by an Apolar Solvent. Similar films were swollen with deuterated benzene or toluene (23% of the film weight), stretched in the swollen state, and observed through neutron difraction. All operations took place in a closed container that was saturated with benzene or toluene vapor. In these conditions as well, the membranes are the only strong component of the composite. The diffraction pattern is shown in Figure 5a; it is similar to that presented in Figure 3a. Comparison with the calculated spectrum (Appendix and Figure 5b) shows that the distribution of intensity along the diffraction ellipse is that expected for an affine deformation of a dodecahedral cell.

Structures of Films With Weak Membranes. This situation in encountered in case of films with neutralized membranes, swollen with water and stretched in the swollen state. In these conditions, strong uniaxial stretching causes plastic deformation of the films (see MECHANICAL PROPERTIES).

Films With Neutralized Membranes, Swollen With Water. These films were obtained from latex dispersions that had been neutralized to pH 10 with NaOH. They were placed in a closed container saturated with D_2O vapor at a

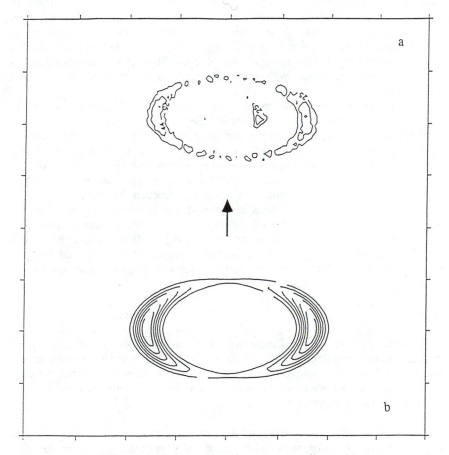

Figure 5. (a) Diffraction from a film that was swollen with benzene and stretched up to λ =1.5. (b) Calculated scattering pattern for an affine deformation λ =1.5 of a polycrystalline sample (see Appendix B for details). The arrow indicates the direction of stretching.

temperature $T = T_g + 25°C$. When the swelling was sufficient, they were stretched in a few minutes and then the container was cooled down while the deformation was maintained. Finally the films were examined through neutron scattering.

For polycrystalline samples, the first diffraction ring is again deformed into an ellipse with the long axis along the direction normal to the stretching. However, the distribution of intensity on this ellipse is opposite to that observed in dry films: in the present case the intensity is weak in the perpendicular direction and strong in all other directions (Figure 6a). With increasing deformations, the first order ring is progressively transformed into a set of 2 stripes that are centered on the stretching direction and elongated in the perpendicular direction. The second order ring deforms in the same way, and merges with the first ring when the stretching is very high (Figure 6b).

These patterns may be analyzed to give a picture of the objects that cause the scattering. In the direction of stretching, the diffraction pattern is periodic; accordingly, the objects that cause the scattering must be repeated in this direction. Since the stripes are located on the ellipse resulting from deformation of the original diffraction ring, the repetition of these objects matches the repetition of the deformed cells. In the other directions of the pattern, there are no repetitions, therefore the objects are not correlated in any other direction. Moreover, the stripes are narrow in the stretching direction and elongated in the perpendicular direction; accordingly, these objects must be quite elongated in the stretching direction and thin in the perpendicular direction. Consequently the objects that produce the stripes may be described as "strings" that are oriented along the stretching; each string is made of thin water pockets that are repeated at each cell; neighboring strings are not correlated in any direction (Figure 7).

Discussion

In this work we have compared the macroscopic and microscopic deformations of cellular materials submitted to uniaxial stress. In all situations that we found, the patterns of deformation are reproducible. Indeed, the mode of deformation appears to be determined exclusively by the geometry and by the relative strengths of the two components of the composite.

"Weak" Dispersed Phase, "Strong" Continuous Phase. This situation was obtained in films that were stretched in the dry state, at high temperature or after swelling by an apolar solvent. These conditions are chosen so that the cores will have a plastic behavior at large deformations. Conversely, the membranes remain elastic because their glass transition temperature is very high and because they are not swollen by apolar solvents.

The diffraction patterns correspond to a perfectly affine deformation of the material. Consequently, the cells and the network of membranes must be deformed according to the macroscopic deformation. Moreover, there is no loss of correlation between neighboring cells, and therefore no slip or shear within the membranes.

This result is consistent with expectations based on the relative strengths of cores and membranes. The membranes form a network which is elastic, regular and fully connected. Under uniaxial stress, this network alone would respond with a

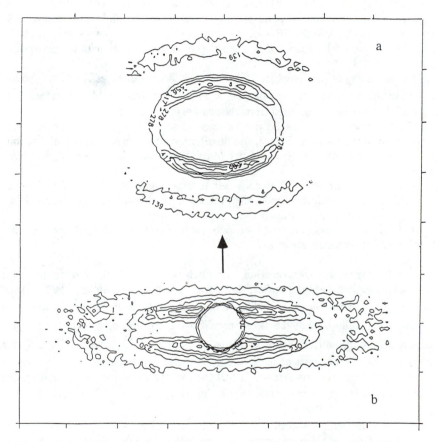

Figure 6. (a) Diffraction from a film stretched at high water content (19% D_2O) and high temperature ($T_g + 25°C$) up to λ =1.25. (b) Diffraction from the same film stretched to λ =3.

homogeneous deformation which would be in every point affine to the macroscopic deformation. Since the cell cores are weaker than the membranes, they do not modify this mode of deformation. Indeed the only strong constraint set by the cores is that the volume of each cell must remain constant; this constraint is satisfied with an affine deformation field.

"Strong" Dispersed Phase, "Weak" Continuous Phase.   This situation was obtained in films swollen by a polar solvent, water or methanol. In this case the PS cores remain rigid, while the PBA outer shells and PAA membranes are softened. The films may then be described as dispersions of hard spheres in a viscoelastic matrix.

Mechanical testing of the rehydrated films (water content in excess of 10%) shows that large macroscopic deformations ($\lambda > 3$) are not recoverable; therefore the mechanical response has changed from elastic to plastic.

The diffraction patterns do not correspond to an affine (i.e. homogeneous) deformation of the material. Indeed, the distribution of intensity on the diffraction ellipse is opposite to that observed in dry films. This loss of affinity may have two origins:

a) Either the deformation within each cell is inhomogeneous, because the weak hydrophilic region is deformed more than the strong hydrophobic core. Meanwhile, the lattice is deformed in an affine way;

b) Or the lattice is not deformed affinely, and therefore some of the original correlations between cells are lost.

Inhomogenous Deformation of Each Cell.   Upon rehydration, the hydrophilic regions become weaker; therefore, they may deform to a larger extent than the dry hydrophobic cores. Let λ_c be the average strain of the cores, and λ_r that of the whole cell; λ_r may be taken equal to the macroscopic strain, since the deformation of the lattice is affine in this case. The ratio λ_c/λ_r is constrained by the condition that the thickness of the hydrophilic region must remain non zero even in the most stretched regions. Assuming that the average thickness of the hydrophilic region is 15% of the outer particle radius, this yields $\lambda_c/\lambda_r < 1.2$ for the stretched state.

Because of this inhomogeneous deformation, the distribution of water in the cell does not follow the deformation of the lattice. In the parallel direction, the water layer is thicker than expected from an affine deformation of the lattice, and in the perpendicular directions it is thinner. Consequently, the form factor that describes interferences in the unit cell is modified. This is most apparent at Q values where the form factor vanishes, because the diffraction orders are cancelled at those values. In some directions, the extinction of form factor will occur at a location that corresponds to the main diffraction lines, while in other directions the extinction may be shifted away from these lines. The total intensity will vanish in the first case and reinforce in the second.

The pattern calculated for a deformation at $\lambda_r = 1.5$ and $\lambda_r = 2.5$ are presented in Figure 8. The intensity of the first diffraction order for $\lambda_r = 2.5$ is located in 4

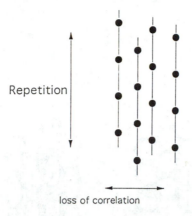

Figure 7. A picture of the defects that cause the scattering in stripes shown in Figure 6. Each "string" is made of thin water pools that are repeated along the stretching direction; neighboring strings are not correlated in any direction.

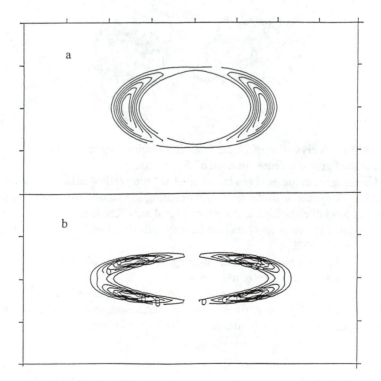

Figure 8. Calculated diffraction patterns for a stretched lattice with inhomogeneous deformation of the unit cell. (a) stretching to λ =1.5. (b) stretching to λ =2.5.

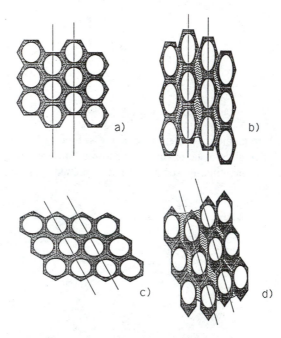

Figure 9. (a) A crystallite of a latex film after swelling by water. The vertical axis of the figure is a dense direction of the lattice.

b) The same crystalite has been stretched along the vertical axis. In addition to the affine deformation of the cells, the stretching has caused some random shifts between rows of cells aligned along the vertical axis. These shifts are accomodated by water "pockets", which cause the "stripes" in the diffraction patterns of Figure 7.

c) A crystallite of a latex film after swelling by water. In this case the vertical axis is not a dense direction of the lattice.

d) The same crystalite has been stretched along the vertical axis. The cells are deformed in the direction of the macroscopic stretching, but the shifts between cells are possible only along a dense direction of the lattice.

spots, instead of the 2 spots obtained for an affine deformation (Figures 4 and 5).

Therefore the deformations that occur with a stiff dispersed phase and a soft continuous phase must involve a change of structure in the lattice and not only in the unit cell.

Loss of Correlations in the Lattice. As explained in STRUCTURES OF DEFORMED FILMS, the stripe patterns correspond to strings of water rich pockets; these water pockets are elongated in the stretching direction and repeated at the lattice spacing in this direction. Since there is no diffraction in the perpendicular direction, the strings are irregularly spaced in this direction.

These strings may be characterized according to where they are and how they are oriented. First, they must occur everywhere in the sample. Indeed, the stripes contain most of the intensity in the diffraction pattern. Therefore they must be produced by objects that contain most of the water. Second, the spacing of these water pockets in a string matches the spacing of cells along a direction where they are nearest neighbours (first diffraction order, [111] and [200] directions). Therefore the structure may also be described as made of columns of cells which are arranged in a direction where successive cells in a column are nearest neighbours in the lattice (Figure 9). In this sense, the lack of scattering in the perpendicular direction indicates that the correlations between columns are lost. This description implies that the stretching of a wet sample causes irregular shifts between columns of cells.

Conclusion

The present latex films are composite materials with a cellular structure. The cell cores are the dispersed phase of the composite, while the membranes form the continuous phase. Through selective swelling of either phase the material can become soft in the dispersed phase or soft in the continuous phase.

The response of these materials to a macroscopic stretching depends on which phase is the weaker phase of the composite. Indeed, in a macroscopic stretching, the dispersed phase tends to respond with a stretching of each cell, whereas the continuous phase tends to respond with a slip between cells.

In the situation where the cell cores are the stiff component, we have demonstrated that the microscopic response to a macroscopic stretching is indeed a homogeneous stretching of all cells. Conversely, in the situation where the membranes are the soft component, the deformations are localized in the membranes, the cells slip past each other, and the lattice is disorganized by these microscopic shear deformations.

Appendix

This appendix presents the calculation of the azimutal variation of the diffracted intensity caused by an affine deformation of the film [10]. It is assumed that the films are organize into polycrystalline structure (**see Diffraction Patterns**), and that the crystallites have random orientations in the undeformed film. An affine deformation of this polycrystal causes each cell to be elongated by a factor λ in the parallel direction, and shrunk by a factor $\lambda^{-1/2}$ in the perpendicular direction.

For a polycrystalline sample, the set of all vectors q that were, before deformation, located within a solid angle $d\Omega$ becomes shifted into a new solid angle $d\Omega'$ located at an angle q' from the stretching direction. Similarly, the intensity located in intensity $d\Omega$ becomes shifted $d\Omega'$; its value may be calculated as:

$$I(\theta') \,/\, I_{iso} = d\Omega'./ d\Omega = \lambda^{3/2} \,/\, (\lambda^3 \cos^2\theta' + \sin^2\theta')^{3/2}$$

This redistribution of the intensity causes a decrease by a factor λ^3 in the parallel direction and an increase by a factor $\lambda^{3/2}$ in the perpendicular direction. The scattering pattern calculated in this way for the 111 diffraction line of a polycrystalline sample has been shown in **Figure 5b**.

References

1. Dobler, F.; Pith, T.; Lambla, M.; Holl, Y. J. Colloid Interface Sci. 1992, 152, 1
2. Winnik, M.A.; Wang, Y. J. Coatings Techno. 1992, 64, 51
3. Distler, D.; Kanig, G. Colloid Polym. Sci. 1978, 256, 1052
4. Zozel, A., Heckmann, W., Ley, G.and Mächtle, W.,
   Makromol. Chem., Macromol. Symp. 1990, 35-36, 423
5. Joanicot, M., Wong, K., Maquet, J., Chevalier, Y., Pichot, C., Graillat, C.,
   Lindner, P., Rios, L., Cabane, B., Progr. Colloid Polym. Sci. 81 (1990) 175
6. Chevalier, Y., Pichot, C., Graillat, C., Joanicot, M., Wong, K., Maquet, J.,
   Lindner, P., Cabane, B., Colloid Polym. Sci. 270 (1992) 806
7. Wang, Y., Kats, A., Juhué, D., Winnik, M.A., Langmuir 8 (1992) 1435
8. Rieger, J., Hädicke, E., Ley, G., Lindner, P. Phys. Rev. Lett. 1992, 68, 2782
9. Wong, K., Joanicot, M., Richard, J., Maquet, J., Cabane B,
   Macromolecules 26 (1993) 3168
10. Rharbi,Y. Thesis of Paris-Sud university, june 1995
11. Pieranski, P. Contemp. Phys. 1983, 24, 25.
12. Guinier, A. Theorie et technique de la radiocristallographie; Dunod: Paris 1964

Chapter 18

Influence of Morphology on Film Formation of Acrylic Dispersions

M. P. J. Heuts[1], R. A. le Fèbre, J. L. M. van Hilst, and G. C. Overbeek

Research and Development Department, Zeneca Resins bv, Sluisweg 12, P.O. Box 123, 5140 AC Waalwijk, Netherlands

Latex drying behaviour and (film) properties as minimum film forming temperature (MFT), dynamic mechanical behaviour, hardness and tensile strength have been compared, for latices prepared according to two stage emulsion polymerisation and by blending low and high Tg single stage latices. Compared to single stage counterparts, having the same overall composition, two stage systems as well as blends remain a significant lower MFT up to high levels of high Tg polymer. All differences, including the ones in dynamic mechanical behaviour, can be ascribed to differences in latex morphology. For all latices, initial dry rates are observed which deviate from established film formation models and from experiments described in literature. It is shown that latex composition and latex morphology have a small, irrefutable influence on the initial water evaporation rate.

Often cosolvents are a substantial part of a water borne paint formulation because of their ability to reduce the minimum film formation temperature (MFT) of acrylic latices. At the moment paints based on these latices are regarded as more environmentally-friendly when they are compared to solvent borne paints. To fulfil future governmental and environmental requirements regarding solvent emissions from paints, water borne paints need to be developed that contain little or no cosolvent. This requires latices that do not need cosolvents to achieve a low MFT. For both interior and exterior paint applications, it is imperative that these low MFT latices do not impair coating properties such as block resistance, hardness, gloss and drying behaviour. There are two main latex preparation methods [1,2] which result in low MFT latices that retain such coating properties. These methods are:
*   blending of high and low MFT latices,
*   using a two-stage or sequential polymerisation technique [3-5].
These preparation methods will result in latex types with different particle morphologies.

[1]Current address: Zeneca Resins, 730 Main Street, Wilmington, MA 01887

Coatings made of these latices will have different morphologies and therefore different coating properties. It is the aim of this paper to compare latices, prepared according to these two methods, not only with each other, but also with the single stage latex having the same overall composition. This will be done by comparing mechanical properties as hardness, block resistance, Young's modulus and maximum elongation. Furthermore the influence of the method of latex preparation on its drying behaviour is studied.

Experimental

Emulsion Polymerisation.   All latices were prepared according to the same semi-continuous procedure. The materials used are listed in Table I.

Table I. Components for the used emulsion polymerisation processes		
Reactor charge :	wt (g)	
demineralised water	812.5	
sodium lauryl sulphate	8.78	
ammonium persulphate	4.39	
Feeds :	stage (1) wt (g)**	stage (2) wt (g)**
demineralised water	$257.5 \times R$	$257.5 \times (1-R)$
sodium laurel sulphate	$8.78 \times R$	$8.78 \times (1-R)$
butyl acrylate	$X \times R*$	$X \times (1-R)$
methyl methacrylate	$Y \times R*$	$Y \times (1-R)$
acrylic acid	$26.3 \times R$	$26.3 \times (1-R)$
Neutralisation/solids adjustment:		
ammonia/demineralised water	30.0	

*X+Y=851.7 g; the X/Y ratio depends on the theoretical Tg of the copolymer according to the Flory-Fox equation. **R is the weight fraction of the first stage monomers on the total amount of monomers; R=1 for single stage or average latices.

The emulsion polymerisations were carried out in a 2 litre baffled glass reactor equipped with a condenser, a mechanical stirrer and a nitrogen inlet tube. During the whole process a nitrogen atmosphere was maintained in the reactor.

The water, surfactant and initiator were charged to the reactor, together with 114.4 g of the pre-emulsified stage(1). The temperature was raised to 85°C and maintained throughout the procedure. About 5 minutes after reaching the reaction temperature the rest of the stage(1) feed was added in 90*R minutes. Directly thereafter the pre-emulsified stage(2) feed was added in 90*(1-R) minutes. After completion of the second feed the reaction temperature was maintained for 30 minutes before cooling to room temperature. Based on the solids content (before adjustment), the monomer conversion was >99.7%. The pH was set to 7.0±0.1 with ammonia and the solids content of the latex was adjusted to 45.0 wt% . After this the latex was filtered over a 200 mesh cloth.

Particle Size Analysis. Particle size analysis was performed on the Otsuka ELS-800 (photon correlation spectroscopy, Stokes-Einstein). Prior to the analysis, samples were diluted in a 1 mM aqueous NaCl solution, with a pH set to 8.3±0.2 with NaOH.

Dynamic Mechanical Thermal Analysis (DMTA). DMTA was performed on 0.8 - 1.0 mm thick free films, which were prepared by pouring the latex into a aluminium petri dish and drying them at 60°C for 18 hours. The analyses were done on a Polymer Laboratories DMTA MkII in dual cantilever bending mode. The temperature range was -40°C to 140°C; heating rate 4°/min; frequency 1Hz; strain x4.

Determination of König Hardness. The König Hardness was determined by using an Erichsen model 299/300, according to ASTM D 4366. Coatings were prepared by casting a 80 µm thick wet film on a glass plate and subsequently dried for 18 hours at 60°C (for latices with a MFT above 60°C, a drying temperature of 100°C was used).

Gravimetric Determination of Evaporation Rate. The water evaporation rate from the latex was recorded as weight loss in time. Therefore a film was cast on a cold rolled steel Q-panel using a 100 µm wire rod applicator. Directly after casting, the Q-panel was placed on an analytical balance. Special care was taken to avoid influences of temperature, relative humidity and air flow.

Determination Minimum Film Formation Temperature. The MFT of the latices was determined with the Sheen MFFT Bar model SS-3000. The wet film thickness used was 60 µm.

Determination of Stress Strain Curves. Stress strain experiments were performed on an Instron 4301 fitted with a 100 N load cell and an optical extensometer. The crosshead speed was 100 mm/min. The required free films were prepared by casting a 400 µm wet film on release paper and subsequently drying them for 6 hours at 50°C. Still being warm, the films were cut into test strips and dried for another 16 hours at 50°C. The stress strain curves were recorded at room temperature (25°C). The measurements were repeated five times.

Determination of Block Resistance. On a testchart (Leneta, type 2C) a film was cast using a 125 µm wire rod applicator, dried for 1 hour at room temperature, followed by 16 hours storage at 50°C. Then the testchart was cut into 1 cm wide strips. Two of these strips were placed perpendicular on top of each other (coating to coating) under a pressure of 1kg/cm^2 at a temperature of 25°C or 50°C for 4 hours. The tested film passed the block resistance test when the two test strips separated without film damage.

Results and Discussion

Latex Synthesis. A series of single stage latices plus two series of sequential latices have been prepared, of which the key parameters and some properties are summarised in Table II. For the sequential (two-stage) latices a low Tg stage of 5°C and a high Tg stage of

Table II. Key parameters of the prepared latices						
Code*	Ratio R**	Tg(1)/Tg(2) (°C)	Overall Tg (°C)***	Particle size (nm)	MFT (°C)	König Hardness (s)
AV5	1	5/-	5	75	0	23
AV10	1	10/-	10	78	10	43
AV15	1	15/-	15	76	13	69
AV20	1	20/-	20	81	16	94
AV25	1	25/-	25	75	21	120
AV30	1	30/-	30	78	27	141
AV36	1	36/-	36	78	35	177
AV41	1	41/-	41	84	44	172
AV60	1	60/-	60	83	66	179
SH20	0.8	5/60	15	84	5	40
SH30	0.7	5/60	20	84	7	51
SH40	0.6	5/60	25	76	13	66
SH50	0.5	5/60	30	81	16	86
SH60	0.4	5/60	36	81	37	102
SH70	0.3	5/60	41	76	55	120
SH80	0.2	5/60	47	78	59	167
HS20	0.2	60/5	15	78	3	40
HS30	0.3	60/5	20	78	7	54
HS40	0.4	60/5	25	79	14	74
HS50	0.5	60/5	30	80	26	100
HS60	0.6	60/5	36	80	37	111
HS70	0.7	60/5	41	82	53	139
HS80	0.8	60/5	47	84	53	167
BL20	-	60/5	14	-	2	21
BL30	-	60/5	19	-	3	28
BL40	-	60/5	25	-	3	40
BL50	-	60/5	30	-	5	52
BL58	-	60/5	-	-	13	-
BL60	-	60/5	36	-	15	79
BL70	-	60/5	41	-	33	106
BL80	-	60/5	47	-	57	132
BL90	-	60/5	54	-	61	157

*AVx = single stage or average latex (x denotes the theoretical overall Tg); SHy = two stage latex; sequence low Tg high Tg (y denotes the high Tg polymer percentage); HSy = two stage latex; sequence high Tg low Tg (y denotes the high Tg polymer percentage); BLy = blend of AV5 and AV60 (y denotes the percentage of AV60). ** see Table I. ***the overall Tg is the theoretical Tg based on the overall composition, according to the Flory-Fox equation.

60°C has been used. Both low Tg-high Tg and high Tg-low Tg feed sequences have been performed.

A fourth series of latices was prepared by blending latices with a Tg of 5°C (AV5) and a Tg of 60°C (AV60) in various ratios. The overall Tg, as presented in Table II, for the sequential latices as well as for the blends shows that the latex has an overall composition similar to the corresponding single stage latices.

DMTA. Figure 1 shows the DMTA storage moduli (E') for the samples coded AV5, AV25, SH40, HS40 and BL40, of which the last four have the same overall composition. For the single stage polymers a large drop in E' is observed when these polymers pass their glass-rubber transition (Tg). Normally, when a polymer passes its Tg, a drop in E' of 2-3 orders of magnitude is seen within a temperature range of 20-30°C. For the blend and two stage polymers a different behaviour is observed. The E' curve of the blend shows two distinct transition regions. One between 10 and 30°C and the other between 70 and 90°C. The presence of two such distinct transitions means that the low and high Tg polymers are not mixed on a molecular level, but that they are present in separate domains. The "plateau" between 30 and 60°C indicates that the high Tg polymer contributes in film toughness [6]. The presence of such a "plateau" in the E' curve indicates that, upon heating, the blend as well as the two stage polymers retain more of their toughness over a wider temperature range than the single stage polymer having the same overall composition.

For the two stage latices it can be observed that the "plateau" is less prominent, a more gradual change in E' is observed. This indicates that in the two stage latices a higher interaction exists between the low and high Tg polymer. This higher interaction can indicate an increased surface-surface interaction between the soft and hard polymer domains, probably due to a smaller domain size of the hard polymer.

MFT. The particle size of the prepared latices is presented in Table II. The variance in particle size is considered to be small enough not to effect the MFT significantly [7]. As can be seen in Figure 2 and Table II, the MFT of the single stage latices increases proportionally with their overall Tg. The blends and the two stage latices show a totally different behaviour.

For the blends the MFT stays below 10°C up to a high Tg polymer content of 50%. Above 50% the MFT starts to rise rapidly. This behaviour is explained as follows. At small fractions of high Tg polymer in the blend, the low Tg polymer is present as the continuous (film forming) phase whereas the high Tg polymer acts like an almost inert filler. When the fraction of high Tg polymer increases, it is not longer possible to form a continuous film at ambient temperatures.

As can be seen in Figure 2, the MFT of the two stage latices lies between the MFT of the blend and the single stage latices with the same overall compositions up to 50% high Tg polymer. This is in line with the DMTA observations. Although the presence of separate low and high Tg polymer domains is evident, the interaction between the low and high Tg polymer is different than for the blends. This is caused by different latex morphologies. In the blends the low and high Tg polymer are present as separate particles, in the two stage latices they are present in the same particle. As they have

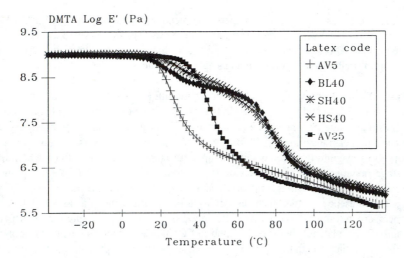

Figure 1. DMTA log E'- temperature curves for latex systems having the same overall composition (AV25, BL40, SH40, HS40) and the low Tg phase of the blend (AV5). For clarity, the curves have been shifted to a storage modulus of 1E9.0 Pa @ -40°C.

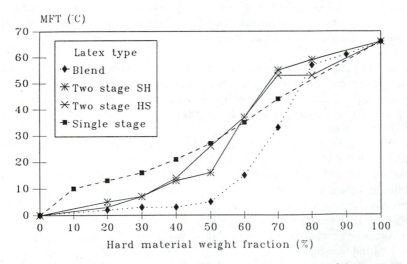

Figure 2. Latex MFT as function of the percentage high Tg material. The single stage latices, with the corresponding overall composition, are added for comparison (see Table II).

comparable latex particle sizes, a significant smaller domain size of high Tg polymer is expected in both types of two stage latices.

König Hardness. The pendulum damping test is sensitive in detecting differences in coating hardness, where hardness is defined as resistance to deformation. The results of the König Hardness determinations are given in Table II and depicted in Figure 3.

For the single stage latices the Tg dependence of the König Hardness presents itself as a S-curve with the steepest inclination where the Tg resembles the measurement temperature (in this case 25°C). The data presented in this paper, show only the middle and upper part of this S-curve. The steepest increase in König Hardness is observed in the Tg range 10 to 30°C. When the coating has a Tg below the measurement temperature, all polymer chains are in their rubbery state and are relatively easy to deform. This results in a low and hardly changing König Hardness. In the region just below or just above the Tg of the film, the glass transition dictates the deformation resistance, in a similar way as visible in the DMTA E' curves of the single stage polymers AV5 and AV25 (Figure 1). There as well, the observed transition is not instantaneous but spans a region of about 20-30°C. For coatings with a Tg (far) above the measurement temperature, all polymer chains are in their glassy state and therefore are not deformable, resulting in high König Hardnesses.

The blends used in this investigation contain different ratios of polymers with Tg's of 5 and 60°C. At low high Tg polymer fractions (10-30%), the resulting film can be described as a continuous film of low Tg polymer with islands of high Tg polymer embedded into it. The deformation behaviour of this film will be quite similar to the pure low Tg material. At increased volume fractions of high Tg polymer (30-50%), the König Hardness starts to rise, whereas the MFT (see Figure 2) hardly rises in this region. If the high Tg polymer fractions come above 50%, a gradual (almost linear) increase in König Hardness is observed. This behaviour is in contrast with the way the MFT changes with increasing fractions of high Tg polymer. At low fractions of high Tg polymer the König Hardness is dominated by the low Tg polymer, at higher fractions there is no explicit region were the high Tg polymer dominates the König Hardness.

The two stage material curves seem an intermediate between a single stage latex and a blend, with the steepest change in their 'S-curve' probably between 70 and 80% high Tg material. This is in agreement with the observations and conclusions regarding DMTA and MFT that two stage prepared latices contain elements observed in blends as well as in single stage latices.

Block Resistance. The block resistances of coatings with the same overall composition (AV25, SH40, HS40 and BL40) and a MFT of ±13°C (AV15, SH40, HS40 and BL58) are shown in Table III. At 25°C only AV15 fails, at 50°C AV15 and AV25 fail. When comparing the E' curves for the materials with the same composition (Figure 1), it is noticed that at a (DMTA) temperature of 50°C, the E'-value for AV25 is about 10 times smaller than for the blend and the two stage polymers. (The E'-value for AV15 is expected to be even lower.) For these experiments there seems a good agreement between the temperature dependence of E', and the observed block resistances.

Stress Strain results. Figure 4 presents the stress strain data obtained for polymers with the same overall composition. Table III shows their elongation at break and their Young's modulus.

The results are in line with the observations described above. The single stage polymer AV25 (theoretical Tg 25°C) is near its Tg and behaves as a tough plastic, resulting in the highest Young's modulus. The two phase systems (BL40, SH40 and HS40) are all based on a low Tg polymer matrix (Tg 5°C) in which high Tg material (Tg 60°) is embedded. For all four latices the elongation to break is comparable.

However, a completely different picture is obtained when latex compositions with a MFT of ±13°C are compared. Figure 5 and Table III show the corresponding data. The soft single stage material AV15 has a very low Young modulus, due to the fact that this material is completely in its rubbery state. It can be stretched up to a higher elongation before break than AV25 (Figure 4). The two stage materials have a considerably higher Young's modulus but have a lower maximum elongation than the blend or single stage material with the same MFT.

The maximum elongations obtained by the two stage materials are similar or higher than the maximum elongation obtained by the single stage material having a Tg of 25°C. The block resistance tests have shown the actual strength (moduli) of the two stage and blended polymers (SH40, HS40, BL40 and BL58) to be comparable at room temperature and even superior at 50°C, when compared with a single stage polymer with a Tg of 25°C.

Table III. Modulus and maximum elongations for the systems with a hard polymer content of 40% and for systems with a MFT of ±13 °C					
Code	MFT (°C)	Young's Modulus (kPa,@25°C)	elongation at break (%)	block resistance @25°C	@50°C
AV15	13	115	380	fail	fail
AV25	21	2040	240	pass	fail
SH40	13	1450	200	pass	pass
HS40	14	1600	180	pass	pass
BL40	3	650	240	pass	pass
BL58	13	220	595	pass	pass

Drying. In literature a number of approaches[8,9] are described to study the drying behaviour of latices. Our approach resembles Eckersley's [8]. However, in stead of pouring the latex in a petri dish, the latex was cast on a cold rolled steel Q-panel. Unexpectedly, this led to a remarkable change in results. When a petri dish was used, all latices tested show a drying behaviour as described by Eckersley. Initially the latices seem to lose water at a constant rate and after some time this rate decreases and approaches zero. This behaviour is consistent with the film formation models as proposed by Vanderhoff et al. [10] and Croll [9,11].

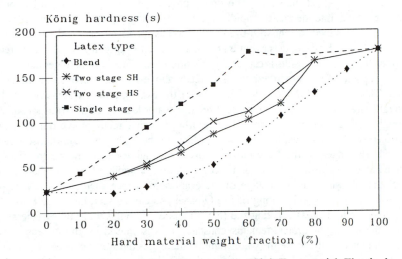

Figure 3. König hardness versus the percentage high Tg material. The single stage latices, with the corresponding overall composition, are added for comparison (see Table II).

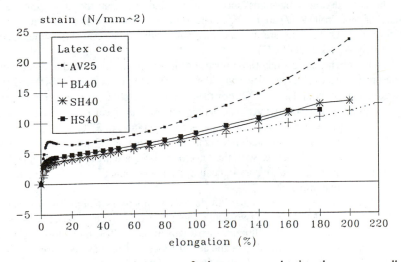

Figure 4. Stress - elongation curves for latex systems having the same overall composition.

When drying latices on a Q-panel, two striking differences are observed. The first is the much shorter drying time (one hour compared to several [8]). Second, the absence of a constant drying rate at the start of the drying process. In Figures 6 and 7 it can be seen that the drying rate decreases rapidly at the beginning of the drying process. This behaviour seems inconsistent with the drying models proposed by Vanderhoff et al.[10] and Croll [9,11]. Both models assume an initial constant evaporation rate. The difference between their and our results is caused by the difference in measurement geometry. When water evaporates from a petri dish the air directly above the latex becomes saturated with water. This results in a reduced driving force for the evaporation of water from the latex. The rate determining factor, with respect to latex water loss is not the transport of water through the latex but the diffusion of water from the air directly above the latex to the environment. As a Q-panel has no vertical edges on the side, water can diffuse sideways as well. The edges of the latex film will dry more quickly than the centre. This was verified by casting a film of a 0.5% aqueous solution of sodium lauryl sulphate on a Q-panel and following the initial stage of the evaporation process. Sodium lauryl sulphate was used to assure good wetting of the Q-panel. As can be seen in Figure 8 the water evaporation rate decreases during the first 5 minutes although the total surface area, of which the water evaporates, remains constant. This is caused by the increasing water concentration directly above the water film. At the edges the water concentration directly above the water film will rise less quickly than in the centre because water can diffuse sideways. This will result in a non-constant dry rate at the start of the drying process. As Figure 8 indicates, the time needed for the dry rate to become constant is larger than 5 minutes.

The latex drying simulations shown above, indicate that the drying models by Vanderhoff and Croll (9-11) are not adequate to describe the aspects of the early stage in film formation. Our results are more comparable with the experiments, simultaneously presented by Winnik et al (12,13).

The latices chosen to study the influence of latex composition on the kinetics of the drying process are two-stage latices with the low Tg-high Tg sequence (SH) and latices having the same overall Tg of 25°C. For this comparative study, special attention has to be given to the drying conditions. All drying experiments have been carried out at a relative humidity between 49% and 50% and at a temperature of 27.0±0.5°C.

Figure 9 presents the cumulative weight loss as function of time for the two-stage latices. The results for the latices with the same overall Tg are presented in Figure 10. To avoid the influence of possible errors in coating thickness, only the first 30 minutes of the drying process are compared.

In Figure 9 it can be seen that with decreasing fraction of high Tg polymer (and therefore decreasing MFT) of the two-stage latex the drying process takes longer. So, the composition of a latex clearly influences its drying process. The more low Tg material present in the latex particles the slower the drying process. This may be caused by the fact that the latex particles with the higher fraction of low Tg polymer are more easily deformed and hence may coalesce to a higher extent or form a denser packing then the latex particles with a lower fraction of low Tg polymer. Both phenomena result in less and smaller pores through which water can be transported.

The influence of latex preparation method on the drying behaviour can be seen in Figure 10. The two-stage latices (MFT 13°C) dry faster than the single stage latex (MFT

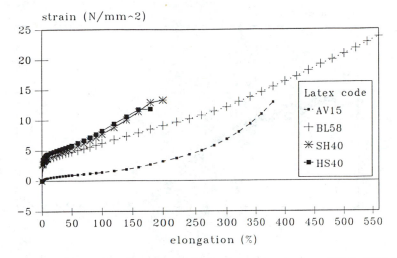

Figure 5. Stress - elongation curves for latex systems having a MFT of 13°C.

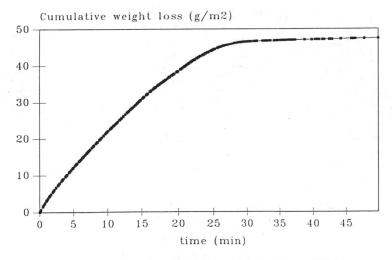

Figure 6. Cumulative weight loss in time of latex AV15.

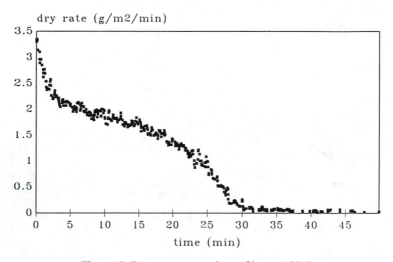

Figure 7. Dry rate versus time of latex AV15.

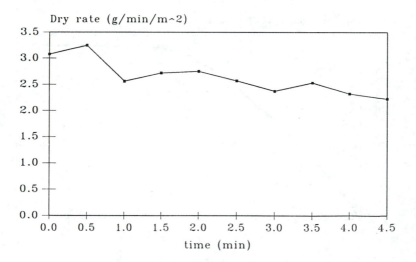

Figure 8. Cumulative weight loss in time of a 0.5% aqueous solution of sodium lauryl sulphate on a cold rolled steel Q-panel.

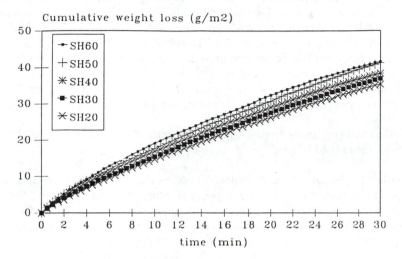

Figure 9. Cumulative weight loss versus time of films from latices prepared according to the two stage procedure.

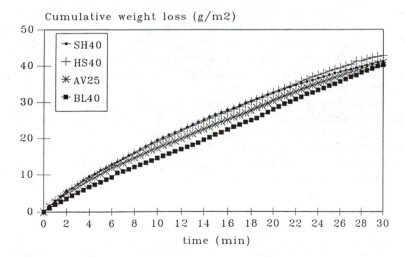

Figure 10. Cumulative weight loss as function of time from latex films, having an overall Tg of 25°C.

21°C). The blend (MFT 5°C) dries considerably slower. The fact that the blend dries slower than the two-stage or the single stage latices is expected, according to the results presented in Figure 9. The results found for the two stage and single stage latices are not in line with this interpretation. Their individual MFT's predict the reverse of what is found in practise. It is expected that the single stage latex with a MFT of 21°C dries faster than the two-stage latices with a MFT of 13°C. Therefore it seems that the presence of both high Tg polymer and low Tg polymer in the same particle limits the kinetics of particle deformation or coalescence. In the blend of the low Tg and high Tg latex the interaction of the low Tg and high Tg polymer is absent prior to the film formation process. The low Tg polymer particles can deform or coalesce relatively unhindered by the high Tg polymer.

The film drying results for the blends and the single stage latices are comparable with Winnik's results (*12,13*).

Particle Morphology. It is interesting to notice the large similarity in mechanical properties between two stage latices, prepared either by a soft-hard or a hard-soft polymerisation sequence. This suggests comparable particle morphologies and indicates that the two stage emulsion polymerisation process applied according to Table I, results in latices which are much alike. This requires further investigations.

Conclusions

The two methods of latex preparation, blending and two stage polymerisation, can produce low MFT latices which have improved mechanical properties over the single stage counterparts having the same MFT. Blended latices have a lower Young's modulus and a higher elongation at break than their two stage counterparts with the same MFT.

Not many differences are observed between the properties of the two stage latices with a low Tg high Tg sequence or a high Tg low Tg sequence, indicating comparable particle morphologies.

All latices show a similar drying behaviour. To obtain comparable data, special attention must be paid to experimental conditions. Not only drying conditions are very important but also the geometry used during the drying experiments. With the method of drying used for this investigation, a clear influence of the latex preparation method on the initial drying rate is observed. These differences in dry rate of the different types of latices are related to latex morphology.

Acknowledgments

The authors wish to thank Mrs M. Westerlink, Mr L. Donders, Mrs M. van Loo, Miss C. van Iersel and Mr P. de Bont for performing most of the experimental work and Mr F. Buckmann for fruitful discussions with respect to film formation of latices.

Literature cited

1.   Friel J.M., EP 466409 (1992); Makati A.C., US 4968740 (1990).
2.   Klesse W., EP 376096 (1990); Frazza M.S., EP 429207 (1991); Larsson B.E., EP 15644 (1980).

3. Arnoldus, R.; Adolphs, R.L; Zom, W.Z.W. *XXth Fatipec Congress*, Nice 1990, 81.

4. Devon, M.J.; Gardon, J.L.; Roberts, G.; Rudin, A. *J. Appl. Polym. Sci.* **1990**, *39*, 2119.

5. Tongyu Cao; Yongshen Xu; Yanjun Wang; Xuesun Chen; Aiqin Zheng, *Pol. International* **1993**, *32*, 153.

6. O'Connor, K.M.; Tsaur, S.L. *J.Appl. Pol. Sci* **1987**, *33*, 2007.

7. Eckersley, S.T.; Rudin, A. *J.Coat. Technol* **1990**, *62*, 89.

8. Eckersley, S.T.; Rudin, A. *Progr. Org. Coatings* **1994**, *23*, 387.

9. Croll, S.G. *J. Coat. Technol* **1986**, *58*, 41.

10. Vanderhoff, J.W.; Bradford, E.B.; Carrington, W.K. *J. Polym. Sci. Symp* **1973.**, *41*, 155.

11. Croll, S.G. *J. Coat. Technol* **1987**, *59*, 81.

12. Feng, J.; Winnik, M.A. "Latex blends and the kinetics of drying of latex dispersion". This ACS Volume.

13. Winnik, M.A.; Feng, J. *J. Coat. Technol* **1996**, *68*,39.

Chapter 19

Mechanical Studies of Film Formation in Waterborne Coatings by Atomic Force Microscopy

A. G. Gilicinski and C. R. Hegedus

Air Products and Chemicals, Inc., 7201 Hamilton Boulevard, Allentown, PA 18195

Atomic force microscopy (AFM) was been used to study film formation in coatings prepared from aqueous dispersions of acrylic/ polyurethane hybrid polymers. Analysis of solvent and low heat bake effects on film formation of latex blends of softer (low T_g) and harder (higher T_g) polymers were limited by the inability of topographic AFM methods to identify domains in the blend coatings. New AFM modes (friction force, force modulation, and phase contrast) were used to map mechanical properties (friction, stiffness, and adhesion) while simultaneously imaging topography. This approach was effective in identifying hard and soft domains by supplementing topographic data with mechanical property maps that reflected phase identity in the blend coatings. Mechanical AFM modes will have tremendous utility in film formation studies of waterborne coatings, especially for blends.

The advent of atomic force microscopy (AFM) has provided coatings researchers with high resolution methods for three-dimensional imaging of coating topography (*1*). Examples include a recent report on analysis of the extent of coalescence of waterborne latex coatings by measurement of the number density of protruding (uncoalesced) polymer particles from the coating surface (*2*). Other examples include analysis of the extent of coalescence in larger latex systems (300 nm diameter) by measuring the ability of the probe tip to measure between the latex particles (*3,4*).

While tremendously useful, topographic AFM methods are limited by their inability to identify different domains on a coating surface. New AFM modes seem ideally for this due to their ability to simultaneously measure topography and local mechanical properties. Three of these were used and compared in a study of solvent and environmental effects on film formation of waterborne latex coating blends.

Atomic Force Microscopy of Coalescence of Waterborne Latex Coatings

Single Component Latex Systems. For single component latex coatings, the topographic information from AFM measurements provides a powerful picture of film formation processes. For example, the use of the AFM to measure the extent of coalescence in film formation in waterborne coatings formed from acrylic/ polyurethane hybrid dispersions (40-60 nm diameter particle size) has been recently demonstrated (2). The basis for this measurement is quantitation of the extent of protrusion of uncoalesced polymer particles from the surface of the dried coating. Correction of image features for the probe tip size (typically 20 nm) is a key to the successful quantitative use of AFM in this application. If it is assumed that the coating surface reflects bulk processes that occur during film formation, then this approach to the analysis of surface topography can be used as a convenient probe of bulk film formation.

The general mechanism for film formation of a waterborne latex coating is shown in Figure 1. Coating topography for a system that has only partially gone through this mechanism should be markedly different from that of a completely coalesced coating. In this way, AFM images of the dried coating surface essentially provide a measure of how far down the mechanistic pathway the coating proceeded. To illustrate this, Figure 2 shows an AFM image of the surface of a coating that did not undergo extensive coalescence during the drying process. In contrast, highly coalesced coatings that have undergone complete film formation are flat at down to nanometer scales. Measurement of the number density per unit area of uncoalesced polymer particles protruding a certain distance from the surrounding coating surface provides a quantitative measure of extent of coalescence for that latex system.

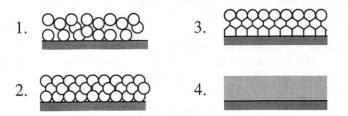

Figure 1. Model for film formation mechanism of a single component waterborne latex coating. After application (1) and initial water evaporation (2), particle close packing is achieved with water filling the interstices. Initial film formation (3) begins with compacted particles, and interparticle diffusion leads to an isotropic coating (4).

Multicomponent Latex Blends. The use of topographic AFM data is straight-forward for studying film formation of single component systems. The study of blends, however, is limited by the lack of chemical information tied to the topographic images. There are many cases where a blend of different polymer latexes can yield advantageous coating properties that combine performance

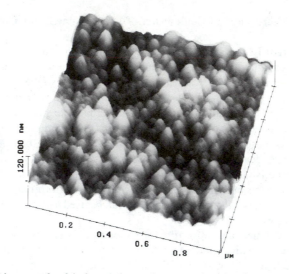

Figure 2. AFM image of a dried waterborne latex coating that has not undergone extensive film formation. Uncoalesced polymer particles are seen as spheres protruding from the surface of the coating.

characteristics of the individual components (5). In these cases, it is not easy to identify the component responsible for features observed in AFM images, and interpretation of the results can be problematic. More general questions about the film formation mechanism, as well as an understanding of the mechanistic effects of additives (such as organic solvents) or environmental effects (such as low temperature bake during drying), cannot be addressed without the ability to identify the chemical nature of the topographic features in the AFM images.

Topographic information can be used to derive some identification of different phases when there is substantial difference in the topography of different blend components and this difference is maintained in the blend coating. An example of this was seen in the coalescence study of the waterborne acrylic polyurethane hybrids (2). The lower T_g acrylic component showed an unusual pattern of shallow depressions 50-100 nm wide and only several nanometers deep. The basis of these depressions was thought to be swelling of the polymer particles (due to residual solvent from polymer synthesis) followed by particle shrinkage after water and solvent evaporation (and particle close packing) was complete.

In some blends with the lower T_g component, these shallow depressions are seen along with other coating morphology, and in these cases chemical identification of these features can reasonably be made. In many cases, however, the phase is too small in size to reproduce the characteristic topography seen in single component coatings, or there is a change in behavior between the single component coating and the phase in a multi-component blend. In such cases, a characteristic topography cannot be used to identify chemically different areas of the surface of a coating blend.

Solvent and Temperature Effects in Waterborne Coatings from Latex Blends

The film formation mechanism of multi-component polymer latexes is more complex than simple single component systems, and the effect of added organic solvent on film formation or environmental effects during cure are more difficult to predict. The study of a blend system with varying organic solvent content and drying conditions is described here to set the context for mechanical studies that will be described next.

Coating Preparation. Waterborne coatings from a blend of acrylic/polyurethane hybrid dispersions having a 1:1 ratio of polymer acrylics with high and low T_g ranges were prepared as described previously (2). The upper end of the T_g range was 100 and 50 °F, respectively. Residual N-methylpyrrolidone from the polymer synthesis was present and contributed a 150 grams/liter organic solvent level to the waterborne formulation. Organic solvent content was varied by the addition of Texanol (Eastman) to the formulation to boost solvent to 200 and 250 grams/liter. Coatings were applied with a 6 mil draw down bar to obtain coatings with a dry film thicknesses of 2 to 2.5 mils. A set of coated panels was dried under ambient lab conditions (70 °F, 50% relative humidity). Two other sets were dried for 15 minutes at ambient conditions and then for 30 minutes in ovens at 120 °F and 150 °F.

Solvent Effects. As found in the previous study, a series of shallow depressions was observed on parts of the coating surface in the blends dried at ambient temperature (2). Their shapes were remarkably uniform at lower solvent levels, varying only by about 10% in size and depth. There was a dramatic change in these depressions as solvent content was increased, with both diameter and depth increasing as solvent was added to the formulation. Size measurements are shown in Figure 3. This is consistent with the hypothesis that the depressions are due to shrinkage of the particles after water evaporation (and particle close packing) is achieved. For solvents that partition into the polymer, greater solvent levels would cause more fully swelled particles to shrink more and create larger depressions.

At higher solvent levels, the spread in diameter and depth increased, and it became more difficult to unambiguously identify the depression features on the hilly topography of the coating surface. This is likely due to the increased coalescence and extent of film formation induced by the solvent in the formulation. These effects lower the differentiation between blend components and make phase identification by topographic features more problematic. Topographic AFM imaging can only be used at best as a semi-qualitative analysis tool in this case.

Low Temperature Bake Effects. Heat applied during the film formation process in simple waterborne latex systems is known to enhance interparticle molecular diffusion and particle coalescence. A similar effect might be expected in multi-component blends as well. This issue was studied with the same set of coatings at three solvent levels in their formulations (150, 200, and 250 grams/liter) dried under ambient conditions (70 °F, 50% relative humidity), 120 °F, and 150 °F.

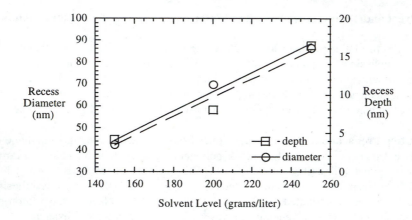

Figure 3. Average recess diameter and depth of shallow depressions in blend coating surfaces as a function of solvent content in the waterborne formulation. Data were averaged over 25 depressions per sample. The sizes were remarkably uniform for the lowest solvent level, with typical depth and diameter ranges of ± 10%. As solvent content increased the range increased as well; deviation for the highest solvent level was over 25%. This is a reflection of greater coalescence in the high solvent coating and a greater uncertainty in assigning topography to a component of the blend.

AFM images of the low solvent coatings dried at three temperatures showed a striking trend. The depressions seen in the low solvent coating at ambient temperature are rarely observed in the 120 °F dried coating, and are absent from the 150 °F coating. Two possible explanations can be proposed. One is that heating causes solvent evaporation and particle shrinkage to occur more rapidly and not to be delayed until particle close packing is achieved. In this case, the shrunken particles would be close packed at the end of water evaporation. A second hypothesis is that timing of events is similar, but that enhanced polymer chain mobility and increased coalescence causes subtle topographic features not to survive the drying process.

At higher solvent levels (250 grams/liter), different behavior is observed. The depressions seen after ambient temperature drying are present after low temperature heat baking, however the nature of the coating topography changes significantly. After ambient drying, the depressions are relatively large and deep, and although there is a relatively large range (± 25%) to their shapes and sizes, they are still identifiable as the same type of depressions observed at lower solvent levels after ambient drying. After a 150 °F bake was performed for 30, however, the depressions lose their distinct character and are very difficult to distinguish from the coating topography. The difference between ambient and 150 °F bakes on the coating topography can be seen in Figure 4.

Topographic AFM can be used to derive general conclusions about effects of solvent and temperature. Only limited interpretation is possible, however, without the ability to identify domains in coating blends. New modes of AFM are now emerging that make identification possible based on mechanical properties.

(A)

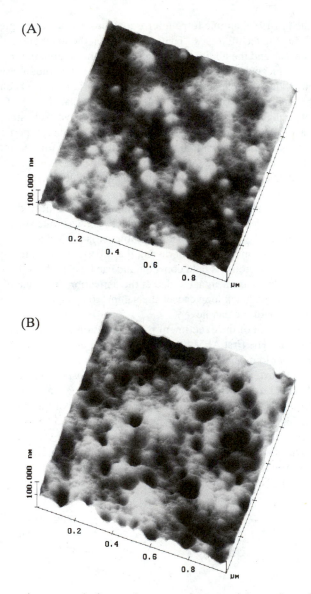

(B)

Figure 4. One micron atomic force microscopy images of the surface of two dried waterborne coatings from a 1:1 blend of high and low T_g acrylic polymer in acrylic/ polyurethane hybrid dispersions. Both coatings had identical formulations, including solvent added to a level of 250 grams/liter. The coatings were dried for 15 minutes after application in a lab atmosphere (70 °F, 50% relative humidity) and were then further dried for 30 minutes in (A) lab atmosphere, and (B) in a 150 °F oven.

New Mechanical Modes of Atomic Force Microscopy

Since its invention in 1986, atomic force microscopy has not been a single technique but a family of imaging technologies. They are based on the interaction of a probe tip with a sample surface and the ability to precisely move the sample in three dimensions using a piezoelectric ceramic scanner (6). "Contact mode" AFM was the original version of the experiment, however a range of variations has been since developed. These allow analysis of softer polymer samples, analysis of liquids, and measurement of a range of sample properties (friction, stiffness, adhesion, electrical charge, capacitance, ionic conductance, magnetism, heat, optical properties, and others) as a function of spatial location. Modes of AFM useful for polymer analysis will be discussed to illustrate ways that the tool can be used for topographic imaging and for mechanical analysis to identify domains in coating blends.

Contact Mode AFM. The most widely used form of AFM has been the "contact mode" experiment, where a tip is in constant contact with a sample surface at a constant applied force. A low force constant cantilever is pushed back (deflected) by the sample surface by a constant amount during the experiment, and the applied force (typically 5-100 nanoNewtons) is held constant and can be calculated by Hooke's law. Changes in topography cause a signal to detect the deflection of the cantilever holding the tip, and a feedback loop causes the sample to change height to reestablish the constant deflection of the cantilever.

A key component of the experiment is the detection scheme used to measure cantilever deflection. The first AFM had a scanning tunneling microscope on the back of the cantilever to measure changes in distance (6). Interferometric and capacitive methods have also been used. Currently, the most widely used detector is the optical reflection sensor, which features a simple design capable of achieving atomic resolution (7). A laser beam is focused onto the back of the cantilever, which is coated with a thin layer of gold for reflectivity. The reflected beam is directed to a split photodetector, and the relative voltage signal on the upper vs. lower halves of the detector provides an inexpensive yet incredibly sensitive monitor of cantilever deflection. This provides tremendous sensitivity in height changes on the sample surface. A schematic for the contact mode experiment is shown in Figure 5.

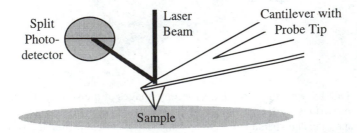

Figure 5. Schematic for contact mode AFM experiment. The sample sits on a piezo-electric scanner that moves it with Angstrom precision under a sharp probe tip.

Tapping Mode AFM. Contact mode AFM involves very low forces (5-100 nanoNewtons) applied to the sample surface. For soft polymeric materials, however, even this low force may be enough to damage the surface or push polymer chains around on the sample. Tapping Mode AFM is a version of an oscillating cantilever mode of the experiment introduced in 1993 (*8*). In this experiment, the cantilever and tip are oscillated at the cantilever's resonant frequency (around 300 kHz for a single crystal silicon integrated cantilever/tip). The same optical detector scheme is used as for contact mode AFM, except that instead of measuring the angle of the reflected laser beam, the total voltage spread on the photodetector is measured. As the sample approaches the freely oscillating cantilever/tip, the largest excursions of the tip oscillations will be dampened by "tapping" contact with the sample. The signal governing the tracking of the sample now becomes the degree of attenuation of the total voltage signal rather than the relative signal above or below the split in the photodetector. This scheme is depicted in Figure 6.

The practical benefit of this mode of AFM is that lateral forces are almost completely eliminated from the experiment. Instead of a constant drag of the tip over the sample, the tip "taps" its way along the surface with multiple taps for each physical point along the scan. The total applied force is reduced as well, making the method amenable to analysis of a range of soft materials including polymer coatings. A range of commercial instrumentation is now available with this mode of AFM.

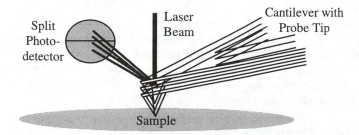

Figure 6. Schematic for tapping mode AFM experiment. The experiment is similar to contact mode AFM except for two changes. First, the cantilever and tip are oscillated so that the probe "taps" along the surface instead of maintaining constant contact. Second, the total voltage sweep is measured on the photodetector. Contact with the surface is monitored using the voltage sweep as a measure of the attenuation of the oscillations by "taps" on the sample surface.

Friction (Lateral) Force Microscopy. The measurement of friction in a contact mode AFM experiment (where a tip is in constant contact with a sample surface) can be easily done by adding a vertical split to the photodetector (*9*). By scanning the surface under the tip at 90 degrees to the axis of the cantilever, frictional effects will cause the cantilever to experience torsional forces based on the magnitude of friction between the probe tip and the sample surface. This method simultaneously monitors the reflected laser beam around the horizontal split in the photodetector to yield the height information, and around the vertical split to yield frictional/ lateral force information. In this manner, topography and friction information are simultaneously obtained by monitoring both data channels during the experiment.

Frictional effects are thought to be primarily surface-related phenomena, and will provide indirect composition information about the coating surface ([10]). Some flexibility in the experiment can be achieved by applying different surface treatments to the probe tip to change the range of frictional interaction observed in the experiment. Qualitative results can be obtained from uncalibrated cantilevers, however the torsional behavior of the cantilever must be calibrated to allow quantitative friction force experiments to be made.

Force Modulation Atomic Force Microscopy. The contact mode experiment can be further modified to allow local stiffness to be measured by applying a small AC modulation to the cantilever during contact mode scanning ([11]). Typical modulation frequencies are in the range of 5-20 kHz. A phase sensitive detector is used to measure the AC signal to essentially measure how the sample responds to the modulated signal. If the sample is softer than the force constant of the cantilever, it will yield to the tip pushing in and out of the surface. If it is stiffer, then the cantilever will not travel into the sample the same distance as the amplitude modulation. For different ranges of sample stiffness, cantilevers with different force constants can be used to provide additional dynamic range for the stiffness measurements.

The force modulation AFM experiment consists of two components: the DC signal that is used to measure topography, and the AC signal used to measure local sample stiffness. Both signals are measured simultaneously, so data for topography and local stiffness are acquired simultaneously. The power of the technique is the ability to map local stiffness and to correlate this mechanical behavior with topographic features on the surface.

Phase Contrast Atomic Force Microscopy. The first mechanical mode of AFM that works with the tapping mode instead of the contact mode AFM experiment is the phase contrast mode. Phase contrast AFM is almost identical to tapping mode except that an additional property of the cantilever oscillation is monitored and changes in that signal can be related to material properties. In addition to measuring the frequency and magnitude of the cantilever oscillation, the imaginary component of the resonant circuit is monitored. The theoretical understanding of the effects that give rise to changes in phase behavior is not yet mature, and reflects the fact that new AFM modes are invented and become commercially available faster than theoretical work can be done to understand them. However, the signal appears to be influenced both by the local stiffness of the surface as well as adhesive interactions between the tip and sample.

The phase contrast method has the great advantage of working with the tapping mode, which is critical for analyzing soft samples. The primary disadvantage is the lack of aids to interpreting the contrast mechanism that may be observed for a given coating sample.

Mechanical Studies of Waterborne Latex Coatings

To evaluate the potential for applying the three available mechanical modes of AFM (friction force, force modulation, and phase contrast microscopies), coatings from the blend studies were analyzed. All three proved useful in distinguishing between the two components of the blend coatings. Different information was obtained from the three modes. Friction force microscopy (FFM) appeared to yield primarily surface-sensitive information about polymer material at the coating surface. Force modulation AFM was more sensitive to subsurface as well as immediate surface material properties; for the same samples analyzed by FFM, force modulation data showed more complex structure and features that could reasonably be explained by sensitivity of up to 50 nm below the coating surface. Phase contrast AFM was most difficult to interpret but appeared to have material sensitivity midway between the other two modes; not as sensitive to subsurface material properties as force modulation, but sensitive to more than just the surface molecular properties that governed the FFM signal.

Friction (Lateral) Force Microscopy Results. Friction data clearly showed differences between the two phases in the blend coating system. An example of the topography and friction data from this experiment are shown in Figure 7 for the coating sample with the most clear differentiation between phases (lowest solvent level ambient cure coating). The topography was similar to that observed in earlier tapping mode AFM studies on these samples, with the difference that the shallow depressions easily seen in tapping mode are more difficult to discern in the contact mode measurement made in friction force microscopy. This illustrates the main drawback to using FFM - restriction of topography maps to contact mode imaging.

Friction data were very revealing. Spherical protrusions reminiscent of uncoalesced particles from earlier studies were found to have lower friction than lower areas between the protrusions. This fit the hypothesis of the lower spots being the lower T_g component of the blend. There was virtually a complete correlation between high topography spots in the height map corresponding to low friction spots on the friction map, indicating that the experiment was primarily probing at-surface frictional behavior. If subsurface properties contributed to the frictional response there would have been low friction spots present at topographically low spots.

The range of frictional response was found to change with solvent level and low temperature bake. Solvent addition decreased the friction range observed, lowering the frictional contrast between different domains. This effect was even more pronounced after low temperature heat bakes, with contrast after 150 °F heating being reduced to less than half the contrast found for ambient drying.

Force Modulation AFM Results. Stiffness data from force modulation experiments also showed distinct domains, but in a different manner than FFM. An example of the results is shown in Figure 8 for the same sample imaged by FFM in Figure 7. Several striking differences are noted. First, the topography data appear to

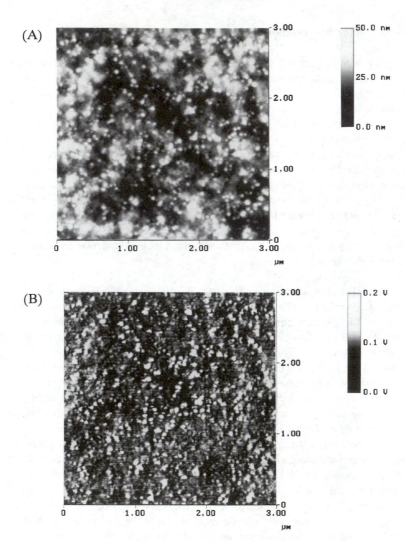

Figure 7. Results from friction force microscopy experiment on a low solvent (150 grams/liter) formulation of a coating blend dried at ambient lab conditions (70 °F, 50% RH). Two three micron images were simultaneously obtained: (A) Topography data obtained in contact mode AFM, and (B) frictional data for the same spatial area as scanned in A. The data are presented in "topview" mode, with grayscales reflecting topography and friction data, respectively. The topography scale in (A) is 50 nm, with topographically high spots being lighter colored and low spots being dark. The friction scale in (B) is a relative scale, with 0.2 Volts signal difference between the vertically split halves of the photodetector (low friction is light, higher friction is dark). Friction results clearly identify the high (less coalesced polymer) spots as being the higher T_g acrylic component of the acrylic/polyurethane hybrid dispersion formulation.

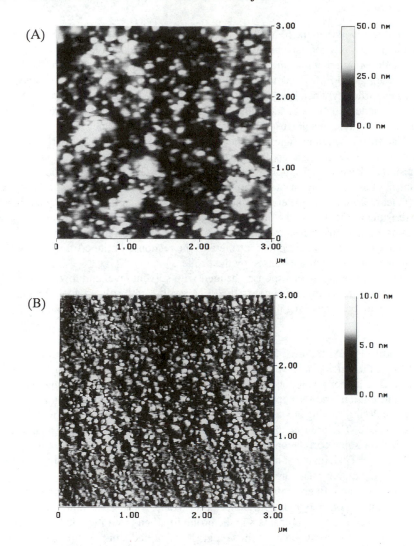

Figure 8. Results from force modulation AFM experiment on low solvent (150 grams/liter) formulation of a coating blend dried at ambient lab conditions (70 °F, 50% RH). Two 3 micron images were simultaneously obtained: (A) Topography data obtained in contact mode AFM, and (B) stiffness data for the same spatial area as scanned in A. The data are presented in "topview" mode, with grayscales reflecting topography and stiffness data, respectively. The topography scale in (A) is 50 nm, with topographically high spots being lighter colored and low spots being dark. The friction scale in (B) is a modulation scale, with 10 nm scale on the modulation travel of the tip relative to the modulation driven signal (high stiffness is light, low stiffness is dark). Stiffness data identify the high (less coalesced polymer) spots but also pick up stiff spots that do not correspond to topography data. This is likely due to sensitivity to subsurface mechanical properties of the polymer coating.

be less well resolved than the FFM results. Part of the reason is due to incidental tip issues, but there was a consistent difficulty in achieving the same resolution with the contact mode topography imaging in force modulation AFM compared to the same mode of imaging in FFM. The modulation of the tip into the sample may be causing slight added difficulty when imaging softer areas of the coating compared to FFM (where this modulation is not added) and especially compared to lower force methods such as tapping mode imaging used in the phase contrast AFM experiment.

The stiffness data in Figure 8B show many more particles than the friction data in Figure 7B, despite the fact that the topographic images (Figures 7A and 8A) show a similar density of protruding spherical particles. This is probably due to the mechanism of obtaining stiffness data in the force modulation mode. The probe tip is driven at least several nanometers into the coating surface, and the polymer material that is resisting or yielding to the probe is likely being influenced by mechanical stiffness extending tens of nanometers into the coating. Subsurface particles of the high T_g component are expected to contribute to the response of the coating if they are within the material volume sensed by the probe.

The range of stiffness response changed with solvent level and low temperature bake in a similar manner to the FFM results, with solvent addition and low temperature bake resulting in lower stiffness contrast between the domains on the coating surface.

Phase Contrast AFM Results. Differentiation based on phase behavior of the cantilever oscillation was also obtained in experiments with the range of coatings in this study. The higher resolution tapping mode imaging method provided markedly improved resolution, and the depressions observed earlier in the low solvent ambient temperature drying coating are seen in Figure 9A (shown at higher resolution than Figures 7 and 8). The protruding spheres are present as before, and the phase data in Figure 9B show some correlation between low phase change areas of the coating surface and high T_g domains of uncoalesced particles. The lower areas of topography show as varying higher phase areas on the phase image.

The possible influence of adhesion interactions between tip and sample may lead to a prediction of topography-influenced phase data. If there is some small adhesion between lower T_g domains and the probe tip, then the tip would be expected to have a higher adhesion to areas of the topography that had a higher surface area of contact to that tip. Depressions that reflected curvature at the end of the probe tip could provide higher contact areas (and therefore adhesive force) than rims of the depressions. Without a greater understanding of the mechanism giving rise to the phase contrast signal, deep interpretation is necessarily speculation, however there is some correlation between the shapes and locations of depressions in the low T_g areas and higher phase areas of similar shape. If adhesions (and its direct relation to the surface area of interaction) were not influencing the signal, the phase signal of low T_g areas with multiple depressions should not reflect the depressions.

The phase contrast signal range did not change as much as the friction and stiffness data did when comparing coatings with different solvent levels and dried with varying low temperature bakes. The reasons for this are not yet clear.

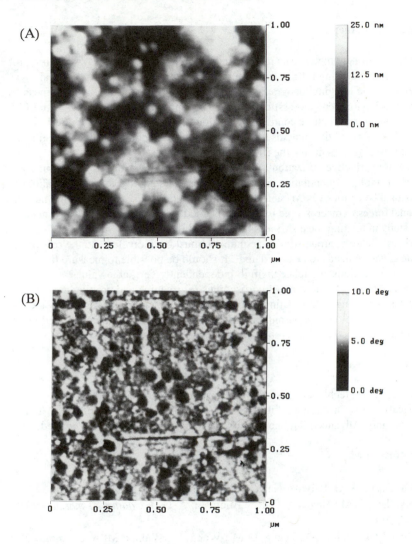

Figure 9. Results from phase contrast AFM experiment on low solvent (150 grams/liter) formulation of a coating blend dried at ambient lab conditions (70 °F, 50% RH). Two one micron images were simultaneously obtained: (A) Topography data obtained in tapping mode AFM, and (B) phase data for the same spatial area as scanned in A. The data are presented in "topview" mode, with grayscales reflecting topography and stiffness data, respectively. The topography scale in (A) is 50 nm, with topographically high spots being lighter colored and low spots being dark. The phase scale in (B) is ranged at 10 degrees difference in phase signal (high phase change is light, low is dark). Phase data change more at the low spots thought to be low T_g domains; this may reflect tip-sample adhesion being measured by the method. Less adhesion would be expected for the high T_g polymer domains.

Future Developments

The promise of nanometer scale materials characterization offered by scanning probe microscopies is nearing fulfillment with new AFM modes that probe material properties while imaging topography at nanometer resolution. Friction, stiffness, and adhesion behavior were successfully mapped to topography in this study with FFM, force modulation and phase contrast AFM. Identification of domains in waterborne coating blends aided the study of film formation as a function of solvent level and environmental factors during the film formation process.

More effective utilization of these tools for coatings studies will occur as further advances in quantitation and theoretical understanding are made. Calibration of torsional behavior of FFM cantilevers and an improved basis for relating nanoscale frictional forces to macroscopic friction is needed. Force modulation AFM needs more study in relating local stiffness data to modulus and other rheological properties. Phase contrast AFM's mechanism must be unraveled and the role of tip-sample adhesion needs to be elucidated. It should be possible to measure the adhesive interactions and relate them to independently verifiable values.

The development of new microscopies for imaging and probing nanoscale properties is coming as the coatings industry faces tremendous new technological challenges in developing environmentally compliant systems. The continued rapid development of methodologies such as AFM will help meet that challenge.

Acknowledgments

We acknowledge technical assistance by Kristen Kloiber (Air Products and Chemicals, Inc.). Discussions with Ken Babcock (Digital Instruments, Inc.) and Dr. Don Chernoff (Advanced Surface Microscopy, Inc.) are gratefully acknowledged.

Literature Cited

1.   Gilicinski, A. G. Polymer News **1996**, in press.
2.   Rynders, R. M.; Hegedus, C. R.; Gilicinski, A. G. *J. Coatings Technology* **1995**, *67*, 59-69.
3.   Goh, M. C.; Juhue, D.; Leung, O-M.; Wang, Y.; Winnik, M. A. *Langmuir* **1993**, *9*, 1319-1326.
4.   Meier, D. J.; Lin, F. *Langmuir* **1995**, *11*, 2726-2733.
5.   Hegedus, C. R., unpublished data.
6.   Binnig, G.; Quate, C.; Gerber, G. *Phys. Rev. Lett.* **1986**, *56*, 930-938.
7.   Ohnesorge, F.; Binnig, G. *Science* **1993**, *260*, 1451-1454.
8.   Strausser, Y. E.; Heaton, M. G. *American Laboratory* **1994**, 20-29.
9.   Noy, A.; Frisbie, C. D.; Rozsnyai, L. F.; Wrighton, M. S.; Lieber, C. M. *J. Am. Chem. Soc.* **1995**, *117*, 7943-51
10. Butt, H.-J.; Kuropka, R. *J. Coatings Technology* **1995**, *67 (848)*, 101.
11. Radmacher, M.; Tillman, R. W.; Gaub, H. E. *Biophysical J.* **1993**, *64*, 735-742.

Chapter 20

Latex Film Formation at Surfaces and Interfaces

Spectroscopic Attenuated Total Reflectance and Photoacoustic Fourier Transform IR Approaches

B.-J. Niu, L. R. Martin, L. K. Tebelius, and Marek W. Urban[1]

Department of Polymers and Coatings, North Dakota State University, Fargo, ND 58105

This chapter explores novel spectroscopic approaches that can be utilized in the analysis of film formation near surfaces and interfaces. Emphasis is given to attenuated total reflectance (ATR) and step-scan photoacoustic (SSPAS) Fourier transform infrared (FT-IR) spectroscopy as both techniques can provide meaningful information from various surface depths of organic films as well as are capable of determining molecular level interactions among film components. This chapter attempts to provide a comprehensive description of internal and external factors that may influence coalescence near the film-air (F-A) and film-substrate (F-S) interfaces of latex films.

In the last few years our efforts focused on understanding mobility and distribution of surfactant molecules during and after latex film formation.[1-13] Although these studies have led to several findings concerning the influence of a latex chemical makeup and coalescence conditions in relation to the behavior of small molecules in polymer networks, it became quite apparent that there are numerous opportunities for learning more about latex film formation. Our particular interest lies in molecular level understanding of processes near the film-air (F-A) and film-substrate (F-S) interfaces, as these zones of latex films have a significant influence on numerous properties. A particular emphasis will be given to a chemical makeup of latex particles, their size and size distribution, and the presence and behavior of low molecular weight species during coalescence. In a context of the film formation, we will focus on the recent developments in attenuated total reflectance (ATR) and step-scan photoacoustic (S^2-PAS) spectroscopies, as utilized to the interfacial and surface studies of latex films. In view of the above considerations, let us first establish the principles governing these measurements.

[1]Corresponding author

0097–6156/96/0648–0301$17.75/0

ATR FT-IR Spectroscopy

When light passes through two media with different refractive indices, and the media are in contact with each other, the path of the light is distorted. This is schematically illustrated in Figure 1. The majority of light is transmitted at a 90° angle of incidence (θ), and the light is partially reflected at $\theta < \theta_c$. A total reflection will occur at $\theta > \theta_c$. When the angle of incidence θ is greater than θ_c, and $n_1 > n_2$, the light is totally reflected. The critical angle is defined by $\theta_c = \sin^{-1} n_2/n_1$, and n_1 and n_2 are the refractive indices of a crystal and a sample, respectively. This is the basis for internal reflection spectroscopy (IRS).

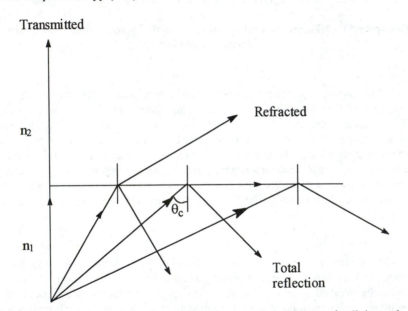

Figure 1. Schematic diagram of the refractive index changes on the light path as a function of incidence angle.

IRS can be used to measure the optical spectrum of a sample that is in contact with an optically denser and transparent medium. Since there is only one reflection during the measurement, sensitivity is limited. Therefore, this technique was modified by increasing the number of reflections, giving rise to attenuated total reflection spectroscopy (ATR). Experimental setups of the internal reflection technique for single reflection and multi-reflection methods is shown in Figure 2, B and C. For comparison, Figure 2 also shows transmission (A), reflection-absorption (D), diffuse reflectance (E) and photoacoustic (F) setups.

One of the advantages of ATR is that it is possible to conduct depth penetration experiments. The penetration depth, d_p, is expressed by the following equation:

$$d_p = \frac{\lambda}{2\pi\left(\sin^2\theta - n_{12}^2\right)^{1/2}} \qquad (1)$$

where: d_p (cm) is the depth penetration into the surface; n_1 and n_2 are the refractive index values of an ATR crystal and a sample, respectively; θ (degree) is the angle of incidence, and λ (cm^{-1}) is the wavelength of electromagnetic radiation in medium n_1.

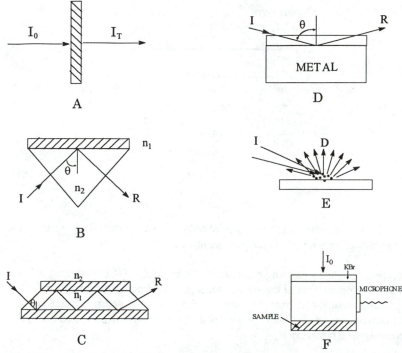

Figure 2. Schematic diagram of IR techniques: A - transmission; B - single reflection IRS (internal reflection spectroscopy); C - attenuated total reflection (ATR); D - reflection-absorption; E - diffuse reflectance setup; F - photoacoustic setup.

As shown by eqn. 1, one disadvantage for ATR FT-IR spectroscopy is that the depth of penetration is wavelength dependent. In an effort to eliminate this effect, a new algorithm was developed.[14] Furthermore, because eqn. 1 was derived with an assumption that the examined specimens are homogeneous, any composition/concentration variations preclude the use of this useful relationship for quantitative purposes. In an effort to utilize this relationship, especially if one is interested in depth profiling experiments, the approach schematically illustrated in Figure 3 was developed.[14] The surface is divided into n layers with each layer thickness, h_j. At the each boundary layer, L_j, the response of the sample to local evanescent waves is characterized by a complex refractive index defined by $\hat{n}_j = n_j - ik_j$, and n_j is refractive index and k_j is the absorption index. By applying developed algorithm to each layer,[14] which is assumed to be homogeneous, eqn. 1 can be utilized. However, the layers among themselves are not homogeneous, and by stacking all layers together, the surface is reconstructed by a step-wise treatment of volumes occupied by each layer. This approach allows quite accurate quantitative analysis of surfaces and its precision is determined by the number of spectra collected at various depths. Based on this principle, quantitative analysis of surfactant concentration near the F-A and F-S interfaces may be examined.

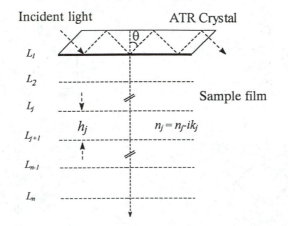

Figure 3. A schematic diagram of numerically slicing a nonhomogeneous surface to form a stack of parallel thin homogeneous layers.

Furthermore, by incorporating polarized IR light in an ATR experiment, it is possible to determine orientation of the surface species. The diagram of polarized ATR FT-IR experimental setup is schematically illustrated in Figure 4, and transverse electric (TE) and transverse magnetic (TM) are parallel and perpendicular polarized components of electromagnetic radiation.

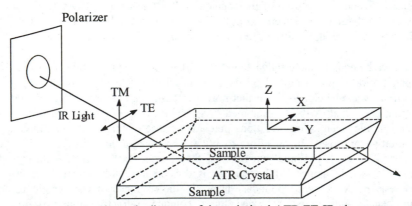

Figure 4. A schematic diagram of the polarized ATR FT-IR elements.

Photoacoustic (PA) Fourier Transform Spectroscopy

Photoacoustic spectroscopy is based on the detection of an acoustic signal emitted from a sample due to absorption of modulated radiation. This is schematically illustrated in Figure 5, A. The sample is placed in an acoustically isolated chamber to which a sensitive microphone is attached. On absorption of modulated light, heat is generated within the sample. Its release leads to temperature fluctuations at the sample surface. These temperature fluctuations at the sample surface cause pressure changes in a surrounding gas, which, in turn, generate acoustic waves in the sample chamber. The pressure changes of the gas are detected by a sensitive microphone, and the obtained electrical signal is Fourier-transformed. In the PAS experiment, the absorbed energy is released in a form of heat that is transferred to the sample surface, and the efficiency of the heat transfer is determined by the thermal diffusion coefficient of the sample, a_s (cm), and the modulation frequency of the incident radiation, ω:[15]

$$a_s = \left(\frac{\omega}{2\alpha}\right)^{1/2} \qquad (4)$$

where: α (cm^2/s) is the thermal diffusivity: ω (s^{-1}) is the angular modulation frequency, and equal to $4\pi v \upsilon$ (v (cm/s) and υ (cm^{-1}), are velocity of mirror and wavenumbers, respectively). The thermal diffusion length μ_{th} (cm) is related to the thermal diffusion coefficient a_s through

$$\mu_{th} = \frac{1}{a_s} = \left(\frac{2\alpha}{\omega}\right)^{1/2} \qquad (5)$$

Based on this relationship, the thermal diffusion length is inversely proportional to the modulation frequency. Therefore, by changing ω, one changes μ_{th}, which in effect result in changing the depth from which the acoustic signal is generated. Thus, the capability of surface depth profiling is one of the most appealing features of PA FT-IR. Urban et al.[16,17] have demonstrated several applications of PAS regarding surface depth profiling. However, the major drawback of this approach is that the thermal diffuse length μ_{th} also depends on wavenumber. As a result, low wavenumber regions result in deeper penetration depths, and at high wavenumbers, the spectral information comes from shallower depths. In an effort to eliminate the wavenumber dependence, and obtain the penetration depth wavenumber independent spectra, step-scan photoacoustic FT-IR spectroscopy can be employed.[18]

In a step-scan interferometry, a mirror of a Fourier transform interferometer is moved incrementally, and the spectra are acquired while the retardation is constant. As a result, the Fourier frequency dependence is eliminated and the photoacoustic sampling depth is constant across the spectrum. Furthermore, using a two-phase lock-in amplifier, in-phase (I) and in-quadrature (Q) components of the signal can be simultaneously acquired, and used to obtain the signal phase. A schematic diagram of a PAS cell is illustrated in Figure 5, A, and a detection of (I) and (Q) spectra are presented in Figure 5, B. While amplitude modulation of the PA signal is accomplished by a chopper to generate PA signal, in the phase modulation the retardation is sinusoidally varied by dithering the mirror at a fixed frequency. The disadvantage for amplitude modulation is that it produces a signal containing a large DC component, placing phase modulation in advantage over the amplitude modulation. The reason behind is that the use of PM rather than AM for a signal generation increases the step-scan signal-to-noise ratio (SNR) by a factor of 2. Secondly, the amplitude of the PM can be used to select the wavelength region most efficiently modulated. Thus, the step-scan mode, particularly in conjunction with PM, provides a constant frequency modulation for all wavelengths.

In this chapter, Fourier transform infrared spectroscopy, especially ATR and step-scan photoacoustic FT-IR will be employed to the analysis of latex film formation at the film-air and film-substrate interfaces. The distribution of surfactant molecules at the interfaces will be followed during the film formation and the structural features, such as orientation of surfactant molecules near the film-air and film-substrate interfaces will be examined. Our particular interest is understanding of molecular level interactions resulting from coalescence conditions, chemical makeup of latex particles, surfactant/copolymer interactions, and their effect on macroscopic film properties.

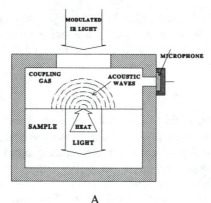

A

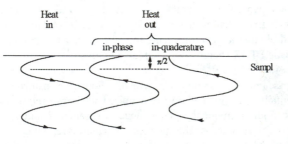

B

Figure 5. A - Schematic diagram of step-scan photoacoustic cell; B - Schematic diagram of two components of in-phase (I) and in-quadrature (Q) in step-scan phase analysis.

Homopolymer Latex Blends

Let us examine a series of ATR FT-IR spectra obtained from the F-A interface of a 50/50 mixture of polystyrene (p-Sty) and poly(n-butyl acrylate) (p-BA) latex homopolymers. Figure 6, Traces A through E, illustrate the spectra recorded from the same specimen using TM polarized light, obtained at the angles of incidence between 60 and 40°. Such choice of the incidence angles was dictated by the fact that such a range allows us to vary the depth of penetration of light into the film from 1.3 to 2.3 μm. Therefore, molecular level information from different depths can be obtained.

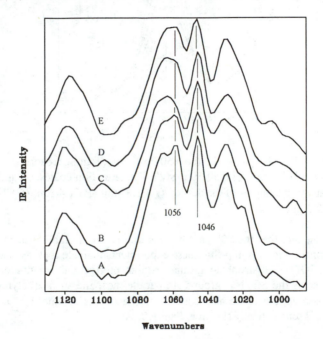

Figure 6. ATR FT-IR spectra in the 950-1150 cm⁻¹ region, recorded at the F-A interface (TM polarization) of a 50/50 p-Sty/p-BA latex film coalesced under ambient conditions, at various angles of incidence: A - 1.3 μm, B - 1.4 μm, C - 1.6 μm, D - 1.9 μm, E - 2.3 μm.

Analysis of the spectra shown in Figure 6 shows that, while going from 1.3 to 2.3 μm depths into the F-A interface, the intensity of the S-O stretching bands, resulting from the presence of the $SO_3^-Na^+$ entities associated with H_2O and acid groups at 1046 and 1056 cm⁻¹ decreases with depth. At 1.6 μm (Trace C), the 1056 cm⁻¹ band is not detected, and the 1046 cm⁻¹ band continues to decrease at a greater depth (Traces D, E). At the same time, the intensity of the 700 cm⁻¹ band changes in such a way that the band reaches its maximum around 1.6 μm from the top surface layer (Figure 7, Trace C). What is even more interesting is that, for the spectra recorded using TE polarization (Figure 7, Traces A-E), the strongest intensity of the

700 cm^{-1} band is also detected around 1.6 μm, and the band is more pronounced in the TE polarization.

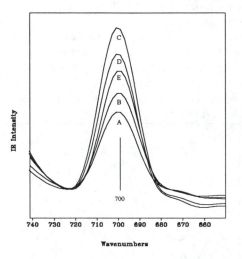

Figure 7. ATR FT-IR spectra in the 650-750 cm^{-1} region, recorded at the F-A interface (TM polarization) of a 50/50 p-Sty/p-BA latex film coalesced under ambient conditions, at various angles of incidence: (A) 1.3 μm, (B) 1.4 μm, (C) 1.6 μm, (D) 1.9 μm, (E) 2.3 μm.

Considering the fact that the 700 cm^{-1} band is due to the aromatic out-of-plane C-H normal deformation modes in p-Sty, these experiments indicate that styrene rings as a well as the SO$_3^-$Na$^+$ hydrophilic groups exhibit preferential orientation near the surface. Whereas the SO$_3^-$Na$^+$ groups are parallel near the surface, styrene rings of the p-Sty phase appear to be also parallel, at the depths around 1.6 μm from the surface. This is schematically shown in Figure 8, A.

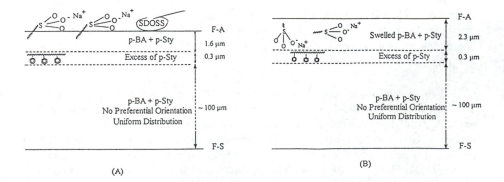

Figure 8. Stratification of p-Sty and orientation changes of surfactant molecules near the F-A and F-S interfaces: A - 40 % humidity; B - 100 % humidity.

In view of the above observations, and considering our previous data pertaining to the behavior of the $SO_3^-Na^+$ groups of SDOSS,[5-7] one of the puzzling issues is the intensity changes of the 700 cm^{-1} band. Since this band can serve as a probe for p-Sty behavior, let us examine if there are changes in the p-Sty content as a function of the coalescence time at various surface depths. For that reason, we followed the intensity changes of the 700 cm^{-1} band over a period of time at various angles of incidence. Figure 9 illustrates Sty intensity changes as a function of depth from the surface from 24 to 72 hours. After 24 hours, the p-Sty intensity increases at greater depths. After 40 hours, at the shallower depths, ranging from 1.3 to 1.6 μm, the 700 cm^{-1} band further increases. The 700 cm^{-1} band reaches its maximum at 1.6 μm from the F-A interface after 56 hours, and at depths beyond 1.6 μm continues to decrease with the increasing penetration depths. The same trend is found in the F-A interface spectra recorded 72 hours after coalescence. However, when the penetration depth reaches approximately 2 μm (corresponding to 43° angle of incidence), the band intensities converge for both

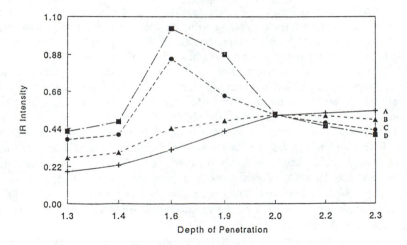

Figure 9. Relative intensities of the 700 cm-1 band at the F-A interface (TE polarization) at various surface depths as a function of time: A - 24 hrs; B- 40 hrs.; C - 56 hrs.; D - 72 hrs.

polarizations. Although from the latex film formation point of view, it is apparent that there are changes in the latex composition across the film, at this point, one should realize that the effect of critical angle needs to be addressed. The question arises if the presented results are not affected by the optical distortions resulting from a proximity of the critical angle.[14,15] For that reason we measured the refractive index of the latex films in the non-absorbing region. It appears that the refractive index is 1.53 ± 0.01. This value of refractive index gives a critical angle of 40°, and solving over the spectral range of interest, gives critical angles between 39.6° and 40.5°. Thus, the data convergence at 43° is not affected by the proximity of the critical angle, and results from coalescence. In contrast, the 700 cm^{-1} band showed no intensity changes at the F-S interface for either the TM or TE polarizations, indicating that stratification of p-Sty occurs only near the F-A interface.

Although based on the presented data it is quite apparent that there is a preferential phase separation near the F-A interface, let us utilize dynamic mechanical thermal analysis (DMTA) and examine the tan δ values as a function of temperature, as their maximum represent a glass transition temperature (T_g) of a polymer.[19] Analysis of the DMTA tan δ curves for the 50/50 p-Sty/p-BA latex mixture showed the presence of two T_gs at -52°C and at 98°C, indicating the presence of two separate phases within the latex film. While the T_g of -52°C is due to the p-BA latex component in a mixture, the 98°C T_g, is that of p-Sty. These observations indicate that, at this stage of coalescence, the two latex homopolymers in the mixture do not coalesce into a uniform network, and although DMTA data indicates that the separation occurs, these, and for that matter any other measurements, are unable to distinguish which portion of the film is phase separated. In contrast, ATR FT-IR surface depth profiling experiments near the F-A interfaces clearly demonstrate non-homogeneity in the direction perpendicular to the film plane. Thus, the presence of two separate T_g's results from stratification near the F-A interface. It should be noted that there were numerous studies dealing with the phase separation in polymers and polymer blends, to our best knowledge, this is the first approach actually showing where, in respect to the rest of the film, the phase separation occurs.

The above data demonstrate that the 50/50 p-BA/p-Sty latex film is a non-uniform composite of soft and hard particles near the F-A interface, and hydrophobic and hydrophilic particles of the composite latex will facilitate different interactions of the individual components. Therefore, the ability of water uptake can be used as a means of altering these interactions. A comparison of the results for the latex exposed to 100% RH (not shown), with the data shown in Figure 9 (40% RH) indicated that, upon exposure to humidity, the bands of 1046 and 1056 cm^{-1}, characteristic of SDOSS, are not detected. Furthermore, the band due to p-Sty at 700 cm^{-1} is significantly weaker, and reaches its maximum intensity at approximately 2.3 μm from the F-A interface. As we recall, a maximum intensity at 40 % RH was detected near 1.6 μm for the F-A interface.

Based on these experiments, it is obvious that when the latex films are exposed to 100% humidity, water uptake occurs. Furthermore, since surfactant molecules are water soluble, their removal from the surface is attributed to the fact that water penetrates the network, thus allowing SDOSS molecules to diffuse into it. This process is facilitated by an excess of the free volume at the temperature of experiment (27°C), as the T_g of the p-BA phase is -52°C, as oppose to the p-Sty phase which exhibits the T_g of 98°C. Thus, p-BA has substantially accessive amount of the free volume and represents a good medium for water uptake and subsequent diffusion of SDOSS molecules. While this behavior of SDOSS under given conditions is anticipated, the diminished intensity of the 700 cm^{-1} band due to the p-Sty phase is somewhat surprising. However, when water diffuses into the film, the top p-BA surface layer can be plasticized by water, which results in swelling it. Thus, the effective thickness of the p-BA layer increases due to water intake, and this phenomenon is believed to be responsible for the hydrophobic p-Sty phase being detected at greater depths. While Figure 8,A schematically depicts a stratification of the p-Sty phase near the F-A interface, Figure 8, B illustrates the case of the top p-BA layer being swelled by H_2O. When such a latex film is exposed to 100% RH, the top layer becomes thicker. This is shown in Figure 8, B.

At this point it is appropriate to mention that, in contrast to our previous studies on 50/50 p-Sty/p-BA copolymer,[12] not a mixture of homopolymers, there was no stratification of polystyrene near the F-A interface. Furthermore, there were no intensity changes of the 700 cm^{-1} band for the TM and TE polarizations, and no intensity differences between various penetration depths. As one would anticipate, the DMTA data showed the presence of a single T_g at 16°C. Thus, the results on the p-Sty/p-BA copolymer indicated no phase separation, which was also accompanied by no preferential orientation of the polystyrene rings near the surface. In addition, only the 1046 cm^{-1} band characteristic of the $SO_3^-Na^+-H_2O$ interactions near the F-A interface was detected. However, the 1056 cm^{-1} band was not present,[12] indicating that the $SO_3^-Na^+$-HOOC latex interactions are not detected. Thus, there are significant differences in the latex surface and interfacial properties, depending upon the original latex makeup. When latex particles are made up of a copolymer, the latex particles are able to coalesce, thus giving a single phase latex film. In a latex film composed of homopolymers, the hard p-Sty and soft p-BA particles display a phase separation, but the separation occurs near the F-A interface, and stratification of the p-Sty layers are detected. The phase separation between the homopolymers during the film formation also influences the mobility and orientation changes of SDOSS within the latex film.

Latex Particle Composition

In an effort to determine how the monomer composition of the latex films influences the surfactant concentration and its orientation throughout the film, the following latex copolymers were prepared: 50% Sty/50% n-BA, 30% Sty/70% n-BA, and 10% Sty/90% n-BA. Traces A, B, C, D, and E of Figure 10a illustrate ATR FT-IR spectra of the 50% Sty/50% n-BA latex copolymer. The spectra were recorded using parallel (TE) polarization. For reference purposes, the spectrum of neat SDOSS surfactant is shown in trace F. The presence of the 1046 cm^{-1} and 1056 cm^{-1} bands indicates that there is a significant amount of water and hydrogen-bonded acid groups associated with the $SO_3^-Na^+$ surfactant groups near this interface.[5] Based on the analysis of the spectra shown in Figure 10a, it appears that there is a decrease in the surfactant concentration of the $SO_3^-Na^+$ groups associated with water and acid entities when going from approximately 1.35 to 2.30 μm. This is demonstrated by the decrease of the 1046 cm^{-1} and 1056 cm^{-1} band intensities. However, when the perpendicular (TM) polarization spectra are examined, the situation changes. Instead of the 1046 cm^{-1} and 1056 cm^{-1} bands, the 1050 cm^{-1} S-O stretching band is detected for the penetration depths ranging from 1.35 to 1.48 μm. This is illustrated by traces E and D, Figure 10b. As the penetration depth increases, the 1046 cm^{-1} and 1056 cm^-[1] bands become more pronounced (traces C, B, and A, Figure 10b).

The data presented in Figures 10a and 10b indicate that the $SO_3^-Na^+$ hydrophilic groups of SDOSS are present in three forms: non-associated (1050 cm^{-1}),

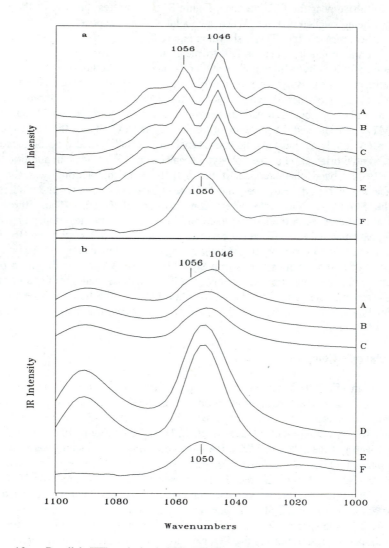

Figure 10a. Parallel (TE) polarized ATR FT-IR spectra, recorded at various depths of penetration from the F-A interface of the 50/50 Sty/n-BA latex cast on PTFE: (trace A) 2.30 μm, (trace B) 1.89 μm, (trace C) 1.65 μm, (trace D) 1.48 μm, (trace E) 1.35 μm, and (trace F) transmission spectrum of SDOSS surfactant.

Figure 10b. Perpendicular (TM) polarized ATR FT-IR spectra, recorded at various depths of penetration from the F-A interface of the 50/50 Sty/n-BA latex cast on PTFE: (trace A) 2.30 μm, (trace B) 1.89 μm, (trace C) 1.65 μm, (trace D) 1.48 μm, (trace E) 1.35 μm, and (trace F) transmission spectrum of SDOSS surfactant.

associated with H_2O (1046 cm^{-1}), and associated with COOH groups (1056 cm^{-1}). Furthermore, while non-associated $SO_3^-Na^+$ groups are perpendicular at shallow depths from the surface, the $SO_3^-Na^+$ associations with H_2O and COOH groups are preferentially parallel. Going deeper into the F-A interface, the $SO_3^-Na^+$ are preferentially parallel, and no free $SO_3^-Na^+$ groups are detected. Although at this point it is too early to asses the origin of the free SDOSS near the surface, the presence of the hydrophobic styrene groups near the F-A interface may repel water, acid groups or other hydrophilic entities, therefore leaving surfactant groups in a non-associated form.

With these data in mind let us examine the same latex films at the F-S interface. For the 50% Sty/50% n-BA latex films cast on PTFE, SDOSS is present at significantly lower concentration levels than that detected at the F-A interface. As shown in Figures 11a and 11b, for both parallel (TE) and perpendicular (TM) polarizations, respectively, the 1050 cm^{-1} band is detected at the penetration depths ranging from 1.35 to 1.48 μm. Furthermore, the band shifts to 1046 cm^{-1}, when penetration depth approaches 1.65 μm. Thus, based on this analysis, it is apparent that not only the concentration of SDOSS is much smaller near the F-S interface, but its environment also changes. This behavior is attributed to the fact that the PTFE surface is hydrophobic, and repels water molecules present near the F-S interface. Therefore, only small fractions of SDOSS are detected as not associated with water molecules. However, when the distance from the PTFE substrate increases, the hydrophobic effect of the PTFE substrate diminishes, making it easier for the surfactant molecules to become associated with water.

Let us now consider the effect of substrate surface tension on the distribution and/or orientation of SDOSS across the 50%/50% Copolymer Sty/n-BA latex films. For that reason latex films were deposited on a liquid Hg, which exhibits a surface tension significantly higher than that of PTFE (approximately 420 mN/m). Figure 12a illustrates ATR FT-IR spectra recorded from the F-A interface using parallel (TE) polarization. It appears that the 1046 cm^{-1} and 1056 cm^{-1} bands are strong, indicating that a high concentration of SDOSS hydrophilic surfactant groups is present at this interface. Furthermore, $SO_3^-Na^+$ surfactant groups associated with water and acid groups are preferentially parallel to the surface and, as observed in Figure 12a, decrease, while penetrating deeper into the film (traces E through A). When, however, TM polarization is employed, the situation changes again. Figure 12b depicts a series of the ATR FT-IR spectra recorded from the F-A interface using perpendicular (TM) polarization. As shown in Figure 12b (traces E through A), an increase of the 1046 cm^{-1} and 1056 cm^{-1} bands is observed up to 1.65 μm into the surface. However, 1.89 μm below the surface, the 1046 cm^{-1} band dominates this spectral region. These data indicate that these are not only orientation changes of the $SO_3^-Na^+$ groups, but there are significantly different environments at various depths. Furthermore, a comparison of traces A and E in Figures 10b and 12b also indicates that there is a significant effect of the substrate on the distribution and environment of the $SO_3^-Na^+$ groups. For example, at approximately 1.35 μm depth (trace E in Figure 10b) near the F-A interface, the SDOSS in a non-associated form is present when PTFE is used as a substrate. However, on a liquid Hg, the SDOSS is detected, but its concentration levels are significantly smaller and moreover, the $SO_3^-Na^+$ environment contains water and acid groups (Figure 12b).

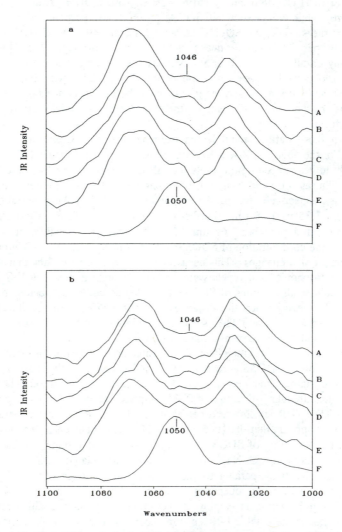

Figure 11a Parallel (TE) polarized ATR FT-IR spectra, recorded at various depths of penetration from the F-S interface of the 50/50 Sty/n-BA latex cast on PTFE: (trace A) 2.30 μm, (trace B) 1.89 μm, (trace C) 1.65 μm, (trace D) 1.48 μm, (trace E) 1.35 μm, and (trace F) transmission spectrum of SDOSS surfactant.

Figure 11b Perpendicular (TM) polarized ATR FT-IR spectra, recorded at various depths of penetration from the F-S interface of the 50/50 Sty/n-BA latex cast on PTFE: (trace A) 2.30 μm, (trace B) 1.89 μm, (trace C) 1.65 μm, (trace D) 1.48 μm, (trace E) 1.35 μm, and (trace F) transmission spectrum of SDOSS surfactant.

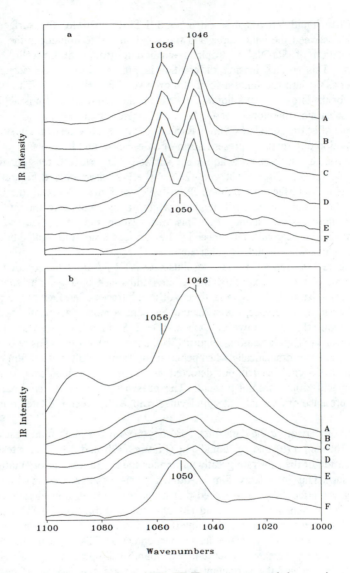

Figure 12a Parallel (TE) polarized ATR FT-IR spectra, recorded at various depths of penetration from the F-A interface of the 50/50 Sty/n-BA latex cast on Hg: (trace A) 2.30 μm, (trace B) 1.89 μm, (trace C) 1.65 μm, (trace D) 1.48 μm, (trace E) 1.35 μm, and (trace F) transmission spectrum of SDOSS surfactant.

Figure 12b Perpendicular (TM) polarized ATR FT-IR spectra, recorded at various depths of penetration from the F-A interface of the 50/50 Sty/n-BA latex cast on Hg: (trace A) 2.30 μm, (trace B) 1.89 μm, (trace C) 1.65 μm, (trace D) 1.48 μm, (trace E) 1.35 μm, and (trace F) transmission spectrum of SDOSS surfactant.

When liquid latex is cast on a solid PTFE substrate, the surface tension differential between the liquid latex and the solid substrate influences the migration and orientation of $SO_3^-Na^+$ hydrophilic surfactant groups at both F-A and F-S interfaces. This surface tension differential is present during coalescence, but it diminishes as the latex becomes a solid film on a solid PTFE surface. The situation is different for the latex cast on a liquid Hg surface: the liquid latex film is deposited on a liquid substrate, therefore creating a surface tension differential as coalescence occurs, and after the solid latex film is formed. For the latex cast on Hg surface, the surface tension differential is present during coalescence, and it increases as the latex becomes a solid film cast on a liquid Hg surface. The surface tension differential effect is illustrated in Figures 10b and 12b for the F-A interface and Figures 11b and 12b for the F-S interface of the 50/50 Sty/n-BA latex films. As seen, the 1046 cm^{-1} and 1056 cm^{-1} bands increase substantially while going deeper into the interfaces for the latex films cast on liquid Hg. This observation indicates that the $SO_3^-Na^+$ surfactant groups are greatly influenced by the surface tension differential between the solid latex film and the liquid Hg substrate.

Figure 13a, traces E through A, illustrate ATR FT-IR data recorded from the F-S interface of the 50% Sty/50% n-BA latex films cast on Hg. The spectra were recorded using parallel (TE) polarization. Although traces of surfactant are present at the F-S interface, in essence, the surfactant present is mostly associated with water. The situation changes, however, when the F-S interface is examined using perpendicular (TM) polarization. Figure 13b illustrates that the intensity of the 1046 cm^{-1} band increases dramatically when penetrating deeper into the latex film (trace A). Again, the 1056 cm^{-1} band is not detected at this interface, indicating that, at this depth, the acid groups are not present. The presence of the 1046 cm^{-1} band results from the presence of residual water molecules trapped between the latex film and the substrate.

A comparison of the data for a 50% Sty/50% n-BA latex films cast on PTFE (Figures 10b and 11b) and on liquid Hg (Figures 12b and 13b) indicates that the surface tension of the substrate greatly influences the migration and orientation of the SDOSS surfactant in a latex film. Although this observation may not be very surprising, the effect is so pronounced that it requires further considerations. When the latex film is deposited on a liquid Hg, the concentration of the $SO_3^-Na^+$ groups associated with water near the F-S interface increases. In addition, the spectra recorded from the F-S interface using perpendicular (TM) polarization show a significant intensity increase of the 1046 cm^{-1} band while penetrating deeper into the tension of Hg causes the surfactant migration toward the F-S interface, and the hydrophilic $SO_3^-Na^+$ surfactant groups are preferentially perpendicular to the surface. This behavior results from the fact that SDOSS lowers the surface tension of the latex, and the latex film, after being deposited on Hg with a high surface tension, film for the film cast on Hg. These observations suggest that the higher surface attracts the surfactant to the F-S interface, in an attempt to lower the surface tension difference between Hg and the latex copolymer. Liquid Hg has also an effect on the F-A interface, as shown by a comparison of the spectra shown in Figures 10b and 12b. The 1050 cm^{-1} band due to non-associated $SO_3^-Na^+$ surfactant groups is not detected when the latex films cast on Hg substrate. Furthermore, the concentration of

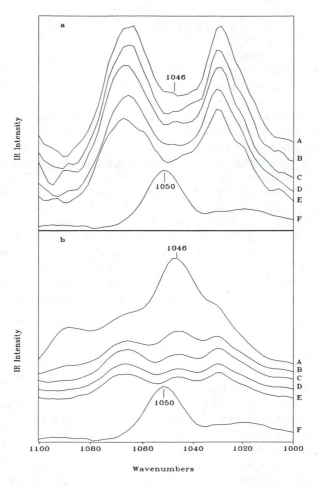

Figure 13a Parallel (TE) polarized ATR FT-IR spectra, recorded at various depths of penetration from the F-S interface of the 50/50 Sty/n-BA latex cast on Hg: (trace A) 2.30 μm, (trace B) 1.89 μm, (trace C) 1.65 μm, (trace D) 1.48 μm, (trace E) 1.35 μm, and (trace F) transmission spectrum of SDOSS surfactant.

Figure 13b Perpendicular (TM) polarized ATR FT-IR spectra, recorded at various depths of penetration from the F-S interface of the 50/50 Sty/n-BA latex cast on Hg: (trace A) 2.30 μm, (trace B) 1.89 μm, (trace C) 1.65 μm, (trace D) 1.48 μm, (trace E) 1.35 μm, and (trace F) transmission spectrum of SDOSS surfactant.

the SO_3Na^+ surfactant groups increases while going deeper from the F-A interface inot the latex film deposited of the Hg substrate.

    One of the factors that also affects distribution and orientation of surfactant molecules near the interfaces is the latex copolymer composition. In essence, compatibility between the copolymer chemistry and surfactants will determine to what extent surface tension at the substrate will influence surfactant mobility. For that reason we modified the Sty/n-BA particle composition. When the concentration of styrene in the latex copolymer is diminished, the SDOSS surfactant concentration near the F-A and F-S interfaces is smaller. For the 30% Sty/70% n-BA and 10% Sty/90% n-BA latex copolymers cast on the PTFE surface, no surfactant is detected at the F-A or F-S interfaces, for either parallel (TE) or perpendicular (TM) polarizations. However, when the same latex copolymers were cast on the liquid Hg substrate, the surfactant concentration and its distribution throughout the latex films changes, especially near the F-S interface. For the 30% Sty/70% n-BA latex films cast on liquid Hg substrate, the F-A interface spectra show no surfactant bands present at this interface for either parallel (TE) or perpendicular (TM) polarizations, but at the F-S interface 1046 cm^{-1} surfactant bands are detected for both polarizations. Furthermore, concentration of the SDOSS surfactant groups parallel to the surface decreases slightly when going deeper into the film for the parallel (TE) polarization. However, for the perpendicular (TM) polarization spectra recorded at the F-S interface, an increase of the 1046 cm^{-1} band while penetrating deeper into the film is detected. These observations indicate that for this monomer composition, the F-S interface of the latex films contains most surfactant groups that are parallel to the surface near this interface, and the $SO_3^-Na^+$ groups become perpendicular to the surface while going deeper into the F-S.

For 10% Sty/90% n-BA latex films cast on liquid Hg substrate the SDOSS intensities remain constant throughout the examined depths near the F-A interface with the parallel (TE) polarization. However, they slightly decrease while penetrating deeper from the F-A interface with a perpendicular (TM) polarization. These observations indicate that the amount of SDOSS surfactant groups parallel to the surface throughout the examined portion of the latex film remains constant, and the amount of the surfactant groups perpendicular to the surface decreases, while going deeper into the latex film.

No surfactant bands were detected over the examined depth near the F-S interface for 10% Sty/90% n-BA composition with the perpendicular (TM) polarization. However, the 1046 cm^{-1} surfactant band is detected for the parallel (TE) polarization, indicating that most of the surfactant present at this interface is parallel to the surface.

When 30% Sty/70% n-BA and 10% Sty/90% n-BA latex films cast on liquid Hg were examined, the 1056 cm^{-1} band was not detected at either F-A or F-S interfaces, indicating that, for this composition, the mobility of surfactant is limited. However, as we recall the data shown in Figures 10a and 12a, the 1056 cm^{-1} band, due to association of the $SO_3^-Na^+$ surfactant groups with the acid groups, was detected only for the 50% Sty/50% n-BA copolymer mixture. Therefore, latex composition clearly affects the mobility of SDOSS. For a higher concentration of styrene, the acid groups were driven towards the F-A interface, but when the styrene concentration is lowered to 30% Sty and 10% Sty, the surfactant groups are found to be only associated with water.

In summary, the presented data allow us to depict a scenario concerning the orientation of SDOSS and the effect of the substrate surface tension for a core/shell latex films. A schematic representation of the effect of surface tension on the surfactant mobility through the latex copolymer film is depicted in Figure 14a. As shown in Figure 14a, the low surface tension of the PTFE substrate drives the hydrophilic $SO_3^-Na^+$ surfactant groups towards the F-A interface of the latex film. However, when the surface tension of the substrate is increased (by using liquid Hg), the surfactant migrates toward the F-S interface of the latex film (Figure 14b). There is also the effect of latex composition which influences the surfactant mobility through the latex film. A schematic representation of the effect of monomer composition on the surfactant mobility through the latex copolymer film is depicted in Figure 15. As the styrene concentration in a latex composition increases, the surfactant tends to migrate towards the F-A interface.

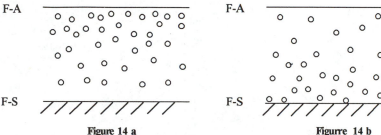

Figure 14 a

Low surface tension substrate
o surfactant

Figurre 14 b

High surface tension substrate
o surfactant

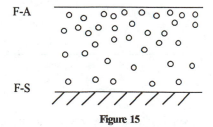

Figure 15

High styrene concentration

o surfactant

Step-Scan Photoacoustic Depth Profiling

By changing modulation frequency of an FT-IR interferometer equipped with a photoacoustic cell, it is possible to obtain information from various surface depths. Step-scan PAS FT-IR spectra recorded from the F-S interface of a 50%/50% Sty/n-BA latex film, using 400, 300, 200, and 100 Hz phase modulation frequencies are shown in Figure 16, Traces A–D, respectively. Trace A recorded at 400 Hz shows two weak bands at 1056 and 1046 cm^{-1} which are due to SO$_3$⁻Na$^+$···HOOC and SO$_3$⁻Na$^+$···H$_2$O associations, respectively. However, when the modulation frequency is decreased to 100 Hz, the 1056 and 1046 cm^{-1} bands decrease to become eventually non-detectable (Trace D). At the same time, a new band at 1050 cm^{-1}, which is due to the symmetric S-O vibrations of free SO$_3$⁻Na$^+$ hydrophilic end groups, is detected at lower modulation frequencies. This band dominates the spectra recorded with 100 Hz (Figure 16, Trace D). Since the depth of penetration of IR light is inversely proportional to the phase modulation frequency,[20] for 400 Hz phase modulation frequency, the

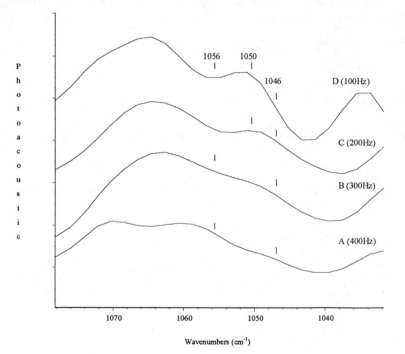

Figure 16. Step-scan PAS FT-IR spectra recorded at different modulation frequency near the F-S interface: A - 400Hz; B - 300Hz; C - 200Hz; and D - 100Hz.

depth of penetration is shallowest. In an effort to establish penetration depths at these modulation frequencies, equation 1 can be used. Using the values of $\alpha = 1.57 \times 10^{-3}$ cm^2/s for PMMA,[21] the penetration depths at all examined frequencies were determined and are summarized in Table I. Based on the data shown in Figure 16, Traces A–D, a higher content of non-bonded SO$_3$⁻Na$^+$ hydrophilic end groups exist at the greater depths near the F-S interface. This is because water evaporates out of the latex film, resulting in less water molecules trapped at greater depths. Therefore, there are less chances for the SO$_3$⁻Na$^+$ hydrophilic end groups to associate with H$_2$O and COOH groups. However, the presence of the 1056 and 1046 cm^{-1} bands is detected at the shallower depths, indicating that there are traces of water molecules near the F-S interface.

Table I. Penetration depths at various modulation frequency

	Modulation frequency (ω)			
	400 Hz	300 Hz	200 Hz	100 Hz
Penetration Depth (μ_{th})	11 μm	13 μm	16 μm	22 μm

The 1056 and 1046 cm^{-1} bands detected at the shallower penetration depths, and the 1050 cm^{-1} band detected deeper into the F-S interface, indicate that the surfactant molecules migrate towards the interface to minimize the interfacial surface tension. Furthermore, their presence detected at different penetration depths in the step-scan IR spectra indicate that there are significant differences in the film formation near the F-S interface. In the case of the F-S interface, it is far more difficult for water molecules to evaporate from the F-S interface because they may coalesced top layers which inhibit this process.

In order to determine the distribution of SDOSS molecules at various penetration depths, the step-scan spectra were also recorded with 0° and 90° phase, which referred to as in-phase (I) and in-quadrature (Q) spectra. Before we analyzed these data, it is necessary to determine relative penetration depths for 0° and 90° spectra because the relative penetration depths for 0° and 90° spectra depend on the phase modulation angle. In an effort to define the shallow and deep depths and their positions with respect to 0° and 90° spectra, a reference carbon black spectrum was collected. Since carbon black absorbs almost all of the IR radiation and it is thermally thick, stronger peak-to-peak signals recorded from the step-scan PA spectra are attributed to the surface of carbon black, and weak signals come from deeper penetration depths. For that reason, a series of the carbon black spectra were recorded at various phase angles and utilized as references to determine the relative depth penetration for 0° and 90° spectra. The results are shown in Figure 17 with curves A$_0$ and A$_{90}$ illustrating the peak-to-peak intensity (Volts) vs. phase modulation angles for 0° and 90° spectra at 1000 Hz modulation frequency, respectively. In this Figure, both A$_0$ and A$_{90}$ are cosine waves, and are shifted by 90°. These data also show that, when the phase angle is 0°, A$_{90}$ has the higher peak-to-peak intensity, thus the signal comes from the surface and it is in-phase (I) with the incident radiation, and A$_0$ comes from deeper penetration depths, thus represents in-quadrature (Q) spectra. On the other hand, when the phase modulation angle is 90°, the 90° spectrum comes

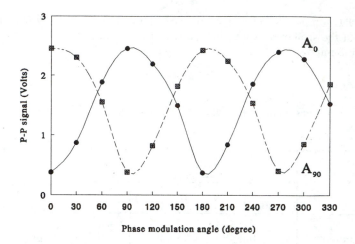

Figure 17. A plot of peak-to-peak intensity vs. phase modulation angle at 1000 Hz modulation frequency: (A_0) - 0° phase and (A_{90}) - 90° phase.

from the bulk spectrum (Q), and the 0° spectrum comes the surface (I). The same trends were detected at modulation frequency of 400 Hz.

With this in mind, let us go back to the main theme and analyze distribution of SDOSS molecules across the latex film. Figure 18 (A-D) illustrates a series of spectra recorded from the F-S interface from 0° to 90°, in 10° increments, at 400, 300, 200 and 100 Hz, respectively. In this experimental setup, the phase modulation angle is 0°, thus the 0° phase (Q) spectrum represents the signal from deeper penetration depths, and a 90° (I) is a photoacoustic signal from shallower depths. As shown in Figure 18 (A), the bands at 1056 and 1046 cm^{-1} gradually increase, as the phase angles increases from 0° to 90°. The phase angle at 0° provides deeper depth of penetration than the phase angle at 90°. Since the 1046 cm^{-1} band intensity is not detected, distribution of SDOSS molecules is not homogeneous within a 11 μm layer from the F-S interface.

In a series of the step-scan spectra shown in Figure 18 (B), a 300 Hz phase modulation frequency from 0° to 90° was used. Although the bands at 1056, 1050 and 1046 cm^{-1} are detected, the band intensities at 1056 and 1046 cm^{-1} decrease to minimum, and the 1050 cm^{-1} band increases, as the phase angles change from 90° to 0°. This observation again indicates that the SDOSS molecules are not uniformly distributed near a 13 μm layer from the F-S interface. Since water evaporation is one of the major driving forces during coalescence, film formation at this depth from the F-S interface will be significantly affected. The amount of water molecules decreases as the penetration depths increase after coalescence. Therefore, the 1050 cm^{-1} band becomes detectable at 13 μm layer and the band at 1046 cm^{-1} due to the SO_3^- $Na^+\cdots H_2O$ association decreases, to become non-detectable.

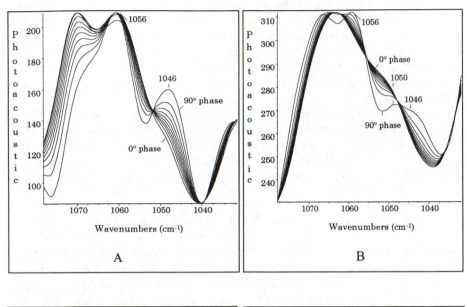

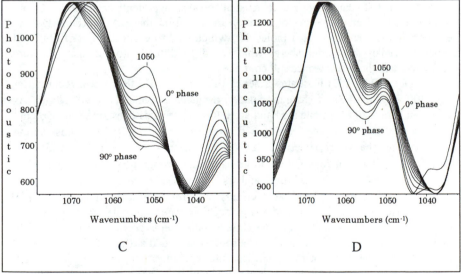

Figure 18. 10 steps of step-scan PAS FT-IR spectra recorded at various modulation frequency with 0° and 90° phase angles near the F-S interface: A - 400Hz; B - 300Hz; C - 200Hz; and D - 100Hz.

When the modulation frequency is set at 200 Hz, the bands at 1056 and 1046 cm^{-1} are not detected, and only 1050 cm^{-1} band is present. This is shown in a series of spectra in Figure 18 (C). The band at 1050 cm^{-1} is strongest for the (Q) spectra, and decreases as the phase angle decreases (I). Similar trends for the band at 1050 cm^{-1} are detected for 100 Hz modulation frequency with the 0° and 90° phase angles at the F-S interface. These spectra are shown in Figure 18 (D). Again, this observation indicates that a fewer of SDOSS molecules are present at the greater depths from the F-S interface.

Having identified spectral features near the F-S interface, let us analyze SDOSS at the F-A interface. The step-scan photoacoustic FT-IR spectra recorded with 400, 300, 200, and 100 Hz of phase modulation frequency at the F-A interface are shown in Figure 19, Traces A–D, respectively. Similarly to the discussion above, the depths of penetration for 400, 300, 200, and 100 Hz of phase modulation frequency are 11, 13, 16, and 22 μm, respectively. The 1056 and 1046 cm^{-1} are the strongest for 400 Hz (Trace A). However, both decrease as the phase modulation frequency decreases, and become almost non-detectable above 200 Hz phase modulation (Traces C and D). Instead, the 1050 cm^{-1} band due to non-bonding $SO_3^-Na^+$ hydrophilic end groups on SDOSS is detectable at 200 Hz. These data suggest that, when going deeper into the F-A interface, there is less H-bonding between $SO_3^-Na^+$ and water, and no associations between $SO_3^-Na^+$ hydrophilic end group and water and COOH groups are present. Therefore, both 1056 and 1046 cm^{-1} bands are non-detectable. Similarly, the 1056 and 1046 cm^{-1} bands are detected at the shallower depths and the 1050 cm^{-1} band is detected at the deeper penetration depths from the F-A interface. Due to equilibrium between inner water molecules, which tend to evaporate out of film, and water molecules from atmosphere, which tend to diffuse into the film, the film formation near the F-A interface still continues.

Figure 20 (A-D) illustrates a series of step-scan FT-IR spectra recorded at 400, 300, 200, and 100 Hz phase modulation frequencies, with 0° and 90° phases. As shown in Figure 20 (A), while the band at 1046 cm^{-1} remains constant, the 1056 cm^{-1} decreases as the phase angle changes from 0° (I) to 90° (Q). This observation indicates that the water molecules are present near a 11 μm depth, and form $SO_3^-Na^+\cdots H_2O$ associations. On the other hand, the strongest band intensity at 1056 cm^{-1} is detected with Q spectrum, and as the phase angle increases, this band decreases,. indicating that more acid groups are present near 11 μm into the F-S interface. This behavior is attributed to the fact that water molecules are trapped near this interface, while the acid groups are closer to the deeper portions of 11 μm depth. Similarly, the same trends are found in the spectra recorded at the 300 Hz (Figure 20, B). However, the band intensity of the 1046 cm^{-1} decreases as the phase angle decreases. Hence, at greater depths, less water molecules are present, then inhibiting $SO_3^-Na^+\cdots H_2O$ associations. Again, a weak band intensity of the 1056 cm^{-1} is detected for Q spectrum. Since water evaporated towards the film surface interface, more water molecules are able to stay in shallower penetration depths. Based on these spectra recorded from 13 μm into the F-S interface, detected concentrations of water molecules increase parallel with the water evaporation direction, and decrease with the penetration depths.

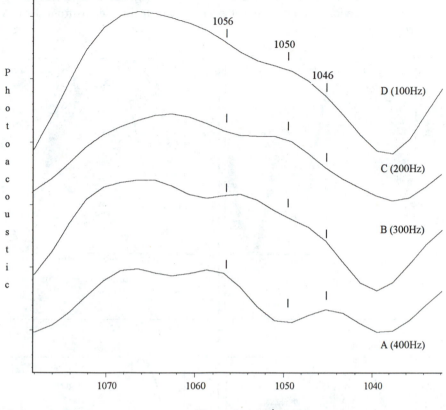

Figure 19. Step-scan PAS FT-IR spectra recorded at different modulation frequency near the F-A interface: (A) 400Hz; (B) 300Hz; (C) 200Hz; and (D) 100Hz.

Similar trends are observed near the F-A interface for modulation frequencies at 200 and 100 Hz. This is shown in Figures 20, C and D, respectively. In this case, the 1050 cm^{-1} band is detected at a 16 μm layer near the surface, and decreases at greater depths, indicating that there are no SDOSS molecules beyond 22 μm boundary. By combining spectroscopic data from the F-S and F-A interfaces, it can be seen that SDOSS molecules exude towards the F-A and F-S interfaces, but they are not homogeneously distributed across the film. Since 50% of Sty is incorporated in this composition, hydrophobicity is relatively high, resulting in that the copolymer expels water molecule from latex particles interstices during coalescence. Therefore, one can detect free SDOSS surfactant near a 13 μm layer from F-S interface. Because only the 1050 cm^{-1} band is detected below 13 μm, coalescence is complete below this thickness from the F-S interface. At the F-A interface, water molecules from the surroundings are able to diffuse into the surface during coalescence, thus SO_3^- $Na^+ \cdots H_2O$ associations are detected at the 11, 13 and 16 μm depths from the F-A interface.

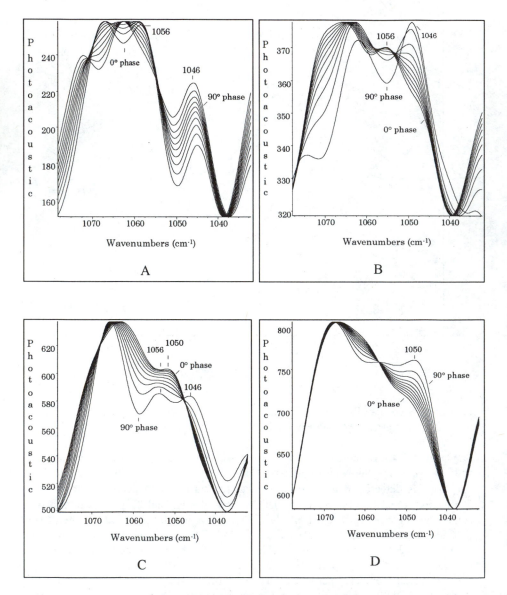

Figure 20. 10 steps of step-scan PAS FT-IR spectra recorded at various modulation frequency with 0° and 90° phase angles near the F-A interface: (A) 400Hz; (B) 300Hz; (C) 200Hz; and (D) 100Hz.

Although we are interested in quantitative assessments of SDOSS distributions near interfaces, at this point such analysis is not available. Therefore, a semi-quantitative analysis will be employed. One of the beneficial features of the step-scan PAS is that the spectra are not wavenumber dependent. Therefore, the relative band areas can be used as a measure of the relative concentrations of SDOSS. However, before the analysis of the relative band areas as a function of penetration can be accomplished, it is necessary to convert the phase angles to penetration depths. Although R-G theory[15] predicts that the $2\pi\mu_{th}$ layer is the deepest detectable penetration depth beneath the surface, for any practical purposes, due to a low signal-to-noise ratio beyond $3\mu_{th}$, $3\mu_{th}$ is the deepest penetration depth that can be detected. In order to obtain relative quantities of SDOSS, the relative areas of the 1046, 1056, and 1050 cm^{-1} bands, normalized to the strongest band at 1068 cm^{-1}, due to the C-C vibrational modes, were utilized. The plots of the band areas as a function of the phase rotation angles are shown in Figures 21 and 22 for F-S and F-A interfaces, respectively. In Figure 21, A, relative quantities of SDOSS molecules forming $SO_3^-Na^+\cdots COOH$ associations (1056 cm^{-1}) remain constant throughout the entire film thickness, whereas the amount of the $SO_3^-Na^+\cdots H_2O$ entities (1046 cm^{-1}) decreases, as the phase angle decreases from 90° to 0°. Thus, there are fewer $SO_3^-Na^+\cdots H_2O$ associations at greater penetration depths. In the case of 300Hz shown in Figure 21, B, the relative amounts of $SO_3^-Na^+\cdots H_2O$ and $SO_3^-Na^+\cdots COOH$ species continuously decrease to become non-detectable. However, relative concentrations of the non-bonding SDOSS molecules (1050 cm^{-1}) gradually increase, while going further away from the F-S interface. Concentrations of the non-bonding SDOSS molecules further increases at approximately 16 μm penetration depths into the film. This is shown in Figure 21, C. Furthermore, concentration of the non-bonding SDOSS (1050 cm^{-1}) decreases as the penetration depth increases. This is shown in Figure 21, D. Similar trends are observed for the relative concentrations of $SO_3^-Na^+\cdots H_2O$, $SO_3^-Na^+\cdots COOH$, and non-bonding SDOSS molecules near the F-A interface. This is shown in Figure 22 (A-D). However, a comparison of the data shown in Figure 21 and 22 indicate that the concentration of SDOSS near the F-A interface is almost as twice as large as compared to the F-S interface.

Using the surfactant distribution data in coalesced films, it is possible to depict how coalescence process may vary at various distances from the F-A and F-S interfaces. Although one could possibly divide the film thickness into three zones: F-A, central, and F-S interfacial regions, it appears that a degree of coalescence chnages across the film is continues. The schematic diagram of the distribution of SDOSS molecules is illustrated in Figure 23. One property of surfactant molecules is that these entities tend to migrate to interfaces in order to minimize the interfacial surface tension. In our previous studies,[10] we illustrated that the surfactant migration is influenced by the surface tension of the substrate after latex coalescence. In this case, latex was cast on polytetrafluoroethylene (PTFE), with the initial interfacial surface tension γ_{int} (H_2O-PTFE) equal to 50 mN/m, and the surface tension of the latex polymer being approximately 32.2 mN/m. As coalescence progresses, water evaporates, leaving the copolymer film in contact the PTFE substrate, and forming solid-solid interface. Using the concept of interfacial surface tension expressed by

$\gamma_{int} = \gamma_{sub} + \gamma_{film} - 2(\gamma_{sub}^{d}\gamma_{film}^{d})^{1/2}$, the interfacial surface tension between the latex film and PTFE substrate changes to 1.9 mN/m. In this equation, γ_{int} is the interfacial surface tension, γ_{sub} and γ_{film} are the critical surface tensions for a substrate and a polymeric film, respectively, and γ_{sub}^{d} and γ_{film}^{d} are the dispersion components of the substrate and polymeric film, respectively. Apparently, the surfactant molecules migrate to the F-S interface due to the higher surface tension in the initial stages of coalescence, and more surfactant molecules migrate toward the F-A interface due to the lower interfacial surface tension between polymeric film and substrate after coalescence. Therefore, surfactant molecules cumulate near F-A and F-S interfacial regions.

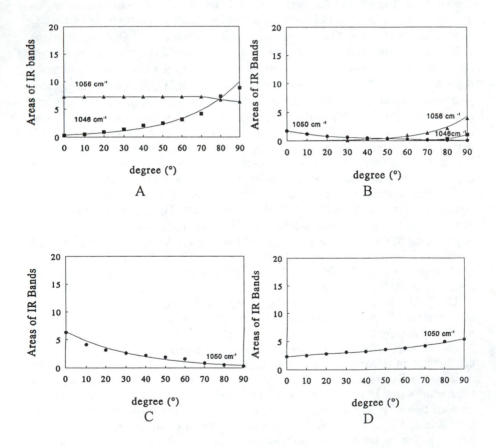

Figure 21. Plots of relative quantities of SDOSS at various modulation frequency vs phase angles near the F-S interface: A - 400Hz; B - 300Hz; C - 200Hz; and D - 100Hz.

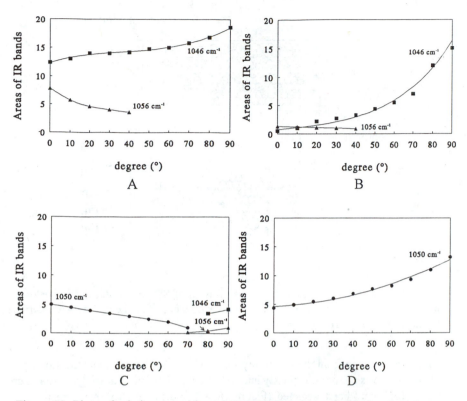

Figure 22. Plots of relative quantities of SDOSS at various modulation frequency vs. phase angles near the F-A interface: A - 400Hz; B - 300Hz; C - 200Hz; and D - 100Hz.

Since there are three different types of SDOSS interactions with the latex components: (1) $SO_3^-Na^+\cdots H_2O$; (2) $SO_3^-Na^+\cdots HOOC$, and (3) free SDOSS molecules, we can follow latex film formation by detecting the SDOSS behavior during coalescence. Let us go back to the step-scan IR spectra recorded with 400 Hz of modulation frequency near the F-S interface, which corresponds to the penetration depths near 11 μm. Because the band at 1046 cm^{-1} is detected, water is present between the particle interstices. Furthermore, the presence of the 1056 cm^{-1} band indicates that the particle inter-diffusion is not complete because the water molecules does not evaporate completely in this layer. This is schematically illustrated in Figure 23, which shows the F-S interface near 11 μm. At approximately 13 μm, it appears to

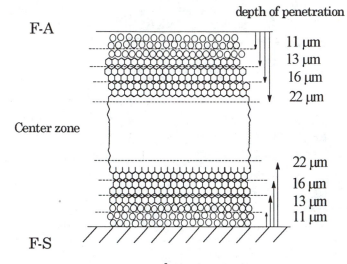

Figure 23. A schematic diagram of the distribution of SDOSS molecules in coalesced film. SDOSS molecules have a twice relative concentration near the F-A interface than near the F-S interface.

be a transition stage from the wet to dry stage of coalescence. In this transition layer, polymer particles are not completely inter-diffused, and at approximately 16 μm, there is only 1050 cm^{-1} band detected (Figure 18, C), indicating that there are no SO_3^- Na$^+$···H$_2$O associations, resulting in coalesced film. This is shown in Figure 23, 16 μm from the F-S interface.

As we recall, Figure 18, D illustrated that at 22 μm from the interface, only 1050 cm^{-1} band was detected, indicating that the film formed continuous phase. Although in the case of F-A interface, the presence of water in the surrounding atmosphere allows diffusion into the F-A interface, thus affecting kinetics of coalescence process, similar trends are detected.

Conclusions

These studies show that ATR and PAS in a step scan mode of operation can be effectively utilized in the analysis of surfaces and interfaces of latexes as well as other films. Our current studies involve urethanes, thermoplastics, as well as silicone elastomers. Although at this point we are able to quantify ATR measurements, quantitative PAS analysis will require further investigations because it requires understanding of the detection processes as well as material's properties. With the current setup, ATR is capable of effective depth profiling up to 2-3 μm, whereas PAS step-scan measurements exhibit significantly wider range, up to 100 μm, or more, when proper modulation frequency ranges are employed.

Acknowledgments: Acknowledgment is made to the donors of The Petroleum Research Fund, administrated by the ACS, ACS-PRF# 29083-AC7 for partial support of this research. The author is also thankful to various industrial sponsors for funding parts of this research.

References:

1. M. W. Urban and K. W. Evanson, Polym. Comm., 31, 279 (1990).
2. K. W. Evanson and M. W. Urban, Surface Phenomena and Fine Particles in Water- Based Coatings and Printing Technology, M. K. Sharma and K. J. Micale, Eds., Plenum Press, New York, 1991. 3. K. W. Evanson and M. W. Urban, J. Appl. Polym. Sci., 42, 2309 (1991).
3. K. W. Evanson and M. W. Urban, J. Appl. Polym. Sci., 42, 2309 (1991).
4. K. W. Evanson, T. A. Thorstenson, and M. W. Urban, J. Appl. Polym. Sci., 42, 2297 (1991).
5. T. A. Thorstenson and M. W. Urban, J. Appl. Polym. Sci., 47, 1381 (1993).
6. K. W. Evanson and M. W. Urban, J. Appl. Polym. Sci., 42, 2287 (1991).
7. T. A. Thorstenson, L. K. Tebelius, and M. W. Urban, J. Appl. Polym. Sci., 47, 1387 (1993).
8. J. B. Huang and M. W. Urban, Appl. Spectrosc., 46, 1666 (1992).
9. J. B. Huang and M. W. Urban, Appl. Spectrosc., 46, 1014 (1992).
10. T. A. Thorstenson, L. K. Tebelius, and M. W. Urban, J. Appl. Polym. Sci., 49, 103, (1993).
11. B.-J. Niu and M.W.Urban, J.Appl.Polym.Sci., 60, 371, (1996).
12. 12. B.-J. Niu and M.W.Urban, J.Appl.Polym.Sci., 60, 389, (1996).
13. B.-J. Niu and M.W.Urban, J.Appl.Polym.Sci., 60, 379, (1996).
14. M. W. Urban, *Attenuated Total Reflection Spectroscopy of Polymers - Theory and Practice*, American Chemical Society, Washington, DC, 1996.
15. A. Rosencwaig, *Photoacoustics and Photoacoustic Spectroscopy*, John Wiley & Sons, New York, 1980.
16. M.W.Urban and J.L.Koenig, Appl.Spectrosc., 1986, 40(7), 994.
17. M. W. Urban, *Vibrational Spectroscopy of Molecules and Macromolecules on Surfaces*, John Wiley & Sons, New York, 1996.
18. R.M. Dittmar, J.L.Chao, R.A.Palmer, Appl.Spectrosc., 1991, 45(7), 1104.
19. M.B.Roller, J.Coat.Technol., 1982, 54(691), 33.

Chapter 21

Rate-Limiting Steps in Film Formation of Acrylic Latices as Elucidated with Ellipsometry and Environmental Scanning Electron Microscopy

J. L. Keddie[1], P. Meredith[2], R. A. L. Jones, and A. M. Donald

Polymer and Colloid Group, Cavendish Laboratory, Department of Physics, University of Cambridge, Madingley Road, Cambridge CB3 0HE, United Kingdom

Our data indicate that evaporation is the rate-limiting step in film formation when the temperature of a latex is about 20 K or more above its glass transition temperature (T_g). When a latex is nearer to its T_g, the rate-limiting step in film formation is deformation of the latex particles, possibly by viscous flow of the polymer driven by the reduction in surface energy. In this latter case, there is evidence that a drying front first creates air voids. Subsequently, a coalescence front moves inward from the periphery in the plane of the film. Evaporation rates are retarded in a latex that is well-above its T_g, probably as a result of the reduced surface area of water, caused by extensive particle deformation. We studied the kinetics of film formation in an acrylic latex using ellipsometry and environmental-SEM, techniques which allow *in situ* observation of wet and partially-wet latices. We fit our data to a model describing the coalescence of voids by viscous flow.

We have employed methodology (similar to what we used previously (*1*)) to determine the effect of *temperature* on the kinetics of film formation. We correlate our experimental findings with the kinetics of the passage of drying and coalescence fronts in the plane of the film and normal to its surface.

In our previous publication (*1*), we studied the kinetics of film formation of acrylic latices, paying close attention to the effects of the *glass transition temperature* (T_g). Using a combination of ellipsometry and environmental scanning electron microscopy (ESEM), we identified the four conventional stages of film formation. We found that

[1]Current address: Department of Physics, University of Surrey, Guildford, Surrey GU2 5XH, United Kingdom
[2]Current address: Health and Beauty Care, Procter and Gamble Ltd., Rusham Park, Whitehall Lane, Egham, Surrey TW20 9NW, United Kingdom

when film formation takes place at temperatures not far above the T_g of the latex, there is an intermediate stage between II and III. In this intermediate stage, which we called II*, water has evaporated from the interstitial voids but the particles have not deformed sufficiently to compact into a void-free, dense array, yet even so the latex is optically transparent. The duration of Stage II* is inversely related to T_g. Figure 1 illustrates the stages of film formation, as we described them previously.

In this work, we will differentiate between drying and coalescence fronts in the plane of a film and normal to it. Although they appear as separate fronts, in reality, the two could be interrelated, as pointed out below. Also note that passage of a drying front (in which the water/air meniscus recedes) does not necessitate the simultaneous passage of a coalescence front (in which particles rearrange and deform to fill all available space). We therefore refer in this paper to drying and coalescent fronts separately.

Interest in, and discussion of, fronts in the plane of a latex film have persisted over three decades. In 1964 Hwa (2) reported his observation that a latex cast on a flat surface first becomes optically transparent near its outer edges. He observed that during film formation, there are three regions in the plane of the film: a transparent. apparently dry region near the edges of a sample; a central turbid, wet region; and a cloudy intermediate region. He proposed that this intermediate region consisted of flocculated particles, and he derived and tested a mathematical relationship between the radii of the transparent and flocculated regions. During film formation, the outer clear region increases at the expense of the central turbid, wet region. Okubo *et al.* (3) studied the same three regions, and they also reported that a thin film-formed layer or "skin" existed on the surface of the central wet region. This skin is indicative of a drying and coalescence front normal to the latex surface. The front observed moving in from the sample edges could be the same as that in the skin, so that a distinction between the two types of fronts could be artificial. Chevalier *et al.* (4), observing similar phenomena with small angle neutron scattering (SANS), used the phrase "*coalescence front*" to describe the growth of the dry, clear region at the expense of the central, turbid, wet region. SANS revealed that the change in volume fraction of polymer is quite abrupt at the point of film formation. Coalescence fronts have been observed with cryogenic SEM (5), and drying fronts have been associated with the transport of surfactant to the film center (6). Most recently, Winnik and Feng (7), drawing upon work on the drying of particle monolayers by Denkov *et al.* (8), have developed a description of drying and film formation in the plane of the film. They propose that as a result of a film being thinner near its edges, evaporation is faster there. As film formation begins at the sample edges, water flows radially outward, carrying particles with it.

Our technique of ellipsometry is highly-sensitive to the passage of fronts in the plane of a latex film. As a matter of fact, Stage II*, that we reported previously, can be considered to be a "flocculated region" between the turbid (Stage II) and transparent (Stage III) latex. The time duration of Stage II* is a direct function of the width of the region and the velocity of the fronts.

Sperry and co-workers (9) have suggested that there are two important components in the process of latex film formation: evaporation of the aqueous solvent and deformation/compaction of particles leading to void closure. They postulated that either of these steps could be rate-limiting, depending on the temperature of the process and other factors. Here, we report experiments that indicate which process is rate-limiting for two acrylic latices at various temperatures. We propose an equation to describe how the time for film formation (defined by the onset of optical clarity) varies as a function of the temperature of the latex, when particle deformation is rate-limiting. Determining the rate-limiting step in the film-formation of a latex is of scientific interest because it can provide insight into the

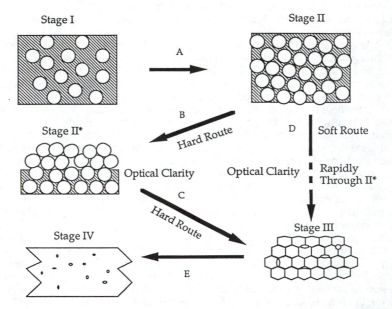

Figure 1. Idealized representation of the four stages of latex film formation plus the intermediate stage called II*. At the onset of Stage II, particles come into contact with each other. During Stage II*, water has evaporated from between the particles but voids still exist in areas where particles are not well-deformed. In Stage III there is a densely-packed array, and in Stage IV there is a continuous coating. Reprinted with permission from ref. 1. Copyright 1995 American Chemical Society.

mechanisms of the process. Understanding latex film-formation has technological relevance because it can lead to control and optimization of latex properties.

Materials and Experiments

We studied the same latices described in our previous publication (*1*). The latices are based on copolymers of methyl methacrylate and 2-ethyl hexyl acrylate. They were emulsified and stabilized with a combination of nonionic and anionic surfactants. Cellulose ether was used as a protective colloid. The latices initially contained 55 weight percent water. Their T_gs vary with composition, as summarised in Table I. At a given temperature, the polymer in Latex 2 - with a higher T_g - is expected to be more viscous than the one in Latex 4.

Because these latices can sometimes film-form within minutes of being cast, many techniques cannot adequately study the process due to their relatively slow data acquisition times. We use two techniques that allow observation of film formation as it occurs. One of these, phase-modulated ellipsometry (*10*) is both fast and non-destructive. We can obtain information about the structure of a latex every one or two seconds without having to manipulate or mechanically probe it. We use the other technique, ESEM (*11*) to image wet and partially-wet materials. The latex sample is studied in its *natural* state, without the need for an electrically- conductive coating or special sample preparation.

Table I. Latex Chemical Compositions[a] and Physical Characteristics

Latex No.	Chemical Composition		Diam. (nm)[b]	$T_g (K)$[c]
	MMA (mol.%)	2-EHA(mol.%)		
2	54.3	44.8	488	286
4	42.9	56.2	554	268

[a] Latices were not purified; all copolymers contain 0.9 mol. % methacrylic acid.

[b] Mean values of particle diameter were determined with photo-correlation spectroscopy. Standard deviation is approximately ± 40 nm.

[c] Determined from differential scanning calorimetry of the dried latices.

Using a commercial spin-coater rotating at 1800 r.p.m., we spun wet latex on the unpolished side of a silicon wafer (diameter of 5.08 cm (2 in.)) for approximately two seconds to create a planar, smooth and wet coating. The thickness of the coating was reproducible when using this casting technique. During most ellipsometry experiments we placed the sample on a stage having a fixed temperature ranging from 278 to 318 K. We immediately conducted *kinetic* scans in air on the drying sample with a phase-modulated ellipsometer, a Uvisel model manufactured by Jobin-Yvon. In this type of scan, the angle-of-incidence and wavelength of radiation were fixed (at 53 ° and 413.3 nm, respectively), and the ellipsometric angles (ψ and Δ) were obtained as a function of time while the latex film-formed. A fixed beam spot (with areal dimension of about 3 mm by 7 mm) was positioned in the center of the sample. The end of the deposition determined the starting point in measuring the time for film formation. After film formation, the latex coatings were in the range of 50 µm thick. All ellipsometry experiments were conducted in the still air of our laboratory. The room temperature was typically 293 K.

Modelling the latex as a semi-infinite solid, we can find the *effective* complex refractive index from ψ and Δ (*1,12*). The complex refractive index consists of a real component, n, and an imaginary component, k (or extinction coefficient). According to the Lorentz-Lorenz equation (*13*), n increases with increasing material density (and decreasing void content). The extinction coefficient increases with increased light scattering from voids and with increased optical absorption (*12*). The complex refractive index of a latex can therefore be used as an indicator of its microstructure. The technique of ellipsometry has been fully discussed elsewhere (*14*).

We observed the latices in all of their stages of film formation in an ElectroScan ESEM (Model E-3). The design and use of the environmental electron microscope has been described in detail elsewhere (*11,15*) Typically we placed a drop of wet latex in a shallow copper cup, inserted it into the microscope at atmospheric pressure, and cooled to 276 K or below. We gradually replaced the air in the sample chamber with water vapor, taking care not to dry the sample (*16*). By keeping the temperature of the sample and the pressure of water vapor in the chamber near the saturated vapor conditions, we evaporated and condensed water on the surface in a controlled fashion. We imaged the sample surface in both hydrated and dry states, and we used water to decorate and highlight particle edges and interfaces. We raised the sample temperature, as needed, to allow film formation to proceed.

Results and Discussion

We note that we sometimes observed three regions in the plane of the film during its film formation: an optically-transparent region near the edge of the sample, a turbid central region, and a cloudy region between the other two regions. As the clear region expanded inward, the turbid region at the sample center decreased in area. The width and distinctness of the intermediate, cloudy region vary with temperature and T_g of the latex, as will be discussed in detail below. In Latex 4, and in Latex 2 at higher temperatures, the intermediate region was not well-defined. Instead, the turbid region gradually faded into the transparent region.

We see distinct and reproducible trends in our ellipsometric scans. Demonstrating these trends, Figures 2-4 show the effective complex refractive index (both the real component, n, and the imaginary component, k) as a function of time for Latices 2 and 4 at three different temperatures: 310, 292 and 285 K. In each of the figures, film formation - signalled by the onset of optical clarity in the latex - is demarcated with "FF." At 310 K (well above T_g for both latices) the behavior of n for the two latices is quite similar (Figure 2), rising with time somewhat gradually from the value of water ($\approx$ 1.33) to the value of the fully-dense polymer ($\approx$ 1.5). This behavior can be interpreted to mean that the volume fraction of water decreases simultaneously as the fraction of polymer increases, as is expected when evaporation is rate-limiting. As water evaporates, the latices film-form without an intermediate stage. The rise in index is abrupt because a relatively sharp, combined drying and coalescence front passes from the sample outer edge to the center. The region between the transparent and turbid regions is narrow. That is, there is not an intermediate cloudy region that would cause the passage of two fronts (one at its beginning and one at its end). Note that both latices film-form after the same amount of time (within the error margin of the experiment).

Despite the similarities, there are some differences in the behavior of the two latices at 310 K. The index of Latex 2 does not reach its final value of $\approx$ 1.5 for several hours, which is much slower than Latex 4 (with a lower T_g). Furthermore, there is more optical scattering in Latex 2, evident in its higher value of k. These lower values of n and higher values of k indicate that some microvoids exist in Latex 2 at this temperature but that Latex 4 is nearly fully dense.

Near room temperature (292 K), film formation once again occurs at about the same time for the two latices, as seen in Figure 3. Whereas the index of Latex 4 rises gradually as it did at 310 K, the index of Latex 2 falls abruptly and then rises slowly, as was also found previously with other analysis at room temperature (*1*). Note that Latex 2 shows two primary optical changes that we associate with two events. We

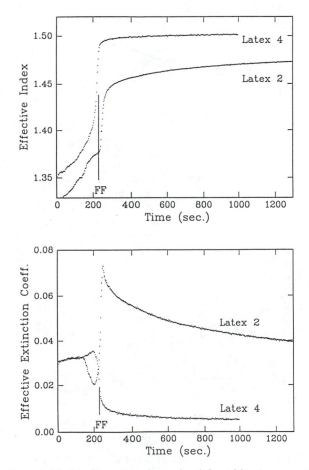

Figure 2. Complex refractive index determined from kinetic scans of Latices 2 and 4 during film formation at a temperature of 310 K. Effective n is shown in the top figure, and k is shown in the bottom one.

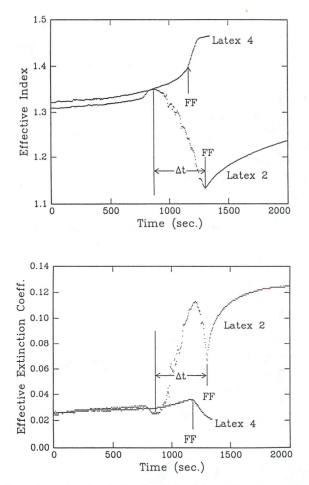

Figure 3. Complex refractive index determined from kinetic scans of Latices 2 and 4 during film formation at a temperature of 292 K. Effective n is shown at the top and k at the bottom. The onset of film formation (optical clarity) is labelled with "FF." The time between the optical change associated with the passage of a drying front and the change associated with film formation (the coalescence front) is identified as Δt.

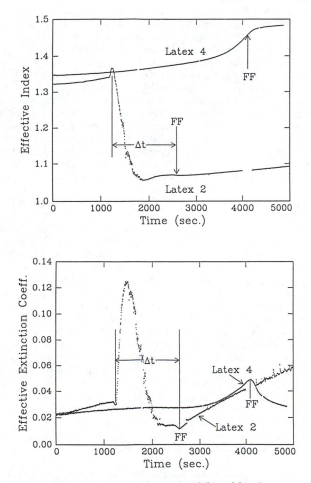

Figure 4. Complex refractive index determined from kinetic scans of Latices 2 and 4 during film formation at a temperature of 285 K. Effective n is shown at the top and k at the bottom. The onset of film formation (optical clarity) is labelled with "FF." The time between the optical change associated with the passage of the drying front and the change associated with film formation is identified as Δt.

associate the first event, signalled as a drop in the effective index, with water receding from the sample surface and concomitantly creating air voids. We reason that only the presence of air voids in the latex will cause its index to drop below the value of water. Seen another way, this is the passage of a drying front that defines one side of the intermediate "cloudy" region. We associate the second event - an increase in n from a minimum - with the average void size decreasing to a critical size, below which light is not significantly scattered, and thereby leading to optical clarity. As the voids continue to shrink, the density of the latex increases and hence so does the index. This second event can be considered to be a coalescence front. We have labelled the time between these two events as Δt on Figures 3 and 4. In the parlance of our previous paper (1), Δt is equal to the time between the onsets of Stages II and II*. Note that the events are also signalled by changes in k, and the second event is more clearly demarcated by k than by n. We attribute the increase in Latex 2's k to enhanced surface roughness. As particles emerge from the water surface, they lead to increased optical scattering and hence an increased value of k. It is useful to interpret our findings in terms of Hwa's "flocculated" or cloudy region between turbid and clear regions. In this context, Δt might correspond to the time between passage of drying and coalescence fronts as both move inward to the center. The changes in n and k for Latex 2 are abrupt at this temperature, corresponding to rather sharp, well-defined fronts.

In Latex 4, on the other hand, we note that k drops and n rises substantially over a period of less than 200 seconds during film formation at 292 K. This latex does not show a pronounced intermediate region at this temperature. The latex passes from a turbid to a transparent state with a single diffuse front. We suggest that the particles deform continuously during the evaporation process, and when the water recedes from the sample surface, a smooth polymer surface emerges. The final low value of k (<0.01) is what we expect for a smooth, optically-transparent solid. There is no evidence for separate drying and coalescence fronts or for a flocculated region with substantial voiding, as was found for Latex 2 at the same temperature.

It is worthwhile mentioning that we have performed kinetic ellipsometry scans on a latex (T_g of ≈ 340 K) that is *non*-film-forming at room temperature. We observe n falling from the value of water (1.33) to less than 1.1 over an interval of about ten seconds. Our previous ESEM study has shown us that the particles in this latex remain spherical after evaporation of the aqueous solvent. Non-deformed, monosized particles packed in a random array will have a volume fraction of solids of about 0.6, and hence a large fraction of the dry latex will be air voids associated with the interstices between particles. Such a material is expected to have a low refractive index. We therefore attribute the abrupt drop in n to the recession of the water meniscus below the latex surface. The index drop is more gradual and less severe in Latex 2 than in the non-film-forming latex, probably because of particle deformation in the former.

At the lowest temperature, 285 K (slightly below the T_g of Latex 2), there are abrupt changes in n and k of Latex 2, as would be expected when water recedes from the sample surface and particles emerge as a drying front passes inward toward the sample center. Since we expect that the polymer is fairly rigid at this temperature, there is a time lag before it deforms sufficiently to close interstitial voids. During the time interval marked as Δt, n is quite low and k is quite high. These optical properties fit in with the idea of a flocculated region of particles with water receded from the sample surface creating air voids. Once again, drying has preceded full particle deformation and coalescence. That is, a drying front appears separately from a coalescence front. In Latex 4, in contrast, the mechanism appears to be similar to that found at higher temperatures. Water remains near the surface throughout the drying process, and its volume fraction decreases with time. Although the sample film-forms from the edges inward, there is no evidence for an intermediate flocculated region. Instead, the latex passes gradually (via a single, diffuse front) from being turbid to clear. We propose that as water evaporates the particles deform more and thereby continuously "squeeze" the water to the surface. This process leads to much more gradual changes in n and k.

Film formation surprisingly occurs sooner in Latex 2 (with a higher T_g) than in Latex 4, as Figure 4 shows. At this and other low temperatures, the "softer" latex 4 takes much longer to film-form in comparison to the "harder" Latex 2. At a temperature of 281 K, the difference in film-forming times for the two latices is striking. Latex 2 film-forms after 5600 seconds, compared to the 8300 seconds taken for film formation of Latex 4. Our finding might, in part, be a result of differences in evaporation rates. This interpretation is consistent with recent evaporation rate studies of hard and soft latex by Winnik and Feng (7). The particles in Latex 4 might continuously deform during film formation and thereby reduce the area of the water meniscus, reducing the overall rate of evaporation. In the harder Latex 2, the particles do not deform to the same extent, and so evaporation is not restricted as much.

The morphology observed with ESEM supports our interpretations of the ellipsometry data. In the micrograph shown in Figure 5, the latex is in the transitional stage, II*. Water is still present in some of the interstitial voids between the particles. The sample surface has been saturated with water, most of it collecting in the necks of adjoining particles. The water is readily apparent because it appears brighter than the polymer (15). There are small regions in which particles have created a densely-packed array. These particles are no longer spherical but have deformed to maximize contact with their neighbors. Even so, voids are seen across the sample surface. The larger voids result from irregularities in either particle size or particle packing, and the smaller voids result from interstitial space between non-deformed particles.

Figure 6 shows a higher magnification view of a surface similar to the one shown in Figure 5. A distribution of voids sizes is apparent. Some voids are less than 100 nm across and appear between partially-deformed particles. As further deformation of the surrounding particles occurs with the passage of time, this type of void should be eliminated. Other voids are in the range of 500 nm, such as the one marked by an arrow in the micrograph. This particular void is hexagonally-shaped and is surrounded by six deformed particles. It appears to result from a packing irregularity. Even if the six surrounding particles were to deform more, the void would not be closed because of its large size. Closure will require flow of the surrounding matrix.

The particle-sized void shown in Figure 6 looks similar to ones observed by Wang and co-workers (17) using atomic force microscopy . They found that holes appear in a PBMA latex after annealing at temperatures well above its Tg for several hours. They attributed their observation to the relaxation of internal defects which cause particles to move away from the surface in order to fill internal voids. In the latices studied here, we do *not* see the appearance of surface voids at later stages of film formation. On the contrary, we observe voids appearing even before evaporation is complete. Such voids subsequently close over a period of hours, according to our ellipsometry and ESEM analysis.

When Latex 2 is held at temperatures near its Tg, we see the particles deforming to a greater extent. Figure 7 shows Latex 2 at a temperature of 285 K in Stage II*. The latex temperature has been raised to this value over a period of about 30 minutes, and water has been observed to have evaporated from the upper portion of the sample. There is no water condensed on the sample surface, and this leads to a rougher topography being observed. Particle-particle boundaries and voids are readily seen in the micrograph. Even after several minutes at temperatures above T_g these boundaries persist. We conclude that particle compaction to a dense array does not immediately follow the evaporation of water in Latex 2. In a latex dried near or

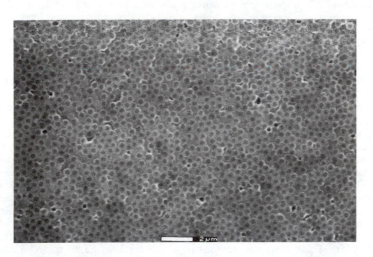

Figure 5: An ESEM micrograph of Latex 2 (T_g = 286 K) in Stage II (or II*) at a temperature of about 276 K. The bright areas are shallow pools of water on the sample surface. Particle boundaries are clearly defined, and voids between the particles are readily apparent. Bar = 2 μm.

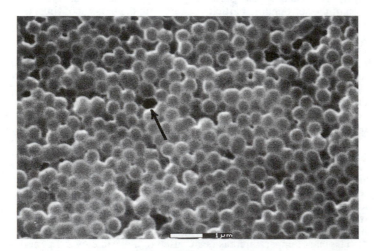

Figure 6: An ESEM micrograph of Latex 2 (T_g = 286 K) in Stage II (or II*) at a temperature of about 276 K at higher magnification than shown in Figure 5. The arrow points to a hexagonal void formed by six particles joining together. Bar = 1 μm.

slightly below its Tg, such as the one shown in the figure, particle deformation is the rate-limiting step in film formation.

We observe something very different in Latex 4. Figure 8 helps to illustrate that evaporation of water is the rate-limiting step in film formation for Latex 4 when at a temperature that is well above its T_g (268 K). On the left side of the micrograph, the sample surface is fully-saturated with water; the sample is at a temperature of about 276 K. The water is visible as the bright, somewhat hazy areas. The darker spots are particles that have emerged from the water surface. Below the surface, the latex is also wet, probably in Stage 2. (In this experiment, we followed a careful pumpdown procedure to ensure that the latex stayed wet.) On the right side, surface water has evaporated very recently (within one minute of capturing the image). Deformed particles are *already* apparent even at these early times and when the latex is still essentially in Stage II. Some of the particles have formed a flat face adjoining their neighbors. Within seconds of the surface water evaporating and after capturing this image, the latex was continuous and void-free. In our previous publication (*1*), we showed Latex 4 only in Stages II and IV since the transformation was so rapid that we were not able to capture a dry, densely-packed array characteristic of Stage III. Even at temperatures close to 273 K, we see the formation of a dense array in Latex 4 immediately following the evaporation of water.

Our ESEM observations, coupled with our study of optical properties, bring us to the following description of the film formation process. When deformation of particles is rate-limiting, such as in Latex 2 close to its T_g, water recedes from the sample surface and moves inward from the sample edges leading to a "drying front." Particles do not deform substantially, however, as evaporation occurs. The latex is not optically clear because air voids develop at the interstices between particles and from irregularities in the packing of particles. Emergence of particles from the wet surface is indicated by the observation of increased optical scattering. This stage of film formation is associated with the cloudy, intermediate region observed between the wet, turbid region and the outer clear regions. At the point of film formation and the passage of a coalescence front, the film becomes optically clear, even though some smaller voids (below the wavelengths of light) persist for much longer. Thus, n is lower and k is higher in a latex under these conditions than in a fully-densified latex.

In other circumstances, such as in Latices 2 and 4 at temperatures well above their T_gs, evaporation is rate-limiting. There is not an intermediate state between a fully-wet (Stage II) latex and a densely-packed (Stage III) latex. As water evaporates, particles can continuously deform. Thus, we observe with ESEM particle deformation at surfaces from which water has recently evaporated. There is no evidence, in this case, for a separate drying front receding from the sample surface. Instead, we suggest that particles continuously deform and thereby fill the space vacated by the water evaporation. Looking at the problem from another point-of-view, it can be said that water is "squeezed" to the sample surface (near the center of sample) as film formation proceeds from the outer edges. Therefore, there is a *gradual* change in optical properties in a single front as deformation accompanies evaporation; the latex gradually becomes 100 percent polymer while the sample surface remains smooth. At the onset of optical clarity, the latex is near its maximum density and free of voids.

When particles deform substantially, they reduce the surface area of the air/water meniscus and thus decrease the amount of water lost to evaporation per unit area of film. As a consequence, a softer latex (in which evaporation is the rate-limiting step of film formation) takes longer to film-form than does a harder latex. This effect is only experimentally-observable at low temperatures (less than about 285 K) when evaporation times are long. These findings are consistent with the assertion made by Sheetz (*18*) over thirty years ago that "the overall rate of water evaporation

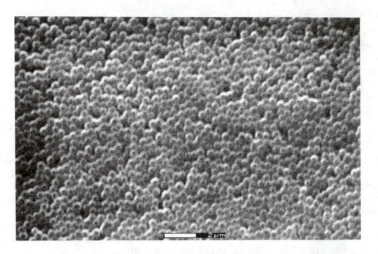

Figure 7: An ESEM micrograph of Latex 2 heated to 285 K, just below its T_g, after being dried at 276 K. The surface is dry. Voids and surface roughness are readily seen. Bar = 2 μm.

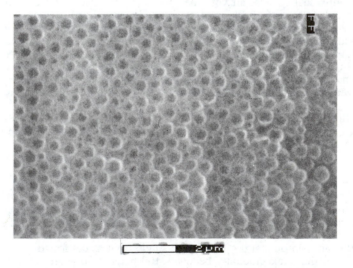

Figure 8: An ESEM micrograph of Latex 4 (T_g = 268 K) in the process of film formation at a temperature of 276 K. On the left side, the sample surface is fully saturated with water. Below the surface, the latex is also wet (Stage 2). On the right side, surface water has recently evaporated. Deformed particles are already apparent, immediately after the evaporation of the water. Within seconds of this micrograph being taken, the latex is continuous and void-free, as was shown in our earlier publication. Bar = 2 μm

will be reduced as the total air-water interface shrinks." Croll (*19*), in contrast, found that drying rates of non-film-forming particles (such as titanium dioxide) were the same as film-forming latex.

Void Closure as a Rate-Limiting Step in Film Formation

In this section, we consider the deformation of particles to eliminate voids. In Latex 2 at temperatures near its T_g, our data indicate that water has evaporated prior to full particle deformation. In this instance, particle deformation is the rate-limiting step in film formation. We will derive a mathematical description of the time for film formation as a function of latex temperature when the deformation step is rate-limiting. We then apply the equation to our data to test if the model provides an adequate description.

When void-closure is the rate-limiting step, then we expect that the water/air meniscus recedes from the sample surface (while moving inward in the plane of the film to the sample center). Then since the particles are too rigid to deform immediately, voids develop near the sample surface (as we have observed with ESEM and deduced from optical properties). The kinetics of the closure of these voids will determine the time for optical clarity and film formation. Mackenzie and Shuttleworth (*20*) have previously considered the problem of spherical voids in a viscous medium decreasing in size under the action of surface energy. They derived an expression to relate the shrinkage of the void radius, r, to the viscosity of the surrounding medium, η:

$$\frac{dr}{dt} = -\frac{\gamma}{2\eta}\left(\frac{1}{\rho(r)}\right) \tag{1}$$

where γ is the surface energy, t is time, and $\rho(r)$ is the relative density which varies with void radius. Note that the term $\rho(r)$ also varies with the microstructural characteristics of the material, such as the number of voids and the initial particle size and packing. Note that this equation is similar to one used by Sperry and co-workers (*9*) to explain the time-dependence of the minimum film-forming temperature in a latex dried below its T_g. Integration of equation 1, assuming that viscosity is constant with time, results in the expression:

$$t - t_0 = \Delta t = \frac{2\eta}{\gamma}[f(r_0) - f(r)] \tag{2}$$

where the void radius is initially r_0 at a time of t_0. The function f varies nearly linearly with r. The temperature dependence of the viscosity can be expressed by the Vogel-Fulcher equation (*21*) :

$$\eta = A\exp\left(\frac{B}{T - T_0}\right) \tag{3}$$

with A, B and T_0 all being material constants. The viscosity of most amorphous polymers near their T_g has been found to be described accurately by the Vogel-Fulcher equation. In latex polymers, however, it has been suggested that water has a plasticizing effect.[Sperry, 1994; Brodnyan, 1964]. If water were to lower the viscosity of a latex polymer, then as water evaporated from within the polymer, the viscosity would be expected to increase with time. Even so, the polymer viscosity should exhibit a temperature-dependence that could be described by equation 3 over narrow ranges of time and temperature.

If we assume that film formation occurs when the void radius decreases below a critical size, r_c, then we can recover an expression in which $\ln(\Delta t)$ varies as the reciprocal of film-forming temperature:

$$\ln(\Delta t) = \frac{B}{T - T_0} + \ln\left[\frac{2A}{\gamma}\left(f(r_o) - f(r_c)\right)\right] \tag{4}$$

by setting t_0 equal to the time when the drying front recedes and t equal to the time when void size equals the critical radius, r_c, that yields optical clarity. We imagine a porous layer of dry, partially-coalesced latex existing during the time interval, Δt. Figures 3 and 4 illustrate how we determine Δt from the ellipsometry data. At temperatures greater than about 306 K for Latex 2 ($T_g = 286$ K), we do not see optical evidence for a dry, porous region during film formation. In this circumstance, we suggest evaporation is rate-limiting, as we found with Latex 4 over a wide range of temperatures.

We have plotted our results to test our model and find good agreement of data for Latex 2 with what is predicted by equation 4 (see Figure 9). When plotted as $\ln(\Delta t)$ versus $(T-T_0)^{-1}$, the data fall on a straight line, with T_0 fit to 257 K. For most glasses, T_0 is typically about 50 K lower than T_g (23). Since T_g for Latex 2 is 286 K, our experimental value is reasonable. Clearly, the length of the time interval, Δt, is thermally-activated, which is what we expect for a process controlled by viscous flow. The slope of the best fit line in figure 10, equal to the effective value of B, is about 200 K. This value is at least one order of magnitude lower than the literature value (24) of B for an acrylic ($\approx 4 \cdot 10^3$ K). One reason for this discrepancy might be the plasticization of the polymer by water. The presence of water in small quantities is known to lower T_g in hydrophilic polymers, and we expect the value of B to be decreased as well. In a related study of void closure in acrylic latices, we have found evidence for a lowering of the viscosity of a latex film that we attributed to plasticization by water (Keddie, J.L.; Meredith, P.; Jones, R.A.L.; Donald, A.M. submitted to *Langmuir.*) Note also, that in this present study we have seen film formation in Latex 2 at temperatures as low as 278 K, even though the Tg (from DSC) of a fully-dried latex is 286 K. Thus, there is additional evidence that the acrylic polymer has been plasticized by the presence of water.

A second factor contributing to a lower effective activation energy for viscous flow could be the presence of water soluble chains in the protective colloid grafted to the particle surfaces and in the latex serum. A third factor could be partitioning of acrylic acid groups at and near the particle surfaces. Consequently, the particle surface, which dominates particle deformation and coalescence, might be less viscous than the bulk of the particle.

Summary

We have used two non-invasive techniques to monitor the stages of film formation in two acrylic latices at temperatures ranging between 278 and 318 K. As film formation takes place, we have followed the changes in water content and in latex microstructure. At higher temperatures, both latices film-form at similar times. At lower temperatures, the lower T_g latex (4) film-forms at *longer* times. We suggest that this is a result of differing amounts of particle deformation and subsequent differences in evaporation rates. The softer latex deforms to a greater extent, reducing the area of the air/water meniscus.

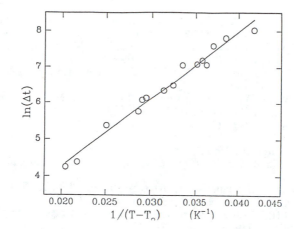

Figure 9. Natural logarithm of Δt (the time between the drying front passage and the onset of optical clarity) for Latex 2 plotted against $1/(T-T_0)$, where T is sample temperature and T_0 is fitted to 257 K. The solid line is the best fit.

Ellipsometry is sensitive to the passage of drying and coalescence fronts normal to the latex surface and in the plane of the film. In a latex at a temperature well above its T_g, a turbid film gradually becomes optically clear. There is no evidence for an intermediate region, described by Hwa as a "flocculated" region. At temperatures closer to the latex T_g, the passage of separate drying and coalescence fronts in the plane of the film are evident. The time interval between the passage of these fronts, Δt, is a function of the temperature of the latex.

In Latex 4, we suggest that the water/air interface (although decreasing in area) remains close to the latex surface throughout the process. Evaporation of water and deformation of particles occur concomitantly. We conjecture that the evaporation of water is the rate-limiting step. In the higher T_g latex (2), by contrast, the drying front recedes inward from the sample edge and produces air voids near the sample surface. These voids shrink, being driven by a reduction in surface energy. The time between the passage of the drying front and the onset of film formation increases with decreasing temperature. This result is in agreement with a model based on the viscous flow of the latex polymer. Our data thus support the notion that at high temperatures (about 30 K above T_g), evaporation is the rate-limiting step, but particle deformation by viscous flow is rate-limiting at lower temperatures.

Acknowledgments

The latices were synthesized at ICI Paints in Slough, UK. We benefited from discussions with Drs. David Taylor, Craig Meekings, Panos Sakellariou, Gerry Meeten and Len Gate. At the time of this work, JLK was supported as an Oppenheimer Associate at the University of Cambridge. Some of this research was conducted through the Colloid Technology Programme, which is funded by the DTI, ICI plc, Schlumberger Cambridge Research, Unilever plc, and Zeneca plc. We are indebted to one of the referees for helpful comments.

Literature Cited

1. Keddie, J.L.; Meredith, P.; Jones, R.A.L.; Donald, A.M. *Macromolecules* **1995**, *28*, 2673.
2. Hwa, J.C. *J. Polym. Sci.: Pt. A*, **1964**, *2*, 785.

3. Okubo, M.; Takeya, T.; Tsutsumi, Y.; Kadooka, T.; and Matsumoto, T. *J. Polym. Sci.: Polym. Chem. Ed.*, **1981**, *19,* 1.

4. Chevalier, Y; Pichot, C.; Graillat, C.; Joanicot, M.; Wong, K.; Maquet, J.; Lindner, P.; Cabane, B. *Coll. Polym. Sci.*, **1992**, *270,* 806.

5. Sheehan, J.G.; Takamura, K.; Davis, H. T.; Scriven, L.E. *Tappi Journ.*, **1993**, *76*(10), 93.

6. Juhué, D.; Wang, Y.; Lang, J.; Leung, O.-M.; Goh, M.C.; Winnik, M.A. *J. Polym. Sci.: Pt. B: Polym. Phys.*, **1995**, *33,* 1123.

7. Winnik, M.A.; Feng, J. *Coatings Tech.*, **1996**, *68*(852), 39.

8. Denkov, N.D.; Velev, O.D.; Kralchevsky, P.A.; Ivanov, I.B.; Yoshimura, H.; Nagayama, K. *Langmuir,* **1992**, *8,* 3183.

9. Sperry, P.R.; Snyder, B.S.; O'Dowd, M.L.; Lesko, P.M. *Langmuir*, **1994,** *10,* 2619.

10. Jasperson, S.N.; Schnatterly S.E. *Rev. Sci. Instr.* **1969**, *40,* 761.

11. Danilatos, G. *Adv. Elecronics & Electon Phys.* **1990**, *78,* 1.

12. Meeten, G.H.; *Optical Properties of Polymers*, Elsevier Applied Science Publishers: London, 1986; p 361.

13. Born, M.; Wolf, E. *Principles of Optics*, Pergamon: New York, 1975; p 87.

14. Azzam, R.M.A.; Bashara, N.M. *Ellipsometry and Polarized Light*; North-Holland: Amsterdam, 1987; p 40.

15. Meredith, P.; Donald, A.M. *J. Microsc (Oxford)* **1996**, *181,* 23.

16. Cameron, R.E.; Donald, A.M. *J. Microsc.* **1994**, *173,* 227.

17. Wang, Y.; Juhué, D.; Winnik, M.A.; Leung, O.M.; and Goh, M.C.*Langmuir*, **1990**, *8,* 760.

18. Sheetz, D.P. *J. Appl. Polym. Sci.* **1965**, *9,* 3759.

19. Croll, S.G. *J. Coat. Technol.* **1986**, *58,* 41.

20. Mackenzie, J.K.; Shuttleworth, R. *Proc. Phys. Soc.*, **1949**, *62,* 838.

21. Vogel, H. *Phys. Z.*, **1925**, *22,* 645; Fulcher, G.S.*J. Am. Ceram. Soc.*, **1925**, *8,* 339.

22. Brodnyan, J.G. ; Konen, T. *J. Appl. Polym. Sci.*, **1964**, *8,* 687.

23. McKenna, G.B. in *Comprehensive Polymer Science Vol. 2*, C. Booth and C. Price, ed.; Pergamon Press: Oxford, 1989, p. 331.

24. van Krevelen, D.W.; Hoftyzer, P.J. Properties of Polymers: Their Estimation and Correlation with Chemical Structure, Elsevier: Amsterdam, 1976; p 343.

Chapter 22

The Relationship Between Film Formation and Anticorrosive Properties of Latex Polymers

R. Satguru, J. C. Padget, and P. J. Moreland

Research and Technology Department, Zeneca Resins, P.O. Box 8, The Heath, Runcorn, Cheshire WA7 4QD, United Kingdom

Attainment of coherent and defect free film formation in latex polymers is essential for achieving good protective properties. Two case studies are reported to highlight this effect in terms of achieving good anticorrosive properties. In the first study, the positive influence of the addition of a non ionic surfactant to a chloro polymer latex in terms of enhanced particle coalescence leading to excellent anticorrosive properties is discussed. In the second case, the deleterious influence of particle pre-crosslinking on coalescence in a styrene-acrylic latex and its consequence on anticorrosive properties is highlighted.

The rate and extent of particle coalescence during latex film forming process has been a topic of interest for many years. Many theoretical and experimental papers have been published and the extent of activity has never been greater than at the present time, with the advent of powerful techniques for studying the process such as neutron scattering (*1*) atomic force microscopy (*2*) and fluorescence spectroscopy (*3*). This work has been driven in part because of the reasonable belief that the rate and extent of particle coalescence will have a profound effect on the properties of the ultimate coating. Nevertheless there has been remarkably little published work on the relationship between coalescence and anticorrosive properties.

The anticorrosive properties of a coating are profoundly influenced by ingress of oxygen and water passing through the coating to the metal surface because corrosion processes depend fundamentally on the presence of oxygen and water. Preventing or at least restricting the ingress of water, oxygen, and/or electrolyte should be the primary consideration for designing an anticorrosive coating. Apart from selecting the optimum polymer type for this condition, coherent film formation is the key for achieving good protection. Attaining coherent film in solvent borne coatings is fairly straight forward. In water borne polymers however, it is more difficult as the film formation has to be derived from dispersed colloidal

0097–6156/96/0648–0349$15.00/0

particles. In this paper we report two related case studies to highlight the importance of coherent film formation to obtain good anticorrosive properties. The first study deals with the influence of post added surfactant on film formation of a chloropolymer latex and the second study looks at the influence of crosslinking on film formation of a styrene-acrylic latex.

Experimental

The chloropolymer and styrene-acrylic latices were synthesised using conventional emulsion polymerisation method.

For water blushing, AC impedance and salt spray testing, films were cast on mild steel test panels using an applicator bar at the specified dry film thickness. Coated films were then dried for 7 days at 25°C and 50% relative humidity before subjecting to the respective testing method.

AC impedance measurement was carried out using a Frequency Response Analyser, Solartron Model 1174.

Atomic Force Microscopy (AFM) measurement was performed using a Nanoscope II instrument. Latex films were cast on acetate sheets and allowed to dry for 7 days at room temperature before subjecting to AFM examination.

Moisture Vapour Transmission Rate (MVTR) values were obtained using the gravimetric weighed cup method at 25°C and 75% relative humidity. Coatings were applied as 1x 12µm dry film thickness on to cascade board and dried at room temperature for 7 days prior to MVTR measurement.

Results and Discussion

Case Study 1 : Chlorine containing vinyl acrylic latex. Chlorine containing polymers exhibit excellent barrier to transmission of both water (liquid/vapour) and oxygen. For this reason solvent borne chloropolymers are well established as anticorrosive coatings. We have found that the attainment of similarly high level of corrosion protection from a waterborne chloropolymer latex is crucially dependent on achieving good particle coalescence. Investigations based on one particular chloropolymer latex (Haloflex* 202) led to some interesting findings. This latex was prepared at a very low anionic surfactant content with colloid stabilisation essentially derived from the initiator end groups (4,5). The average particle diameter of the latex was 230nm and the Minimum Film Forming Temperature (MFFT) was 15°C. It was found that clear films cast from the latex on to steel panels and room temperature dried for 7 days gave excellent performance in a conventional salt spray test. However formulation of the latex into a stable paint formulation (pH=4.5) required the addition of a non ionic surfactant in order to provide sufficient colloid stability. For this reason we studied the effect of surfactant addition on coalescence and coating properties. The chosen surfactant was Synperonic PE 39/70 (ICI), a polyethlene oxide - polypropylene oxide block copolymer surfactant of nominal composition $(EO)_{64}(PO)_{39}(EO)_{64}$.

Latex samples containing different proportions of non ionic surfactant were cast on to mild steel panels at a dry film thickness of 60 µm. After drying for 7 days at 25°C / 50% relative humidity the panels were immersed for 7 days in distilled water and then inspected. A qualitative assessment of film blushing at varying levels of surfactant addition is given in Table I.

Table I. Water blushing as a function of post added non ionic surfactant

Wt% Surfactant	Extent of Blushing
0	Severe
1	Moderate
2	Slight
3	None
4	Very Slight

The film containing no added surfactant exhibited worst blushing ie. a white haze developed. Progressively increasing the surfactant content in the range of 0 - 3 % (wt% on latex solids) gave rise to a progressive decrease in water blushing, with blushing being virtually absent at 3% concentration. A further increase to 4% caused reintroduction of blushing, but significantly less than for the surfactant free film. Evidence that the observed blushing was indeed due to the ingress of water was provided by measuring the capacitance increase of the coatings by an AC impedance technique (6). The rate of capacitance increase during immersion in 5% aqueous sodium chloride decreased as the concentration of non ionic surfactant increased in the range of 0 to 3% on latex solids. From the capacitance measurements the 'notional water uptake' of the films was calculated using the expression derived by Brasher and Kingsbury (7) and presented in Figure 1. The results showed that the capacitance measurements exactly paralleled the visual water blushing observations.

The adsorption isotherm of Synperonic PE 39/70 on to latex was determined gravimetrically and found to be indicative of a strong adsorption characteristic as shown in Figure 2. At concentrations of up to 2.5 - 3.0 % by weight of surfactant on latex solids the majority of the surfactant was adsorbed on to the particle surface. At concentrations of greater than 3%, the excess surfactant was unadsorbed and present in the latex aqueous phase.

Transmission electron microscopy of replicated films containing 0 to 5% post added surfactant after a drying period of four hours, clearly showed the increase in the extent of coalescence with increasing surfactant concentration (6). The sample containing 5% surfactant exhibited surface exudations which were presumed to be excess unadsorbed surfactant present in the aqueous phase of the parent latex.

The above results indicate that post addition of non ionic surfactant to this latex strongly influenced the particle coalescence process in providing a coherent and defect free film at an optimum concentration of 3 wt% on latex solids. Further evidence for this was obtained by measuring the tensile properties of the respective films. It was seen that increasing concentration of non ionic surfactant gave rise to a decrease in modulus and tensile strength and an increase in extension to break. These property changes demonstrate that the surfactant was an effective plasticiser

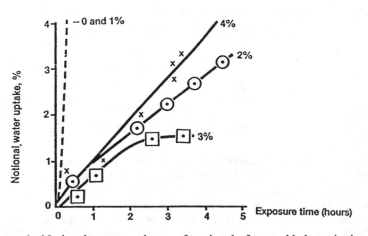

Figure 1. Notional water uptake as a functional of post added non ionic
surfactant onto chlorine containing vinyl acrylic latex at different exposure
times. (---) 0 and 1 wt%; (⊙) 2 wt%; (▢) 3 wt% and (x) 4 wt%.
(Reproduced with permission from reference 6. Copyright 1983 *J. Coatings Tech.*)

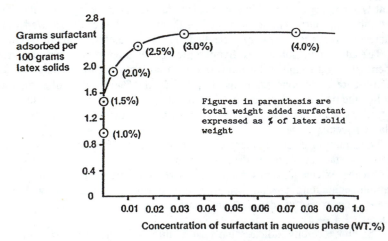

Figure 2. Adsorption isotherm of non ionic surfactant on to chlorine
containing vinyl acrylic latex.
(Reproduced with permission from reference 6. Copyright 1983 *J. Coatings Tech.*)

for the polymer, a fact which presumably explains the observed ability of the surfactant to increase the rate of particle coalescence. An additional possible role of the surfactant is to increases the inherent colloid stability of the parent latex, which in turn allows the particle coalescence to take place in a uniform manner by maintaining its particle integrity (ie. free from flocculation) through out the drying process, resulting in a defect free film. It is however important to note that the optimum protective properties of the film was observed at the monolayer coverage concentration of the surfactant   ie. 3 wt% , and at this concentration the water vapour barrier properties of the latex cast film was comparable to that of the corresponding solvent cast film based on the same copolymer.

Case Study 2 : Styrene acrylic latex. This study highlights the deleterious influence of particle pre-crosslinking on coalescence and its consequence on anti corrosive properties of the derived coatings. Two latices were chosen for this study: Latex AC1 - a styrene acrylic latex, where deliberate crosslinking was introduced during the polymerisation via the use of specific monomer choice. Latex AC2 - also a styrene acrylic latex composed of the same backbone as AC1 but no crosslinking monomer was employed ie. uncrosslinked. Particle diameter of both latices was 75 nm and MFFT of AC1 was 20°C while the MFFT of AC2 was 16°C. The study was based on assessing the respective anticorrosive performance of the coatings as a function of coalescent amount required to form coherent films. Dowanol DPnB and Dowanol PnB (Dow Chemical) combination at a ratio of 3:2 was used as the coalescing agent for this study.

Latices containing 0 to 25 wt% (based on polymer solid content) of coalescent were prepared and 45 µm d.f.t. films were cast on mild steel panels at room temperature (22°C) and allowed to dry for 7 days. The panels were then subjected to ASTM hot salt spray testing. The results showed (see Plates 1 and 2), that latex AC1 required a coalescent level of 20 to 25 wt% while latex AC2 required only 10 wt% for good performance up to 1000h salt spray exposure. Considering the MFFT difference of the two latices is not very large, the above result was significant.

In an attempt to obtain insight into the film formation process of latices AC1 and AC2, films containing different levels of coalescent were subjected to atomic force microscopy (AFM) and moisture vapour transmission measurement (MVTR) studies. AFM provides information on the topology of the film and hence the coalescence process, while MVTR values give information on the barrier to water vapour of the film and hence the coherency of the film. AFM results (see Plates 3 and 4), showed that a coalescent level of approximately 20 wt% was required for latex AC1 while 10 wt% coalescent was sufficient for Latex AC2 to form smooth films. This result was confirmed by measuring the root mean square roughness value from the respective micrographs. The   MVTR values presented in Figure 3, also showed similar results indicating that lowest MVTR value for latices AC1 and AC2 were obtained at coalescent concentrations of approximately 15 -20 wt% and 10 wt% respectively. These results clearly supports the salt spray performance of the respective latex films, and confirms the necessity of higher level of coalescent for pre-crosslinked latex particles to achieve good anti corrosive performance.

HOT SALT SPRAY V COSOLVENT LEVEL (430 HOURS)

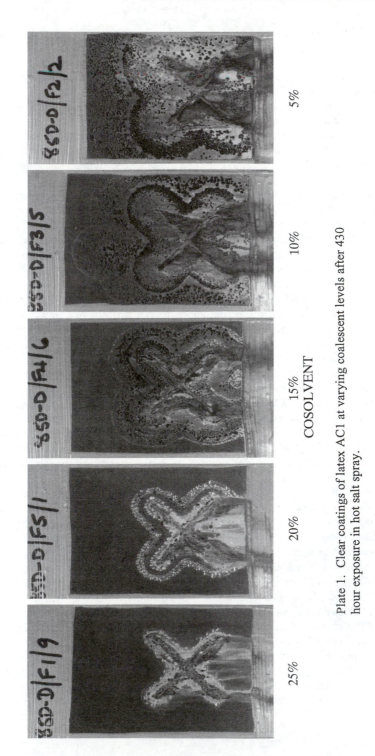

| 5% | 10% | 15% COSOLVENT | 20% | 25% |

Plate 1. Clear coatings of latex AC1 at varying coalescent levels after 430 hour exposure in hot salt spray.

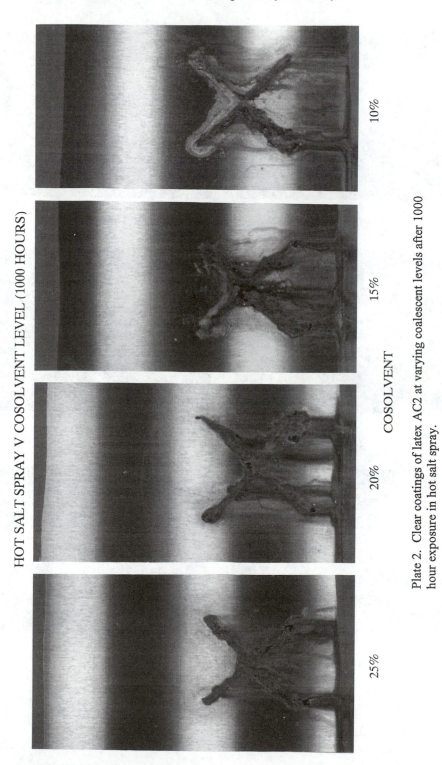

HOT SALT SPRAY V COSOLVENT LEVEL (1000 HOURS)

10%   15%   20%   25%

COSOLVENT

Plate 2. Clear coatings of latex AC2 at varying coalescent levels after 1000 hour exposure in hot salt spray.

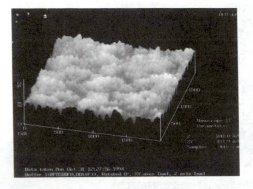

0

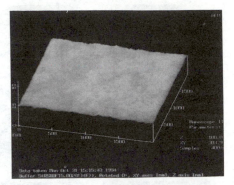

15%

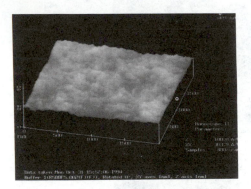

5%

20%

10%

25%

Plate 3. Atomic Force Micrographs of Latex AC1 films containing varying levels of coalescent.

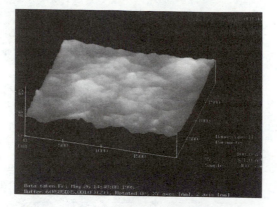

5%

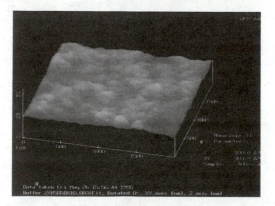

10%

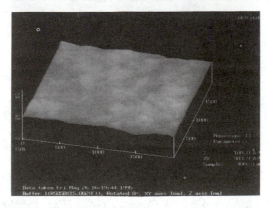

15%

Plate 4. Atomic Force Micrographs of Latex AC2 films containing varying levels of coalescent.

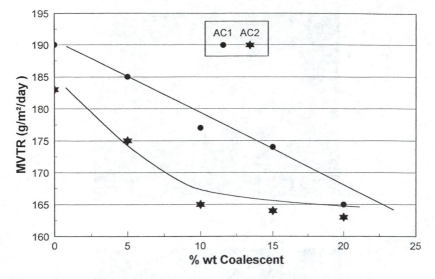

Figure 3. Moisture vapour transmission rates of latex films AC1 and AC2 as a functional of coalescent concentration. (Adapted from ref. 6).

In addition, this study also showed the usefulness of AFM and MVTR measurements to assess the degree of coalescence in latex polymers in relation to its protective properties.

* The word "Haloflex" is a registered trademark of ZENECA Ltd.

Acknowledgments

The authors wish to thank Mr J Farrar Mrs E Jones and Mr N Ormesher of Zeneca Resins for their contributions to this work.

Journal References:

1.   Yoo J N, Sperling L H, Glinka C J, Klein A ., Macromol. 1990, 23, 3962-3967.

2.   Juhue D, Lang J., Langmuir 9, 792, 1993.

3.   Winnik M A, Wang Y, Haley F., J. Coatings Tech., 1992, 64, 811, 51-61.

4.   Burgess A J, Caldwell D, Padget J C., J. Oil Colour Chemistry., 1981, 64, 175.

5.   Moreland P J, Padget J C, Lim Yoo Keng., Proc. Corrosion Asia 1994, No 1042.

6.   Padget J C, Moreland P J., J. Coatings Tech., 1983, 55, 698, 39.

7.   Brasher D M, Kingsbury A H., J. Appl. Chem., 1954, 4, 62.

Chapter 23

Permeation and Morphology in Polymer Latex Films Containing Leachable Additives

P. A. Steward[1], J. Hearn[2], M. C. Wilkinson[3], A. J. Wilson[4], and B. J. Roulstone[5]

[1]185 Stony Lane, Smethwick, Warley, West Midlands B67 7BA, United Kingdom
[2]Department of Chemistry and Physics, Nottingham Trent University, Nottingham NG11 8NS, United Kingdom
[3]OakApple House, Great Wishford, Salisbury SP2 0PA, United Kingdom
[4]Centre for Cell and Tissue Research, University of York, Heslington, York YO1 5DD, United Kingdom
[5]ICI Paints Division, Wexham Road, Slough, Berkshire SL2 5DS, United Kingdom

This paper reviews recent research, performed at The Nottingham Trent University, into the enhancement of the permeability properties of polymer latex films. Both leachate-free model colloid latices, made by surfactant-free emulsion polymerisation, and commercially available latices, containing surfactants and often requiring plasticiser addition to achieve good film formation, have been studied. Water soluble leachable additives such as sucrose, HPMC and a soluble polymer latex, have been employed to deliberately enhance solute permeant flux, and the transport properties of the films are discussed with reference to film morphology. The use of increasing levels of leachable additive allows the solute transport mechanism to be systematically changed from one in which permeation occurs predominantly by solution-diffusion to a mechanism in which convective diffusion predominates and solute permeation tends to become independent of penetrant solubility in the polymer. The results are discussed in terms of the increased hydration of the polymer resulting from additive leaching.

Whenever polymer latex films are used as coatings, then their permeabilities to gases, water vapour and solutes are likely to be of significance. In the case of barrier coatings for substrate protection, a minimisation of the flux would often be desirable, but where coatings are used to aid sustained or controlled release of an active ingredient then an ability to modify the flux is needed. In the latter application, water soluble leachates play an important role.

    This paper utilises free polymer latex films to investigate the transport properties of such films to a range of permeants, and the effects on film permeability of leachable additives ranging from endogenous surfactants, essential plasticisers and molecular sized highly water soluble additives to much larger (latex particulate) leachable additives. In the course of this work, it was found to be necessary to relate the changes in transport mechanism to the effects of additive leaching on film morphology.

Experimental.

Materials. Homopolymer latices used included poly(butyl methacrylate) (PBMA), poly(amyl methacrylate) (PAMA), and poly(hexyl methacrylate) (PHMA). Commercially available copolymers used – typical of those employed in pharmaceutical coatings – were part of the Eudragit range (Table I – data is from the Röhm Haas Data Book (3) unless otherwise stated) based on acrylic-methacrylic acid esters (Röhm Pharma GmbH, Weiterstadt: supplied by Dumas UK Ltd., Tunbridge Wells, Kent, UK). Latices were typically diluted to 5% W/W solids content using analytical grade water (Fisons PLC, FSA Lab. Supplies, Loughborough, UK).

A core-shell latex was prepared from a PBMA core, shot with acrylic acid (AA) (Aldrich Chemical Company Ltd., Gillingham, Dorset, UK)

Films were plasticised, when essential for ease of handling, using glycerol triacetate (triacetin) (supplied by Aldrich Chemical Co. Ltd.).

Film additives used to enhance solute permeability included sucrose (Aldrich Chemical Co. Ltd.), hydroxypropyl methylcellulose (Celacol HPM 450 BP – supplied by Courtaulds Chemicals, Spondon, Derby, UK), Eudragit L 30 D (Table I), sodium chloride (Aldrich Chemical Co. Ltd.), and triacetin (glycerol triacetate, a plasticiser – supplied by Aldrich Chemical Co. Ltd.).

4-nitrophenol (Fisons PLC) was used as a probe of solute permeability. Other permeants included carbon-14 labelled sucrose, and tritiated water (THO) (both supplied by Amersham International PLC, Amersham Laboratories, Amersham, Buckinghamshire, UK).

A series of anilines (aniline, methyl aniline, and ethyl aniline – all supplied by Fisons PLC) were used as a probe of porosity, as was potassium chloride (Aldrich Chemical Co. Ltd.) and either sodium-22 or chlorine-36 labelled sodium chloride (Amersham International PLC).

Solute permeation experiments were buffered to pH 6 using orthophosphate buffer – prepared as described in the C.R.C. Handbook of Chemistry and Physics (4).

Gases used for permeability measurements were helium and carbon dioxide (BOC Ltd., Brentford, Middlesex, UK).

Procedures.

Polymer Latex Polymerisation. Homopolymer latices were prepared surfactant-free (5, 6), with the apparatus sited in a fume cupboard due to the harmful nature of the monomers (e.g., irritating to skin, eyes and respiratory system, harmful by ingestion and inhalation). Latex particle diameters were determined by photon correlation spectroscopy (PCS) using a Malvern Instruments (Spring Lane, Malvern, Worcs, UK) Zetasizer3. Latex solids content (% W/W) was determined gravimetrically from a comparison of the wet and dried weights.

A PBMA-AA core-shell surfactant-free polymer latex was prepared by a shot-growth technique (7). This technique polymerises a core polymer (PBMA), as in a surfactant-free homopolymer emulsion polymerisation (5, 6), and then adds a 'shot' of

Table I Eudragit copolymers: structures and properties

NE 30 D (i.e., Neutral Ester, 30% Dispersion)	L 30 D (Leichtöslich i.e., freely soluble)	RL 30 D[†] (Leichtdurchlässig i.e., freely permeable)	RS 30 D[†] (Schwerdurchlässig i.e., slightly permeable)
poly(ethyl acrylate-methyl methacrylate) (2:1)	poly(methacrylic acid, ethyl acrylate) (1:1)	poly(ethyl acrylate, methyl methacrylate) trimethylammonio-ethyl methacrylate chloride (20:1)[‡]	poly(ethyl acrylate, methyl methacrylate) trimethylammonio-ethyl methacrylate chloride (40:1)[‡]

Polymer molecular weight			
800,000	250,000	150,000	150,000
Polymer T$_g$ /K			
265 (*1*)	303 (*2*)	328 (*2*)	328 (*2*)
z-average latex particle diameter (by PCS) /nm			
163.0	130.7	177.3	151.9
Commercial additives			
(i) iso-nonyl phenol poly(ethylene glycol) (*1*)	(i) sodium dodecyl sulphate (0.7%) (ii) polysorbate 80 (2.3%) (iii) requires triacetin plasticiser (15%)	(i) sorbic acid (0.25%) (ii) requires triacetin plasticiser (15%)	(i) sorbic acid (0.25%) (ii) requires triacetin plasticiser (15%)

† A pseudo-latex: prepared by emulsification of the bulk polymerised polymer.
‡ Ratio of neutral ester groups to ammonium groups.

a second monomer (the AA shell) when the core monomer is at high percentage conversion. Like the core monomer, the AA was initially cleaned of inhibitor by distillation at reduced pressure, in a nitrogen atmosphere. The rate of progress of the core monomer polymerisation was pre-determined by monitoring a polymerisation gravimetrically (5) to determine the time at which 90% conversion was reached – i.e., the point at which it was intended to add the shot of AA (8) of mass equal to 2% of the mass of the PBMA core monomer. The final product was cleaned by microfiltration (9), and this included an acid washing stage to convert the carboxyl groups to their acid form.

Film Preparation. All Eudragits (e.g., RL 30 D, RS 30 D, and L 30 D), with the exception of Eudragit NE 30 D, required plasticising with 0.15 g g^{-1} triacetin, in order to make the free films manageable. This, like the other film additives (i.e., the sucrose, HPMC, etc.), was added to the water used to dilute the latex to 5% W/W solids content – before addition to the latex – in preparation for casting. (Although not an essential additive, the effectiveness of triacetin plasticiser as a permeability enhancer when added to Eudragit NE 30 D was also tested in a series of experiments.) Film additives were used at loadings measured as a fraction per gram of latex solids content e.g., 0.2 g (g.latex solids)$^{-1}$, i.e., a 20% load. When Eudragit L 30 D was used as an additive, it was initially still in the form of an aqueous based latex (at 5% solids content). Addition of Eudragit L 30 D to another latex such as Eudragit NE 30 D therefore effectively diluted the solids content of each individual latex. E.g., a loading of 100% Eudragit L 30 D in Eudragit NE 30 D was a 1 : 1 mixture of 5% total solids content, but with the Eudragit NE 30 D diluted to 2.5%.

Films were cast from a fixed volume of latex onto a Pyrex glass substrate. The wet latex was contained with a glass cylinder – sealed to the substrate by a smear of high melting point silicone grease.

Casting took place on a levelled platform in a fanless laboratory oven. The oven contained fixed amounts of desiccant (per film) to remove the humidity resulting from the evaporation of water. PBMA films were cast at 353 K for 72 hr; PAMA and PHMA films at 313 K for 15 hr; and the Eudragit latices at 313 K for 24 hr. These differences in times and temperatures were due to the various glass transition temperatures (T_g) of the polymers, and the time required to dry the films at the differing temperatures. The T_g of PBMA is quoted (10) as 308 K, and that of PAMA and PHMA as 263 $\rightarrow$ 265 K (Polysciences (UK) Ltd., Moulton Park, Northampton, UK, personal communication). Values for the Eudragit polymers are given in Table I.

After casting, Eudragit films could be removed from the substrate by careful peeling, following gentle heating of the back of the substrate, being careful to avoid stretching the film. The homopolymer latices required soaking in water to assist removal. In the case of PBMA, the water needed to be hot (343 K) because the film was brittle at ambient temperature.

Solvent cast surfactant-free homopolymer latex films were prepared by dissolving the freeze-dried polymer at 5% W/V in butan-2-one (BDH Chemicals Ltd., Poole, Dorset, UK) and films were cast onto a PTFE substrate. Drying was initially carried out in a fume cupboard at ambient temperature, with the film partially shielded to prevent too fast an evaporation rate. Once solidified (ca 4 hr), the film was transferred to a vacuum oven, where it was kept under vacuum, again at ambient temperature, for a further 96 hr, to remove any residual solvent.

Films were always used within 24 hr of casting unless otherwise stated.

Determination of Solute Permeability Coefficients. The 4-nitrophenol permeability of the films was determined with the film located to form the interface of two chambers of a 'permeability cell.' Films were always oriented with the polymer-substrate interface (with respect to casting orientation) facing the permeant charge to avoid uncertainty resulting from the 'permeability side-difference phenomena' often discussed in the literature (*11*, *12*, *13*) on free-film properties. The 'donor' and 'receiver' chambers initially contained the charge of permeant and a buffer solution, respectively. The 4-nitrophenol was recrystallised from toluene and dissolved in orthophosphate buffer (pH 6) such that it was it was in its undissociated form (⟩ 99%). (Note that 4-nitrophenol is a skin irritant and is harmful by ingestion.) The film was sealed between flanges on the permeability cell chambers by means of silicone grease, and clamped. The permeability cell was then placed in a constant temperature bath (303 K), and each chamber of the cell filled (65 ml) simultaneously – to avoid stress on the film.

Each chamber of the permeability cell was stirred at 100 rpm using glass propeller-type stirrers, driven by a common motor.

Measurement of the transport of permeant through the film into the receiver chamber was done spectrophotmetrically, using a semi-automated system. The receiver chamber solution was continuously pumped, using a multichannel peristaltic pump (Gilson Medical Electronic – supplied by Anachem, Luton, Bedfordshire, UK) to flow-through compact quartz cells (80 μl capacity, 1 cm pathlength: supplied by Hellma England Ltd., Westcliff-on-Sea, Essex, UK) positioned in a UV spectrophotometer (model number PU8730, Philips Scientific, Philips Analytical, Cambridge, UK), from where it was returned to the permeability cell. All tubing, with the exception of that around the rollers of the peristaltic pump, was PTFE.

Five films could be monitored simultaneously using five flowcells in a computer controlled automatic cell changer (Philips cell programmer, part number PU8737). Absorbance data was collected at (typically) 15 min intervals over the course of 24 hr, and was initially stored in the spectrophotometer (controlled by PU8714 User Programming Software). Data was then transferred (via a PU8702 computer interface) to a personal computer, and a spreadsheet was used to convert the absorbance data i, to concentration data by means of a pre-determined Beer-Lambert coefficient, and ii, to a permeability coefficient.

Permeability coefficients (*P*) were calculated from an integrated form of Fick's first law of diffusion (relative to the boundary conditions of the system):

$$Pt = \frac{2.303 \, lV}{2A} \log_{10} \left[\frac{C_0}{C_0 - 2C_t} \right] \qquad (1)$$

where:
l = film thickness (m);
V = receiver cell volume (= donor cell volume) (m^3);
A = area of film exposed to permeant (m^2);
C_0 = initial permeant solution concentration (g dm^{-3});
C_t = receiver solution permeant concentration (g dm^{-3}), at time t (hr).
(Derivation of a similar equation has been given by Flynn (*14*).)

The permeability coefficient can therefore be calculated by least squares regression analysis of the data once the steady state flux has been achieved, following permeant equilibration within the film.

In the case of the additive-containing films, solute permeability coefficients were typically determined whilst the additive was simultaneously leaching from the film. However, films that had been pre-leached of additive were also investigated. Films were pre-leached whilst *in-situ* in the permeability cell – which was filled with buffer and typically left for a period of 96 hr whilst the additive was leached. Conditioning of these pre-leached films in the permeability cells either involved the cell being i, emptied and the film dried (24 hr) before the start of the permeation experiment, or ii, the film being left wet before the start of the permeation experiment. In the latter case, the leachant was left in the cell, and the 4-nitrophenol was added to the donor chamber in the form of 4-nitrophenol crystals.

The aniline permeants were used at a concentration of 8 mM made-up in orthophosphate buffer (pH 6). (Note that the anilines are all flammable and toxic by inhalation, ingestion and skin absorption.) The permeability coefficients of the anilines were determined in a manner similar to that of the 4-nitrophenol. Since the anilines were found to adsorb onto the short section of flexible tubing within the peristaltic pump, automated sampling had to be replaced by a manual procedure.

Determination of Gas and Water Vapour Permeability Coefficients. CO_2 and He gas permeability coefficients were determined using a non-overflowing Daventest apparatus and the British Standard Method (*15*). With this procedure, the film acts to separate the gas from a vacuum created beneath it. The flux of gas is then measured manometrically.

Water vapour permeability coefficients were determined gravimetrically with the film at the interface of a relative humidity (RH) difference of 81% – achieved by sealing the film to the top of a sample bottle containing a saturated solution of ammonium sulphate, and placing the film in a desiccator (containing silica gel) kept at a constant temperature of 298 K. The seal between the film and sample bottle was achieved using a 'glue' prepared from a sample of the film dissolved in butan-2-one. Saturated ammonium sulphate solution gives an RH of 81% (corresponding to a water vapour pressure of 1.92 cmHg) at 298 K, and this varies little (±0.1%) over a temperature range of ±5 K (*4*). The weight of the sample bottle (*W*) was measured at regular intervals, and the water vapour permeability coefficient was calculated from:

$$ P = \frac{dW}{dt} \frac{l}{1.92 \times A} \qquad (2) $$

where dW/dt = rate of change of weight of the sample bottle (g hr^{-1}).

Other Techniques. KCl (0.2 M) permeability was determined with the films in the permeability cells and the transport of electrolyte into the receiver chamber being monitored conductometrically.

Carbon-14 labelled sucrose permeation and the permeation of THO – both β-emitters – was also measured with the films in the permeability cells. Samples were periodically removed (typically 250 μdm^{-3}) from the receiver chamber, and the activity monitored by liquid scintillation counting (using a Canberra Packard model 2250CA Tri-Carb liquid scintillation counter) following addition of Canberra Packard Ultima Gold Scintillation Cocktail (10 ml). The long half-lives of the isotopes meant that this decay did not need to be accounted for over the comparatively short periods of the permeation experiments. The use of the radio-isotopes as permeants presents numerous health and safety problems (*16*) and radio-isotopes are typically subject to local legislation or codes of practice which should be consulted before use.

Sodium-22 labelled NaCl is a gamma emitter (and therefore requires greater care (*16*) when 'handled' than do the other radio-isotopes), and was analysed using a solid scintillation method, using a Canberra Packard 500 Auto Gamma Counter, with a 3" crystal. Chlorine-36 labelled NaCl was detected using the aforementioned liquid scintillation counting procedure. Both were used as permeants with the film mounted in the standard permeability cell, and like the sucrose and THO, small samples were removed from the cell for measurement, and discarded (in accordance with local regulations).

Leaching of additives (sucrose, HPMC, and the Eudragit L 30 D) from the films was determined gravimetrically, as a function of time: the additive leaching into a volume of water (e.g., 250 ml), or buffer, far greater than that which would become saturated with the additive.

The partition of the permeants (*4*-nitrophenol and the anilines) into films of known weight was determined spectrophotmetrically – until solution equilibrium was achieved – with the film in a constant temperature bath (303 K), and agitated by an orbital motion bed.

Pore distributions were determined by the technique of mercury porosimetry (using a Carlo-Erba Porosimeter 2000 – imported by Fisons PLC). Such an instrument operates over a range of 3.7 $\rightarrow$ 7,500 nm pore radius. (The use of this method and apparatus involves the handling of mercury which is very toxic by ingestion, inhalation and skin contact.)

Scanning electron microscopy (SEM) was performed both by i, CBDE (Porton Down Salisbury, Wiltshire, UK) using a Hitachi field emission SEM, and ii, The Institute of Polymer Technology and Materials Engineering (University of Technology, Loughborough, UK). All fracture cross-sections of films were performed with the film frozen in liquid nitrogen to prevent deformation of the film.

Transmission electron microscopy was performed using a tensile snap method (FFTEM) to fracture cross-sections. This specialised method freezes the film sample at the triple point of nitrogen whence it is both fractured and replicated. This work was performed at the Centre for Cell and Tissue Research (University of York, Heslington, York, UK) using a (Oxford Instruments) Cryotech CF-4000 to perform the fracture, and a JEOL 1200 Ex TEM.

PBMA-AA core-shell latex samples of known weight were titrated conductometrically versus freshly prepared sodium hydroxide (nominally 0.1 M). The latex, in a Pyrex beaker, was initially outgassed with N$_2$, to reduce the effects of atmospheric CO$_2$ contamination, and was kept under a blanket of N$_2$ during the titration. The latex was stirred using a PTFE magnetic follower, and the NaOH added by means of a microlitre syringe. The conductivity was determined using a Jenway 4010 conductivity meter (supplied by Fisons): stirring was halted during the conductivity reading. Core-shell latex films (prepared from an equivalent weight of latex as used in the aforementioned titrations) were also titrated.

Results and Discussion.

Additive-Free Polymer Latex Films.

Surfactant-Free Polymer Films Containing No Endogenous Additives. The additive-free homo-polymers differed only in the length of their alkyl groups, and their aforementioned polymer T_g's, and this was reflected in their respective permeability properties (Table II and Figure 1). The permeability coefficient for PBMA was on the limit of instrument resolution, and the film was found to effectively act as a barrier over a period of ca 72 hr (i.e., the longest period monitored). 4-nitrophenol was originally chosen as a permeant, by Wicks (17) for its solubility in both water and polymer, and it was apparent in this work that the 4-nitrophenol permeability coefficient increased with increasing hydrophobicity (i.e., carbon chain length) of the films. This was the expected result for a hydrophobic permeant in which it would be expected that solubility in the film would increase with polymer hydrophobicity. I.e., such permeation in a non-porous membrane is a three step process in which a penetrant must i, dissolve (partition) into the polymer, ii, diffuse through the membrane, and iii, desorb from the polymer. The partition of permeant into the polymer was immeasurably low in the case of PBMA, which also showed little opacity due to swelling with water: the opacity being only slightly greater for PAMA and PHMA.

Contributory to permeant transport is the motion of the polymer chains. Diffusional motion in any medium is dependent upon the dynamic formation and elimination of free-volume (i.e., 'holes' of no fixed size or shape) into which a penetrant molecule may move. Hole formation is a random process dependent on the thermal motion of the molecules constituting the medium. When a penetrant travels through the medium, its rate of diffusion is dependent on the probability of the penetrant being able to 'jump' from one hole to an adjacent hole. This is itself dependent on the hole having sufficient volume to accommodate the penetrant molecule. The diffusional jump direction is random, but bulk flow occurs due to statistical probability arising from the concentration gradient. If the medium is a polymer, penetrant diffusion is dependent upon the cooperative motion of the polymer chain segments. Various theories for the diffusion of a permeant in a polymer exist (e.g., Fujita's free-volume theory (18), and the activated zone theory (19)), and attempt to rationalise a diffusional jump in terms of either the activation energy, E, of the formation of a suitably sized hole, or the availability of a distribution of suitably sized holes. Generally, the diffusion coefficient, D, is expressed by the following proportionality (20, 21):

$$D \propto \exp -\left[\frac{V_p}{V_f}\right] \exp -\left[\frac{E}{RT}\right] \tag{3}$$

where:
V_p = volume required by the diffusant;
V_f = sample free volume;
R = universal gas constant;
T = temperature.

Table II 4-nitrophenol solute permeability coefficients, and properties, of homopolymer films

Polymer Type	Alkyl Group Length	Polymer T$_g$ /K	Permeability Coefficient /1×10^{-10} m^2 hr^{-1}	
			Latex Cast Film	Solvent Cast Film
PBMA	4	308 (*10*)	1	1
PAMA	5	263 → 268†	25	—
PHMA	6	263 → 268†	80	55

† Polysciences (UK) Ltd., Moulton Park, Northampton, UK, personal communication.

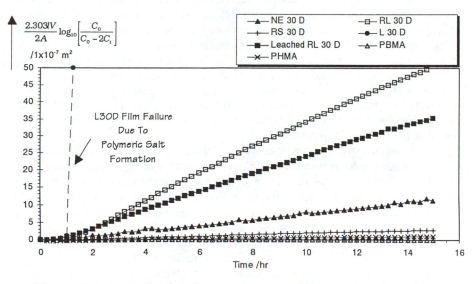

$$\frac{2.303lV}{2A} \log_{10}\left[\frac{C_o}{C_o - 2C_t}\right]$$

/1×10^{-7} m^2

Legend: —▲— NE 30 D   —□— RL 30 D   —+— RS 30 D   —●— L 30 D   —■— Leached RL 30 D   —△— PBMA   —✕— PHMA

L30D Film Failure Due To Polymeric Salt Formation

Time /hr

Figure 1 Relative 4-nitrophenol permeation rates of films cast from the various Eudragit polymer types.

Therefore, it was evident that in the case of PBMA (but not PAMA or PHMA), the polymer's T_g which was close to the temperature at which the permeation experiment was performed (i.e., 303 K), was likely to contribute to its extremely low permeability coefficient due to stiffening of the polymer chains.

The effect of polymer T_g was also readily apparent in the morphology of the polymer latex films. By FFTEM, the softer PAMA and PHMA appeared structureless (Figures 2 and 3, respectively). In contrast, PBMA showed vestigial particulate boundaries (Figure 4) in the fracture cross-section – albeit grossly deformed from their original spherical shape and PCS particle size of ca 500($\pm$50) nm (cf. PHMA and PAMA of 650($\pm$50) nm).

SEM's of the PBMA film surface showed (Figure 5) particles which were far less deformed than those at the film interior – presumably indicating the 'power' of the process of wet sintering: a mechanism by which latex particle coalescence has been reported (22) to form continuous films.

It has previously been reported (23) that PBMA films cast for 3 hr at 353 K undergo an aging process in which permeability decreases as a function of storage time – presumed to be due to further gradual coalescence, as described by Bradford and Vanderhoff (24, 25, 26), or the process of autohesion as described by Voyutskii (27, 28, 29). Such films showed a greater level of structure (by FFTEM) than those films shown here. The extra annealing time (i.e., casting for 72 hr) has therefore had a beneficial effect on film structure, and the film appears visually more amorphous. It might therefore be expected that film permeability would be more reproducible over a prolonged period of storage and less likely to exhibit the aforementioned effects of aging – of obvious benefit for a controlled release coating. However, because the film was already showing barrier like properties, it was not possible to show whether the permeability coefficient decreased with time, or not. In this respect, the structureless nature of the softer PHMA and PAMA polymers might prove even more advantageous. However, such an advantage was irrelevant as the polymers were far too tacky to use as a coating – tending to adhere and cohere at the least opportunity.

Although no particle boundaries were visible in the latex cast PHMA and PAMA films, the 4-nitrophenol permeability coefficient of PHMA was significantly greater than for an equivalent solvent cast film (Table II) indicating that the permeability of the latex cast film could possibly decrease during prolonged storage as further polymer chain interdiffusion leads to increased densification of the film. Further evidence of increasing film densification as a function of film age was provided by He gas permeability coefficients decreasing from $3.02(\pm0.48)\times10^{-9}$ cm^3 cm s^{-1} cm^{-2} cmHg^{-1} for the unaged PHMA film to 2.36×10^{-9} cm^3 cm s^{-1} cm^{-2} cmHg^{-1} for a film aged 18 days.

None of the additive-free, surfactant-free films were permeable to electrolyte.

Commercial Polymer Latex Cast Films. In contrast to the surfactant-free polymers prepared specifically for the above work, the commercially available Eudragit polymers specifically designed as pharmaceutical coatings (2, 3) contain endogenous components residual to their manufacture, or added for the sake of particle stability during storage, etc.

The effects on the 4-nitrophenol permeation of the various polymer types, combined with their endogenous additives and, in the case of all Eudragits with the

Figure 2 Featureless FFTEM of a PAMA film's cross-section.

Figure 3 Featureless FFTEM of a PHMA film's cross-section.

Figure 4 FFTEM of a PBMA film's cross-section showing vestigial particles, albeit highly deformed.

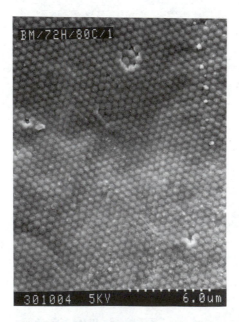

Figure 5 SEM of a PBMA film's polymer-air interface showing particulate close packing.

exception of Eudragit NE 30 D, of 15% (g.polymer)$^{-1}$ triacetin plasticiser are shown in Figure 1. As expected, the carboxylated Eudragit L 30 D dissolved at the experimental pH, due to the formation of polymeric salts. This occurred within the first 2 hr of the start of permeation – before the attainment of permeant (partition) equilibrium within the film and, hence, before a permeability coefficient could be measured.

Eudragit RL 30 D showed the expected greater permeability than Eudragit RS 30 D (3.81($\pm$0.08)$\times10^{-7}$ m^2 hr^{-1} versus 0.153($\pm$0.011)$\times10^{-7}$ m^2 hr^{-1}) as a result of its greater hydrophilicity due to its functionality. Also, from Figure 1, it was apparent that the magnitude of the permeability increased considerably due to the plasticiser added to the Eudragit RL 30 D film. Whereas a Eudragit RL 30 D loaded with 15% triacetin yielded a permeability coefficient of 3.81($\pm$0.08)$\times10^{-7}$ m^2 hr^{-1}, a comparable film pre-leached (48 hr) of its triacetin yielded a permeability coefficient of 2.68($\pm$0.18)$\times10^{-7}$ m^2 hr^{-1}. (Note, the typical plasticiser load of 0.15 g g^{-1} was typically 97% leached from the film within 1 hr of insertion into water – based on gravimetric and spectrophotometric evidence.)

The effect of a leachable endogenous additive was apparent in the case of Eudragit NE 30 D in which the film typically yielded a 4-nitrophenol permeability coefficient of 4.00($\pm$0.36)$\times10^{-8}$ m^2 hr^{-1}. By comparison, a Eudragit NE 30 D film prepared from an extensively dialysed sample of the latex yielded a permeability coefficient of 0.884($\pm$0.025)$\times10^{-8}$ m^2 hr^{-1}. This was presumably due to the removal of the nonyl phenyl poly(ethylene glycol) surfactant, which otherwise possibly acts to either plasticise the polymer or allows the imbibition of water into the film through which a permeant may travel.

As might be predicted from the above permeability coefficient results, Eudragit RL 30 D showed a much lower permeation activation energy at 40.7 kJ mol^{-1} (by means of an Arrhenious-type plot), than Eudragit NE 30 D (50.5 kJ mol^{-1}). The equilibrium 4-nitrophenol partition coefficients of the polymers (Figure 6) were also ca 3-fold greater in the case of Eudragit RL 30 D (278.6$\pm$21.6) compared to Eudragit NE 30 D (94.2$\pm$12.7).

Endogenous or essential additives were thus seen to have a marked effect on film permeability, and this can be correlated to the transport mechanism through the film and, hence, film morphology. By the technique of mercury porosimetry, Eudragit NE 30 D showed some evidence of a pore structure (Figure 7), whereas Eudragit RL 30 D showed no sign of any significant porosity. SEM's of Eudragit NE 30 D did show some structure (Figure 8), but no porosity. The structure possibly exists between vestigial latex particles whose complete coalescence was inhibited by the presence of grafted surfactant. The differential-type plot (Figure 7) for Eudragit RL 30 D, whilst apparently showing some structure in the film at high pressure, shows no real pore distribution, and the 'structure' was possibly due to compression of the film at pressures approaching ca 2$\times10^8$ N m^{-2} (2000 Bar). Equally, the resolution of SEM showed no porosity in the case of Eudragit RL 30 D. However, permeants can also be used as probes of porosity. In the absence of an ion exchange mechanism, or porosity of a diameter greater than that of the probe molecule, electrolyte will not permeate the film unless there is the presence of a solvent for both the permeant and the polymer (*20*). In the case of non-ionic Eudragit NE 30 D, KCl failed to be detected on the receiver side of the film – indicating no porosity of a radius greater than the hydrated radii of K$^+$ (3.31 Å (*30*)) or Cl$^-$ (3.32 Å (*30*)).

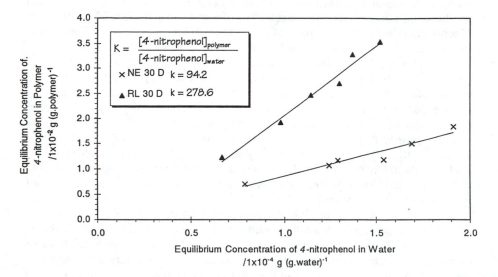

Figure 6 Equilibrium *4*-nitrophenol partition plots for Eudragit NE 30 D and Eudragit RL 30 D films.

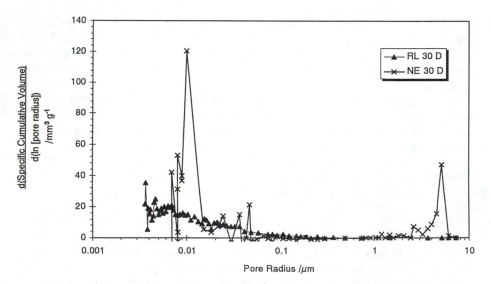

Figure 7 Pore distribution exhibited by Eudragits RL 30 D and NE 30 D films – determined by mercury intrusion porosimetry.

Corresponding to this, the transport of the series of anilines (aniline, methyl and ethyl aniline) through Eudragit NE 30 D showed significant correlation with the uptake of the anilines by the polymer (Table III): the permeability coefficients increasing in the same rank order as their uptakes (Figure 9). This was indicative of the aforementioned three stage process of activated diffusion through an amorphous polymer, in which the permeability coefficient is defined by:

$$P = DS \tag{4}$$

where:
D = diffusion coefficient;
S = solubility coefficient.

The above results are complimentary to those of Göpferich (*1*) who, following the leaching of endogenous surfactant from Eudragit NE 30 D (used in the preparation of a pharmaceutical matrix-type device), found an increase (of up to 50%) in the flux of drug compared to a device in which the surfactant had been removed before preparation of the matrix. Göpferich ascribed this to increased hydration of the polymer due to void space left by the leached surfactant – but not due to the formation of continuous pores.

In contrast to the above properties of Eudragit NE 30 D, Eudragit RL 30 D showed aniline permeability coefficients which tended to a constant value, despite varying permeant uptakes (Figure 9 and Table III). This was indicative of permeant transport through water-filled pores (convective transport) where the diffusion coefficients would be expected to approach those of the anilines in water. Transport in water is generally less impeded than in polymer due to the greater degree of thermal motion of the molecules of a liquid when compared to a (visco-elastic) solid. This convective transport is not, however, necessarily unimpeded, and the porosity may be of such a scale that it is sufficient to allow water into the film to plasticise the polymer – and therefore increase the rate of penetrant transport – without being of sufficient size to allow unimpeded aniline permeation. The molecular volume of each aniline may be calculated using the atomic contributions of the components, as determined by Le Bas (*31*) (data cited by Wilke (*32*)). From this data, and assuming that the molecules are spherical, the anilines' radii may be calculated as 3.52 Å, 3.74 Å and 3.94 Å for aniline, methyl aniline and ethyl aniline, respectively. This implies a pore of diameter of at least 7.88 Å is required for unimpeded convective transport.

Because of the cationic nature of Eudragit RL 30 D and the dissociation of its quaternary ammonium chloride, its permeability to electrolyte could not easily be determined by conductivity measurements: hence, the use of either i, $^{36}Cl^-$ labelled NaCl, or ii, $^{22}Na^+$ labelled NaCl. The $^{36}Cl^-$ labelled electrolyte was found to be gradually lost during the course of the experiment – presumably due to exchange with the chloride ions of the polymer. (Its presence was not detected in the environment of the experiment.) Eudragit RL 30 D was, however, found to be permeable to the $^{22}Na^+$ ions (Figure 10), indicating the presence of continuous pathways through the film. (*I.e.*, whereas in the case of the anilines permeation can occur simultaneously by both activated diffusion and convective diffusion, only convective diffusion will allow the passage of electrolyte in the case of $^{22}Na^+$.) The effective permeability coefficient of this $^{22}Na^+$ transport calculates to be 1.26×10^{-7} m^2 hr^{-1}. Since the flux (J) is effectively given by:

Figure 8 FFTEM of Eudragit NE 30 D film's cross-section showing some evidence of film structure.

Table III Aniline permeability coefficients, and uptakes of Eudragits NE 30 D and RL 30 D films

Aniline	Permeability Coefficient $/1\times10^{-7}$ m^2 hr^{-1}	Aniline Uptake by Film /g (g.film)$^{-1}$
Eudragit NE 30 D		
Aniline	0.91	0.004
Methyl Aniline	1.47	0.0575
Ethyl Aniline	2.84	0.1115
Eudragit RL 30 D (+ 15% triacetin)		
Aniline	3.05	0.0057
Methyl Aniline	3.17	0.0381
Ethyl Aniline	3.16	0.0671

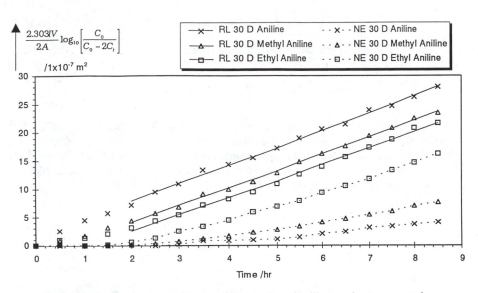

Figure 9 Plot showing the tendency of the series of anilines to be transported at a similar rate through Eudragit RL 30 D films, but at different rates through the Eudragit NE 30 D films.

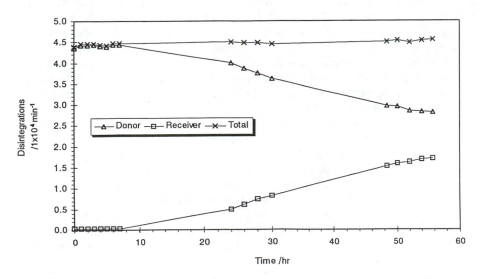

Figure 10 Plot showing the increase in sodium-22 ions on the receiver side of the permeability cell, and decrease of donor ion concentration, as NaCl is transported through continuous porosity existing in Eudragit RL 30 D film leaching triacetin plasticiser. (Reproduced with permission from reference 6. Copyright 1995 John Wiley.)

$$J_{Total} = J_{Activated} + J_{Convective} \qquad (5)$$

and since the flux is proportional to the diffusion coefficient (*i.e.*, Fick's law), then from equation (4), it is seen that:

$$P_{Total} \propto D_{Total} = D_{Activated} + D_{Convectional} \qquad (6)$$

In the case of electrolyte, $D_{Activated} = 0$, and the measured permeability coefficient is effectively the diffusion coefficient. A diffusion coefficient of 1.26×10^{-7} m^2 hr^{-1} through the aqueous channels within the polymer compares to a value of ca 4.6×10^{-6} m^2 hr^{-1} for Na$^+$ ions in free aqueous solution. The diffusional path is therefore found to be highly tortuous, implying that the film has an effective thickness (L) given by:

$$L = \tau \times l \qquad (7)$$

where:

l = measured film thickness;

τ = tortuosity factor.

The transport of water (in the form of THO) across the film was investigated both in the presence of a 4-nitrophenol flux, and without (Figure 11). When measuring 4-nitrophenol permeability coefficients, films were normally oriented with the donor solution against the polymer-substrate side (with respect to casting orientation) of the film. This rule was followed when investigating the flux of THO in the presence of 4-nitrophenol, but with reference to the 4-nitrophenol as opposed to the THO. The orientation of the film with respect to the THO donor therefore changed dependent on whether the THO was travelling with the 4-nitrophenol or against the 4-nitrophenol flux. (I.e., when no flux of 4-nitrophenol was present, the THO 'dissolved' *into* the polymer-substrate side of the film, and this was also true when the THO travelled with the flux of 4-nitrophenol. When travelling against the 4-nitrophenol flux, the THO donor was against the polymer-air side of the film.)

Permeability coefficients of water (calculated assuming a concentration of THO in H$_2$O) showed that transport rates were comparable to those of the 4-nitrophenol. However, the permeability coefficient of THO travelling with the flux of 4-nitrophenol was of greater magnitude than that when travelling against the 4-nitrophenol flux. This was therefore evidence that no osmotic transport mechanism was in operation, and is of pharmaceutical significance for reservoir type devices. Although controlled release devices have been built (*33*) based on the principle of osmotic pressure controlling the release of active agent, in the case of a reservoir device, the build-up of osmotic pressure can lead to rupture of the reservoir skin leading to effective loss of the usability of the device, and potential overdose.

Polymer Latex Films Containing Additions of Water Soluble, Leachable Additives.
The three main additives investigated were generally typical of those that might included in sustained release formulations.

The first additive, sucrose, is of relatively low molecular weight (342 g mol^{-1}), is highly water soluble (i.e., 225 g (100 g.water)$^{-1}$ at 303 K (extrapolated from data in Kaye & Laby (*34*)), and might be expected to be potentially molecularly dispersable.

HPMC (shown below) is described (by Courtaulds Chemicals, Spondon, Derby, UK, personal communication) as having limitless aqueous solubility – but the solubility is effectively limited by the solution viscosity – and it was most easily dissolved in cold water. It is a non-ionic cellulosic ester with a degree of polymerisation of 250 (M.W. 58,5000). HPMC is widely used in the food industry, and also pharmaceutically where it is crosslinked to form highly water-swellable hydrogels.

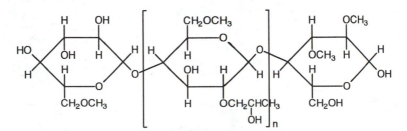

Details of Eudragit L 30 D are given in Table I. It is an anionic copolymer, and dissolves at pH $\rangle$ 5.5 forming polymeric salts. It is therefore used as an enteric pharmaceutical coating. Its use in this study was based on the principle that because of its particulate nature, then it would dissolve and leach to form porosity of greater diameter than the molecular additives.

The effect on the permeability of Eudragit NE 30 D by two other additives – triacetin and sodium chloride – was also investigated.

Triacetin plasticiser was considered as an essential additive for handling hard polymers such as Eudragit RL 30 D, but had been found to increase the permeability coefficient of such a film. Its effect on the *4*-nitrophenol permeability of Eudragit NE 30 D was therefore briefly investigated.

NaCl, like sucrose, could be used as a highly water soluble additive, but its ionic strength has the additional effect of tending to destabilise the polymer latex. It has been reported by Okubo (*12*) *et al.* that, in the case of a surfactant-free latex, a porous, flocculated layer was apparent at the film interfaces. For a latex of only marginal stability, the film-air interface was observed to be porous. Addition of electrolyte to the latex, before casting, further destabilised the latex such that the film-substrate interface was also found to be porous. However, addition of surfactant to increase the latex stability led to a close packed, non-porous structure in agreement with the findings of Isaacs (*35*). This ability to flocculate latices of differing particle-particle separations has also been used by Okubo (*36*) and He to produce films which show asymmetric properties. For the purpose of the permeability experiments performed as part of this study, NaCl was added, at a range of concentrations, to extensively dialysed Eudragit NE 30 D latex.

Film Solute Permeability. The effect of the sucrose, HPMC and Eudragit L 30 D additives on the *4*-nitrophenol permeability coefficient is displayed in Figure 12, in which the additives were leaching from the film during the course of the permeation

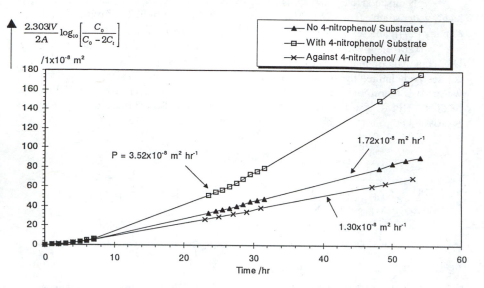

Figure 11 Permeability of THO (i) with, (ii) against, and (iii) without a flux of 4-nitrophenol, through Eudragit NE 30 D film. († Substrate, for example, in the legend indicates which film face is against the donor solution.)

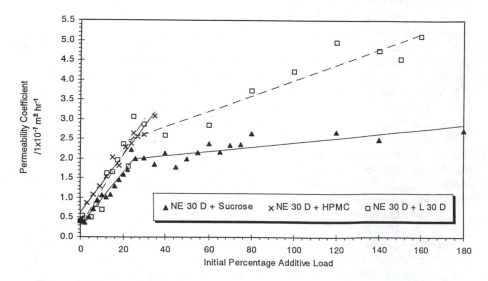

Figure 12 The trends in the 4-nitrophenol permeability coefficients – as a function of initial additive load – for Eudragit NE 30 D films.

experiment. For each additive there was an initial linear increase in permeability which was directly proportional to the load of additive. The maximum load of HPMC was limited by the viscosity of the latex, whilst both sucrose and Eudragit L 30 D were loaded into Eudragit NE 30 D up to the point at which the film became structurally fragile and could not be lifted from the casting substrate.

In understanding these results, it was necessary to investigate the mechanism of transport through the film, and the use of KCl electrolyte permeability measurements give some insight into the problem. In the case of the additive-free Eudragit NE 30 D film, no electrolyte was found to be transported across the film, and the same was true of the sucrose loaded film at low levels of addition. As the sucrose load was increased, and whilst the 4-nitrophenol permeability coefficient was showing the initial linear increase, then still electrolyte failed to permeate the film. There then occurred a point (Figure 13) at which KCl did start to permeate sucrose loaded Eudragit NE 30 D, and this was found to occur at a level of sucrose addition almost corresponding to the point at which the 4-nitrophenol permeability coefficient showed a break in its initial linear increase. Beyond this level of sucrose addition (hereafter termed 'C_s' ($\approx 25\%$)), whereas the 4-nitrophenol permeability coefficient increased at a lesser rate, the KCl permeation rate started to increase with increasing amounts of sucrose. A calculation of the pore radii determined by comparison of the effective K^+ diffusion coefficients through the film and in free aqueous solution showed that the effective pore radius in the film was tending to plateau at $\rangle$ 40% sucrose load (*37*). At a sucrose loading of 40%, the series of three anilines demonstrated permeability coefficients which were greater than those of the sucrose-free Eudragit NE 30 D film, but which were still dependent on the rank order of there solubility in the film.

The addition of sucrose to the already more hydrophilic Eudragit RL 30 D film showed a much poorer degree of 4-nitrophenol permeability enhancement (Figure 14) compared to that of Eudragit NE 30 D + sucrose. However, the overall magnitude of permeability coefficient was greater than that of Eudragit NE 30 D.

The permeability coefficients of the films containing HPMC were generally greater than those for an equivalent load of sucrose. At 25% HPMC addition to Eudragit NE 30 D, the 4-nitrophenol permeability coefficient was 2.65×10^{-7} m^2 hr^{-1}. This compared to a value of 1.99×10^{-7} m^2 hr^{-1} for an equivalent load of sucrose. This therefore implies a far greater ease of transport in the HPMC loaded film when compared to the sucrose loaded film. That the transport mechanism changed from one which was dominated by solution-diffusion in the case of the sucrose loaded film, to one which was dominated by a convective mechanism in the case of the HPMC loaded film was shown by the permeation of the anilines. Whereas the anilines showed diverging fluxes even at 40% sucrose loading, the aniline fluxes all tended to the same rate for 25% HPMC loaded Eudragit NE 30 D films (Figure 15). In the case of Eudragit L 30 D loaded Eudragit NE 30 D films, the aniline fluxes at 25% addition still showed some variation despite a 4-nitrophenol permeability coefficient of similar magnitude to that when convective aniline transport was evident in the case of HPMC addition. However, the fact that the 4-nitrophenol permeability of the film was still increasing showed that the Eudragit L 30 D additive was still having a marked effect on the void-volume of the film. At 100% Eudragit L 30 D addition, the 4-nitrophenol permeability coefficient was

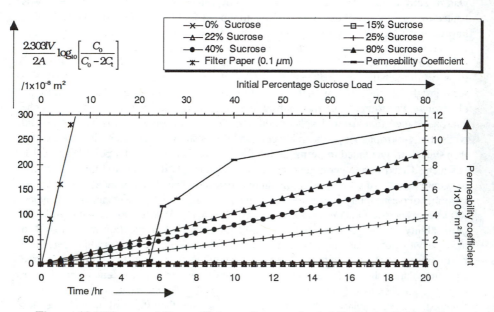

Figure 13 KCl permeability coefficients of sucrose loaded Eudragit NE 30 D film, as a function of initial sucrose load.

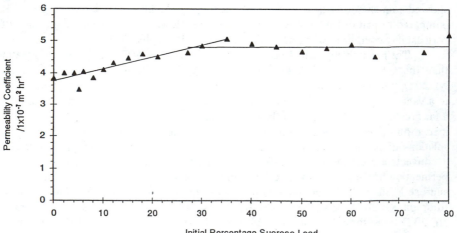

Figure 14 Trend in the *4*-nitrophenol permeability coefficient of Eudragit RL 30 D (+ 15% triacetin) film with increasing sucrose load.

4.23×10^{-7} m^2 hr^{-1} , and the anilines again showed similar permeability coefficients (Figure 15) reflecting convective diffusion through water filled pores.

The relative effectiveness of the sucrose, HPMC, and Eudragit L 30 D water soluble additives on the 4-nitrophenol permeability of Eudragit NE 30 D was shown by the respective permeation activation energies at additive loadings above where the break in the initial linear permeability increase occurred (i.e., on the plateau side of the permeability coefficient versus load curves). Table IV indicates that all three additives greatly reduced the permeation activation energy. Whilst sucrose was the least effective additive, HPMC was by far the most effective – providing an activation energy lower than for Eudragit L 30 D addition when the load of HPMC was far less (and limited by the HPMC solution viscosity).

It has already been stated that PBMA films were very brittle at ambient temperature. Because of these handling difficulties, which were exacerbated by the presence of sucrose, and the necessity of soaking the films in water to remove them from their casting substrate (which would therefore leach some of the additive), very few permeation experiments were performed on such films. However, although the additive-free PBMA films were effectively barriers to 4-nitrophenol over the time scale of a typical permeation experiment, it was shown that the permeability of sucrose loaded PBMA films increased by some three orders of magnitude (i.e., from $\sim 0.001 \times 10^{-7}$ m^2 hr^{-1} for the sucrose-free PBMA, to 0.15×10^{-7} m^2 hr^{-1} for a 5% sucrose loaded film, to 6.68×10^{-7} m^2 hr^{-1} for a 40% loaded film).

The addition of triacetin to Eudragit RL 30 D has already been shown to enhance the film's permeability when compared to the unplasticised film. Table V gives the 4-nitrophenol permeability coefficients of Eudragit NE 30 D loaded with various levels of triacetin up to the point at which the films become too tacky to handle. Rather than increase the permeability coefficient, it was found that low levels of triacetin first reduced it, implying greater film coalescence. The tabulated permeability results show a minimum at 2% addition, after which the permeability starts to increase – presumably as the leaching of the plasticiser (during the course of the permeability experiment) starts to become influential in leading to channels of lowered resistance through which the permeant can travel. The initial decrease in film permeability due to a plasticiser has been observed elsewhere, for example, by Goodhart (*38*) *et al.*

NaCl was added at levels from 0 to 20% per gram of Eudragit NE 30 D solids content, and the resultant 4-nitrophenol permeability coefficients are given in Table VI. The results are not easily interpreted since any increase in film permeability may arise from i, porosity in the particle packing of the destabilised latex film, or ii, the leaching from the film of the NaCl leading to the formation of voids. It was evident, however, that whilst the 4-nitrophenol permeability was greatly increased by the NaCl (when compared to the NaCl-free film) there was no trend of increasing permeability with increasing load of additive. The greater scatter in these results is ascribed to the poorer quality of the films resulting from the destabilised latex.

When examined visually, dialysed Eudragit NE 30 D films containing NaCl showed evidence of the additive on the film surface in the form crystalline 'flowers,' which were never seen on the additive-free film (or films containing any of the other additives which exuded from the film.) These are seen in greater detail in the SEM in Figure 16. SEM's of the fracture cross-section (Figure 17) showed crystals of the additive

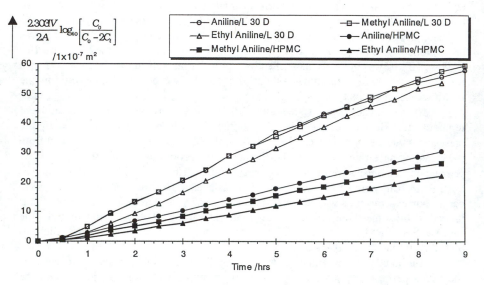

Figure 15 Permeation of the series of anilines through Eudragit NE 30 D film loaded (as indicated in the legend) with either (i) 100% Eudragit L 30 D, or (ii) 25% HPMC.

Table IV Arrhenius activation energies for the permeation of 4-nitrophenol through Eudragit NE 30 D film

Film Additive	Load /% (g.polymer solids)$^{-1}$	Activation Energy /kJ mol^{-1}
None	—	50.5
Sucrose[†]	25%[†]	50.3
Sucrose	40%	34.6
Eudragit L 30 D	100%	18.24
HPMC	25%	10.67

† Film pre-leached of sucrose and dried.

Table V The effect of triacetin plasticiser on the 4-nitrophenol permeability coefficient of Eudragit NE 30 D film

Triacetin Load /% (g.NE 30 D)$^{-1}$	Permeability Coefficient /1×10^{-8} m^2 hr^{-1}
0	4.00
1	3.07
2	2.74
5	3.10
10	3.12

Table VI 4-nitrophenol permeability coefficients of films cast from extensively dialysed Eudragit NE 30 D loaded with sodium chloride

NaCl Load /% (g.NE 30 D)$^{-1}$	Permeability Coefficient /1×10^{-7} m^2 hr^{-1}
0	0.0884
0.25	1.01
0.5	1.22
1	1.32
2	1.27
5	1.10
10	1.13
20	1.17

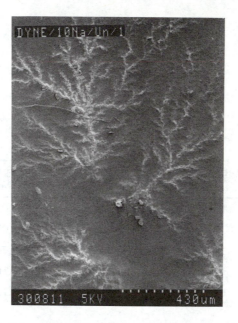

Figure 16 SEM of the polymer-air surface of a dialysed Eudragit NE 30 D film containing NaCl (10%), showing crystal-like growth of the exuding additive.

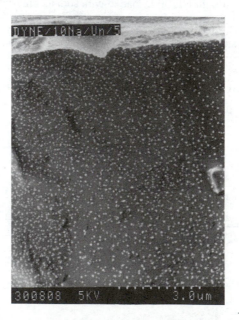

Figure 17 SEM of the fracture cross-section of a dialysed Eudragit NE 30 D film containing NaCl (10%) showing the dispersion of NaCl crystals within the film interior.

implying that the NaCl was not molecularly dispersed (although clustering may result due to incompatibility between the polymer and additive causing it to exude). Because of the soft nature of the Eudragit NE 30 D polymer, which led to a highly coalesced film, the effects of the electrolyte on particle packing were not evident by this means.

In contrast to NaCl, when sucrose was used as the additive in Eudragit NE 30 D, exudations appeared as hemispherical 'blisters' on the film surface. Similar blisters were also observed on sucrose loaded PBMA films (Figure 18). The number of such blisters increased with increasing sucrose load up to the point where they merged to form a complete glossy layer (at ca 150% addition). The formation of such exudations was dependent on the environment of the film. Whilst a film with a low sucrose loading would typically only start to visibly show such blisters after ca 24 hr, a similar film stored in a desiccator would not show any exudation even after two weeks – although exudations would appear after removal from the desiccator. This indicated that humidity enhanced the rate of exudation, and was possibly the cause of it. The distribution of sucrose within the film is indicated in Figures 19 and 20 which show SEM's of the upper and lower surfaces of a PBMA film containing 15% sucrose. The upper surface (Figure 19) shows close packed PBMA particles together with dispersed sucrose exudations. The lower surface (Figure 20), however, whilst showing larger exudations, also shows that the PBMA particles are not in close contact with their nearest neighbours, but are dispersed in sucrose. This is in keeping with observation of latex film drying, which showed that a skin was formed initially, trapping wet latex underneath, and implies that the films dry from the upper open surface downwards. In the wet latex, the sucrose is dissolved in the aqueous phase. As the film dries, it remains preferentially in the aqueous phase until supersaturation leads to its desorption predominantly towards the base of the film. However, wet latex which was trapped beneath the initial skin must obviously lose its water through that skin (which becomes increasingly thick as the film dries). Therefore, the evaporating water provides an upwardly moving flux to carry and ultimately deposit sucrose towards the film's upper surface.

Additive Leaching and Film Porosity. The leaching of sucrose, Eudragit L 30 D, and HPMC from Eudragit NE 30 D was monitored gravimetrically as a function of both leaching time and initial load. Figure 21 shows the results for both sucrose and HPMC leaching, and Figure 22 for Eudragit L 30 D leaching. The rates of leaching are found to be dependent on the additive properties.

The sucrose leaching data show a change dependent on the initial load (e.g., apparent as the increase in amount leached at loads $\rangle 15\%$), and dependent on the time allowed for leaching. The following points (where C_s was identified earlier in relation to the 4-nitrophenol permeability increase) summarise the conclusions based on the sucrose data:

- Below C_s and at short leaching time ($\langle$ 6 hr): sucrose was not easily leached – presumably as a result of the percolating water having insufficient time to penetrate the polymer and reach the dispersed sucrose. This difficulty in leaching showed that sucrose did not simply reside at the film surface.
- Below C_s and at long leaching times ($\rangle$ 6 hr): the water penetrated the film to a greater extent and the sucrose was leached more easily.

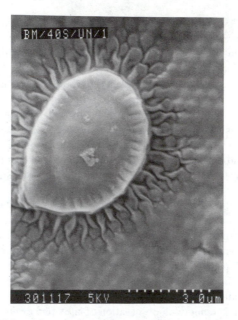

Figure 18 SEM of the polymer-air surface of a surfactant-free PBMA film loaded with 40% sucrose showing an example of a sucrose exudation. (Reproduced with permission from reference 37. Copyright 1995 John Wiley.)

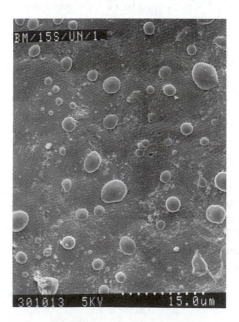

Figure 19 SEM of the polymer-air surface of a PBMA film containing 15% sucrose, and showing close packed PBMA particles and exudations of sucrose.

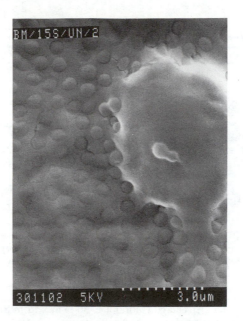

Figure 20 SEM of the polymer-substrate side of a PBMA film containing 15% sucrose showing sucrose exudations, and PBMA particles dispersed in sucrose. (Reproduced with permission from reference 37. Copyright 1995 John Wiley.)

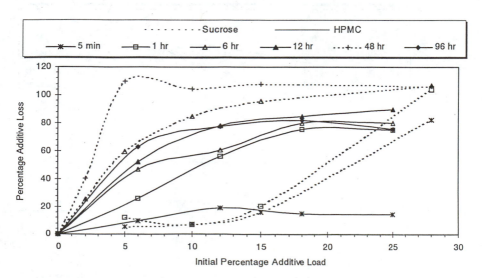

Figure 21 Plot of the percentage sucrose and HPMC loss from Eudragit NE 30 D film as a function of initial load.

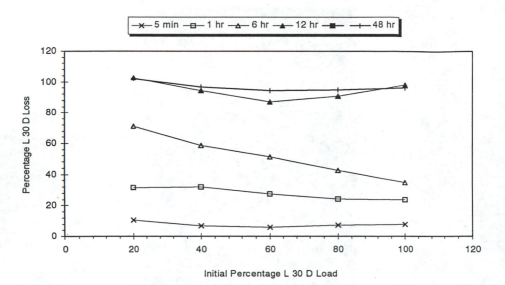

Figure 22 Plot of the percentage Eudragit L 30 D loss from Eudragit NE 30 D film as a function of initial load.

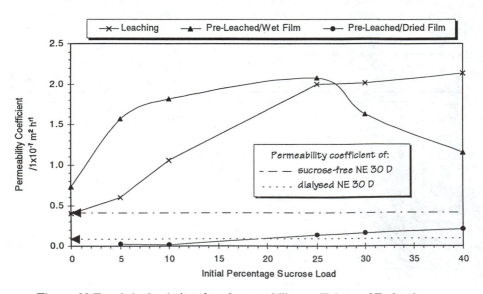

Figure 23 Trends in the 4-nitrophenol permeability coefficients of Eudragit NE 30 D film either (i) leaching sucrose, (ii) having been pre-leached of sucrose and kept wet, or (iii) having been pre-leached of sucrose and dried.

- Below C_s, the dispersed nature of the sucrose yielded no significant porosity linking the two sides of the film. Since carbon-14 labelled sucrose permeation experiments indicated that films were barriers to a sucrose flux, sucrose in the bulk of the film could only leach if an exit channel existed to one or other of the film's surfaces.
- Above C_s, there was a marked increase in the amount of sucrose leached from the film at short leaching times. This result was probably a combination of i, higher levels of sucrose residing on the film's surfaces, from whence it was more easily removed, and ii, the greater porosity resulting from increased amounts of sucrose progressively leached on the front of the percolating water as it moved towards the film's interior. (The fact the amount leached was ⟩ 100% in some instances may be ascribed to i, the leaching of endogenous surfactant, combined with ii, the fact that in some instances, polymer particles had been seen by SEM to be dispersed in additive therefore allowing polymer to be lost from the film.)

The HPMC leaching data was somewhat different. Never was the full load of HPMC leached from the film. However, like sucrose and presumably for similar reasons (i.e., the creation of increasing numbers of hydrophilic pathways on leaching), higher levels of HPMC addition were more easily leached than lower levels. The inability to leach the full load HPMC on the timescale of the experiment was attributed to film swelling and to the very viscous nature of the additive (having a tendency to form a hydrogel).

The leaching of Eudragit L 30 D from Eudragit NE 30 D (Figure 22) suffered from different problems to the other two additives due to it being present in particulate form as opposed to molecular form. Whilst this may have been advantageous in that it may lead to the formation of particulate sized holes (as opposed to molecular sized holes), leaching times were prolonged due to the necessity of having to initially dissolve the particles.

Whilst additive leaching is of obvious importance in creating the porosity necessary to enhance film permeability, the physical properties of the polymer are important for the films ability to sustain that porosity. Despite the enhanced solute permeability of the sucrose loaded Eudragit NE 30 D films, mercury intrusion porosimetry did not show the presence of a pore distribution in a pre-leached (40% sucrose) and dried film, and the same was true of Eudragit NE 30 D films leached of either HPMC or Eudragit L 30 D. The reason for this was again indicated by the use of 4-nitrophenol as a solute permeant. Films were pre-leached of sucrose and then either dried or kept wet before the start of a 4-nitrophenol permeation experiment. Both sets of films were leached (96 hr) *in situ* in the permeation cell, but one set was then oven dried and the permeation experiment performed by the normal method, whilst to the other set of films, 4-nitrophenol was added to the leachate in the donor permeability cell in the form of crystals. This latter film was therefore never out of the water. Resultant permeability coefficients are given in Figure 23. The graph shows that the sub-ambient T_g Eudragit NE 30 D polymer could not retain porosity if the film was allowed to dry, whereas keeping the film wet did allow the film to retain an enhanced level of permeability. A number of other conclusions can also be drawn from the graph:

- At low initial levels of sucrose addition (⟨ C_s), the permeability coefficients of the pre-leached/wet film were greater than those of the film leaching sucrose during the course of the permeation experiment. This presumably results from the extended leaching times of the pre-leached film (96 hr) compared to the film leaching whilst permeating. Low levels of sucrose have been shown to be difficult to leach, and the extended soak time would further advance this leaching. Films leaching at the same time as undergoing permeation would not have completely leached their sucrose before the end of the permeation experiment.

- At low initial levels of sucrose addition ($\langle$ C_s), the pre-leached/dried film yielded a lower permeability coefficient than a dialysed Eudragit NE 30 D film. This can be attribute to the extended annealing time given to the film on drying after leaching.

- At high initial levels of sucrose addition ($\rangle$ C_s), the pre-leached/wet film yielded permeability coefficients which were less than those of the film leaching sucrose during the permeation experiment. This is again ascribed to the length of time that the film was under water. Higher levels of sucrose have been shown to be readily leached quite quickly; the reduced permeability coefficient therefore implies that some of the resultant porosity/void space yielded by the sucrose is lost – presumably by a process of wet sintering (i.e., healing due to the polymer-water interfacial tension).

- At high initial levels of sucrose addition ($\rangle$ C_s), the pre-leached/dried film yielded permeability coefficients which were greater than that of the dialysed Eudragit NE 30 D film. This implied that for a film which once demonstrated porosity, then the drying/annealing process did not 'heal' the film to the extent that was shown by the coalescence of casting initially. I.e., regions of reduced film densification existed (which reduced resistance to permeability) in this pre-leached/dried film compared to a dialysed Eudragit NE 30 D film as cast from the latex.

The ability of a hard polymer (i.e., one whose T_g is above ambient temperature) to retain porosity after leaching was shown by both mercury porosimetry and SEM evidence of sucrose-leached PBMA. Figure 24 shows the pore distribution of a PBMA film leached of 40% sucrose, whilst Figures 25 and 26 show the porosity in the upper and lower faces. Figure 27 (the fracture cross-section) shows that the porosity was present through the full thickness of the film. This evidence correlates with the aforementioned increased 4-nitrophenol permeability coefficient.

Eudragit NE 30 D films loaded with Eudragit L 30 D were unique with respect to this work in that they consisted of both a soft polymer and a hard polymer – either of which could be separately leached. Eudragit L 30 D is soluble at pH $\rangle$ 5.5, whereas Eudragit NE 30 D is not, whilst Eudragit NE 30 D is soluble in toluene when Eudragit L 30 D is not. Figure 28 shows the polymer-air interface of a 1 : 1 ratio Eudragit NE 30 D : Eudragit L 30 D film which has been leached of the hard Eudragit L 30 D fraction. This compares to Figure 29 which shows a film cast from the same polymer ratio, but leached of the soft Eudragit NE 30 D fraction. Both polymer types are of similar latex particle size (Table 1), and so the effects of film healing are immediately apparent since the film which was leached to leave just the softer polymer (i.e., Eudragit NE 30 D) does not display porosity. Interestingly, if the load of Eudragit L 30 D was increased to 150% (i.e., 2 : 3 ratio of Eudragit NE 30 D : Eudragit L 30 D) then the hardness of the Eudragit L 30 D, together with its high level of dispersion prevented complete deformation of the Eudragit NE 30 D particles during casting. Thus, leaching the Eudragit L 30 D fraction of the film resulted in some porosity and a film in which the Eudragit NE 30 D particles could still be identified (Figure 30) – a feature not observed in any of the other Eudragit NE 30 D/additive combination films investigated.

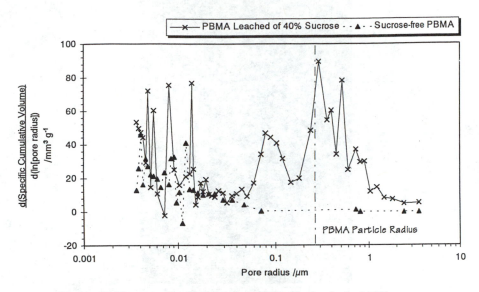

Figure 24 The pore distribution of a PBMA film leached of 40% sucrose, compared with the sucrose-free film.

Figure 25 SEM of the polymer-air interface of a PBMA film leached of 40% sucrose.

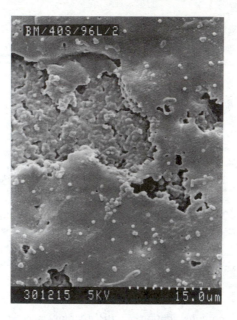

Figure 26 SEM of the polymer-substrate interface of a PBMA film leached of 40% sucrose.

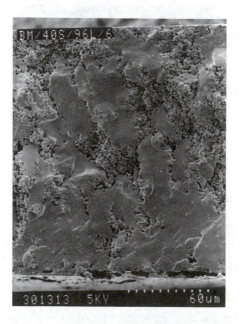

Figure 27 SEM of the fracture cross-section of a PBMA film leached of 40% sucrose.

Figure 28 SEM of the polymer-air interface of 1 : 1 ratio film of Eudragit NE 30 D and Eudragit L 30 D leached of Eudragit L 30 D.

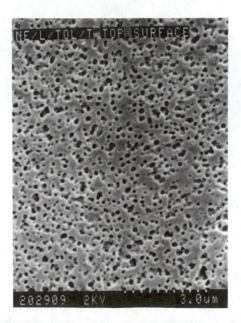

Figure 29 SEM of the polymer-air interface of 1 : 1 ratio film of Eudragit NE 30 D and Eudragit L 30 D leached of Eudragit NE 30 D.

Figure 30 SEM of the fracture cross-section of a 2 : 3 ratio film of Eudragit NE 30 D and Eudragit L 30 D leached of Eudragit L 30 D.

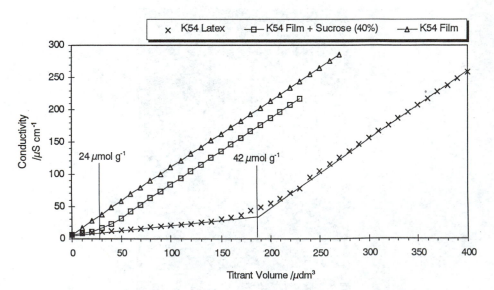

Figure 31 End-group analysis of (i) carboxylated AA-PBMA core-shell latex (coded K54), (ii) K54 latex film, and (iii) K54 latex film leaching an initial load of sucrose (40%).

In the case of mixed Eudragit NE 30 D and Eudragit L 30 D films, differences in thickness were also apparent in the leached films. Whilst films leached of Eudragit NE 30 D contracted to fill the void volume, films leached of Eudragit L 30 D remained close to their original thickness as measured after casting. This latter point also indicates that the films were not stratified: if the two latex types had separated into layers, then the leaching of either would have left a film whose thickness correlated with that of the unleached fraction.

Availability of Sites of Hydrophilicity. The permeability enhancement of Eudragit NE 30 D by the presence of HPMC resulted from the greatly increased hydrophilicity of the film. This hydrophilicity allowed water to penetrate the film such that it was visibly swollen (therefore allowing a penetrant to move relatively unhindered through an aqueous environment, as opposed to the resistive environment of the polymer). Eudragits RL 30 D and RS 30 D are similarly sold, as coatings, on the basis of their relative degrees of hydrophilicity, and this is reflected in the Röhm Pharma coding policy (Table 1).

Evidence from previous work on the morphology of PBMA polymer latex films cast for 3 hr, compared to PBMA films used in this study cast for 72 hr (both at 353 K) implied that the permeability of such films depended on the availability of particle boundaries. I.e., those films cast as part of this study were effectively barriers over the course of the permeation experiment, and showed less evidence of particulate boundaries than those film cast for 3 hr – which also showed higher 4-nitrophenol permeability coefficients (*13*).

Further evidence that particle boundaries were important in film transport arose from the fact that sucrose (when used as a permeant) was unable to diffuse through, for example, a Eudragit NE 30 D film – and therefore was presumably unable to penetrate an individual latex particle – but did enhance the permeability of a film when used as a film additive. This implied that in order that the sucrose existed in the film (to create porosity upon its leaching), then it must reside at the particle boundaries.

It was therefore thought that for the enhancement of permeability by an increase in hydrophilicity of the latex, then that hydrophilicity would be maximised if it existed at the particle interfaces: i.e., attached to the latex particle surfaces in the form of a shell. Hence, the use of PBMA-AA core-shell latex.

The availability of the carboxyl groups of the AA shell was therefore determined by conductometric titration (Figure 31). In the case of the AA-PBMA latex (coded K54), an end point was found at 186 μdm^3 (for 20 ml of latex), equivalent to 42 μmol g^{-1}. For a film cast for 72 hr at 353 K, from 5 ml of the latex, no accessible carboxyl groups were found within the film or on its surface. However, for a film containing 40% sucrose, cast as before and then leached for 96 hr in water, a carboxyl end point was found at 27 μdm^3 of titrant. This corresponded to 58% of the total end point to be expected for 5 ml of latex – a result compatible with the permeation studies in which the sucrose film additive opened hydrophilic pathways for ions to enter the film.

When latex films are fractured, they tend to do so across particle diameters rather than between interparticle boundaries – an observation which has been interpreted (*39*) in terms of functional group hydrogen bonding at the vestigial latex particle boundary regions. Thus at the level of functionality used in this study, and with it being confined to the interparticle regions, then the lack of access to those groups by titrations was not unexpected. In contrast, where functionality was at a higher level and dispersed throughout the polymer phase, extensive hydration of the film results as, for example, has been shown to happen for Eudragit RL 30 D polymer.

Gas and Water Vapour Permeability. Gas permeability measurements proved to be a more sensitive probe of film permeability than solute permeability measurements. For example, gas permeability measurements were able to detect film aging (5) (due to, for example, the effects of further gradual coalescence (24, 25, 26) or autohesion (27, 28, 29)) which could not be detected by solute permeability measurements (due to the effects being hidden by the experimental uncertainty).

Sucrose, HPMC and Eudragit L 30 D all reduced the CO_2 permeability of Eudragit NE 30 D films (Table VII) – a feature that is advantageous for many commercial uses where such films are used as encapsulants and are required to act as barriers to preserve the film contents. Surprisingly, considering the evidence of the SEM's (Figures 28 to 30), mixed films of Eudragits NE 30 D and L 30 D (1 : 1) leached of one or other component, did not exhibit high CO_2 permeability coefficients (Table VII). (The Daventest apparatus used to measure gas permeabilities would normally be unusable to measure the gas permeability of a porous film since it would be unable to retain a vacuum for sufficient time for a measurement to be made.) Whilst leaching the hard Eudragit L 30 D polymer may allow the film to heal, leaching the soft Eudragit NE 30 D polymer left a highly porous film. The evidence of the gas permeability results implies, however, that the porosity evident in the SEM's was not contiguous through the full thickness of the film. I.e., sufficient film remained to act as a partial barrier to the CO_2, and to retain the vacuum in the Daventest apparatus.

The CO_2 permeability coefficients result from the additives having lower permeabilities than the Eudragit NE 30 D, due to their either being crystalline in nature (e.g., sucrose), or having a higher T_g than Eudragit NE 30 D (e.g., HPMC and Eudragit L 30 D).

The permeability of the films to water vapour was affected by the hydrophilicity of the film. This was shown by comparing the water vapour permeability coefficients of Eudragits RL 30 D and RS 30 D (Table VIII). Despite the functionality of Eudragit L 30 D (see Table I), it yielded the lowest water vapour permeability (Table VIII) of any of the four Eudragits investigated. This is tentatively ascribed to polar repulsion from the functionality limiting the polymer chain motion such that the polymer's diffusivity towards permeants is also reduced. The water vapour permeability coefficients of the Eudragit L 30 D-containing Eudragit NE 30 D films were reduced compared to the additive-free Eudragit NE 30 D film (Figure 32) – the water vapour permeability coefficient showing a decreasing trend with an increasing fraction of Eudragit L 30 D.

The water vapour permeability coefficients of sucrose loaded Eudragit NE 30 D films were, in general, higher than the sucrose-free film (Figure 32 and Table VIII). This is possibly due to the hydrophilicity of the sucrose providing a route for the water vapour, together with the inability of the Eudragit NE 30 D polymer chains to interdiffuse into the sucrose such that voids or regions of low chain density existed at the sucrose-polymer interfaces. However, increasing loads of sucrose tended to reduce the water vapour permeability of the sucrose-loaded Eudragit NE 30 D film (Figure 32), indicating that

Table VII CO_2 permeability coefficients of Eudragit NE 30 D films

Additive	Load /% (g.NE 30 D)$^{-1}$	CO_2 Permeability Coefficient /1×10^{-9} cm^3 cm s^{-1} cm^{-2} cmHg^{-1}
None	—	1.99
Sucrose	25%	1.59
HPMC	25%	1.37
Eudragit L 30 D	100%	0.37
Pre Leached Films		
Additive	**Initial Load** /% (g.NE 30 D)$^{-1}$	CO_2 **Permeability Coefficient** /1×10^{-9} cm^3 cm s^{-1} cm^{-2} cmHg^{-1}
Eudragit L 30 D	100%; then leached of L 30 D fraction	3.36
Eudragit L 30 D	100%; then leached of NE 30 D fraction	5.47

Table VIII Eudragit water vapour permeability coefficients

Eudragit Film Type	Permeability Coefficient /1×10^{-6} g hr^{-1} cm^{-1} cmHg^{-1}
RL 30 D	2.65
RS 30 D	2.14
NE 30 D	1.65
L 30 D	1.39

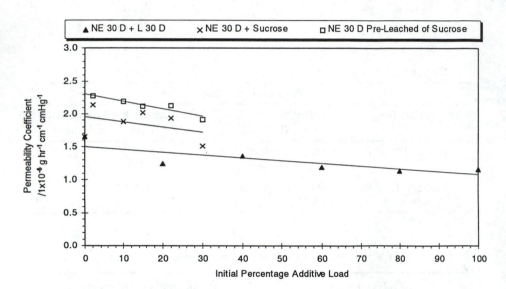

Figure 32 Trends in the water vapour permeability coefficient of Eudrgait
NE 30 D film loaded with either (i) sucrose or (ii) Eudragit L 30 D, and Eudragit
NE 30 D film pre-leached of sucrose.

whilst the increased hydrophilicity aided permeability, excess dry sucrose itself hindered permeant transport.

The water vapour permeability of the HPMC loaded Eudragit NE 30 D film was of similar magnitude to that of the sucrose loaded films – again presumably related to the highly polar nature of the HPMC.

Conclusions.

The solute transport properties of a polymer latex film have been found to be enhanced, in a systematic manner in permeation experiments by the addition of water soluble hydrophilic additives.

The influence of the physical properties of the polymer, such as T_g, and additive-polymer compatibility are also of great importance, and the properties of a polymer film depend on the storage temperature and/or the environment in which the film is utilised. For example, soft polymers coalesced to a degree whereby they appeared featureless on a particulate scale (by both SEM and FFTEM) and were unable to sustain porosity following the leaching of an additive. Hard polymers, however, formed films which showed (by SEM and FFTEM) vestigial latex particles, and which were able to sustain porosity following additive leaching.

The ability to prepare a porous film (especially for use as a coating) has contradictory requirements. The retention of porosity requires a hard polymer, but the use of such a polymer is least suitable for film formation, at worst simply forming a friable powder. If the polymer does coalesce into a film, then the film is likely to show extended aging properties (*13*) such as increased densification of particle boundaries and further gradual coalescence. These effects are undesirable and lead to reduced predictability of film properties, but have, however, been shown to be minimised by the thermal annealing of a film (e.g., PBMA films cast for 72 hr as opposed to 3 hr, at 353 K). The use of a softer polymer can minimise these effects (due to polymer chain motion producing sufficient interdiffusion during film casting to the extent that particle boundaries no longer exist and any further chain motion is no different from that of the bulk polymer) to the point where they are negligible for larger (e.g., solute) permeants, but they may still be apparent for smaller (e.g., gaseous) permeants.

For solute permeation, the influence of water on the film is of utmost importance. A hydrophobic permeant which is soluble in the polymer will diffuse through it due to the random cooperative molecular motions of the polymer chains (independent of the presence of water). However, diffusants are typically required to be hydrophilic to some extent, and it is their ability to diffuse into and out of the polymer, typically into an aqueous environment that is important.

The presence of water in the film can have a number of effects on its solute transport properties. Not only will it allow the permeation of penetrants which are insoluble in the polymer alone, it may also act to plasticise the polymer, albeit localised in the hydrophilic regions of the film (possibly at the particle boundaries). Such plasticisation would presumably increase the rate of transport of both hydrophilic and hydrophobic permeants. At the lowest levels of addition of molecular additives, such as sucrose or HPMC, leaching does not result in porosity of a size suitable to allow unimpeded permeant transport. The enhancement of permeability may therefore be considered to result from the greater void space within the films yielding a greater degree of polymer chain motion due to a loosening of the film structure.

The mode of permeation through the films used in these studies may be interpreted as i, pure solution-diffusion-type transport, ii, permeant transport through hydrophilic and/or plasticised pathways, and iii, (unimpeded) transport through pores in which the pore diameter is greater than the permeant size, and which pass from side-to-side of the film. Each of these may occur singularly or in combination. Modes 'ii' and 'iii' differ in that in the latter case the pores might be considered as permanent and of fixed position, whereas in the former case the plasticised channels may be transient and dependent on random chain motion. The change from plasticisation to porosity is very gradual, and this is seen as the initial linear increase in permeability as the load of additive increases.

4-nitrophenol does not readily penetrate surfactant-free PBMA, but does permeate when the film contained, for example, sucrose. Similarly, the increased 4-nitrophenol permeability of surfactant-present Eudragit NE 30 D compared to dialysed Eudragit NE 30 D is purely an enhancement due to the presence of surfactant. If, however, the additive fails to increase the gas permeability of a film, it must be assumed that the additive itself does not plasticise the polymer. An increase in the permeability of an aqueous solute is therefore, predominantly dependent on the additive allowing a greater amount of water into the film.

Following the initial linear increase in solute permeability with additive load, the permeability coefficients tended to plateau as the film transport mechanism changed from being predominantly by solution-diffusion to one where convective transport occurred (as evinced by electrolyte transport through a film which was previously a barrier). The measured permeability coefficients would be expected to be a function of all modes of film transport since their respective fluxes are additive (14). It might therefore be expected that the measured permeability coefficient would increase, up to the point where a convective flux occurs, after which the permeability would increase to an even greater extent due to the higher diffusivities found in water as opposed to polymer. However, there was no sudden increase in flux since convective transport was not a sudden occurrence in the film. In the immediate region of a pore, whose length can be increased by increased additive load, the decrease in film thickness is gradual up to the point when an infinitely thin film exists. Further addition of additive then simply leads to only a small increase in permeability.

If the additive is clustered within the film, then non-continuous pores will form, although the ability to fully leach the sucrose additive (albeit slowly at low levels of addition) implies that the additive does have access to one or other film surface. If the additive is well dispersed, then the access of water will be through only narrow hydrophilic pathways whose length increases as water (and therefore leaching) progresses into the film – which explains the slow kinetics of leaching. It is the increase in the number and size of these pathways, from interparticle boundary plasticisation up to true porosity, which leads to the systematic increase in permeability with increasing additive load. However, the restricted transport of electrolyte showed that true unimpeded porosity was never achieved, by the addition of an additive, before the film became structurally fragile.

The three main additives investigated each had their own advantages with respect to their ability to enhance film permeability. Sucrose is a relatively small molecule which dissolves quickly, allowing it to be well dispersed and easily leached once water has access to it. HPMC does not dissolve so readily and forms a viscous solution – both of which hinder leaching. The viscosity of trapped HPMC would also hinder the transport of a penetrant, but in mitigation the trapped HPMC also prevents the film from healing, and imbibes water through which the permeant can travel. The trapped HPMC also allows the film to be leached and dried and then re-used where it still shows enhanced permeation rates compared to the HPMC-free film. Eudragit L 30 D is possibly the least effective additive. The gain which may result from its large size (and, hence, large pore size) is in reality lost due to the time it takes for such a large bodies to dissolve, and due to the decreased ability to disperse a given weight of the additive when compared to a molecular additive.

Each of the additives investigated can effectively be used to increase film solute permeability. Low levels of addition tend to allow control of permeation via a solution-diffusion mechanism of transport due initially to increased access of water into the film. Higher levels of additive lead to convective film transport, allowing the permeation of electrolyte and tending toward transport which is independent of solute solubility in the polymer, but which shows little increase in actual transport rate presumably due to the fact that there is a high degree of tortuosity in the porous network, and the narrowness of the pores. Increasing the additive load still further leads to an increase in electrolyte transport through the film implying an increase in the pore number density, or a widening of the existing pore diameter. Whilst such an increase allowed greater electrolyte transport, the effect on the other permeants was minimal compared to the initial permeability increase. This was presumably a result of the limiting amount of additive which could be contained by the film. Hence, despite the increasing porous network, the tortuosity of that network becomes a limiting factor, and no further significant increase in permeant transport is seen before the additive loading is such that the film fails structurally.

Literature Cited.

1. Göpferich, A.; Lee, G. *J. Controlled Release.* **1992**, *18* 133-144.
2. Lehmann, K.O.R. *Chemistry and Application Properties of Polymethacrylate Coating Systems.* Ch. 4. In *Drugs and Pharmaceutical Sciences, 36, Aqueous Polymeric Coatings for Pharmaceutical Dosage Forms*; Editor J.W. McGinity; Marcell Dekker: NY, 1989, 153.
3. *Eudragit Product Brochure*, Röhm Pharma GMBH, Weiterstadt. A folder of product data sheets supplied by Dumas UK Ltd., Tunbridge Wells, Kent TN2 5TT, UK.
4. *C.R.C. Handbook of Chemistry and Physics;* R.C. West ed.; 62nd Ed., C.R.C. Press Ltd., 1981-1982.
5. Steward, P.A. *Modification of the Permeability of Polymer Latex Films.*, Nottingham Trent University PhD Thesis, 1995.
6. Steward, P.A.; Hearn, J.; Wilkinson, M.C. *Polym. Int.* **1995**, *38* (1), 1-12.

7.   Chainey, M.; Wilkinson, M.C.; Hearn, J. *Ind. Eng. Prod. Res. Dev., N° 2, Am. Chem. Soc.* **1982**, *21* (*2*), 171-176.
8.   Sakota, K.; Okaya, T. *J. Appl. Polym. Sci.* **1977**, *21*, 1035-1043.
9.   Wilkinson, M.C.; Hearn, J.; Cope, P. *Br. Polym. J.* **1981**, *13*, 82-89.
10.   *Polymer Handbook.*, Brandrup, J.; Immergut, E.H., 2nd Ed., Ch. IV, 1975, 157.
11.   Abdel-Aziz S.A.M.; Anderson W.; Armstrong P.A.M. *J. Appl. Polym. Sci.* **1975**, *19*, 1181-1192.
12.   Okubo M.; Takeya T.; Tsutsumi Y.; Kadooka T.; Matsumoto T. *J. Polym. Sci., Polym. Chem. Ed.* **1981**, *19*, 1-8.
13.   Roulstone, B.J. *Permeation Through Polymer Latex Films.*, Trent Polytechnic Ph.D. Thesis 1988.
14.   Flynn, G.L.;Yalkowsky, S.H.; Roseman, T.J. *J. Pharm. Sci.* **1974**, *63* (*4*), 479-510.
15.   British Standard BS 2782 (ISO 2556-1974), part 8, method 821A, British Standards Institute, 1979.
16.   *Guide for Users of Labelled Compounds*, Amersham International PLC.
17.   Wicks, S.R. *Competitive Sorption Studies with Nylon 6 and Active Carbon.*, University of Bath PhD Thesis, 1982.
18.   Fujita, H. *Adv. Polym. Sci.* **1961**, *3*, 1-47.
19.   Barrer, R.M. *Diffusion In and Through Solids.*, Pub. Cambridge University Press, 1941.
20.   Yasuda, H.; Lamaze, C.E.; Ikenberry, L.D. *Makromol. Chem.* **1968**, *118* (*2858*), 19-35.
21.   Yasuda, H.; Lamaze, C.E.; Peterlin, A. *J. Polym. Sci.* **1971**, *Part A-1, 9* 1117-1131.
22.   Brown, G.L. *J. Polym. Sci.* **1956**, *22*, 423-434.
23.   Roulstone, B.J.; Wilkinson, M.C.; Hearn, *J. Polym. Int.* **1992**, *27*, 51-55.
24.   Bradford, E.B.; Vanderhoff, J.W. *J. Macromol. Chem.* **1966**, *1*, 335.
25.   Bradford, E.B.; Vanderhoff, J.W. *J. Macromol. Sci. Phys.* **1972**, *B6*, 671-694.
26.   Vanderhoff, J.W. *Br. Polym. J.* **1970**, *2*, 161-172.
27.   Voyutskii, S.S. *J. Polym. Sci.* **1958**, *32* (*125*), 528-530.
28.   Voyutskii, S.S. *Autohesion and Adhesion of High Polymers.*, Polymer Reviews, 4, Pub. Wiley Intersci. NY, 1963.
29.   Voyutskii, S.S.; Vakula, V.L. *Rubber Chem. Technol.* **1964**, *37* 1153-1177.
30.   Nightingale jr., E.R. *J. Phys. Chem.* **1959**, *63*, 1381-1387.
31.   Le Bas, G. In *The Molecular Volumes of Liquid Chemical Compounds.*, Pub. Longmans, London & NY, 1915.
32.   Wilke, C.R.; Chang, P. *A. I. Ch. E. J.* **1955**, *1*, 264-270.
33.   Theeuwes, F. *J. Pharm. Sci.* **1975**, *64* (*12*), 1987-1991.
34.   Kaye, G.W.C.; Laby, T.H. *Tables of Physical and Chemical Constants.*, 13th Ed., Pub. Longman, 1971.
35.   Isaacs P.K., *J. Macromol. Chem.* **1966**, *1* (*1*), 163-185.
36.   Okubo M.; He Y., *J. Appl. Polym. Sci.* **1991**, *42* (*8*), 2205-2208.
37.   Steward, P.A.; Hearn, J.; Wilkinson, M.C. *Polym. Int.* **1995**, *38* (*1*), 13-22.
38.   Goodhart F.; Harris M.R.; Murphey K.S.; Nesbitt R.U. *Pharm. Technol.* **1984**, *8* (*4*), 64-71.
39.   List P.H.; Kassis G. *Acta Pharmaceutica Technologica.* **1982**, 28 (1), 21.

Chapter 24

Film Formation and Morphology in Two-Component, Ambient-Cured, Waterborne Epoxy Coatings

Frederick H. Walker[1] and Olga Shaffer[2]

[1]Air Products and Chemicals, Inc., 7201 Hamilton Boulevard, Allentown, PA 18195–1501
[2]Emulsion Polymers Institute and Polymer Interfaces Center, Lehigh University, Bethlehem, PA 18015–4732

Film formation from two component, waterborne, ambient cured coatings is a complex process involving a number of steps that must occur in the correct order. Iodocarboxylic acids such as 3-iodopropionic acid were used to stain amine containing domains in polymeric films based on amine cured water-borne epoxy resins. A film cast 0.5 hr. after mixing a solid epoxy dispersion and a polyethyleneamine type hardener shows phase-separated amine containing domains in an epoxy continuous phase. Two hours after mixing the morphology changes to show individual particles with amine-rich areas at the interparticle boundaries. In contrast, films cast from a water-borne combination of liquid epoxy resin and a highly epoxy-compatible, self-emulsifying amine hardener are shown to be of a much more uniform nature with this technique.

The development of two component, waterborne epoxy coatings is driven by environmental and worker safety regulations requiring ever lower usage of solvent, and consumer preference for coatings with low odor and water cleanup. Generally, these products are employed in markets that require ambient cure and certain advantages, such as enhanced chemical resistance, of a crosslinked polymer over a thermoplastic polymer. When considering the film formation process in these systems, it is convenient to divide the technology into two categories, depending on the molecular weight and physical nature of the epoxy resin employed.

Type I, which has been in use for over 30 years *(1)*, utilizes liquid epoxy resin (epoxy equivalent weight, EEW ca. 190) and a modified poly(ethylene amine) hardener. Usually, the hardener serves as the emulsifier for the epoxy resin, although sometimes the epoxy resin is pre-emulsified in water with surfactants, primarily to adjust package ratios. Typical hardeners are made by reacting a poly(ethylene amine) such as triethylene tetramine or tetraethylene pentamine with: i) fatty acids or dimer acids; and/or ii) epoxy resin, and then modifying that product in some way to reduce the 1° amine content. The strong catalytic effect of water on the amine epoxy reaction *(2)* results in too short a time period over which the mixture can be successfully applied to a substrate (i.e. pot life) without such modification. The amines are partially neutralized with a volatile carboxylic acid such as acetic acid, which increases the water solubility of the hardener and the colloidal stability of the combined, emulsified system. Nevertheless, pot lives are (with some exceptions) usually less than three hours. The humidity and corrosion resistance are generally inadequate for application directly to steel substrates with these older products, though some newer refinements of this technology show significant improvement in that regard. Chemical resistance is acceptable for many uses. The volatile organic content (VOC) achievable with this approach is very low or even zero, and the major market is for the protection of concrete substrates.

The advent of solid epoxy resin dispersion technology (Type II) gave much improved humidity and corrosion resistance, and longer pot life. These products are based on much higher molecular weight resin (EEW about 500 to 650). Because the process for preparation of dispersions of these resins is complicated, they are always supplied in a pre-dispersed form in water. The surfactant technology used to create stable dispersions can be sophisticated, and frequently involves modification of the surfactant for chemical incorporation into the resin. *(3)* The principal hardeners used in this case are the partially neutralized epoxy/poly(ethylene amine) adduct variety described above, modified to optimize performance with the epoxy dispersion. Pot lives of about five to eight hours can be achieved, and with skilled formulation corrosion and humidity resistance is sufficient for light duty maintenance of steel substrates. The VOC of a clearcoat is typically in the 240 - 360 g/L (2 - 3 lb/gal) range. The major market is for use over masonry and related substrates, where the very fast tack-free times (<1 hr.) are particularly attractive for coatings in schools, hospitals, and the like.

Film Formation. Film formation is the process whereby a liquid substance is applied to a substrate and transformed into a thin, coherent, solid coating possessing the requisite barrier and mechanical properties for its intended use. In ambient cured coatings, the process depends on the type of binder system (see Table I). For solvent-borne thermoplastics, solvent evaporation is the only process that need occur. *(4)* Difficulties are encountered in solvent-borne thermosets when the crosslinking process raises the T_g of the film above room temperature. At this point the very slow diffusion of growing oligomers and polymers becomes the rate limiting factor for the curing reactions. *(5)*

In water-borne systems, film formation is generally complicated by the need to transform the binder from a two-phase colloidal dispersion into a single polymeric phase. With thermoplastic latexes, the problem involves achieving sufficient interparticle diffusion and chain entanglement of the high molecular weight emulsion polymers to occur without reducing the T_g of the copolymer to the point where the necessary degree of hardness cannot be obtained. *(6)* In practice, this is accomplished with slowly evaporating coalescing solvents. Sometimes, the latexes particles also contain core-shell and other complex morphologies which lower the minimum film forming temperature (MFT) while maintaining adequate hardness. *(7)*

Film formation from water-borne thermosets is arguably the most complicated of these processes. For optimum properties, all of the steps shown in Table I must occur in the correct order: film application, evaporation of water, particle coalescence, solvent evaporation, and crosslinking. If, for example, the film is applied under more humid conditions than the system was designed to accommodate, the solvent evaporates before the water. The requisite solvency that allows the constituents to diffuse into a uniform mixture is never achieved, and poor barrier and mechanical properties result.

Table I. Film Formation In Ambient Cured Coatings

Solvent-borne Thermoplastic	Solvent-borne Thermoset	Water-borne Thermoplastic	Water-borne Thermoset
(i) Apply Solution	(i) Apply Solution	(i) Apply Dispersion	(i) Apply Dispersion
(ii) Evaporate Solvent	(ii) Evaporate Solvent	(ii) Evaporate Water	(ii) Evaporate Water
	(iii) Crosslink	(iii) Coalesce	(iii) Coalesce
		(iv) Evaporate Solvent	(iv) Evaporate Solvent
			(v) Crosslink

A further complication of these ambient cured systems is the ongoing polymerization reactions, which begin to increase the molecular weight of the materials as soon as they are mixed. This decreases solubility in the coalescing solvent, raises the MFT, and increases the thermodynamic and kinetic barriers to diffusion of the ingredients into a uniform film. Thus, barrier and mechanical properties can also change as the pot life progresses. *(8)*

In this work we describe a newly developed staining technique to probe the structure of ambient cured water-borne epoxy films with transmission electron microscopy. It is shown that the film morphology in some systems can change with time after mixing the components. In addition, the structure of the film can be manipulated through the choice of raw materials.

Experimental

Film Based on Solid Epoxy Dispersion. The following published clearcoat formulation *(9)* was employed. In a container was mixed 125.2 g of a 55% nonvolatile (NV) solid epoxy dispersion (EPI-REZ 5522-WY-55, Shell Chemical Co.; EEW = 625), 7.7 g of 2-propoxyethanol (Ektasolve EP, Eastman Chemical Co.), and 7.5 g of a commercial blend of aromatic solvents with a flash point of 140°F (Hi Sol 15, Ashland Chemical Co.). In a second container was mixed 20.3 g of a 60% NV poly(ethylene amine) adduct curing agent (EPI-CURE 8290-Y-60, Shell Chemical Co.; AHEW =

163) and 39.3 g of deionized water. The stoichiometry is calculated to be 1.5:1 epoxy groups per amine hydrogen, and the calculated VOC of the formulation was 310 g/L (2.6 lb/gal), with the final solvent blend calculated to be 90.6% EP and 9.4% aromatic blend, due to EP that was present in both the hardener and curing agent. Films (ca. 1.3 mil dry film thickness) were cast using a #50 roll bar (Paul N. Gardner Co.) on a partially fluorinated polyethylene film (Tedlar, E.I. DuPont de Nemours & Co., Inc.) 0.5, 2, 4, and 8 hr. after adding the contents of the second container to the first and hand mixing. They were cured for at least 2 weeks in a climate controlled room at 25°C and 50% relative humidity. In some other experiments, the same formulation and procedure as above was employed except that the aromatic solvent was removed from the formulation, as noted below.

Film Based on Liquid Epoxy Resin. A mixture of 83.3 g of liquid epoxy resin (Epon 828, Shell Chemical Co.), 16.9 of nonylphenol, and 1.67 g of a commercial blend of aromatic solvents with a flash point of 104°F were combined in one container, and 33.7 g of the cycloaliphatic amine hardener described below (87.6% NV in EP, AHEW = 92.2 as is), 14.2 g of EP, and 10.0 g of deionized water were combined in a second container. The stoichiometry was 1.2 epoxy groups per amine hydrogen. The cycloaliphatic amine adduct curing agent is based on polycycloaliphatic polyamines (PCPA, Air Products and Chemicals, Inc.), a small portion of which have been covalently grafted in a proprietary process to a water soluble polymer chain consisting mostly of poly(ethylene oxide). The components were hand mixed as above, and after standing for 30 min., the viscosity was reduced to 25 sec., Zahn #2 cup with 20.3 g of DI water. Films (ca. 2 mil dry film thickness) were cast 0.5, 1, 2, and 4 hours after initial mixing, and cured for >2 weeks.

Film Based on Solvent Based Solid Epoxy Resin and Polyamide Hardener. A mixture of 86.6 g of solid epoxy resin (D.E.R. 671-PM75, Dow Chemical Co.; 75% NV in propylene glycol monomethyl ether), 20.1 g of polyamide hardener (Ancamide 350A, Air Products and Chemicals, Inc.; AHEW = 100), 9.5 g of n-butanol, and 9.5 g of xylene were thoroughly hand mixed, and after a 15 min. induction period, films were cast as described above.

Film Based on Solvent Based Solid Epoxy Resin and Amine Adduct Hardener. A mixture of 67.4 g of D.E.R. 671-PM75, 15.4 g of a 60% NV (in an n-butanol/aromatic solvent blend) isolated 2:1 molar adduct of ethylene diamine and liquid epoxy resin (AHEW = 85.5 based on solids), 9.5 g of n-butanol, and 9.5 g of xylene were thoroughly hand mixed, and after a 15 min. induction period, films were cast as described above. The isolated adduct was prepared using a large excess of ethylene diamine, and removing the excess under vacuum until the residual ethylene diamine was less than 1% by weight as measured by GC.

Film Staining and Microscopy. After removing the substrate, the films were cross-sectioned with a cryoultramicrotome to about a 100 nm film thickness. After mounting on a TEM grid which has a polymer support film, the sections were section-side floated in a 2% solution of the staining reagent for 10 minutes, then rinsed several times with DI water, dried, and examined with a transmission electron microscope (TEM) operated at 100 kV accelerating voltage. The stain most commonly employed was 3-iodopropionic acid, but 2-iodoacetic acid, 3-iodobenzoic acid, and phosphotungstic acid were also used as noted.

Electrochemical Impedance Spectra. The impedance spectra were measured on 2 mil films cast on grit blasted steel using a roll bar, and cured under the same conditions as the free films (above). The EIS coating cell exposed a 1 cm diameter circle of the coating to 1.0 M NaCl in water. A platinum gauge counter electrode was used, and potentials were referenced to the aqueous Ag/AgCl electrode. A Princeton Applied Research (PAR) Model 273A potentiostat and model 5210 lock-in amplifier was used, and PAR model 398 impedance software (version 1.11) was used for data acquisition. Experiments were run in 2 parts: single sine waves were used from 100 down to 5 kHz, and a composite waveform was used from 10 Hz to 0.05 Hz. Data were combined into one spectrum for analysis. Sine wave excitation voltages of 10 and 15 mV were used for the single and multi-sine experiments, respectively. Spectra took 13 min. to acquire. Testing was done after 1 and 24 hr. of immersion in NaCl solution.

Results and Discussion

The use of various stains such as osmium tetroxide, *(10)* ruthenium tetroxide, *(11)* and phosphotungstic acid *(12)* to examine the morphology of polymeric films is a common technique. To our knowledge the use of iodocarboxylic acids to stain amine containing domains within a polymeric matrix is new. *(13)* The stain was designed to work by the reaction of the carboxylic acid with the amine groups to form the corresponding salts. The iodine atom, which has a very high nuclear density, scatters the incident electron beam, with the result that amine containing areas within the film appear dark in a photograph.

A cross-section of a solvent borne clearcoat prepared from a solid epoxy resin with an EEW of about 475 crosslinked with a typical medium solids polyamide hardener is shown in Figure 1. The cross-section was stained for 10 minutes with a 2% solution of iodopropionic acid. This is a typical epoxy resin and curing agent combination that has seen widespread use in coatings for the protection of metal substrates in highly corrosive environments for decades. There are some diagonal stripes visible in the photograph which are an artifact caused by the microtome's diamond knife as it sliced the cross-section, and a few specks of dust are also visible. Otherwise, the film is homogeneous at this level of resolution.

Type II Film Late In Pot Life. The morphology of films based on a Type II solid epoxy dispersion was studied using a published clearcoat formulation, employing an epoxy dispersion with an EEW of 625 (based on solids) and the recommended amine/epoxy adduct hardener with an AHEW of 163. A noteworthy feature of this formulation is that it utilizes a large stoichiometric excess of 1.5 epoxy groups per amine hydrogen. This is about the minimum excess necessary to improve the water resistance and corrosion resistance of the coating for direct-to-metal applications, *(14)* and excesses as large as 2:1 are sometimes recommended. This formulating practice is

Figure 1. Cross-section of a film from a solvent borne solid epoxy resin cured with a polyamide hardener, stained for 10 min. with iodopropionic acid.

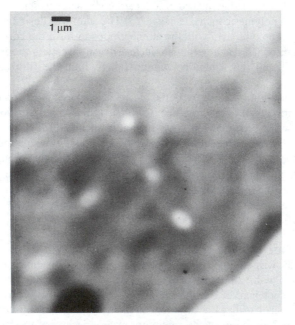

Figure 2. Cross-section of a film prepared from the solid epoxy dispersion with no curing agent, stained 2 hr. with iodoacetic acid.

common with solid epoxy dispersions, and is in contrast with solvent borne epoxy formulations, where stoichiometries closer to 1:1 are normally employed.

In Figure 2, a cross-section of film made from a solid epoxy dispersion and stained for 2 hours with 2% iodoacetic acid is shown. Note that this is a control film that contains no amine hardener. The film was very soft and difficult to cut. We ascribe the vague shadowy pattern present to nonuniformity in the cross-sectioning caused by the softness of the film.

Figure 3 shows the unstained cross-section of the cured, formulated clear coat obtained from a film that was cast 2 hours after mixing the amine and epoxy dispersion, and Figure 4 shows a cross-section of the same film that was stained for 10 min. with 2% iodopropionic acid. In the absence of stain, a cell pattern is barely visible, but with the stain, a core-shell type structure becomes clearly visible. Individual particles are clearly present. The particles are about the same size as the starting dispersion, which was measured by TEM to have a particle size of about 500 nm. The particles appear to have been somewhat elongated by the cutting process. They stain darker on the outside of the particles than in the interior regions, indicating that amine concentrations are higher at the particle boundaries than in the interiors. Unfortunately this is only a qualitative technique, and we have no way to measure what the difference in concentration is. A stained cross-section obtained 4 hours after mixing is virtually indistinguishable from Figure 4.

In Figure 5, the stained film obtained 8 hours after mixing is shown. The structure is similar to the film obtained 2 hours after mixing, except that now some of the largest particles appear to be deformed into dumbbell-like shapes.

The core-shell type structure seen in Figures 4 and 5 probably arises as follows (see Figure 6). The hardener is a step-growth oligomer formed by the reaction of multifunctional epoxides with a polyethylene amine, followed by further adduction with aromatic monoepoxides, which is then partially neutralized with acetic acid. *(15)* Thus it contains both hydrophobic and hydrophilic regions in the same molecule, and the number average molecular weight would be anticipated to be on the order of 1000. When mixed into an aqueous medium containing some glycol ether, it would be anticipated that such an amine adduct would partition into the particle cores, into the aqueous phase, and onto the particle surface. This situation would tend to cause the amines to react preferentially with epoxy groups at or near the surface of the particle, based simply on where the amines are located, as well as the catalytic effect of water on the reaction. Reaction near the surface will result in the formation of a crosslinked shell. This shell would in turn raise the barrier to diffusion of amine into the particle interiors.

If this model is correct, one would expect that small particles, with a larger surface-to-volume ratio than larger particles, would achieve a higher degree of cure, at least during early stages of the curing process when film formation occurs. Therefore the T_g and hardness of small particles would also be greater. Small particles could then act as a mandrel around which larger particles could be deformed by capillary and other forces *(16)* present during the film formation process. This explains the formation of the deformed particles seen 8 hours after mixing.

The need for large stoichiometric excesses of epoxy to reduce water sensitivity may be related. With more amine present in a formulation, more will be partitioned into the aqueous phase, and thus more could be trapped at the particle boundaries in

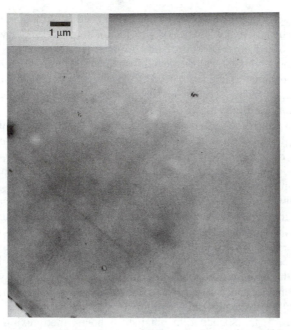

Figure 3. Cross-section of a film from a solid epoxy dispersion and hardener, film cast 2 hour after mixing, no stain.

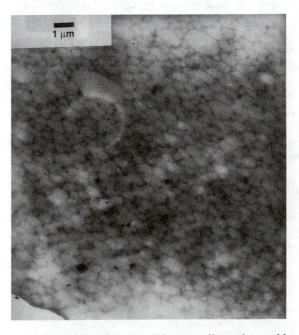

Figure 4. Cross-section of a film from a solid epoxy dispersion and hardener, film cast 2 hour after mixing, stained 10 min. with iodopropionic acid.

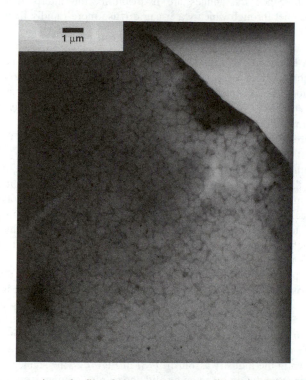

Figure 5. Cross-section of a film from a solid epoxy dispersion and hardener, film cast 8 hour after mixing, stained 10 min. with iodopropionic acid.

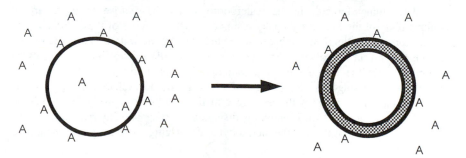

Figure 6. Process for the formation of core-shell type morphology from a waterborne epoxy coating. 'A' represent unreacted amines, and the circle represents an epoxy resin particle. The crosslinked shell is represented by the hash pattern.

the cured film. Since the amines are hydrophilic, the interparticle regions may represent a channel through which water can diffuse more rapidly than it can through other parts of the film.

Type II Film Early In Pot Life. Figure 7 shows the stained cross-section of a film obtained much earlier in the pot life of this formulation, only 1/2 hour after mixing. The morphology is completely different. There are amine containing domains embedded in an epoxy continuous phase. Some of the amine domains are quite large, exceeding 1 μm in diameter.

Numerous modifications to the staining technique were run to check for the possibility that these results were due to some artifact of the experiment. Staining in ethanol and ethyl acetate, instead of the usual aqueous medium, gave the same result. To eliminate the possibility that an S_N2 reaction with amine was eliminating iodide which might migrate to some other area of the film, staining was conducted with 3-iodobenzoic acid in ethanol, also with the same result. Finally, staining was reduced from 10 min. to only 10 sec. Since we believe the stain functions through acid-base reactions which are generally considered to occur at diffusion controlled rates, *(17)* this change should make no difference, as was the case.

These photomicrographs indicate that films formed shortly after mixing the amine and epoxy components have phase separated. It is well known in solvent borne coatings that amine curing agents based on polyethylene amines are not highly compatible with epoxy resins. This is why it is generally recommended that coatings based these curing agents be mixed with the epoxy for about 30 min. to 1 hr. before application, in order to avoid the exudation of amines to the surface of the coating, which is known in the industry as 'blush'. *(18)* In Figure 8, the stained cross-section of a solvent borne formulation based on a solid epoxy resin and an isolated amine adduct hardener is shown. This film was cast only 15 min. after mixing. Similar amine-rich domains appear to be present, although they are not as prevalent as was the case in the waterborne system.

The amine domains in the waterborne system also arise from the initial incompatibility of the amine and epoxy resin. The fact that more of these domains are present probably results from the phase separated state of the waterborne system when the films are cast. In the solvent borne systems, incompatibility develops as the solvent evaporates, since the formulations are initially clear or at most slightly hazy. We believe that the change in morphology of the waterborne system results from the reaction of the curing agent with the surface of the solid epoxy dispersion. Once the surface has been modified sufficiently, the dispersion surface and any remaining amine are no longer incompatible, and the thermodynamic driving force for phase separation is eliminated.

Novel Type I Film. Operating under the premise that the morphologies discovered above may be responsible, at least in part, for the generally poorer performance of waterborne epoxy coatings relative to their solvent borne counterparts, we tried to develop a waterborne epoxy system that would have a more uniform film morphology. In doing so we employed three principles.

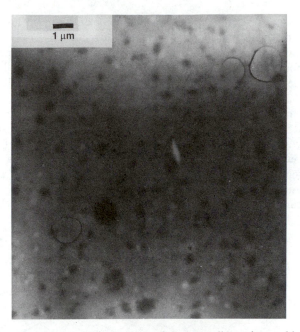

Figure 7. Cross-section of a film from a solid epoxy dispersion and hardener, film cast 0.5 hours after mixing, stained 10 min. with iodopropionic acid.

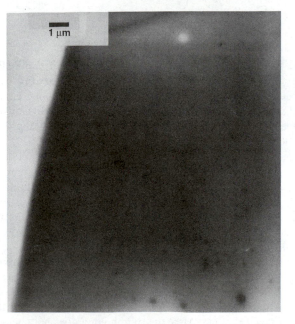

Figure 8. Cross-section of a film from a solvent borne solid epoxy resin and amine adduct hardener, film cast 0.25 hours after mixing, stained 10 min. with iodopropionic acid.

The first was to base the system on liquid epoxy. Its much lower molecular weight and viscosity should reduce the barriers to coalescence.

The second was to utilize a hydrophobic, epoxy-compatible amine as the basic building block for the curing agent. For this purpose PCPA appeared to be ideal. PCPA is a complex amine mixture made up of numerous components. Representative constituents are **I** and **II**. In contrast to the polyethylene amines, considerable experience with the formulation of PCPA has shown it to be highly compatible with epoxy resin, and its water solubility is very low. This type of amine structure also contains primary amines on a secondary carbon. The rate of reaction of such amines with a glycidyl ether is slower than that for the primary amine on a polyethylene amine, which would then extend the time after mixing the amine and epoxy during which the viscosity and molecular weight would stay low enough to afford good application and film formation.

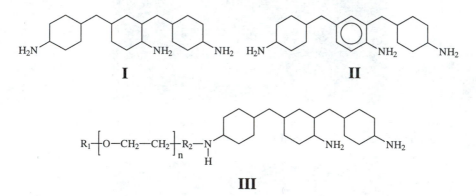

The third principle was to modify a small portion of the PCPA stream by attaching water soluble polyethylene oxide chains in a proprietary process, resulting in a mixture of products typified by **III**. In this way, **III** serves as the emulsifier for the unmodified PCPA and the liquid epoxy resin. In addition, the emulsifying chains eventually become part of the crosslinked network. This is generally thought to be a way to reduce water sensitivity. *(19)*

The formulation given above was developed using experimental design techniques. It contains a plasticizer (nonylphenol), as is usually required in the formulation of an epoxy resin and cycloaliphatic amine hardener. In the absence of a plasticizer, the T_g of the network exceeds room temperature at an early stage in the cure process, at which point the amine-epoxy reaction essentially stops. The formulation also contains a small amount of glycol ether coalescing solvent, and has a calculated VOC of 144 g/L (1.2 lb/gal).

The stained cross-sections of films cast 0.5 hr. and 4 hr. after mixing are shown in Figures 9 and 10, respectively. A few light regions are visible in some areas of the film, and may indicate the presence of some undispersed epoxy resin. However, the vast majority of the films appear to have a uniform morphology within the limitations of the resolution of this technique. Also, the morphology does not change through the useful pot life of the system, which is about 4 hours as measured by the 60° gloss of the clearcoat.

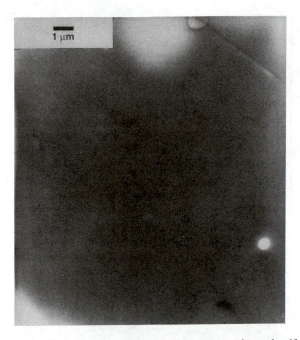

Figure 9. Cross-section of a film from liquid epoxy resin and self-emulsifying curing agent, film cast 0.5 hours after mixing, stained 10 min. with iodopropionic acid.

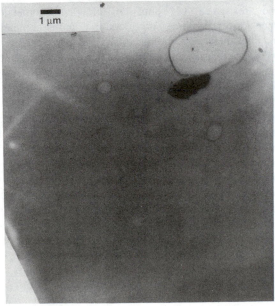

Figure 10. Cross-section of a film from liquid epoxy resin and self-emulsifying curing agent, film cast 4 hours after mixing, stained 10 min. with iodopropionic acid.

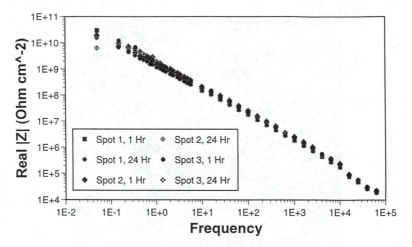

Figure 11. Electrochemical impedance spectrum of a 2 mil. film cast on grit blasted steel based on liquid epoxy resin and self-emulsifying curing agent. Three spots on the panel were tested after immersion in 5% NaCl for 1 hr. and 24 hr. periods. Reproduced with permission from reference 13.

That this approach can yield films with a high degree of water resistance is confirmed by the electrochemical impedance spectrum (EIS) shown in Figure 11. *(13)* A 2 mil film on grit-blasted steel tested in three separate 1 cm^2 spots with a 1M NaCl solution gives a pore resistance >10^{10} ohms/cm^2 after 1 hr. immersion. This is a very high pore resistance for a waterborne epoxy coating, and is comparable to the initial pore resistance of high quality traditional solvent based maintenance coatings. *(20)* Only a slight drop in pore resistance is observed after 24 hr. of immersion.

Conclusions

Iodocarboxylic acids have been utilized to stain amine containing areas in cross-sections of an epoxy network. The technique shows that films prepared from an aqueous solution of a solid epoxy dispersion and amine adduct curing agent have a heterogeneous morphology that changes as a function of the time between mixing of the reactive components and the preparation of the film. At short time intervals, the amine phase separates within an epoxy continuous phase, whereas at later times the film is composed of particles that have amine rich shells. The change in morphology was attributed to compatibilization of the amine with the epoxy dispersion surface due to reactions between them. By grafting emulsifying polymer chains onto a hydrophobic and highly epoxy resin-compatible mixture of amines, a curing agent capable of emulsifying liquid epoxy resin was prepared. Films made from this system exhibited a much more uniform morphology, similar to that obtained from traditional solvent borne epoxy resin formulations. These films also exhibited a very high initial pore resistance as measured by electrochemical impedance spectroscopy.

Acknowledgments. The authors thank Dr. A. Gilicinski for obtaining and assisting in the interpretation of the electrochemical impedance spectra. The assistance of Mrs. K.E. Everett, who prepared most of the coatings for this work is gratefully acknowledged.

Literature Cited

1.   Bolgar, L. Br. Patent 1 108 558.
2.   Rozenberg, B.A. *Adv. Polym. Sci.*, **1986**, *75*, 113.
3.   (a) Elmore, J.D.; Cecil, J.D. U.S. Patent 4 315 044 , 1982. (b) Williams, P.R.; Burt, R.V.; Golden, R. U.S. Patent 4 608 406, 1986. (c) Becker, W.; Godau, C.; U.S. Patent 4 886 845, 1989. (d) Dreischoff, H.; Geisler, J.; Godau, C.; Hoenel, M. U.S. Patent 5 236 974, 1993.
4.   Wicks, Z.W.; Jones, F.N.; Pappas, S.P. *Organic Coatings: Science and Technology,* John Wiley and Sons, Inc.: New York, NY, 1992, Vol. 1, pp 35 - 48.
5.   Wisanrakkit, G.; Gillham, J.K. *J. Coatings Tech.* **1990**, *62*, (783), 35 - 40.
6.   (a) Winnik, M.A.; Wang, Y.J. *J. Coatings Tech.*, **1992**, *64*, (811), 51 - 61. (b) Linne, M.A.; Klein, A.; Miller, G.A.; Sperling, L.H.; Wignall, G.D. *J. Macronol. Sci.-Phys.,* **1988**, *B27*, 217 - 231.
7.   Wicks, Z.W.; Jones, F.N.; Pappas, S.P. *op. cit.*, Vol. 1, p 47.
8.   Wegman, A. *J. Coatings Tech.*, **1993**, *66*, (827), 27.
9.   Clear Enamel Formulation 24-294, Hi-Tek Polymers, Inc., 1988.
10.   (a) Kato, K. *Polymer Letters*, **1966**, *4*, 35.   (b) El-Aasser, M.S.; Vanderhoff, J.W.; Misra, S.C.; Manson, J.A. *J. Coatings Tech.*; **1977**, *49* (635), 71.
11.   Trent, J.S.; Scheinbeim, J.I.; Couchman, P.R. *Macromolecules*, **1983**, *16*, 589.
12.   Hess, K.; Gutter, E.; Mahl, H. *Naturwissenschaften*, **1959**, *46*, 70.
13.   Walker, F.H.; Everett, K.E.; Kamat, S. *Proc XXII High Solids, Waterborne, and Powder Coatings Symposium,* **1995**, 88.
14.   Galgoci, E. *Proc. XXII High Solids, Waterborne and Powder Coatings Symposium,* **1995**, 119.
15.   Degooyer, W.J. U.S. Patent 4 539 347, 1985.
16.   Eckersley, S.T.; Rudin, A. *J. Coatings Tech.*, **1990**, *62*, (780), 89 - 100.
17.   Lowrey, T.H.; Richardson, K.S. *Mechanism and Theory in Organic Chemistry*, Harper & Row: New York, NY, 1976, p 406.
18.   Hare, C. *Protective Coatings*; Technology Publishing Co.: Pittsburgh, PA, 1994; p 210.
19.   Wicks, Z.W.; Jones, F.N.; Pappas, S.P. *op. cit.*, Vol. 1, p 70.
20.   (a) Monetta, T.; Bellucci, F.; Nicodemo, L.; Nicolais, L. *Prog. Org. Coatings*, **1993**, *21*, 353; (b) Rammett, V.; Reinhard, G. *ibid.*, **1992**, *21*, 205.

Chapter 25

Stripe Patterns Formed in Particle Films: Cause and Remedy

Eiki Adachi[1,3], Antony S. Dimitrov[1,4], and Kuniaki Nagayama[2]

[1]Nagayama Protein Array Project, ERATO, JRDC 5–9–1 Tokodai, Tsukuba, Ibaraki 300–26, Japan
[2]Department of Life Sciences, Graduate School of Arts and Sciences, University of Tokyo, Komaba, Meguro-ku, Tokyo 153, Japan

When a droplet of suspension of particles dries on a glass surface, particles collect near the edge of the droplet (contact line) and often leave a striped pattern of particles as water evaporates. During drying processes, the motion of the droplet contact line resembles stick-slip motion and it shrinks towards the center of the droplet with an oscillatory motion. To explain the oscillatory motion and the mechanism of the stripe formation, we formulated a mathematical model that includes a friction force which the contact line feels when particles flow from the inside of the droplet to the droplet edge. As a result of competition between this friction force and surface tensions at the contact line, the droplet oscillates as it dries and generates a striped film composed of particles.

When a suspension droplet dries on a solid substrate, we often observe striped patterns of particles, which remain on the substrate after evaporation of the water. Natural and artificial stripe patterns are shown in Figure 1 for examples. Natural strips, on a car body and on a cup wall, are composed by dusts and coffee powders, respectively. The stripe of dusts was found after a rainy day on a waxed car body, which means droplets were formed on it. The coffee powder stripe was found on a Monday morning, which was left on a desk on the last Friday night. These kind of patterns, of course, can be obtained by using artificial particles such as polystyrene particles. They imply existence of common principle for striping.

When we examine the drying process, we observe that the droplet contact line shrinks towards the center of the droplet with an oscillatory motion (1), causing generation of a particle-array film at the contact line. The shrinking motion of the droplet can be broadly classified as a "stick-slip" motion (2), but the motion we

[3]Current address: National Institute for Physiological Sciences, Okazaki, Aichi 444, Japan
[4]Current address: L'Oreal Tsukuba Center, 5–5 Tokodai, Tsukuba 300–26, Japan

0097–6156/96/0648–0418$15.00/0

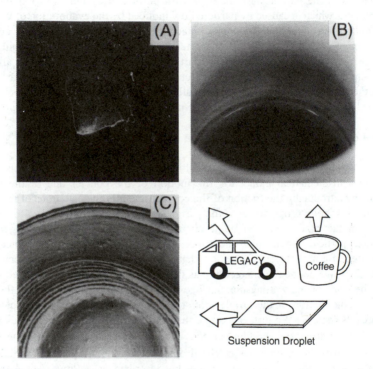

Figure 1. The stripes formed in nature. (A) the stripe of dusts on a car body are some time found after a rainy day when the car body is waxed, which means droplets can be formed on the body. (B) the stripe of coffee powders on a cup wall is formed when one leaves his cup with coffee on a desk for several night. We can find other stripes, for examples, in a kitchen, in a bath room, and in a toilet, where water is there. (C) when a suspension droplet including mono-dispersed particles dries on a glass plate, regular stripes can be obtained.

observed is different from the stick-slip motion that has been observed by other researchers (3-5). In general, an object undergoing stick-slip motion periodically converts its kinetic and internal energy to thermal energy.   The periodic energy dispersion is caused by coupling of the friction at the contact surface with the object motion.   In addition to the energy dispersion, the motion of the suspension droplet accompanies discharges of suspended particles from the inside to the boundary of the droplet (6).   The particles assemble to form particle-array films (7,8) at the droplet contact line, which is caused by particle flow induced by the evaporation of water from the film surface near the receding contact line.   We call this process, convective self-assembly of particles, which characteristically occurs in the thin liquid layer of particle suspensions (9).   Since the particle flow is viscous, the flow affects the stick-slip motion of the droplet as a source of friction at the droplet contact line.   On the other hand, the shrinking motion affects the self-assembly of particles.   This coupled system must show a pattern or structure in the final particle-array films.

Model of Droplet Edge for Formulation of Striping

To investigate the pattern resulting from these competing processes, we derived a mathematical model for the motion of the contact line of the shrinking droplet by introducing the friction of the contact line into the equations of motion, and setting this friction proportional to the particle flow at the contact line.

In the model we assume a circular, sessile suspension droplet containing spherical particles on a solid substrate.   We further assume the droplet edge is composed of a monolayer of particles that wets the surface with the suspension (see Figure 2).   We define the contact line as the intersection of the extrapolated meniscus with the plane of the wetting-film surface.   To derive a relation between the particle flow and the motion of the droplet contact line, we first assume the droplet shrinks towards its center.   We further assume that a water flow (J_W) induced by evaporation (J_e) from the array and the film surface, coupled with viscous drag on the particles, carries the particles from inside of the droplet to the boundary of the particle arrays. Therefore, the total particle motion can be described as a flow, J_P.   Although J_e causes energy transfer, we simply consider J_e is the cause of inducing of J_W and J_e. Since the suspension flow (a mixture of J_P and J_W) is viscous, a friction force, s, acts on the wetting-film surface close to the contact line so as to prevent shrinkage of the droplet.   The surface tension of the wetting film (γ_f) also tends to prevent droplet shrinkage, but the surface tension of the droplet (γ_L) tends to cause droplet shrinkage. As a result, γ_L competes γ_f with and s at the droplet contact line.

Motion of Contact Line and Stripe Number Depending on Particle Volume Fraction

Let h, h, and v be the shear viscosity, the wetting-film thickness, and the velocity of the suspension flow at the film surface, respectively.   Since a velocity of the suspension flow, u, is written as $u=zv/h$ (at z=0, u=0; z=h, u=v) at low v (laminar flow) (10), σ is written as

$$\sigma = \eta \frac{\partial u}{\partial z}\Big|_{z=h} = \eta \frac{v}{h} \tag{1}$$

When the suspension is treated as a continuous fluid, v is defined from the averaged momentum of the water and particles as

$$\frac{1}{h}\int_0^h dz\rho\frac{zv}{h} = \frac{1}{2}\rho v = m_P J_P + m_W J_W \tag{2}$$

where m_P and m_W are the masses of the particle and the water molecules, respectively. In eq 2, ρ is the average density of the suspension, defined as

$$\rho = \rho_P \phi + \rho_W (1 - \phi)$$

where ϕ is the particle volume fraction in the wetting film and ρ_P and ρ_W are the mass density of a particle (m_P/V_P), and the mass density of a water molecule (m_W/V_W), respectively (V_P and V_W are the volumes of a particle and of a water molecule, respectively). The flows, J_P and J_W, are defined as

$$J_P = \phi \frac{v_P}{V_P}, \quad J_W = (1 - \phi)\frac{v_W}{V_W}$$

Where, v_P and v_W are the velocities of particle and water molecules, respectively. From eq 2, v is expressed by J_P. Therefore, eq 1 takes the form

$$\sigma = \frac{2\eta}{h}\frac{\beta\phi + \varepsilon(1-\phi)}{\beta\phi[\phi + \varepsilon(1-\phi)]}V_P J_P \tag{3}$$

where, $\beta = v_P/v_W$ and $\varepsilon = \rho_W/\rho_P$. This equation relates the particle flow, J_P, to the motion of the contact line through an equation of motion of the droplet.

When ρ^*, R, and θ are a effective mass density of the contact line (adjustable parameter), the droplet radius, and the effective contact angle of the droplet, respectively, the equation of motion is written by using eq 3

$$\rho^*\frac{d^2 R}{dt^2} = \gamma_f - \gamma_L\cos\theta + \frac{2\eta}{h}\frac{\beta\phi + \varepsilon(1-\phi)}{\beta\phi[\phi + \varepsilon(1-\phi)]}V_P J_P \tag{4}$$

where t is time. The first and second terms on the right side of eq 4 are the wetting film surface tension and the horizontal projection of the droplet surface tension, respectively. The third term on the right side of eq 4 represents the friction force. Since the solutions that represent the motion of the droplet contact line, $J_P = J_P(t)$ and $R = R(t)$, are necessary, we need another relation between J_P and R. This relation can be obtained by considering two conservation laws for the particle number and for the total volume of the particle and water molecules during the array formation and film wetting. If we assume that the wetting film thickness at the contact line is equal to the particle diameter, the particle number conservation law can be written as

$$2\pi \int_R^{R_0} drrh\phi_P(r) = 2\pi V_P \int_0^t dtrh J_P \tag{5}$$

where $R_0 = R(t = 0)$.   The left side represents the total particle number in the particle array and in the wetting film.   The right side represents the particle number transferred to the particle array and the wetting film from the droplet across the contact line from time 0 to t.   In eq 5, ϕ_P is the volume fraction of the particles in the particle array and in the wetting film.   The total volume of particles and water molecules, which move from inside the suspension droplet to the wetting film, should be balanced with the volume of water evaporating from the particle array and the wetting-film surface and with the volume of the new wetting film formed at the contact line.   This conservation law can be expressed as

$$2\pi Rh(V_P J_P + V_W J_W)$$

$$= 2\pi J_e V_W \int_R^{R_0} drr[1 - \phi_P(r)] + 2\pi R \frac{dR}{dt}h \tag{6}$$

Combining eqs 5 and 6 then yields

$$\frac{d(RJ_P)}{dt} = -\frac{J_e V_W \beta \phi}{hV_P[1+(\beta-1)\phi]}\left[R\frac{dR}{dt} + V_P RJ_P - \frac{h}{J_e V_W}\frac{d}{dt}\left(R\frac{dR}{dt}\right)\right] \tag{7}$$

By combining eqs 4 and 7, we derive a non-linear differential equation for the motion of the contact line as

$$\frac{d^3 R}{d t^3} + 2\lambda\frac{d^2 R}{dt^2} + (\lambda^2 + \mu^2)\frac{dR}{dt} - V_S(\lambda^2 + \mu^2)$$

$$+ \frac{1}{R}\frac{dR}{dt}\left[\frac{d^2 R}{dt^2} - \frac{h}{J_e V_W}(\lambda^2 + \mu^2)\frac{dR}{dt} - \frac{\gamma_f - \gamma_L \cos\theta}{\rho^*}\right] = 0 \tag{8}$$

where

$$V_S = \frac{(\gamma_f - \gamma_L \cos\theta)h}{2\eta}\frac{\beta\phi}{1+(\beta-1)\phi}, \quad \lambda = \frac{1}{2h}\left(\frac{J_e V_W \beta\phi}{1+(\beta-1)\phi} - \frac{2\eta}{\rho^*}\right), \quad \mu^2 = \frac{2\eta J_e V_W}{\rho^* h^2} - \lambda^2$$

and ε is assumed to be one for simplicity.   Note that the fifth term on the left side of eq 8 is nonlinear.   When this term is sufficiently small (i.e., R is large compared with dR/dt), eq 8 simplifies to a third-order linear differential equation.   If the contact angle, θ, and the volume fraction, ϕ, are assumed to be constant during the formation of stripes, an oscillatory solution of eq 8 is obtained as a function of t as

$$R = R(t) = R_0 + V_S\left\{t - \frac{1}{\mu}\left[e^{-\lambda t}\sin(\mu t + 2\alpha) - \sin 2\alpha\right]\right\} \tag{9}$$

Where

$$\alpha = - \tan^{-1} \frac{\lambda}{\mu}$$

The corresponding boundary conditions for eq 9 are

$$\frac{dR}{dt} = 0, \quad \frac{d^2 R}{dt^2} = 0 \quad (t = 0)$$

The number of stripes per unit length, N_S, along the droplet radius is estimated by the particle density in a particle-array film. Since the particle density is proportional to $J_P(t)/V(t)$ [$V(t) = dR(t)/dt$], an area becomes distinguishable from other areas when $V(t)$ tends to zero with finite $J_P(t)$. The area is then recognized as a dense stripe. Therefore, the number of real roots of the equation, $V(t) = 0$, is equivalent to the number of stripes. It is clear from eq 9 that the real roots exist only in a limited range, $0 \le t \le T$. Therefore, the number of roots is equal to $\mu T/\pi$, which is the phase of sine function in eq 9 divided by π. The number of stripes per unit length is estimated from the number of roots and the linear receding velocity, V_S, in eq 9 as

$$N_S \approx \frac{\mu T}{\pi V_s T} = \frac{\mu}{\pi V_s} \tag{10}$$

Eq 10 can be written in terms of ϕ by substituting for μ and V_S from eq 8 to yield

$$N_S \approx - \frac{\eta}{\pi(\gamma_f - \gamma_L \cos\theta)h} \frac{\sqrt{c_0 + c_1\phi + c_2\phi^2}}{\phi} \tag{11}$$

where,

$$c_0 = \frac{4 P_1 P_2 - P_2^2}{\beta^2}$$

$$c_1 = \frac{2 P_1 P_2}{\beta} + 2(\beta - 1)c_0$$

$$c_2 = - P_1^2 + (\beta - 1)(c_0 - c_1)$$

and

$$P_1 = \frac{J_e V_w}{h}$$

$$P_2 = \frac{2\eta}{\rho^* h}$$

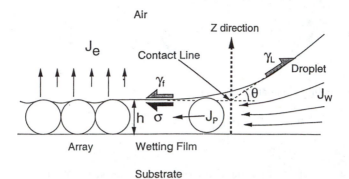

Figure 2. A schematic of the droplet edge.   The particle flow (J_P) is generated by water flow (J_W) that is induced by water evaporation (J_e) from the particle array and the wetting-film surface.   At the contact line region, the droplet's surface tension, γ_L is in competition with the wetting-film's surface tension, γ_f, and the friction force, σ.   The wetting film thickness, h, is assumed to be equal to the particle diameter at the contact line.   (Reproduced with permission from ref. 12. Copyright 1995 American Chemical Society.)

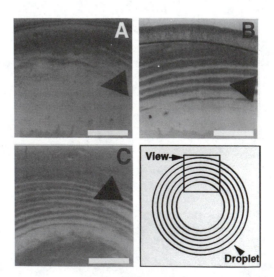

Figure 3. Typical stripe patterns formed by using 144 nm PS particles.   One stripe is pointed by the arrow in each picture.   The initial particle volume fraction of droplets in A, B, and C was 0.01, 0.005, and 0.002, respectively.   The initial droplet volume was 0.5 µl.   The position of each picture in the films is shown by the "VIEW" box in the drawing.   The films are about 3 mm in diameter for each of the samples, and the scale bar is 0.5mm.

Experimental Results and its Agreement with The Theory

We produced striped films of polystyrene particles (PS-particles) by dropping particle suspensions on glass plates and waiting for the water to evaporate. The droplets had a volume of about 0.5 µL, the particle diameter was 144 nm (JSR, STADEX), and borosilicate glass slides (Matsunami microslide glass) were used. To demonstrate the stick-slip motion, the droplet shrinking motion was recorded using a bright-field optical microscope with a CCD video camera. The monolayer of 144 nm PS-particles could be recognized by its blue color due to the interference effect (11).

Typical stripe patterns are shown in Figure 3. These stripes were formed when the droplet contact line started to move with an oscillatory motion. The motion of the contact line, $R(t)$, was determined by plotting its time-dependent position (see Figure 4). The number of monolayer stripes per unit length in the films is plotted as a function of the initial particle volume fraction in Figure 5. The film areas were almost the same (3 mm in diameter) in all experiments. The solid lines in Figure 4 and 5 represent theoretical curves calculated according to eqs 9 and 11 by adjusting c_0, c_1, and c_2, etc. The dashed line shows how N_S is affected by small changes of J_e.

We used PS particles 144 nm in diameter because they gave us the clearest monolayer stripes. This is because the thickness of large-size water film on borosilicate glasses was about 120 nm (confirmed by ellipsometry), which is a little smaller than the particle diameter. Since our theory assumes monolayer-film formation, this coupling between the water film thickness and particle diameter was the most suitable for stripe formation. Actually, it was difficult to obtain monolayer stripes when we used particles either larger or smaller than 144 nm. The characteristics of N_S, such as the maximum number of stripes and the particle volume fractions for $N_S = 0$ (it means perfect monolayers), can be determined by parameters c_0, c_1, and c_2 (see the simple form of $N_S[\phi]$ given in eq 11). Although ϕ in Figure 5 is the initial volume fraction of the droplets and not the particle volume fraction of the suspension in the wetting film, we think this result is acceptable because the volume fraction in the wetting films must be proportional to the initial volume fraction. The position of the droplet contact line motion, $R(t)$, is characterized by μ, λ, and V_S, which determine the spatial frequency of the stripes and the linear receding velocity. These parameters have clear dependence on J_e, h, η, etc., as shown in eqs 9 and 11. We have not yet verified these dependencies due to the lack of sufficient experimental accuracy. However, J_e will be determined by observing evaporation of a water droplet. h and η are going to be determined by an ellipsometry and viscosity measurement of balk solution, respectively. V_W is known from water mass density. Although we have only one adjustable parameter, $\rho*$, the parameter will be expressed by more precise theory. In this study, our objective was to understand the mechanism of the patterning of particle-array films. We therefore focused on the ϕ–dependency of the stripe frequency to compare with the theoretical results.

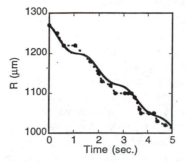

Figure 4. Oscillatory shrinking motion of a droplet contact line, $R(t)$, that contains PS particles 144 nm in diameter. The initial particle volume fraction of the droplets was 0.003. The dots (●) represent experimental data and the solid line is the theoretical curve. The calculated data was obtained by using normalized parameters, $R_0=1$, $V_S=1$, $\lambda=1.5$, $\mu=23$, and $\alpha=0$. (Reproduced with permission from ref. 12. Copyright 1995 American Chemical Society.)

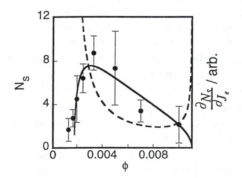

Figure 5. Averaged number of monolayer stripes as a function of the initial particle volume fraction of the droplets. The numbers are normalized by dividing counted numbers with 1.5, which is the half of the droplet diameter. The dots (●) represent experimental data and the solid line is the theoretical curve. The dashed line represents a sensitivity of N_S to J_e (dN_S/dJ_e). Parameters for fitting are $c_0=1.98\times10^{-5}$, $c_1=1.28\times10^{-2}$, and $c_2=-1$ when the amplitude of N_S is 11. (Reproduced with permission from ref. 12. Copyright 1995 American Chemical Society.)

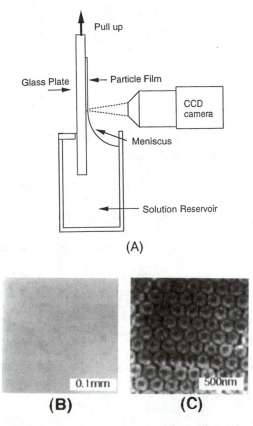

Figure 6. (A) a schematic of an apparatus avoiding stripe patterning of particle films. The glass plate is mechanically pulled up from the solution reservoir at a constant speed. The motion of contact line is monitored by CCD camera attached to an optical microscope. (B) the image of obtained particle array taken by using an optical microscope (OLYMPUS, BHM). (C) the image taken by using a scanning electron microscope (HITACHI S5000H).

Remedy of Striping

From our studies, we now understand that the mechanism of stripe formation results from competition between the droplet's surface tension, the wetting-film's surface tension, and the friction force at the contact line. Since stripes are formed when the contact line moves digitally (stick-slip motion), the motion of contact line must be controlled so as to smoothly move to avoid striping of particle films. Our apparatus realizing the controlling is schematically shown in Figure 6 with the results of uniform particle film formation. Used particles were polystyrene particles (STADEX, JSR) that diameter was 144nm. The glass plate is mechanically pulled up in vertical direction with monitoring of the motion of contact line. When the contact line jumps to backward, the motion of the glass plate is stopped to cancel the jumping, which makes the motion of contact line smooth.   As a result, we could get the particle films without stripes.

References

1.   Dushkin, C. D.; Yoshimura, H.; Nagayama, K. *Chem. Phys. Letters* **1993**, *294*, 455.
2.   Dussan, V. E. B. *Annu. Rev. Fluid Mech.* **1979**, *11*, 371.
3.   Yoshizawa, H.; Israelachvili, J. *J. Phys. Chem.* **1993**, *97*, 11300.
4.   Palberg, T.; Streicher, K. *Nature* **1994**, *367*, 51.
5.   Baumberger, T.; Heslot, F.; Perrin, B. *Nature* **1994**, *367*, 544.
6.   Fitzgerald, S.; Woods, A. W. *Nature* **1994**, *367*, 450.
7.   Denkov, N. D.; Velev, O. D.; Kralchevsky, P. A.; Ivanov, I. B.; Yoshimura, H.; Nagayama, K. *Langmuir* **1992**, *8*, 3183.
8.   Dimitrov, A. S.; Dushkin, C. D.; Yoshimura, H.; Nagayama, K. *Langmuir* **1993**, *10*, 432.
9.   Nagayama, K. *Phase Transition* **1993**, *45*, 185.
10.   Landau, L. D.; Lifshitz, E. M. *Theoretical Physics Vol. 6, Hydrodynamics*, 2nd ed.; Governmental Publisher for Theoretical-Technical Literature: Moscow, 1953 (in Russian); pp71-81.
11.   Dushkin, C. D.; Nagayama, K.; Miwa, T.; Kralchevsky, P. A. *Langmuir* **1993**, *9*, 3695.
12.   Adachi, E.; Dimitrov, A. S.; Nagayama, K. *Langmuir* **1995**, 11, 1057.

NOVEL CHEMISTRY, PROCESSES, AND FILM STRUCTURES

Chapter 26

Cross-Linking Water-Reducible Coatings by Direct Esterification

Guobei Chu[1], Frank N. Jones[1], Rahim Armat[2], and Stacy G. Bike[2]

[1]National Science Foundation Industry/University Cooperative Research
Center in Coatings, Eastern Michigan University, Ypsilanti, MI 48197
[2]Department of Chemical Engineering, University of Michigan,
Ann Arbor, MI 48109

It is demonstrated that direct esterification reactions
can be used to crosslink coatings. An acrylic resin with
high levels of carboxyl and hydroxyl groups crosslinks to
hard, solvent resistant films within 30 minutes at
temperatures as low as 140 $^{\circ}$C when catalyzed by titanates
or strong acids. Tetraisopropyl titanate (TPT) is the
most effective catalyst on a weight basis, with p-
toluenesulfonic acid (p-TSA) equal or close behind. When
small amounts (3 to 6 weight per-cent) of methylolated
melamine formaldehyde (MF) resin are added to p-TSA
catalyzed formulations, crosslinking is slightly
accelerated and surface properties are enhanced.
Presumably, such formulations cure partly by MF resin
crosslinking and partly by direct esterification. Acrylic
resins that crosslink by direct esterification are
potentially economical, do not form crosslinking by-
products that are hazardous air pollutants (HAP) or
volatile organic compounds (VOC), and are readily
adaptable to water-reducible formulation. Rheology of
such formulations is predictable. Cure rates of films
cast from aqueous formulations are similar to those cast
from organic solvent.

For decades most thermosetting coatings have been crosslinked by
one or more of four types of chemical reactions: [1]
transetherification reactions of alcoholated amino-formaldehyde
resins, [2] reactions of unblocked or blocked polyisocyanates with
polyols to form urethanes, [3] a variety of ring-opening reactions of
oxirane (epoxy) groups, and [4] autoxidation, for example of alkyd
resins (1). Many potential alternatives to these well-entrenched
crosslinking chemistries have been studied, and some have found
modest commercial use (2). However, these four chemistries remain
the workhorses of the coatings industry.

0097–6156/96/0648–0430$15.25/0
© 1996 American Chemical Society

Formulators have adapted these four chemistries to satisfy today's stringent environmental and workplace safety requirements. However, future regulations will be even more stringent and may constrain the use of one or more of these chemistries. For example, there is concern about method (1), transetherification of amino resins, because traces of formaldehyde along with larger levels of alcohols are formed as crosslinking by-products and emitted to the atmosphere as Volatile Organic Compounds (VOC) unless incinerated. Methylolated melamine formaldehyde (MF) resins, the type of amino resin most commonly used in water-reducible coatings, emit substantial amounts of methanol and smaller amounts of formaldehyde. Beside being VOCs, both substances are listed as a Hazardous Air Pollutants (HAP) in the U.S. Methanol emissions will be limited by 1997 to 10 tons per year per facility, posing a problem for large operations. MF resins made from alcohols that are not HAP listed can be substituted, but they are difficult to formulate in water-reducible coatings. Clearly it behooves the coatings industry to redouble the search for "greener" crosslinking methods.

This paper will describe preliminary experiments suggesting that esterification of resins containing carboxyl and hydroxyl groups (**Equation 1**) may be adaptable as a crosslinking chemistry for coatings. Lacking a better term, we call it "direct esterification."

$$RCOOH \ + \ HOR' \ \xrightarrow{\text{catalyst}} \ RCOOR' \ + \ H_2O \qquad (1)$$

Direct esterification is one of the oldest and most widely studied reactions in organic chemistry, yet computer literature searches back to 1967 revealed no references to its use as a method for crosslinking coatings. It is known that hydroxyl-rich polymers such as starch (3,4) and polyvinyl alcohol (5) can be crosslinked with dicarboxylic acids, although in some cases the first step in the reaction may be formation of a cyclic anhydride of the diacid. It is also well known that titanates, used as catalysts in this study, can function as crosslinkers (6-8).

It should be noted that other types of esterification reactions are used to crosslink coatings. A common example is reaction of resins bearing carboxyl and oxirane groups (**Equation 2**).

$$RCOOH \ + \ CH_2\!\!-\!\!\overset{\displaystyle O}{\overset{\displaystyle /\,\backslash}{}}\!\!CHR'' \ \longrightarrow \ RCOOCH_2\overset{\displaystyle OH}{\overset{|}{C}}HR'' \qquad (2)$$

Crosslinking of coatings by nucleophile catalyzed transesterification was recently described by Craun (9).

Crosslinking by direct esterification offers potential advantages: It is green, and it is relatively cheap. On the other hand, potential disadvantages, such as slow cure rates and moisture sensitive films, can be envisaged. We will show that, with further research and development, it might be possible to make crosslinking by esterification feasible for certain types of coatings, thus realizing the potential advantages.

Experimental Details

Materials. Materials were the best available grades obtained from the following sources:

Methyl methacrylate (MMA)	Aldrich
Butyl acrylate (BA)	Rohm & Haas
Acrylic acid (AA)	Aldrich
2-Hydroxyethyl acrylate (HEA)	Rohm & Haas
Azobisisobutyronitrile (AIBN)	Polysciences
2-Butoxyethanol (2-BE)	Union Carbide
N,N-Dimethylamino ethanol (DMAE)	Aldrich
p-Toluenesulfonic acid monohydrate (p-TSA)	Aldrich
Tetraisopropyl titanate ("Tyzor TPT")	du Pont
Tetra-n-butyl titanate ("Tyzor TBT")	du Pont
Methyl ethyl ketone (MEK)	Aldrich
Diethanolamine (DEA)	Aldrich
Hexakis(methoxy methylol) melamine resin ("Cymel 303," HMMM)	Cytech
Type R-36 CRS panels	Q-Panel
Double distilled water	

All materials were used as received.

Polymer Synthesis. A dual functional acrylic resin (DFAR) was synthesized under monomer starved conditions using the recipe in **Table 1**. The first portion of 2-butoxyethanol was placed in a 500-mL four neck breakaway flask equipped with a heating mantle with a thermostatic temperature controller, a mechanical stirrer, a thermometer, a nitrogen inlet and a dropping funnel. A nitrogen atmosphere was maintained throughout the process. The reaction solution was maintained at 100 $^{\circ}$C with the controller. A solution of all other ingredients except for the second portion of AIBN was added dropwise with continuous stirring during 3 hrs. Stirring was continued at 100 $^{\circ}$C for another hr and then the second portion of the AIBN was added. Stirring was continued for 2 more hrs at 100 $^{\circ}$C, and the transparent, colorless resin solution was cooled to 25 $^{\circ}$C. Characteristics of DFAR are shown in **Table 1**.

Acid-functional and hydroxyl-functional acrylic resins (AFAR and HFAR, respectively) were prepared by a similar procedure using the recipes shown in **Table 2**.

Hydrolyzed Titanium Chelate (HTC) Catalyst. This material was prepared following Example 2 of Deardorff's patent (10). Tetraisopropyl titanate (35.5g, 0.125 mol) was placed in a 100-mL flask equipped with a stirrer, a thermometer, an addition funnel, a reflux condenser, and a nitrogen inlet and a heating mantle with a temperature controller. The flask was purged with

Table 1. Composition and Properties of Dual Functional Acrylic Resin (DFAR)

Components	Charge g	Charge mol	wt %
Methyl methacrylate	61.1	0.616	30.8
Butyl acrtlate	60.0	0.469	30.0
Acrylic acid	30.0	0.417	15.0
2-Hydroxyethyl acrylate	48.4	0.417	24.2
			100.0
AIBN			
first portion	2.0	0.012	
second portion	0.3	0.002	
2-Butoxyethanol			
first portion	44.4	0.377	
second portion	88.9	0.753	

Properties

"Tg" (Fox eq.) (^{0}C)	10
Nonvolatile (wt%)	64
Viscosity (Brookfield)(Pa s)	51
Acid number (mgKOH/g)	
calculated	117
measured	105
Hydroxyl number (mgKOH/g)	
calculated	117
Molecular weight (Mn; GPC)	10300-15100
Mw/Mn (GPC)	3.6-6.3

Table 2.   Compositions and Properties of Acrylic Resins HFAR and AFAR

	Hydroxyl-Functional Acrylic Resin (HFAR)		Acid-Functional Acrylic Resin (AFAR)	
Components	Charge		Charge	
	g	%	g	%
Methyl mathacrylate	120	30	160	40
Butyl acrylate	160	40	120	30
Hydroxethyl methacrylate	120	30		
Acrylic acid			120	30
		100		100
AIBN first portion	5.1		28	
second portion	1.2		4	
Methyl ethyl ketone	266.7			
Methyl isobutyl ketone			266.7	

Properties

"Tg" (^{0}C) (Fox Eq.)	10		38	
Solids (wt.%)	60.28		60.83	
Hydroxl number (mgKOH/g)				
calculated	129			
OH equivalent weight (solids)	435			
Acid number (mgKOH/g)				
calculated			234	
COOH equivalent weight (solids)			240	
Molecular weight (GPC)				
Mn	15000		3800	
Mw/Mn (GPC)	2.2		2.6	

nitrogen and heated to 80 °C. A solution of 13.25 g (0.125 mol) of diethanol amine (DEA) and 2.25 g (0.125 mol) of water was added with vigorous stirring during 1 hour at 80 °C. A gel formed initially and then dissolved when about 40% of the solution had been added. The product was 49 g of colorless, viscous oil (HTC/isopropanol). A 20-g portion of this material was concentrated on a rotary evaporator (to remove by-product isopropanol) to yield 8.2 g of white solid HTC. HTC was used in two forms, as HTC/isopropanol for nonaqueous formulations and as a 3 weight % solution of solid HTC in water for aqueous formulations.

DFAR Coatings Cast from Solution in Organic Solvent. Catalyst solutions were prepared in isopropanol (for TPT, p-TSA and HTC) or in n-butanol (for TBT). The catalyst solutions were added to the above DFAR resin solution in amounts calculated to give the catalyst levels shown in **Table 3**. These solutions were then diluted with isopropanol or n-butanol to a viscosity that would afford the desired film thickness after application. The resulting coating solutions were aged overnight and then were drawn down on Type R-36 panels obtained from the Q-Panel Company using a #60 wire-round bar. R-36 panels are untreated, matte surface, cold-rolled steel panels of thickness 0.8 mm (0.032 in). Target dry film thickness was 20 um; measured dry film thicknesses varied from 15 to 21 μm. Panels were baked for 30 minutes at various temperature settings in a forced air oven. In each experiment temperature sensitive tape was placed on the front of each panel to measure the temperature reached by the panel. The tape was obtained from Paper Thermometer Co., Box 129, Greenfield, NH 03047. These tapes contain a series of sectors which turn dark when a certain temperature is reached. Both the oven setting temperatures and the readings noted on the temperature sensitive tape were noted. A notation of a single temperature for the tape means that the tape section for that temperature had partially darkened. A notation of a range, for example 127-132 °C, means that the 127 °C sector had fully darkened and the 132 °C sector was unchanged. In general the tape readings ran significantly below the oven readings.

The data reported in **Table 3** are from experiments performed in August during a period of relatively high humidity in the laboratory. Duplicate experiments were performed in February, a period of low relative humidity; the February and August results are compared in **Table 4**.

DFAR Coatings Cast from Aqueous Formulations. Water-reduced coatings were prepared by adding 10 g of N,N-dimethylamino ethanol (DMAE) and catalyst solution [see **Table 5** for amounts, given as parts per hundred (phr) on a solids basis] to 112 g of concentrated DFAR resin solution with stirring. The amount of DMAE added was sufficient to neutralize 75% of the carboxyl groups on the resin based on the theoretical acid number. De-ionized water (78 g) was gradually added with stirring. At first the viscosity increased, and as more water was added it decreased to give a transparent, colorless dispersion or solution at 35% NVW. The aqueous coatings were stored overnight and then drawn down and baked as described above. Dry film thicknesses were 18-23 μm.

Table 3.   Properties of DFAR Coatings Cast from Solution in Organic Solvent
(August)

Catalyst	TPT			TBT			P-TSA			HTC	None
Catalyst level(phr)	1.0	1.5	2.0	1.0	1.5	2.0	1.0	1.5	2.0	2	0
121 ^{0}C(tape)											
129 ^{0}C(oven setting)											
MEK rub resistance	28	44	60	6	11	8	30	40	52		6
Pencil hardness	F	F	H	HB	HB	HB	F	F	H		2B
127-132 ^{0}C(tape)											
140 ^{0}C(oven setting)											
MEK rub resistance	61	82	132	19	38	78	50	81	126	84	6
Pencil hardness	H	H	3H	HB	HB	F	H	H	2H	H	2B
138-143 ^{0}C(tape)											
153 ^{0}C(oven setting)											
MEK rub resistance	200	200	200	152	168	172	200	200	198		20
Pencil hardness	3H	3H	4H	F	F	H	2H	3H	3H		B
148-154 ^{0}C(tape)											
167 ^{0}C(oven setting)											
MEK rub resistance	200	200	200	200	200	200	200	200	200	200	50
Pencil hardness	5H	5H	6H	2H	2H	3H	3H	3H	4H	3H	B

Dry film thickness: 18-20 μm.

Table 4. MEK Rub Resistance of DFAR Coatings Cast from Organic Solution in Summer and Winter

Temperature (°C)			
tape	121	127-132	149-154
oven setting	129	140	167
February			
TPT (phr)			
1.0	17	93	200
1.5	38	192	200
2.0	79	200	200
TBT (phr)			
1.0	17	37	200
2.0	29	200	200
p-TSA (phr)			
1.0	24	67	200
1.5	30	113	200
2.0	38	200	200
August			
TPT (phr)			
1.0	28	61	200
1.5	44	82	200
2.0	60	132	200
TBT (phr)			
1.0	6	19	200
2.0	8	78	200
p-TSA (phr)			
1.0	30	50	200
1.5	40	81	200
2.0	52	126	200

Dry film thickness: 18-25 μm.

The data reported in **Table 5** are from experiments performed in August. Results of experiments performed in February and August, are compared in **Table 6**.

AFAR/HFAR Solution Coating Formulation. AFAR and HFAR resins were mixed in a weight ratio of 40:60 to give a solution with a 1:1 equivalent ratio of carboxyl:hydroxyl groups. Coatings were prepared and drawn down essentially as described above; results are reported in **Table 7**.

Aqueous DFAR Formulations with Added Melamine-Formaldehyde Resin. Aqueous coating formulations were prepared as described above from 50 g (32.3 g solids) of DFAR, 4.5 g (extent of neutralization 75%) of DMAE, and 38.3 g of DI water. HMMM resin ("Cymel 303") and p-TSA were then dissolved in the formulations with stirring. The formulations were kept overnight and then drawn down and baked. Film properties were tested described above. Results are shown in **Table 7**.

DFAR Rheology. DFAR was characterized rheologically using a Bohlin VOR rheometer with a concentric cylinder geometry. The geometry consists of an outer rotating cup of radius 15.4 mm and an inner stationary bob of radius 14 mm. The sample temperature was maintained at 25 ± 0.2 °C. A solvent trap was used to provide a partially vapor-saturated space above the sample to greatly reduce the evaporation rates of the 2-BE and water. The rheological measurements consisted of steady shear viscosity measurements over the shear rate range of 0.185 to 185 sec^{-1}. Since Newtonian flow behavior was observed for all compositions at low shear rates, the viscosity at 0.185 sec^{-1} was chosen as the viscosity shown dilution curves. To obtain dilution curves of DFAR, the resins were partially neutralized with dimethylamino ethanol (DMAE). They were then progressively diluted with water, and the viscosity was measured after each portion of water was added. The mixing was effected by an air driven stirrer. Double distilled water was used to minimize effects of ionic contaminants. Two levels of amine neutralization were tested, one in which the amount of amine was sufficient to neutralize 75% of the carboxylic acid groups in DFAR (extent of neutralization, EN = 75%) and the other at EN = 24%. The results are shown in **Figures 1** and **2**.

For comparison, the dilution curve of a water reducible acrylic resin, designated WRAC, having conventional levels of carboxyl and hydroxyl groups is shown in **Figure 1**. This resin is composed of MMA, BA, AA, and HEA in a 45/36/8/11 mol ratio. It was synthesized at 62 weight % solids in 2-butoxyethanol by a procedure similar to the procedure described for DFAR, as detailed in a forthcoming publication (11).

Panel Heat-up Rate Experiment. Temperature sensitive tapes were placed on uncoated Type R-36 panels and the panels were baked in a forced air oven set at 140 °C for varying times. The tape reading rose to 132-138 °C within 10 min and remained at that level for 24 hr. In repeat experiments small deviations from these results were noted. They were attributed to variations in oven air flow at different locations in the oven.

Table 5. Properties of DFAR Coating Cast from Aqueous Formations (August)

Catalyst and Catalyst Level (phr)

	TPT 1.0	TPT 1.5	TPT 2.0	TBT 1.0	TBT 1.5	TBT 2.0	P-TSA 1.0	P-TSA 1.5	P-TSA 2.0	HTC 1.0	HTC 1.5	HTC 2.0	NONE
121° C(tape)													
129° C(oven setting)													
MEK rub resistance	24	46	71	8	10	14	12	33	57	11	13	20	13
Pencil hardness	HB	HB	HB	B		B	HB	B	HB	HB	B	HB	2B
Impact resistance (in-lb)													
D	100	100	90	160	160	160	160	100	100	160	160	160	160
R	40	40	40	160	160	160	160	40	40	160	160	160	160
127-132° C(tape)													
140° C(oven setting)													
MEK rub resistance	81	102	111	61	86	94	35	66	87	24	27	35	8
Pencil hardness	H	H	H	HB	HB	F	F	F	F	HB	HB	HB	2B
Impact resistance (in-lb)													
D	100	80	80	100	100	80	100	100	80	160	160	160	160
R	40	20	20	40	40	20	40	40	10	160	160	120	160
138-143° C(tape)													
153° C(oven setting)													
MEK rub resistance	200	200	200	96	132	143	198	200	200	62	78	93	29
Pencil hardness	2H	3H	3H	H	H	2H	H	2H	2H	F	H	H	B
Impact resistance (in-lb)													
D	80	80	80	100	100	100	100	100	100	100	100	100	160
R	10	10	10	10	10	10	10	10	10	10	10	10	160
149-154° C(tape)													
167° C(oven setting)													
MEK rub resistance	200	200	200	200	200	200	200	200	200	200	200	200	60
Pencil hardness	2H	4H	4H	2H	3H	3H	3H	2H	3H	2H	2H	4H	HB
Impact resistance (in-lb)													
D	80	60	60	80	80	80	80	80	80	80	80	80	160
R	10	10	10	10	10	10	10	10	10	10	10	10	160

Dry film thickness: 18-23 μm.

Table 6. MEK Rub Resistance of DFAR Coating Cast from Aqueous Formulations in Summer and Winter

Month	February		August	
Catalyst (1 phr)	TPT	p-TSA	TPT	p-TSA
127-132° C (tape) [140° C (oven setting)]	66	31	81	35
149-154° C (tape) [167° C (oven setting)]	200	200	200	200

Dry film thickness: 18-23 μm.

Table 7. Properties of Two-Component (AFAR and HFAR) Acrylic Coatings Cast from Solution in Organic Solvent

Catalyst Catalyst Level (phr)	TPT		TBT		p-TSA	
	1.0	2.0	1.0	2.0	1.0	2.0
127-132° C (tape) 140° (oven setting)						
MEK rub resistance	18	35	9	17	18	20
Pencil hardness	B	B	2B	B	2B	2B
149-154° C (tape) 167° (oven setting)						
MEK rub resistance	99	121	28	61	47	81
Pencil hardness	HB	H	F	B	B	B
171-177° C(tape) 190° C (tape)						
MEK rub resistance	200	200	122	200	184	200
Pencil hardness	3H	4H	2H	2H	H	2H

Dry film thickness: 25-27μm.

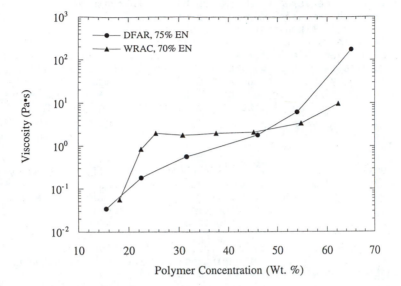

Figure 1. Water Dilution Curves of DFAR at EN 75% and WRAC at 70% EN.

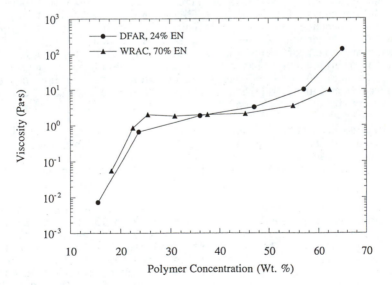

Figure 2. Water Dilution Curves of WRAC at 70% EN and of DFAR at 24% EN.

Coating Testing. Film thicknesses were measured using a
DeFelsko "MIKROTEST" magnetic film thickness gauge. Solvent
resistance was measured by double rubbing with nonwoven paper
saturated with MEK using firm hand pressure. The failure criterion
was when the first penetration through the coating to the panel was
noted. Tackiness of the coating surface was noted during the test,
and some irreversible disturbance of the surface occurred with all
coatings except the ones with 6% HMMM resin added. Pencil hardness
was measured by the method of ASTM D 3363-92a. Impact resistance
was measured using a Gardner Impact Tester having a maximum impact
of 1.84 m-kg (160 in-lb) after 1-3 days aging of the panels.

Results

Dual functional acrylic resin (DFAR) was easily prepared by a
conventional polymerization method using "monomer starved"
conditions to promote polymer uniformity. It was made with high
hydroxyl and acid numbers to increase its functionality (see
Discussion Section). Its only abnormality was its unusually broad
molecular weight distribution, M_w/M_n, which ranged from 3.6 to 6.3.
This broad distribution suggests that esterification reactions may
have occurred to some extent during polymerization or that the HEA
was contaminated with ethylene glycol diacrylate.

The rheological response of DFAR, with its abnormally high
levels of carboxyl and hydroxyl functionality, was compared to WRAC,
a conventional water-reducible acrylic resin. As shown in **Figure 1**,
the dilution curve of DFAR at 75 % EN (extent of neutralization)
lacks the pronounced viscosity plateau in the dilution curve of WRAC
at 70 % EN. As shown in **Figure 2**, when DFAR is neutralized at 24 %
EN using an amount of amine that would be equivalent to 75 % EN for
WRAC, the dilution curve has somewhat more inflection, but not as
much as WRAC at 70 % EN. These dilution curves are similar in shape
to curves reported by Hill and Richards (12) for a family of acrylic
resins containing 10 to 50 mol % of acrylic acids.

Unpigmented coatings were formulated with various catalysts and
films were cast on steel panels and baked. Transparent, glossy
films were formed except for films catalyzed by HTC, which had
uneven, granular surfaces and a tendency to discolor when used on
steel.

Two experimental problems were encountered in evaluation of the
coatings' cure response. The first was the 9 to 15 °C discrepancy
between the oven settings and the temperature tape readings noted in
Tables 3-8. Heat-up rate experiments indicate that tape readings
are consistently lower than oven settings and that variation of the
temperatures actually experienced by panels varies somewhat in this
oven. The tape temperature readings in **Tables 3-8** are judged to be
more reliable than the oven setting readings.

A second problem was poor reproducibility of impact resistance

Table 8. Properties of DFAR Water Reducible Coating Containing HMMM Auxillary Crosslinker

HMMM (phr)	3		6	
P-TSA (phr)	1	2	1	2
121⁰ C (tape)				
129⁰ C (oven setting)				
MEK rub resistance	130	151	177	200
Pencil hardness	H	H	H	3H
Impact resistance				
(in-1b)D	20	20	20	20
R	10	10	10	10
127-132⁰ (C (tape)				
140⁰ C (oven setting)				
MEK rub resistance	200	200	200	200
Pencil hardness	3H	3H	3H	3H
Impact resistance				
(in-1b)D	30	30	60	60
R	10	10	30	40
138-143⁰ C (tape)				
153⁰ C (oven setting)				
MEK rub resistance	200	200	200	200
Pencil hardness	4H	4H	4H	5H
Impact resistance				
(in-1b)D	50	50	60	70
R	10	10	10	30

Dry film thickness: 20-25 μm.

results. While impact resistance tests were performed on all
panels, the results are not all detailed here. In general, films
with lowest MEK resistance had highest impact resistance, but direct
impact resistance fell to the 40 - 80 in-lb range for films that had
200 MEK rubs and reverse impact resistance fell even lower.

Despite these problems, it is thought that the crosslinking
potential of DFAR in solvent-borne formulations is clearly indicated
by the results shown in **Table 3**, and that the results in **Table 5**
represent the cure response of the water-reduced formulations.
Solvent rub resistance is probably the best indicator of relative
extent of cure of the three tests performed in this study. Solvent
rub resistance increases from 6 - 15 double rubs for the least cured
formulations to > 200 double rubs for the most cured. Hardness
tests follow a parallel course, rising from 2B for the least cured
coatings to 2H-3H when 200 MEK rubs are attained and further to 5H
at higher baking temperatures.

The results in **Tables 3** and **5** show that DFAR coatings can
crosslink sufficiently to produce hard, solvent resistant films when
baked for 30 minutes at 138-143 $^{\circ}$C (oven setting 153 $^{\circ}$C) with 1,
1.5, or 2 parts per hundred by weight (phr) of TPT and p-TSA
catalysts. Films catalyzed by TBT and HTC required higher
temperatures to reach 200 MEK rubs. There was little difference
between the cure responses of coatings cast from nonaqueous and
aqueous formulations -- in view of the poor reproducibility of the
solvent rub and hardness tests the small differences noted may be
within experimental error. The apparent order of effectiveness of
the catalysts on a weight basis is TPT > p-TSA > TBT > HTC >> no
catalyst. The differences between TPT and p-TSA are modest and may
not be statistically significant.

The data in **Tables 3** and **5** are from tests performed in August,
a period of relatively high relative humidity in the laboratory.
Results are compared with results obtained in February, a period of
low relative humidity, in **Tables 4** and **6**. Films cast from
nonaqueous formulations appeared to have a slightly better cure
response in February than in August (**Table 4**). There was little
difference when films were cast from aqueous formulations (**Table 6**).
These results suggest that cure may be slightly inhibited by
moisture in humid air but is not further inhibited by moisture in
aqueous formulations. Again, there is a question of whether the
observed differences are larger than experimental error.

A blend of a carboxylic acid functional acrylic resin (AFAR)
with a hydroxyl functional acrylic resin (HFAR) had a relatively
poor cure response, as shown in **Table 7**. The blends cured, but they
required temperatures of about 175 $^{\circ}$C to reach 200 MEK rubs <u>vs</u>.
about 140 $^{\circ}$C for the resin which incorporates both types of
functional groups (DFAR). The large difference in cure response may
be explained by the ability of DFAR to undergo inter- and
intramolecular crosslinking, while the HFAR/AFAR blend is limited to
intermolecular crosslinking.

It should be noted that the surfaces of DFAR coatings were

irreversibly disturbed by the solvent rub test. The surface felt "tacky" during the test, and afterwards the gloss was reduced. In this respect they are inferior to coatings crosslinked with MF resins, which often pass 200 MEK rubs without surface marring.

Addition of a fully methylolated MF (HMMM) resin to the aqueous DFAR formulations was investigated with the results shown in **Table 8**. p-TSA catalyst was used. Comparison of **Tables 8** and **5** shows that addition of 3 weight per-cent of HMMM reduces the bake temperature required to reach 200 MEK rub resistance by about 10 $^{\circ}$C. Addition of 6 weight per-cent of HMMM further improves the cure response. Moreover, addition of 6% HMMM eliminated the problem of surface marring during the solvent rub test.

Discussion

Use of direct esterification for crosslinking of coatings seems obvious. Why have there been so few published reports of it in the recent literature? One reason might be that resins designed for this type of crosslinking are unsuitable for high solids coatings because they need high functionality (carboxyl and hydroxyl groups per molecule) and, therefore, high molecular weights and high viscosities.

However, resins designed for use in water-reducible coatings can have, within limits, high viscosities. As shown in **Figures 1** and **2**, the water-dilution behavior of DFAR is different than that of conventional water-reducible resins like WRAC. These differences can be attributed to different extents of dipole-dipole interactions and of aggregation between the two resin systems. DFAR is more soluble in water due to the greater concentration of hydrophilic groups on the chain, including salted carboxyl groups, free carboxyl groups, and hydroxyl groups. The higher concentration of acid groups in DFAR and the resulting increase in dipole interactions between ionic acid-amine salt groups as compared to WRAC leads to greater chain expansion and interchain interactions. These effects give rise to the initially high viscosity of DFAR in the absence of water (**Figure 1**). In addition, these salt groups impart a high degree of water solubility to DFAR which accounts for the solution-like dilution behavior of DFAR as compared to WRAC. The plateau in the dilution curve of WRAC, due to electrostatic expansion of the polymer chains (<u>11</u>), is not observed with DFAR. In addition, the expansion of these amphiphilic chains in an increasingly water-rich medium leads to chain aggregation followed by a sharp decrease in viscosity. Such a sharp decrease in viscosity is not seen with DFAR at 75% EN; the inflection in viscosity is probably due to a moderate amount of electrostatic chain expansion. The degree of aggregation of polymer chains can be adjusted by varying the extent of neutralization. Indeed, when DFAR is neutralized to only 24%, its dilution behavior approaches that of WRAC at 70% EN (**Figure 2**), suggesting some aggregation of DFAR at higher water contents.

While the water-dilution curves of DFAR are different than those of conventional water-reducible resins, these appear to be no

major obstacles to formulating and applying DFAR resins in aqueous
systems. If rheological characteristics pose no major problems, the
perceived obstacles must be in the crosslinking chemistry. Perhaps
potential investigators assumed that direct esterification reactions
are too slow to be useful. Coatings chemists are familiar with
polyesterification processes to make polyester and alkyd resins,
which generally require temperatures of 220-260 °C and times of 4-10
hours to complete. Coatings, on the other hand, are typically baked
at temperatures below 160 °C in 10 to 30 minutes.

       However, polymerization processes have different kinetic
requirements than crosslinking processes. In polyesterifications
the reactions must be driven to high conversions, generally above
98%. In the case of polyesterification this is difficult because
esterification reactions are reversible and are often of a high
kinetic order, usually somewhere between second order and third
order (8). A consequence of high kinetic order is that
polyesterification reactions that are relatively facile at low
conversions require forcing conditions to drive the reactions to
high conversion. In contrast, crosslinkable polymers can be, and
often are, designed so that adequate crosslinking can be attained at
less than high conversion; if enough functionality is built into the
resins, conversions below 50% may give optimum crosslink density.

       Another difference is that physical removal of water is
difficult in large scale polyesterifications. Sometimes it is the
rate limiting step. But water can easily escape from thin coating
films, probably orders of magnitude faster than it can escape from a
mass of polymer in a large reaction kettle. An indication of the
importance of water removal rate can be found in a recent report by
Jong and Saam (13), where it was shown that polyesterification with
a strong acid catalyst can be affected at reasonable rates at 50 to
80 °C by manipulating a two-phase reaction system so that water is
quickly removed from the polyester.

       The catalyst is critical. Without it a much higher temperature
is required for a given level of cure (See **Table** 3.). In addition
to strong acid (p-TSA) catalyst, titanate catalysts were studied
because the literature (8) indicates that they are more effective as
esterification catalysts than catalysts based on Sn, Bi, Zn, Pb,
etc. TPT, TBT, and p-TSA catalysts are commercially available.
The fourth, HTC, is easily made (10). It is said that its structure
can be represented as:

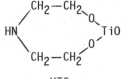

                                    HTC

HTC was studied because it is reported (10) to have exceptional
catalytic activity for esterification and to form stable solutions

in water. However, as described above, it did not perform as well as the other catalysts in aqueous formulations, perhaps because of phase separation. Proprietary commercial titanium chelate catalysts are available but were not investigated. They are reported (6) to be more stable in water than TPT and TBT.

It is unlikely that tetraisopropyl and tetra-n-butyl titanates (TPT and TBT) are the actual catalytic species. As soon as they are added to the coating formulation they probably begin to undergo exchange reactions with the hydroxyl and carboxyl groups of the resin; a complex equilibrium is presumably established. Subsequent addition of water likely perturbs this equilibrium, as an assortment of hydrolysis reactions is possible (7,8). One possible reaction is partial hydrolysis to produce titanoxanes -- species having Ti-O-Ti partial structures. In view of the near certainty that exchange and hydrolysis reactions occur, it is surprising that addition of water has negligible affect on the catalyst activity (Compare **Tables** 3 and **5.**). Hydrolysis could lead to deactivation of the catalyst by precipitation of TiO_2. No precipitation from aqueous TPT and TBT formulations was observed, although long-term stability tests were not performed. With further study it may be possible to find catalysts that are more effective than TPT without the shortcomings of the hydrolyzed titanium chelate catalyst (HTC).

The dual functional acrylic resin (DFAR) was deliberately designed with high functionality, with 21.7 mol per-cent each of AA and HEMA. A typical water reducible resin has 10 mol percent of HEMA and 7.8 mol per-cent of AA.(1) The theoretical hydroxyl number of DFAR is 117 mg-KOH/g-resin, roughly twice the levels commonly used for water reducible resins designed to be crosslinked with melamine formaldehyde (MF) resins. Its theoretical acid number, also 117 mg-KOH/g-resin, is about three times as high that for as such resins. Based on theoretical hydroxyl and acid numbers and on the M_n measured by HPLC its number average functionality is about 30 hydroxyl groups and 30 carboxyl groups per molecule; thus DFAR is capable of forming a highly crosslinked network. If all carboxyl and hydroxyl groups were esterified (probably a physical impossibility because of spatial constraints as high conversion is approached), about 40 per-cent of the monomer units would be connected by crosslinks. Acrylic resins for crosslinking by MF resins with comparable M_n are typically designed so that 10 to 15 per-cent of the monomer units can be connected (1). While quantitative comparisons are problematic because of the different chemistries involved, it seems evident that DFAR can attain a crosslink density comparable to that of contemporary acrylic thermoset coatings if only a moderate fraction (perhaps half or even less) of its carboxyl and hydroxyl groups are esterified.

An effort to measure the extent of crosslinking (i.e the conversion of the esterification reaction) by FT-IR did not yield satisfying results. The hydroxyl and carboxyl groups in the resin give rise to a broad, nondescript peak in the 3100 to 3500 cm^{-1} region; it changes during crosslinking, but no quantitative data could be derived. The FT-IR spectral changes and the observed catalyst effects indicate that crosslinking of DFAR occurs primarily by direct esterification, but the possibility that transesterification reactions contribute to its crosslinking cannot be ruled out.

As reviewed by Fredet and Marechal (8), different authors have drawn varying conclusions about the kinetics and mechanisms of catalyzed direct esterification reactions. It is generally agreed that kinetic order is high; thus the reaction will slow sharply as conversion (extent of reaction) increases. Further, as conversion approaches high levels a second factor will also slow the reaction: restricted mobility within the crosslinked network. In view of these considerations and of the large number of reactive groups in DFAR it seems probable that conversion is well below 100% in all films made in this study. If so there is a substantial concentration of unreacted carboxyl and hydroxyl groups in the films even after high solvent resistance and hardness have been reached.

There are potential disadvantages to designing resins such as DFAR with excess functionality and curing them at fractional conversion. One potential problem is that the coating may have a narrow cure window and be quite sensitive to overbake and underbake. ["Cure window" is a term coined by Bauer to describe the range of bake temperatures at a given time or bake times at a given temperature what will yield satisfactory properties of a given coating (14).] There is even the possibility that crosslinking might continue at a slow rate during service, leading to eventual embrittlement; for this reason coatings based on the DFAR concept will be best suited to applications where they are baked at temperatures high above their maximum service temperatures. Present data show that cure window is a problem but do not enable us to estimate its severity. It is encouraging to note that many coatings, for example most coatings formulated with alcoholated melamine-formaldehyde resins, are deliberately formulated to achieve optimum properties at less than full conversion, and the industry has learned to deal with their rather narrow cure windows and the resulting sensitivity to overbake.

A second potential disadvantage is that unreacted hydroxyl and carboxyl groups in the films may cause the cured films to be sensitive to moisture, bases, and detergents. The degree of this problem will depend on the concentration of unconverted functional groups in the film and on the physical characteristics of the film. High crosslink density and high T_g can be expected to minimize the problems. The severity of this problem is best evaluated in the context of the requirements for a specific application.

A further concern is that ester crosslinks might be too vulnerable to hydrolysis for extended service outdoors or in humid environments. However, ester groups often have sufficient hydrolysis resistance for demanding applications. Powder coatings crosslinked by esterification of oxirane functional acrylic resins with carboxylic acids have sufficient resistance to hydrolysis to be used as automotive topcoats in Japan. Liquid coatings crosslinked by this method are reported to have good weatherability and resistance to acid etching (15). Ester groups present in many weatherable polyester coatings have sufficient hydrolytic stability when highly crosslinked.

The experiments in which HMMM resin was added to aqueous DFAR formulations point toward a potentially useful formulating method: A coating could be formulated to cure partly by a conventional crosslinking reaction and partly by direct esterification. This approach could be used to reduce VOC and HAP emissions while partly retaining the benefits of conventional crosslinkers.

Results of this study raise a question about current baking coatings that are presumed to be crosslinked by other mechanisms: Might esterification play a role in the crosslinking process? Examples of coatings where this might occur are water-reducible acrylic coatings crosslinked with alcoholated melamine-formaldehyde resins and epoxy coatings crosslinked with anhydrides or carboxylic acids. In both cases the catalysts used are also esterification catalysts. Thus it seems possible that, in addition to the recognized crosslinking mechanisms, esterification might play a role, probably secondary but perhaps enough to influence film properties. Recognition of this possibility might be useful to designers of conventional coatings.

Summary and Conclusions

It is demonstrated that an acrylic resin with high levels of carboxyl and hydroxyl groups ("DFAR") can crosslink by direct esterification within 30 minutes at temperatures as low as 140 $^{\circ}$C when catalyzed by certain titanates or strong acids. While only a fraction of the functional groups react, crosslinking is sufficient to give hard, solvent resistant films. This relatively unexplored approach seems especially well suited to water-reducible coatings, where resins can have higher molecular weights, and therefore higher functionality than resins for high solids coatings. Rheology of DFAR indicates more solution-like behavior and much less aggregation than with conventional water-reducible resins.

This approach, i. e. the use of direct esterification to crosslink coatings, has the potential advantage of elimination of the toxic hazard associated with some crosslinkers and elimination of crosslinking by-products which are VOCs or are listed as Hazardous Air Pollutants (HAPs). It is also potentially economical. Thus it merits exploration by the coatings industry in view of impending VOC and HAP regulations.

Many questions remain to be answered before the commercial feasibility of this approach for a given application can be assessed. Potential problems include the possibility of narrow cure windows and moisture sensitive films. The purpose of this paper is not to solve these problems but simply to call to the attention of the coatings industry the possibility of using direct esterification reactions to crosslink coatings.

Acknowledgements

    This project was supported by the National Science Foundation
Industry/University Cooperative Research Center in Coatings at
Eastern Michigan University, North Dakota State University and
Michigan Molecular Institute. Helpful discussions with John Saam and
Kenneth G. Hahn, Jr. are appreciated.

References

(1) Wicks, Jr., Z.W., Jones, F.N., Pappas, S.P. Organic Coatings:
Science and Technology; Vol. I; John Wiley and Sons, New York, 1992;
pp 83-211.
(2) Ibid. pp 212-228.
(3) Yang, C.Q. J. Appl. Polym. Sci. 1993, 50, 47-53.
(4) Yang, C.Q. J. Appl. Polym. Sci., Part A: Polym. Chem. 1993, 31,
1187-93.
(5) Chen, D.; Yu, X. Gongneng Gaofenzi Xuebao; 1991, 4, 231-9.
Chem. Abstr. 117:28193.
(6) Anon. TYZOR^R Organic Titanates; du Pont Specialty Chemicals
Bulletin H-51765 (1993), and references therein.
(7) Rondestvedt, Jr. C.S., Kirk-Othmer Encyclopedia of Chemical
Technology, 3rd ed., Vol. 23, 176-245, John Wiley & Sons, 1983.
(8) Fradet, A.; Marechal, E. Adv. Polym. Sci., 1982, 43, 51-142.
(9) Craun, G.P. J. Coat. Technol. 1995, 67(841), 23-30.
(10) Deardorff, D.L. U.S. Patent 4 788 172, 1988.
(11) Armat, R., Bike, S.G., Chu, G., and Jones, F.N. J. Appl. Polym.
Sci., in press.
(12) Hill, L.W.; Richards, B.M. J. Coat. Technol., 1979, 51(654),
59-67.
(13) Jong, L.; Saam, J.C. Polym. Prepr., 1995, 36(1), 751-752.
(14) Bauer, D.R.; Dickie, R.A. J. Coat. Technol., 1982, 54(685), 57.
(15) Gregorovich, B.V.; Hazan, L. Proc. XIXth Intl. Conf. Org. Coat.
Sci. Tech., Athens, Greece, 1993, 205-218.

Chapter 27

Film Formation in Hexafluoropropylene Plasmas

M. S. Silverstein and R. Chen

Department of Materials Engineering, Technion-Israel Institute of Technology, Haifa 32 000, Israel

The gas phase dominated plasma polymerization (PP) of hexafluoropropylene (HFP) produces 0.2 to 0.5 µm particles and 1 µm spherical particle agglomerates which are all incorporated into a transparent and highly adhering PPHFP film. An energy ratio plasma parameter, E, related to the plasma energy and strongly dependent on flow rate (or residence time) was derived. The maximum and plateau in deposition were successfully described when polymerization and etching were exponentially related to E. The chemical structure of the amorphous, crosslinked PPHFP consists largely of similar amounts of C^*-CF, CF, CF_2 and CF_3 units. Plasma etching was inhibited by fluorine scavenging and HF formation on the addition of hydrogen to the feed. The deposition and coalescence of smaller particles into a smoother and more uniform surface on adding nitrogen reflects the incorporation of nitrogen into the polymer, the increase in the polar component of surface energy and, perhaps, the decrease in dominance of gas phase versus surface polymerization. The low $PP(N_2/HFP)$ electrical breakdown strength may be attributed to an alternate conduction mechanism in the relatively polar plasma fluoropolymer.

Thin fluoropolymer films can have many advantages including a low coefficient of friction, a low surface energy, thermal stability, biocompatibility, and chemical resistance. Unfortunately, the techniques used to deposit thin films of conventional fluoropolymers can involve etchants, solvents and temperatures that limit their applicability. Plasma polymerization is an ambient temperature, solvent-free process that can be used to deposit highly adhering, pinhole-free and crosslinked thin fluoropolymer films from a variety of compounds including those not polymerizable by standard techniques (1-3). These films would have potential applications in the microelectronics, biomedical, membrane, aerospace and automotive fields (4-9).

0097–6156/96/0648–0451$15.00/0

In plasma polymerization the neutral species entering the reactor are fragmented into reactive species as energy is transferred by the electrons in the low pressure plasma environment. The substrate molecules are activated by exposure to the plasma environment and chemical bonds are formed with the polymerizing reactive monomer fragments. The structure and properties of plasma polymer (PP) thin films depend on the plasma environment, specifically the pressure (P), monomer mass flow rate (F_m), power (W) and feed composition, all of which interact in a complex manner. The plasma polymerization of fluorocarbons is even more complex than that of hydrocarbons since the rate of the competing plasma etching reaction is more significant. Hexafluoropropylene (HFP) ($CF_2=CF-CF_3$) is an unsaturated fluorocarbon which tends to polymerize in a plasma rather than etch (*10-12*). The objective of this research is to develop insight into the formation of PPHFP films synthesized under various plasma environments (W, P, F_m and feed composition).

Experimental

Plasma Polymerization. The central part of the plasma reactor system illustrated schematically in Figure 1 is a parallel-plate electrode radio frequency (13.56 MHz) plasma reactor (March Instruments Jupiter III) with an automatic impedance matching network. The anodized aluminum parallel-plate electrodes were 3 cm apart. To increase anisotropy the bottom electrode (21 cm diameter) is smaller, producing a negative DC bias, and is encircled by a ceramic ring, concentrating the plasma. The reactor could be evacuated to 2.5 Pa with a vacuum pump (Alcatel AC-2012) and the temperature of the anodized aluminum parallel-plate electrodes was maintained at 20°C with a circulating liquid cooler (Neslab RTE-100). One inlet rotameter was calibrated for HFP (Matheson) while the other inlet rotameter was calibrated for argon, nitrogen and hydrogen. Glass optical microscopy slides were used as substrates and were centered on the bottom electrode. A detailed description of the plasma polymerization methodology is found elsewhere (*10-12*).

The molar flow rate (F_n) is commonly expressed in units of ml/min at STP (sccm). The basic plasma conditions were 100 W, 187 Pa and 18.3 sccm HFP. Only one plasma parameter, the power, the HFP flow rate or the amount of additional gas, was varied in each set of experiments. The PP films based on HFP and an additional gas will be referred to as PP(gas/HFP). The molar flow rate ratio (F_nR), the ratio of the molar flow rates of gas added to HFP, is used to describe the PP(gas/HFP) feed composition.

The rate of deposition (R_d) was calculated both on a mass and thickness basis. The thickness was measured in several different places using a stylus apparatus with an accuracy of 0.05 μm (Tencor α-step-100) yielding an average deposition rate in terms of thickness (μm/min). The mass gained was measured with an accuracy of 0.1 mg (Mettler AE-163 microbalance) yielding the deposition rate in terms of mass per area (g/(m²min)).

Chemical Structure and Topography. The chemical structure of PPHFP was measured using two complementary methods. Electron spectroscopy for chemical analysis (ESCA) was performed on the PP coated slides using an Al K_α X-ray source, a 40° angle of incidence and an energy of 25 eV for high resolution (Perkin Elmer Physical Electronics 555 ESCA/Auger). The atomic concentrations were determined with an accuracy of 1% in a low resolution scan and the nature of the bonds with carbon in a high resolution C_{1s} scan. The C_{1s} spectra, described as the sum of several mixed Gaussian/Lorentzian peaks, were deconvoluted in order to more accurately describe the contributions of the different bonds to the overall spectrum (*12*). Prior to deconvolution the background was subtracted from the spectrum using one iteration of the Shirley method (*13*).

Transmission Fourier transform infrared spectroscopy (FTIR) (Unicam Mattson 1000 FTIR) provided chemical information that reflected the bulk PPHFP more accurately than the surface specific ESCA. In general, the PPHFP films for FTIR characterization were separated from the substrates by soaking in acetone overnight and drying at room temperature for several days. The similarity of the resulting spectra to those from PPHFP deposited on KBr pellets demonstrated that artifacts were not introduced using this methodology.

Properties. The influence of the feed flow rate ratio on the dispersive and polar components of surface energy (γ^d and γ^p, respectively) and their sum, the total surface energy (γ), was examined. A contact angle (θ) apparatus with an accuracy of 1° (Kernco) and an advancing droplet technique were used (*14*). The relationship between the components of the surface energy and contact angle is expressed by:

$$(1+\cos\theta) = \frac{2}{\gamma_{lv}}\left[\left(\gamma_s^d \bullet \gamma_{lv}^d\right)^{0.5} + \left(\gamma_s^p \bullet \gamma_{lv}^p\right)^{0.5}\right]$$

(1)

where *s* represents solid and *lv* represents liquid in the presence of its vapor. Equation 1 has two unknowns, γ_s^d and γ_s^p, and can be solved by measuring the contact angles of two different liquids (*11*). Bromonaphthalene ($\gamma_{lv}^d = 44.6$ mN/m, $\gamma_{lv}^p = 0$ mN/m) and distilled water ($\gamma_{lv}^d = 22.0$ mN/m, $\gamma_{lv}^p = 50.2$ mN/m) were used (*14*).

The breakdown strengths of the PP films were determined using the capacitor circuit illustrated schematically in Figure 2. The capacitor consists of a PP film on a metallic substrate which serves as one electrode. The metallic substrate was a thin copper sheet that had been polished and then washed with hydrochloric acid to remove the oxide layer. The other electrode was formed by evaporating 0.1 μm thick, 2 mm diameter gold dots on the PP film. Over forty such capacitors were tested for each film using the 'short time test', ASTM-D149. The voltage was ramped from zero at 3.6 V/s (Keithley-487 picoammeter voltage source) and the breakdown voltage was taken as the voltage at which the current jumped by two orders of magnitude (usually reaching 0.1 mA). The breakdown strength is the breakdown voltage divided by the film thickness. As breakdown strength tends to increase with decreasing film thickness the plasma polymerization time was adjusted to yield films of a similar thickness (approximately 2 μm).

The breakdown strength results were interpreted using Weibull statistics. The cumulative failure ($F(x)$) at a given breakdown strength (x) is:

$$F(x) = 1 - \exp\left[-\left(\frac{x}{\sigma}\right)^b\right]$$

(2)

where σ is the mean breakdown strength and *b* is the Weibull shape parameter associated with the distribution breadth.

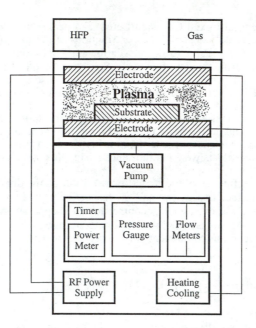

Figure 1. Schematic illustration of the parallel-plate plasma reactor.

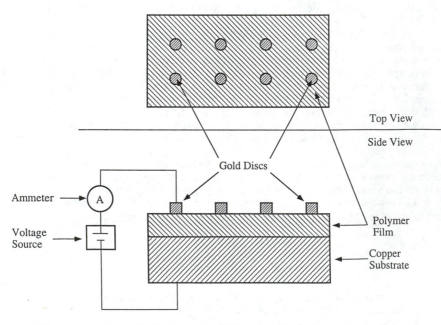

Figure 2. Schematic illustration of the circuit for breakdown strength measurements.

Results and Discussion

Deposition. The basic PPHFP synthesis conditions were 100 W, 187 Pa and 18.3 sccm HFP. Plasma exposure time has little effect on deposition rate, as seen from the linear variation of thickness and mass per unit area with deposition time in Figure 3. The deposition rates calculated from the slopes in Figure 3 are 0.35 μm/min or 0.77 g/(m²min). The ratio of the two deposition rates yields a PPHFP density of 2.2 g/cm³. In spite of the differences in structure, the density of the crosslinked amorphous PPHFP is similar to the densities of commercial semi-crystalline fluoropolymers such as polytetrafluoroethylene (PTFE) and the copolymer of TFE and HFP, fluorinated ethylene-propylene copolymer (FEP).

The variation of the deposition rate with W/F_m (a parameter commonly used to describe the energy to mass ratio in plasma polymerization) is seen in Figure 4. In polymerizations at a constant flow rate the deposition rate increases rapidly at low powers, reaches a maximum and then decreases to a plateau at high powers. The transition from an energy-starved plasma to a monomer-starved plasma occurs near this maximum (*10*). In polymerizations at a constant power the deposition rate increases with increasing flow rate (decreasing W/F_m) reflecting the effect of increasing monomer throughput to a monomer-starved polymerization. The strong dependence of the deposition rate on flow rate, even at a constant W/F_m, is seen in Figure 4.

One convenient expression that can be used to take the strong dependence on flow rate into account is the kinetic energy associated with flow. The kinetic energy per unit mass of the flowing monomer (E_k, J/kg) (equation 3), calculated from the reactor cross-sectional area (A) and the volumetric flow rate (F_v), is inversely proportional to the residence time squared (*12*). The ratio of W/F_m to E_k (equation 4) yields a dimensionless plasma parameter (E) that is proportional to W/F_n^3 (based on the flow rate relationships in equation 5 where M_w is the monomer molecular weight, T is the temperature and R is the gas constant). This energy ratio is more strongly dependent on flow rate than simply W/F_m and thus may provide a better means of normalizing the influences of power and flow rate (*12*).

$$E_k = \frac{1}{2}\left(\frac{F_v}{A}\right)^2 \tag{3}$$

$$E = \frac{W}{F_m}\frac{1}{E_k} = \frac{2W}{M_w(F_n)^3}\left(\frac{PA}{RT}\right)^2 \tag{4}$$

$$F_n = \frac{F_m}{M_w} = F_v\frac{P}{RT} \tag{5}$$

The deposition rate is simply the difference between the rates of polymerization (R_p) and etching (R_e). The dependence of the deposition rate on the plasma environment reflects the effects of the environment on both the polymerization and etching reactions. The ratios of the rates of polymerization, etching and deposition to the monomer flow rate can yield a measure of the polymerization, etching and deposition efficiencies (C_p, C_e and C_d, respectively). Equation 6 expresses an exponential relationship that can be used to describe the

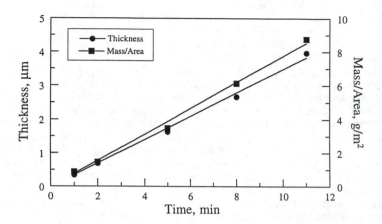

Figure 3. Variation of PPHFP film thickness and mass per unit area with deposition time.

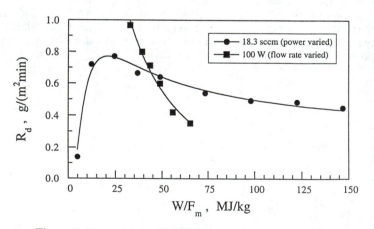

Figure 4. Dependence of PPHFP deposition rate on W/F_m.

dependence of the polymerization and etching efficiencies (where $i=p$ for polymerization and $i=e$ for etching) on the energy ratio plasma parameter (E, from equation 4) (12). The relationship in equation 6 is characterized by a plateau at high E ($C_{i\infty}$) and an activation energy ratio (E_{ai}). The deposition efficiency can therefore be expressed by equation 7.

$$C_i = R_i / F_m = C_{i\infty} \exp\left(-\frac{E_{ai}}{E}\right)$$

(6)

$$C_d = \frac{R_d}{F_m} = \frac{(R_p - R_e)}{F_m} = C_p - C_e = C_{p\infty} \exp\left(-\frac{E_{ap}}{E}\right) - C_{e\infty} \exp\left(-\frac{E_{ae}}{E}\right)$$

(7)

The very different sets of data in Figure 4 are united in one master set for the variation of deposition efficiency with E in Figure 5. The plateau at high E observed in Figure 5 ($C_{d\infty}$) is therefore the difference between the polymerization efficiency and etching efficiency plateaux (equation 8). The curve in Figure 5 was generated from equations 7 and 8 with $C_{d\infty} = 3$ m^{-2} taken from the experimental data and E_{ap}, E_{ae}, and $C_{p\infty}$ as unknowns. The validity of equation 7 is supported by the good fit of the curve in Figure 5 to the experimental data. This attests to the suitability of E for describing the plasma environment. A suitable fit to the data results when E_{ap} (7.0 X 10^{11}) is less than one half of E_{ae} (1.6 X 10^{12}).

$$C_{d\infty} = C_{p\infty} - C_{e\infty}$$

(8)

The sharp increase in deposition at low E indicates that polymerization increases more rapidly than etching ($E_{ap} < E_{ae}$) in the energy-starved plasma. Increasing E means increasing the plasma energy and/or the residence time and thus increasing polymerization and etching. A smaller polymerization activation energy ratio is also consistent with the maximum in Figure 5 since polymerization reaches a plateau while etching continues to increase thus yielding a decrease in deposition. At high E both reactions reach plateaux and the deposition is unaffected by further increases in E.

Chemical Structure. HFP has a CF:CF$_2$:CF$_3$ ratio of 1:1:1 and an F/C ratio of 2. The F/C ratio (from ESCA) of a typical PPHFP (100 W, 187 Pa, 18.3 sccm) is 1.5 (Table I). The deconvoluted peaks (light lines) in the ESCA C$_{1s}$ spectrum in Figure 6 combine in an overall spectrum (heavy line) that is quite similar to the experimental results (light line). The area contributions of C-C, C*-CF, CF, CF$_2$ and CF$_3$ to the overall spectrum area are 7, 25, 22, 26 and 20%, respectively (12). The slightly greater amount of CF$_2$ is typical of plasma fluoropolymers and reflects the more active role of CF$_2$ in the polymerization reaction ($15-18$) The slightly smaller amounts of CF and CF$_3$ and the formation of C-C and C*-CF groups can be associated with monomer fragmentation and etching. The release of atomic fluorine ($1, 15$) and the participation of the CF$_3^+$ ion in etching have been observed (2).

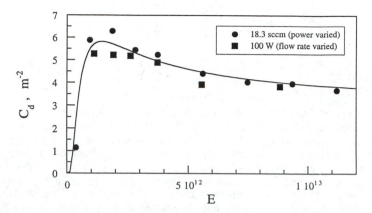

Figure 5. Experimental data from Figure 4 fit with a curve using equation 7 (an exponential master curve relating the deposition efficiency to the energy ratio plasma parameter).

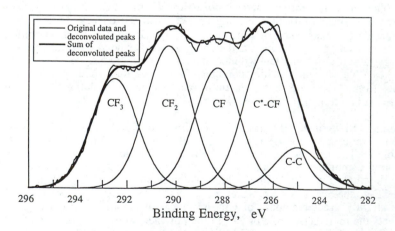

Figure 6. Deconvolution of the ESCA C_{1s} spectrum of PPHFP.

Table I. Structure and Properties of PP(gas/HFP) Films

Name	Symbol	Gas Added to HFP		
		None	H_2	N_2
Feed Flow Rate Ratio	F_nR	0.00	1.04	0.68
Elemental	F/C	1.50	1.46	1.63
Composition	N/C	0.00	0.04	0.28
Ratios	O/C	0.04	0.04	0.10
Dispersive and Polar	γ^d, mN/m	18.5	21.1	19.5
Components of	γ^p, mN/m	0.02	0.07	1.3
Surface Energy	γ, mN/m	18.5	21.2	20.8
Breakdown	σ, MV/m	61	72	23
Voltage	b	1.3	1.7	1.8

The changes in deposition rate on the addition of gases to an HFP plasma with a constant HFP flow rate are seen in Figure 7. The decrease in deposition rate on addition of nitrogen to the feed can be understood in terms of the decrease in HFP concentration, the decrease in residence time and the increase in etching rate (a nitrogen plasma etches PPHFP at 0.03 g/(m²min)). The increase in deposition rate on the addition of hydrogen to the feed can be related to the energetically favored interaction of hydrogen and fluorine to produce HF (*2*). The hydrogen in the plasma scavenges fluorine and by doing so reduces the etching rate. In addition, hydrogen can also participate in the polymerization reaction and be incorporated into the polymer. The deposition rate increases up to 14% and then decreases as the dilution of HFP becomes more significant. These results emphasize the significance of the chemistry involved in the interaction of molecular fragments within the plasma.

The chemical structure of the PP films can be affected both through the incorporation of fragments of the added gas into the polymer and through the effect of the added gas on HFP fragmentation. The surface oxygen content of the PPHFP films in Figure 8 (from ESCA) can result from the reaction of atmospheric oxygen with long lived radicals following polymerization (*1, 19*). As ESCA is a surface sensitive technique it is difficult to know whether the oxygen is found solely on the surface or throughout the bulk. The slight decrease in F/C on the addition of hydrogen reflects a decrease in the amount of CF₃ (*11*). The reaction of hydrogen with the monomer fragments and its incorporation into the polymer produces an increase in the amount of C*-CF (*11*). The addition of nitrogen to the plasma yields the significant increases in F/C, N/C and O/C seen in Figure 8 and Table I. The increase in F/C, the decrease in the amounts of C*-CF and C-C, may result from the formation of volatile carbon-nitrogen products or from a change in the reaction mechanism (*11*). The incorporation of nitrogen also enhances the reaction of atmospheric oxygen with long-lived radicals.

The FTIR spectra in Figure 9 for PP(N₂/HFP) at various flow rate ratios include a broad peak at 1229 cm⁻¹ representing CF₃, CF₂ and CF and a peak at 737 cm⁻¹ associated with C-F bending which has also been shown to be proportional to the degree of crosslinking (*15*). There are no significant peaks between 2000 cm⁻¹ and 4000 cm⁻¹ indicating a relative absence of C-H absorption. The peaks at 1718 cm⁻¹ and 1781 cm⁻¹ represent unsaturation, CF=CF and CF=CF₂, respectively (*20*).

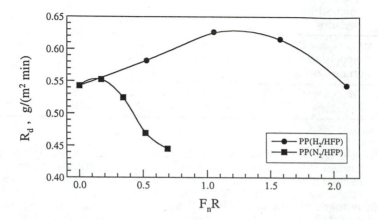

Figure 7. Effect of the ratio of feed flow rates on the deposition rate for PP(gas/HFP).

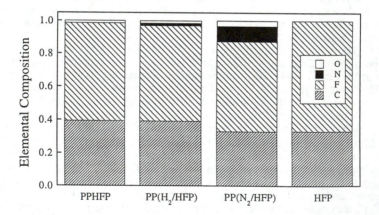

Figure 8. Elemental compositions of PP(gas/HFP) (ESCA) and HFP.

There is a significant change in the carbon unsaturation peaks with the amount of nitrogen in the feed. The peak area increases in size and breadth and the dominant part of the peak shifts from 1718 cm^{-1} to 1748 cm^{-1}. The peak at 1748 cm^{-1} is associated with C=N bonds and reflects the incorporation of nitrogen into the polymer. The incorporation of nitrogen in the polymer is reflected by the change in the relative size of the peaks, the ratio of the area under the peaks near 1748 cm^{-1} (A_{1748}) to the area under the peaks near 1229 cm^{-1} (A_{1229}). The variation of the ratio of the FTIR peak areas with the ratio of feed flow rates is seen for both PP(H$_2$/HFP) and PP(N$_2$/HFP) in Figure 10. While the FTIR peak area ratio remains relatively unchanged for PP(H$_2$/HFP) it increases with the amount of nitrogen in the feed reflecting nitrogen incorporation in the polymer.

Topography. Optical microscopy has revealed spheres on the order of 1 μm on all the film surfaces (*11*). The SEM micrographs in Figure 11 provide more detail as to the film topography. The grainy PPHFP film in Figure 11a is constructed of 0.2 to 0.5 μm particles, and the 1 μm spheres on the film are agglomerates of these same submicrometer particles. This particulate structure is indicative of a gas phase dominated plasma polymerization. Both the particles and the spheres are deposited from the gas phase onto the substrate and are incorporated into the film with further polymerization (*10*).

The smoother PP(H$_2$/HFP) surface in Figure 11b exhibits a submicrometer texture but on a significantly smaller scale. The particles that make up both the film and the spheres are 0.1 μm or less in diameter and exhibit a relatively seamless coalescence when compared to those in Figure 11a. The deposition of smaller particles which coalesce more completely reflects a significant change in the polymerization reaction and/or in the polymer molecules.

No texture at the submicrometer level can be discerned for the PP(N$_2$/HFP) film and spheres in Figure 11c. The formation of spheres indicates that some polymerization does occur in the gas phase. The dominance of gas phase polymerization is also indicated by the significant incorporation of nitrogen since such a high concentration is unlikely to be included through surface absorption (*11*). The spheres that merge into the film and into each other in Figure 11c are indicative of the processes that produced the smooth surface. The deposition of smaller particles which coalesce more completely reflects a significant change in the polymerization reaction and/or the polymer molecules. The decrease in particles size may also be related to an increase in polymer surface energy (*21, 22*). An increase in surface energy would also tend to enhance particle coalescence since the driving force to minimize surface area would be stronger.

Surface Energy. The dispersive component of surface energy for PPHFP, 18.5 mN/m (Table I), is typical of fluoropolymers. The surface energy is greater than that of conventionally polymerized HFP (PHFP) (11.7 mN/m) and similar to that of PTFE (18.6 mN/m), reflecting the loss of CF$_3$ during plasma polymerization among other factors (*14*). The PPHFP polar component of surface energy is on the order of 10^{-2} mN/m and therefore $\gamma_s \approx \gamma_s^d$ (Table I). This relatively small γ_s^p is typical of non-polar polymers such as PHFP and PTFE (*14*). The dispersive component of surface energy increases with both hydrogen and nitrogen content (Table I). This increase in surface energy is the opposite of what might be expected given the decrease in film roughness since a rougher surface has a larger true surface area and therefore exhibits a smaller apparent contact angle (*14*). The increase in the dispersive component of surface energy is, however, similar to that seen for decreasing fluorine contents in conventional fluoropolymers (*14*).

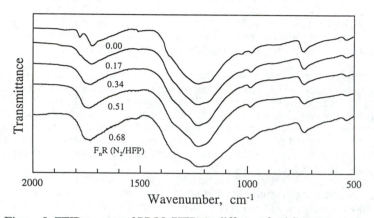

Figure 9. FTIR spectra of PP(N$_2$/HFP) at different feed flow rate ratios.

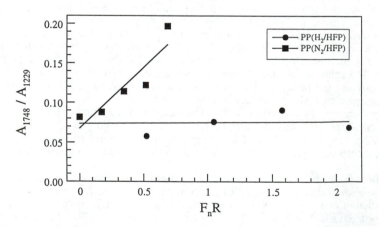

Figure 10. Relative incorporation of nitrogen into PP(gas/HFP) from the ratio of FTIR peak areas.

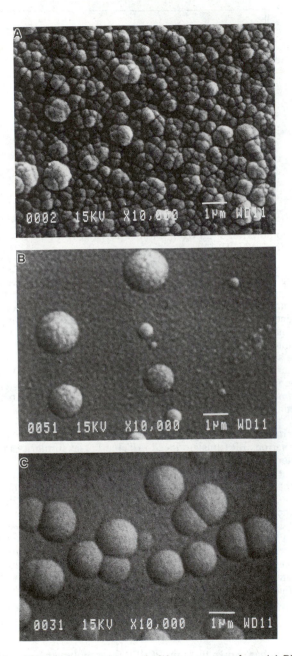

Figure 11. SEM micrographs of film topography: (a) PP(HFP); (b) PP(H$_2$/HFP); (c) PP(N$_2$/HFP).

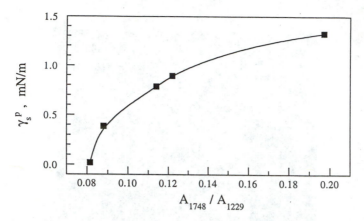

Figure 12. Correlation of the polar component of surface energy with the relative incorporation of nitrogen into the polymer from the ratio of FTIR peak areas for PP(N$_2$/HFP).

The strong correlation between the increasing incorporation of nitrogen and the increasing polar component of surface energy is seen in Figure 12. There is a two order of magnitude increase in the polar component of surface energy with nitrogen content (Table I) although the contribution of the polar component to the total surface energy remains small. For water, the polar term on the right hand side of equation 1 is no longer negligible, reaching 40% of the dispersive term and yielding a significant increase in wetting. The deposition and coalescence of smaller particles into smooth films in PP(N$_2$/HFP) and PP(H$_2$/HFP) can reflect the change in chemical structure, the change in surface energy and, perhaps, the change from a dominantly gas phase polymerization towards a surface polymerization.

Breakdown Strength. The average breakdown strength and Weibull shape parameter for various PP films are listed in Table I. The similarity in the Weibull shape parameter indicates that the breakdown strength distribution, albeit large, is similar in all the PP films. PP(N$_2$/HFP) and PP(H$_2$/HFP) have a slightly narrower distribution that may reflect their smoother, more uniform and less flawed surfaces. The mean breakdown strength for PPHFP is 61 MV/m. PP(H$_2$/HFP) has a 20% larger mean breakdown strength but within the error associated with such a wide distribution. The breakdown strength of PP(N$_2$/HFP) is approximately one third that of PPHFP. This decrease in breakdown strength can be related to the chemical structure of the polymer molecule. The introduction of nitrogen into the polymer molecules and the increase in unsaturation can produce enhanced conductivity and, perhaps, an alternate conduction mechanism in the relatively polar plasma fluoropolymer (*1, 11*).

Conclusions

This investigation has shown that thin PPHFP fluoropolymer films can be readily deposited in an ambient-temperature solvent-free process. The plasma polymerization of HFP has revealed several points of interest:

- The density of the transparent, yellow, highly adhering PPHFP film is 2.2 g/cm^3, similar to those of commercial fluoropolymers.
- The deposition rate (typically 0.35 μ/min, 0.77 g/(m^{2}min)) is independent of deposition time.
- The plasma environment generated by the reaction parameters is strongly influenced by the flow rate and can be conveniently described by E, an energy ratio plasma parameter related to plasma energy and residence time.
- The maximum and plateau in deposition were successfully described by the difference between polymerization and etching, each related exponentially to E through an activation energy ratio. The activation energy ratio for polymerization was less than one half that for etching.
- PPHFP (F/C=1.5) largely consists of similar amounts of C*-CF, CF, CF$_2$ and CF$_3$ and approximately 7% C-C. The abstraction of F and the preferential scission of the C-CF$_3$ bond can explain this structure and the significant amount of unsaturation. The dispersive component of surface energy, 18.5 mN/m, and the negligible polar component of surface energy are typical of fluoropolymers.
- The plasma polymerization of HFP is a gas phase dominated process that produces 0.2 to 0.5 μm particles. These particles and 1 μm spherical particle agglomerates are deposited on the surface and incorporated into the film with further polymerization.
- The addition of hydrogen to HFP can yield a 14% increase in deposition rate resulting from the inhibition of etching through fluorine scavenging and HF formation. The incorporation of hydrogen into the polymer structure yields an increase in the dispersive component of surface energy and produces a smoother surface.
- The addition of nitrogen to HFP yields a PP fluoropolymer with a relatively large F/C. The deposition and coalescence of smaller particles into a smoother, more uniform surface can reflect the incorporation of a relatively large amount of nitrogen, the significant increase in the polar component of surface energy (although the increase in the total surface energy is relatively small) and, perhaps, the decrease in dominance of gas phase versus surface polymerization. The relatively low breakdown strength (23 MV/m) was approximately one third that of PPHFP and may be attributed to an alternate conduction mechanism in the relatively polar plasma fluoropolymer.

Legend of Symbols

A	-	reactor (electrode) area, m^2
b	-	Weibull shape parameter
C_d	-	deposition efficiency, m^{-2}
C_e	-	etching efficiency, m^{-2}
$C_{i\infty}$	-	C_i at high E, (where $i = d, e$ or p), m^{-2}
C_p	-	polymerization efficiency, m^{-2}
d	-	deposition
E	-	energy ratio plasma parameter
e	-	etching
E_{ai}	-	activation energy ratio for C_i (where $i = e$ or p)
E_k	-	kinetic energy associated with flow per unit monomer mass, J/kg
$F(x)$	-	cumulative failure
FEP	-	fluorinated ethylene-propylene copolymer
F_m	-	mass flow rate, g/s
F_n	-	molar flow rate, sccm
F_v	-	volumetric flow rate, m^3/s

HFP	-	hexafluoropropylene
lv	-	liquid in the presence of its vapor
P	-	steady-state plasma pressure, Pa
p	-	polymerization
PHFP	-	poly(HFP)
PPHFP	-	plasma polymerized HFP
PTFE	-	polytetrafluoroethylene
R	-	ideal gas constant, 8.314 J/(mol K)
R_d	-	deposition rate, g/(m^{2}min)
R_e	-	etching rate, g/(m^{2}min)
R_p	-	polymerization rate, g/(m^{2}min)
s	-	solid
T	-	temperature, K
TFE	-	tetrafluoroethylene
W	-	power, W
x	-	breakdown strength, MV/m
γ_i	-	surface energy (where $i = s$ or lv), mN/m
γ_i^d	-	dispersive component of surface energy (where $i = s$ or lv), mN/m
γ_i^p	-	polar component of surface energy (where $i = s$ or lv), mN/m
θ	-	contact angle
σ	-	mean breakdown strength, MV/m

Acknowledgments

The authors gratefully acknowledge the support of the Technion V.P.R. Fund and the Fund for the Promotion of Research at the Technion.

References

1. Yasuda, H. *Plasma Polymerization;* Academic Press, New York, NY, 1985.
2. Biederman H.; Osada, Y. *Plasma Polymerization Processes;* Elsevier Science Publishers, New York, NY, 1992.
3. Boenig, H.V. *Fundamentals of Plasma Chemistry and Technology;* Technomic Publishing, Lancaster, PA, 1988.
4. Morita, S.; Hattori, S. In *Plasma Deposition, Treatment, and Etching of Polymers;* R. d'Agostino, Ed.; Academic Press, New York, NY, 1990; p 423.
5. Ratner, B.D.; Chilkoti A.; Lopez, G.P. In *Plasma Deposition, Treatment, and Etching of Polymers;* R. d'Agostino, Ed.; Academic Press, New York, NY, 1990; p 463.
6. Johnson, S.D.; Anderson J.M.; Marchant, R.E. *J. Biomedical Materials Research* **1992,** *26,* 915.
7. Hoffman, A.S. *J. Appl. Polym. Sci.: Appl. Polym. Symp.* **1988,** *42,* 251.
8. Inagaki, N.; Tasaka S.; Takami, Y. *J. Appl. Polym. Sci.* **1990,** *41,* 965.
9. Takehara, Z.; Ogumi, Z.; Uchimoto, Y.; Yasuda K.; Yoshida, H. *J. Power Sources* **1993,** *43-44,* 377.
10. Chen, R.; Gorelik, V.; Silverstein, M. S. *J. Appl. Polym. Sci.* **1995,** *56,* 615.
11. Chen R.; Silverstein, M. S. *J. Polym. Sci.: Polym. Chem. Ed.* **1996,** *34,* 207.
12. Silverstein, M.S.; Chen R.; Kesler, O. "Hexafluoropropylene Plasmas: Polymerization Rate - Reaction Parameter Relationships", *Polym. Eng. Sci.* accepted.

13. Sherwood, P. M. In *Practical Surface Analysis by Auger and X-ray Photoelectron Spectroscopy;* Briggs D.; Seah, M .P., Eds.; John Wiley & Sons, New York, NY, 1983; p 445.
14. Kinloch, A. J. *Adhesion and Adhesives;* Chapman and Hall, New York, NY, 1987.
15. d'Agostino, R.; Cramarossa, F.; Fracassi, F.; Illuzzi, F. In *Plasma Deposition, Treatment, and Etching of Polymers;* d'Agostino, R., Ed.; Academic Press, New York, NY, 1990; p 95.
16. Golub, M. A.; Wydeven, T.; Cormia, R. D. *J. Polym. Sci., Part A: Polym. Chem.* **1992,** *30,* 2683.
17. Momose, Y.; Takada, T.; Okazaki, S. *J. Appl. Polym. Sci.* **1988,** *42,* 49.
18. Samukawa S.; Furuoya, S. *Jpn. J. Appl. Phys. - Part 2* **1993,** *32,* L1289.
19. Gengenbach, T. R.; Vasic, Z. R.; Chateleir, R. C.; Griesser, H. J. *J. Polym. Sci. Part A: Polym. Chem.* **1994,** *32,* 1399.
20. Conely, R. T. *Infrared Spectroscopy;* Allyn and Bacon, Boston, MA, 1970.
21. Grebowicz, J.; Pakula, T.; Wróbel A. M.; Kryszewski, M. *Thin Solid Films* **1980,** *65,* 351.
22. Wróbel A. M.; Wertheimer, M. K. In *Plasma Deposition, Treatment, and Etching of Polymers;* d'Agostino, R., Ed.; Academic Press, New York, NY, 1990; p 163.

Chapter 28

Fabrication and Application of Particle-Crystalline Films

Kuniaki Nagayama[1] and Antony S. Dimitrov[2]

[1]Department of Life Sciences, Graduate School of Arts and Sciences, University of Tokyo, Komaba, Meguro-ku, Tokyo 153, Japan
[2]L'Oreal Tsukuba Center, 5–5 Tokodai, Tsukuba 300–26, Japan

The formation of thin film organizations of small particles on a substrate has been studied since the beginning of this century. The main efforts were directed to studying the interparticle forces, which are responsible for the structure organization, but not to controlling these forces during the film formation process. Here, we present a novel approach to growing particle arrays, which allows the formation of centimeter-sized particle-crystalline monolayer films. We have specified three factors controlling the rate of an array growth, namely, the diameter of the particles, the particle volume fraction in the suspension, and the water evaporation rate, which depends on the temperature and the relative atmosphere humidity. We have also discriminated the forces arranging the particles into two-dimensional array, namely, the lateral capillary forces between the particles and the hydrodynamic pressure forces pressing each particle to the array's main body. Then, we proposed a simple way to adjust the rate of monolayer production to the growth rate of the densely-packed monolayer or bilayer two-dimensional (2D) array. The obtained large-sized monolayers consist of closely packed domains and, interestingly, in reflected light, mimic the morpho-coloring of the wings of rare types butterflies. In some scale, we are able to regulate the size of the obtained domains by controlling the thickness of the spreading suspension film. Enlarging the domain size can serve future technologies to producing new types of optical gratings, interferometers, antireflection coatings, selective solar absorbers, and especially mass-storage media and microelectronical units.

Ordered arrays of colloidal particles coated on surfaces can be used either as diffraction gratings, optical storage media, or interference layers. Monolayer or thicker layers of random or ordered colloidal particles has shown usage as lithographic masks for preparation of precisely controlled surface textures (1). Textured surfaces of controlled

0097–6156/96/0648–0468$15.50/0
© 1996 American Chemical Society

periodicity are of growing importance for several fields of science and technology. Surface roughness with a periodicity of 10 to 100 nm are responsible for enhanced Raman scattering (*2*). Random and periodic roughness on a submicron scale are parts of optical elements as interferometers (*3*), antireflection coatings (*4,5*), and optical gratings. Selective solar absorbers (*6*) utilize surfaces textured on a micron scale periodicity. Textured surfaces can play an important role in photovoltaics (*7*), and those with a perfect periodicity promise novel technologies for data storage, optics (*8*), and microelectronics.

Especially promising are surfaces textured by layers of two-dimensional (2D) arrays of small particles. The simplest but uncontrolled way to form particle 2D arrays is to spread the particle suspension in a thin layer onto the substrate and leave the solution to evaporate (*9*). The spin-coating technique to form thin particle layers of 2D array, in which the suspension rapidly spreads on rotating substrates, has been developed by Deckman and his group (*1,10-12*). These approaches and the recently developed ones for formation of 2D particle array in wetting films in small cylindrical cells (*13,14*) do not allow to control the growth of the 2D array, which commonly results in a variety of defects, size restriction, and instabilities as, for example, sequences of uncontrollable voids and multilayers. The continuous growth of particle array on large surface areas has not been well documented. In this article, we propose a novel approach to control the formation of large-sized monolayer 2D arrays from fine particles. The controlling mechanism is based on a quantitative analysis of the rate of the array growth. The obtained centimeter-sized monolayer 2D arrays illuminated by white light exert brilliant morpho-coloring as those of morpho butterfly wings.

This report can be formally discriminated into three major focuses: 1) the idea for the continuous array growth and its analysis; 2) experimental realization for the formation of large-sized monolayer arrays; and 3) comparison of the structure and optical properties of monolayer arrays with those of *Morpho*-butterfly wing surface. We would like to note that the experiments were carried out using negatively charged polystyrene particles. Our further experiments with silica particles confirmed that the phenomenon of a thin solid layer deposition on the substrate is quite general and does not depend on the particle internal properties. It is mainly determined by the physical properties of the suspension, particle-particle and particle-substrate interactions, and the particle geometry. The shape and uniformity of the particles are of primary importance for the size of the obtained ordered domains and, hence, for the color properties of the monolayer array.

The essence of our fabrication of two-dimensional assembly of particles stands in the use of a stable wetting film that is made on the substrate as shown in Figure 1 (*15*). The wetting film plays two important roles: 1) it is a 2D liquid medium, where particles can be carried by water flow toward the array's boundary for growth (convective assembly shown in Figure 1C) (*13-18*), and 2) the deformation of its free liquid surface induces attractive force between the particles, when they are pressed between the film surfaces (lateral capillary force, Figure 1D) (*19-21*). In parallel to the usual 3D crystallization, these processes can also proceed according to the free energy difference before and after the reaction in the non-equilibrium state. First, particles undergo the Brownian motion in the thick film (Figure 1A). When the wetting film becomes as thin as the particle size by removing water, ordered 2D domains start to

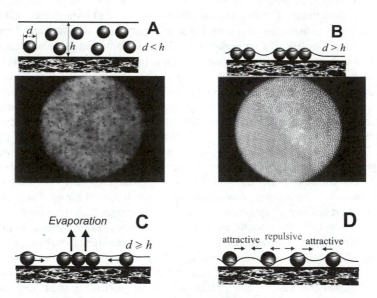

Figure 1. Two-dimensional assembly of particles in a wetting film. (**A**) Particles undergo the Brownian motion in the liquid layer, which has thickness much larger than the particle size. (**B**) Particles start to assemble in the wetting film as the thickness of the film becomes comparable to or slightly smaller than the particle diameter. (**C**) Convective flow responsible for the assembling of particles stimulated by the liquid evaporation. (**D**) Lateral capillary forces responsible for the hexagonally closed packing of the particles. (Adapted from ref. *15*.)

grow (Figure 1B). The removal of water by evaporation is the only one explicit non-equilibrium condition in our system. The most difficult task in this fabrication is not to control the evaporation but to create stable wetting films that are appropriate for the submicron or nanometric particles. At common conditions, the thinning process on solid substrates usually leads to rupturing the film at nanometer-scaled thickness (*22*). Cleaved mica or acid-rinsed glass provides a very wettable surface and, consequently, is suitable for the array formation of colloidal particles when they are larger than 50 nm in diameter. The main our efforts were spent to find experimental conditions, at which the thickness of the spreading suspension films can be gradually changed.

For a wettable solid plate dipped vertically in a suspension of fine particles, we found that monolayers and successive multilayers of 2D particle array spontaneously start to form on the plate surface from the plate-suspension-air contact line downward toward the bulk suspension. The successive formation of monolayers, bilayers, trilayers, etc. (*23,24*) was due to the increase in the wetting film thickness from the plate-suspension-air contact line toward the bulk suspension and to the continuous flux from the bulk suspension toward the film filling it up with densely packed particles. We were able to control the formation of a monolayer or multilayer by carefully withdrawing the substrate (the solid plate) together with the already formed particle arrays from the suspension. Theoretically, when the withdraw rate equals the rate of 2D array formation the array can be continuously formed to any size.

The mechanism for an array formation on substrate plates dipped in water suspensions or solutions is the same as those behind scum formation on the walls of swimming pools, kitchen sinks, etc. Everywhere water contacts wettable surfaces and evaporates, the substances dissolved in the water accumulate in the spreading films at the vicinity of the three-phase contact line. The phenomenon is quite general and depends on the wettability of the surfaces and the properties of the wetting films, rather than the individual chemical properties of the substances dissolved or dispersed in the water.

Materials, Apparatuses, and Procedures

Materials. We used Matsunami microslides (76×26×1 mm) as the solid substrates. The water used in the experiments was purified by using MILLI-Q SP.TOC reagent water system. To control the properties of the spreading suspension film on the glass plates, we added to the suspensions no more than 0.001 mol/l sodium dodecyl sulfate (SDS), 10^{-4} mol/l sodium chloride (NaCl), 0.001 mol/l octanol (Wako Pure Chemical Industries, LTD), and 0.01 mg/ml protein (milk casein, Chameleon Reagent, Japan, or ferritin, Boehringer Mannheim GmbH, Germany). The total amount of the additives was kept to not exceed 5 *vol.%* of the particle volume fraction in the suspension. This restriction appeared to reserve the brilliant coloration of the obtained particle-crystalline films. The additives were mixed together in a closed 20-ml flask. Their concentrations were as follows: 0.01 mol/l SDS, 0.01 mol/l octanol, 0.1 mg/ml protein, and 0.001 mol/l NaCl. This solution was being mixed and slightly warmed up to 40 °C until a transparently clear solution is formed. After cooling to the room temperature, portions of this solution were added to the particle suspension having in mind the restrictions above.

Table 1. Specifications of the latex particles given by the manufacturer

Latex code	Diameter [nm]	Polydispersity [nm]
SS-021-P	2106	± 17
SC-171-S	1696	± 47
SC-108-S	1083	± 10
SC-953-S	953	± 9
SC-081-S	814	± 23
SC-051-S	506	± 10
SC-048-S	479	± 5
SC-032-S	309	± 4
SC-015-S	144	± 2
SC-008-S	79	± 2

Source: Reprinted with permission from ref. *27*. Copyright 1996.

The specifications of the polystyrene particles (Stadex, Japan Synthetic Rubber) that were used are given in Table 1. Originally, the suspensions contained 1 *vol.%* of particles. To exclude the eventual aggregates, the suspensions were filtered through cellulose acetate membrane filters, which had a pore size ranging from 2 to 4 times greater than the particle diameter. We assumed that the particle concentration did not change significantly during the filtration process. To concentrate the particles up to 6 *vol.%* in some of the experiments we used Millipore molecular filtration units with a cut-off pore size of 300 KDa.

Experimental Setups. The simplest experimental setup designed for the preliminary experiments was, in fact, an inclined precleaned microslide plate (*25*). We placed a 100-μl drop of suspension onto the microslide plate and spread it uniformly along the plate surface. Then, we placed the plate on the goniometer stage, which was inclined at a desired angle. Due to gravity the suspension started to flow downward along the microslide plate and the layered array remained behind the receding drop-substrate-air three-phase contact line. We used the above method to figure out the approximate array growth rate when changed the experimental conditions.

Quantitatively detailing the array growth, we have designed and prepared a laboratory setup for withdrawing substrates from particle suspensions. The schematic of this setup and a photograph of the prototype are shown in Figure 2. The working cell was made in-house from pieces of microslides connected to each other by a polymeric glue. It has a narrow gap at the top for withdrawing the substrate plate. A 1-mm gap was chosen to minimize solvent evaporation from the meniscus and confine the evaporation to the particle array film. The substrate plate hangs on a nonelastic string, which is attached to the periphery of a 2-cm in diameter wheel. A stepper motor driver rotates the wheel using a gear box. Thus, the substrate can be withdrawn from the suspension at a rate from 0.1 to 30 μm/s. The array growth process is observed and

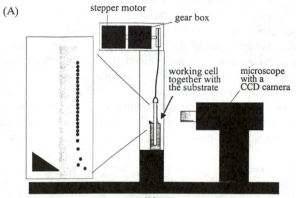

(A)

stepper motor

gear box

working cell together with the substrate

microscope with a CCD camera

rail base

(B)

Figure 2. The setup for fabrication of 2D particle array on microslide substrate plates: (**A**) Schematics and (**B**) Photograph.

recorded using a horizontally placed video-microscope having a resolution power of about 350 nm. In fact, simultaneously keeping constant values for the humidity, temperature, and the particle concentration at the array's leading edge is quite difficult task. Furthermore, our controller of the stepper motor driver allowed only a manual stepwise change of the velocity in 300 steps. Both these inconveniences forced as to select a slightly higher than the calculated substrate withdrawal rate. Keeping this high withdrawal rate for about 1-2 min causes a rupture in the regular array growth and formation of striped particle films – see ref. *26* for the mechanism of the striped pattern formation. The state before rupturing the regular array formation is characterized with a rough leading edge of the array and an accelerated particles' speed with about 30%. We were continuously observing the array growth process and held on the substrate withdrawal (switch off the motor driver) just before the array to rupture. When the regular growth was reestablished we continued the substrate withdrawal (switch on the motor driver). We were able to apply this off/on switch procedure when the array's growth rate did not exceed 3 to 5 μm/s.

Array Formation Procedure. When a protein was not added to the suspension, the glass plates were kept in a chromate cleaning solution at least overnight, then rinsed with water, soaked for more than one hour in 0.1M SDS solution (when particles larger than 506 nm particles were used) or in pure ethanol (when particles smaller than 479 nm were used). The plates soaked in the SDS solution were rinsed again with water and placed under a beaker to dry. The plates soaked in ethanol were placed directly under a beaker. The difference in the washing procedure was necessary to manage the thickness of the spreading suspension film. The SDS molecules adsorbed on the glass substrate kept the film thicker than the ethanol molecules did. The desorption process was slow enough and the surface properties were generally preserved for the time of the experiment – 2 to 3 hours. Because in most of the experiments the laboratory setup and the cell were opened to the room atmosphere, the existence of organic traces in the air often led to local dewettings on the substrate plates. Adding protein molecules significantly improved the plates' wettability. When a protein, ferritin or casein, was added to the suspension, the glass plates were used directly from the original packaging, without additional treatment. However, these additives hampered in some aspects the brilliant coloring of the obtained monolayer arrays.

The experimental cell was sonicated for about 10 sec in a Ney 300 Ultrasonic water bath before each experiment. Then, the cell was washed using a soap solution, rinsed with plenty of water, and dried.

After drying, the experimental cell and a glass substrate were mounted in the apparatus. The stepper motor driver for withdrawing the substrate was set at a rate 3 to 5 times higher than that estimated for a monolayer formation. Then, we filled the cell with a suspension and started to withdraw the substrate plate. To obtain an uniform thickness at the leading edge of the growing particle arrays, we monitored the array growth and gradually decreased the substrate withdrawal rate accordingly. We further proceeded the continuous formation of a monolayer 2D array as described above in the setup description.

To initiate a bilayer formation, we reduced the withdrawal rate by half. When the bilayer started to grow we readjusted the withdrawal rate to obtain uniform thickness at the leading edge of the growing arrays.

All the experiments were done in a room at 25 °C and relative humidity of 48 %RH, measured using a thermo-hygrometer (TRH-10A, Shinyei, Japan).

Color enhancement. The array from micrometer-sized particles exhibit brilliant coloring when illuminated by white light in their native state, after forming and drying. We enhanced this brilliance by coating the particles with silver or gold (up to 10 nm thick) using a vacuum coating technique. An added benefit of this coating is the stabilization of the arrays on the substrate, which, in fact, are fragile before the metal coating.

Characterization of the Obtained 2D Particle Array. The obtained centimeter-sized monolayer particle arrays show interesting color properties resembling natural beauties. We have studied the structure of the 2D particle array in comparison to the structure of butterfly wings, which in some aspects exhibit similar coloring due to the scattering of white light from a periodic surface texture. We used different metric-scale approaches in this investigation. The macroscopic outside view and scattering of white light were observed by naked eyes. The surface structure was figured out by using an optical microscope (Olympus BH, Japan) in reflected light with a dark-field option. To get details in the structures of both the obtained particle arrays and the wings of a butterfly we performed electron microscopy by using a field emission scanning electron microscope (S-5000H, Hitachi, Japan).

We measured the thickness and the refractive index of densely packed monolayers of 144-nm particles by using an ellipsometer (Gaertner Scientific Corporation, Chicago). The incidence angle of the helium-neon laser beam was set at 70°. Using the same apparatus, we also measured the thickness of horizontal wetting water films formed on the microslides by spreading of large water drops.

Schematics of the Array Growth Process

The profile of growing 2D particle array from bulk suspension onto a withdrawing substrate plate, water evaporation flux, j_e, and water and particle fluxes, j_w and j_p, in the vicinity of the array's leading edge are schematically shown in Figure 3 – see also refs. *27* and *28*. The width of the plate is large enough and the growth disturbances at the edges can be neglected, namely, in our model the array's leading edge is a straight line parallel to the plane of the horizontal suspension surface. The formation of layered 2D array can be conveniently split in two main stages: 1) convective transfer of particles from the bulk of the suspension to the thin spreading film (upward in our experiments) due to water evaporation from the film surface and 2) interaction between the particles that lead to specific textures.

Regular Formation of Particle Array Films. The primary driving force for the convective transfer of particles is the water evaporation from the freshly formed particle array. In an atmosphere saturated with water vapor, and after establishing

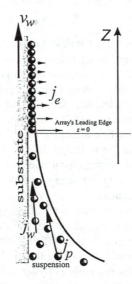

Figure 3. Schematics of the particle and water fluxes at the leading edge of the monolayer particle array growing on a substrate plate that is being withdrawn from a particle suspension. Here, v_w is the substrate withdrawal rate, j_w is the water influx, j_p is the respective particle influx, and j_e is the water evaporation flux. (Adapted from ref. 27)

mechanical equilibrium, the pressure balance in an infinitely small bulk volume inside the spread suspension film is:

$$\Pi + P_{cp} = P_c + P_h , \quad \text{where} \quad P_h = \Delta\rho g \, h_c \tag{1}$$

where Π is the sum of van der Waals and electrostatic disjoining pressures for the suspension wetting films on the substrate plate, P_{cp} is the capillary pressure due to the curvature of the liquid surface between neighboring particles in the particle film, P_c is a reference capillary pressure, P_h is the hydrostatic pressure in a vertical film, h_c is the relative height, $\Delta\rho$ is the density difference between the suspension and the surrounding gas atmosphere, and g is the gravity acceleration.

When the evaporation of water starts, the right-side terms in eq 1 stay almost constant for a given h_c. If we assume that P_c is related to the horizontal suspension surface, i.e., $P_c = 0$, then due to the decrease in the total suspension volume as the water evaporates, h_c slowly increases. We minimized the change in h_c by adding an amount of suspension to compensate the suspension volume decrease. The left-side terms in eq 1, however, increase due to the increase in the curvature of the menisci between the particles, namely, P_{cp} increases, and due to the thinning of the film Π also increases. In some cases, $\partial\Pi/\partial h > 0$, where h is the film thickness, but these films are not stable and not suitable for our technique. Therefore, in an atmosphere, unsaturated with water vapor, a pressure gradient, ΔP, from the suspension toward the wetting film arises due to the water evaporation from the freshly formed particle array (the flux j_e in Figure 3). The pressure gradient

$$\Delta P = (\Pi + P_{cp}) - (P_c + P_h) \tag{2}$$

produces then a suspension influx from the bulk suspension toward the wetting suspension film. This influx consists of a water component, j_w, and of a particle flux component, j_p. The water part of the suspension influx j_w compensates for the water evaporated from the film j_e, and the particle complementary part, j_p, causes particles to accumulate in the film, thus forming dense structures. Naturally, the particle structures thus formed follow the film geometry (*23,29*). The thickness of vertical wetting films increases from the plate-suspension-air contact line downward toward the bulk suspension due to the hydrostatic pressure. Then, successive monolayers, bilayers, trilayers, etc. are expected to be formed as the continuous particle flux j_p fills up the space between the substrate and the film surface. In fact, we observed the formation of successive multilayers when a wettable plate dipped in a suspension of fine particles was kept stationary.

To model the process of regular array growth, we applied the approach of the material flux balance at the array's leading edge (*14,25*). Thus, we calculated the rate of the substrate withdrawal, which must actually be equal to the rate of the array growth. In the developed model we assume two idealized conditions; 1) complete wetting: the suspension wets the substrate by forming a stable wetting film and 2) frictionless: the particles do not stick onto the substrate, if they are not strongly pressed by the film surface as, for example, the larger particle pointed in Figure 4. Both assumptions can be validated for many systems with practical applications.

For a regular and continuous formation of 2D particle array onto a substrate plate, which is schematized in Figure 3, the total water evaporation flux from the particle arrays per unit length of the array's leading edge, J_{evap}, is the integral of water

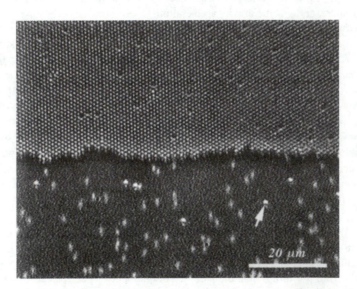

Figure 4.   A part of the leading edge of a growing monolayer particle array. In the upper-half of the photograph, a forming single domain of ordered 953-nm particles is shown. The lower-half shows the suspension meniscus, where the water influx presses the particles toward the forming monolayer. Because of the high velocity in the microscale, $v_p \sim 100$ μm/s, the particles are seen as short fuzzy lines. The few particles seen as bright spots (one is indicated by an arrow) have diameters larger than the average one. They are wedged into the wetting film. The hydrodynamic force is not sufficient to overcome the capillary resistance near the growing arrays boundary. (Adapted from ref. 27)

evaporation flux, $j_e(z)$, along the axis Z, $J_{evap} = \int_0^\infty j_e(z)\, dz$. For practical treatment, we introduce an evaporation length, $l = J_{evap} / j_e$. Here, the evaporation flux from a pure water surface, j_e, depends on the temperature and humidity of the surrounding atmosphere. Both parameters can be experimentally determined. For the steady state process of an array growth the water evaporation, J_{evap}, is exactly compensated by the water flow from the bulk suspension into the arrays, J_w, that is, $J_w = J_{evap}$ or

$$h_f j_w = l j_e, \tag{3}$$

where $J_w = h_f j_w$ and $J_{evap} = l j_e$. Here, h_f is the thickness of the wetting film at the height of the array's leading edge, and is usually slightly smaller than the thickness of the particle arrays, h.   The compensation water influx, j_w, is defined as $j_w = N_w V_w v_{wt}$, where N_w is the number of water molecules per unit volume, V_w is the water molecule volume, and v_{wt} is the macroscopic mean velocity of the water molecules.   The macroscopic mean velocity of the suspended particles, v_p, is proportional to v_{wt}, namely, $v_p = \beta v_{wt}$.   The value of the coefficient of proportionality, β, depends on the particle-particle and particle-substrate interactions and should vary from 0 to 1. The stronger the interactions, the smaller the value of β. For non-adsorbing particles and dilute suspensions β approaches 1. Then, the particle flux, j_p, which is defined as $j_p = N_p V_p v_p$ (where N_p and V_p are the number of particles per unit volume and the volume of a single particle, respectively), is proportional to j_w,

$$j_p = \frac{\beta\varphi}{1 - \varphi} j_w, \tag{4}$$

where $\varphi = N_p V_p$ is the particle volume fraction in the suspension and $1 - \varphi = N_w V_w$ is the water volume fraction.

The particle flux, j_p, drives the growth of particle arrays because the particles attach to the array's leading edge and remain there. The stock of the particles at the array leading edge, $h_f j_p$, is equal to the increase in the total particle volume in the array, namely, the product of the array growth rate, v_c, thickness of the array, h, and array density, $1 - \varepsilon$, (ε is the porosity of the array)

$$v_c h(1 - \varepsilon) = h_f j_p. \tag{5}$$

By substituting j_w from eq 3 into eq 4, and the resulting expression for j_p into eq 5, the rate of the array growth is:

$$v_c = \frac{\beta l j_e \varphi}{h(1-\varepsilon)(1-\varphi)} \tag{6}$$

In the derivation of eq 6, the values of h and $(1 - \varepsilon)$ are not important. What is important is their product, which shows the total volume of particles per unit area. To connect h and $(1 - \varepsilon)$ with the real geometry of the particle array we assume that h is the distance from the substrate to the tops of the particles, namely, for a particle monolayer $h = d$ (d is the diameter of the particles). Then $(1 - \varepsilon)$ is geometrically calculated from the conditions for densely (hexagonally) packed spheres, namely, $1 - \varepsilon = 0.605$. Taking into account that the dense particle multilayers are, in fact, monolayers displaced one over another, we substitute $h(1 - \varepsilon)$ with $0.605kd$ for k-layer particle array – see for example ref. *17* for the particle film thickness. Then, for the formation of dense two-dimensional particle array (monolayers and multilayers as well), we estimate the withdrawal rate of the substrate plate, v_w, by rewriting eq 6

$$v_w = v_c^{(k)} = \frac{\beta l}{0.605} \frac{j_e \varphi}{kd(1 - \varphi)} . \tag{7}$$

Here, $v_c^{(k)}$ is the growth rate of the k-layer array. Equation 7 shows that the growth rate of the dense array depends on the particle volume fraction, φ, water evaporation rate, j_e, diameter of the particles, d, number of layers, k, and an experimentally determined constant, the product βl.

Initiation of the Array Growth. We have already regarded the process of regular array growth accepting that the array growth has been successfully initiated. The origin of the array growth, however, remained to be clarified. Below, we propose an idea how the particle array starts to grow in terms of the water evaporation from a thermodynamically stable wetting film.

When the wettable plate is dipped into the suspension a wetting film arises upward along the plate. Due to the hydrostatic pressure in such vertical wetting films the film thickness, h_f, decreases as the relative height, h_c, increases. Thus, the water-air interface may actually have a relatively high inclination towards the substrate (water-plate interface) and causes a capillary force, which pushes the particles out of the film regions thinner than the diameter of the particles (*18,21*). Hence, the lateral capillary forces are not able to initiate the array growth process in our experiments. Furthermore, at this initial stage of the array formation process, because of the absence of particles in the wetting film, P_{cp} in eq 2 is equal to zero and cannot increase to produce the pressure difference ΔP, which can create suspension influx toward the film. However, the dependence of the film thickness, h_f, on the disjoining pressure, Π, for wetting films can become very steep when h_f is on the order of hundreds nanometers, i.e., the derivative $(\partial \Pi / \partial h)$ can approach extremely large negative values for h_f less than 1 μm. At such a thickness the vertical films exhibit a slight dependence of h_f on the height h_c. Furthermore, when water evaporates from such a film the disjoining pressure, Π, increases thus creating pressure difference ΔP nevertheless that $P_{cp} = 0$ (see eq 2). The created ΔP produces suspension influx, which carries on particles and stores them into the film. We must note here that the wetting film can rupture at high evaporation rates due to the hydrodynamic friction in the compensation influx or at a thickness less than some value h_0, at which the derivative $(\partial \Pi / \partial h)$ can change its sign and becomes positive. Usually, h_0 has a value little less than the thickness of steep $(\partial \Pi / \partial h)$. Depending on the relation between the diameter of the particles and the thickness h_0, particle array or irregularly placed bumps of particles can start to be formed or no particles remain on the substrate (*28*). The onset of a particle array is schematically illustrated in Figure 5 for the case when h_0 is slightly smaller than the diameter of the particles, d. When d is less than h_0 irregularly placed bumps of particles usually remain after drying the film and when d is much greater than h_0 no particles remain after the film drying: they are pushed towards the bulk by the inclined meniscus (see ref. *28* for details).

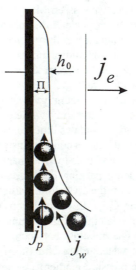

Figure 5. Schematics of the initial state of the array formation in a wetting film with a thermodynamically stable thickness h_0. The particle diameter d is slightly larger than h_0. The water evaporation forces the film thinning, thus increasing the disjoining pressure, Π, and causing suspension influx, $j_w + j_p$, from the bulk toward the film.

Forces Assembling the 2D Array. Completing the schematics of the 2D array formation onto solid surfaces, we propose here our hypothesis for the forces between the particles inside a wetting film leading to the formation of layered particle array. Recent works have suggested that the interactions between particles confined in thin films can be attributed to electrostatic (*30,31*) and lateral capillary forces (*19,21*). As it was suggested (*14,16*) and later proven by direct observations (*18*) the lateral capillary forces are able to gather particles into layered arrays. The phenomenon was observed *in situ* using polystyrene spheres in wetting films on mercury, where the thickness of the film was controlled by injecting suspension into or withdrawing it from the cell (*18*). It was observed that in an atmosphere of high humidity the assembling of the particles into array nuclei starts slowing down. However, when the atmosphere in the cell had low humidity the particles were assembling rather quickly. Hence, the rate of the assembling, which is caused by the water evaporation from the film, becomes much higher than those caused by the lateral capillary forces. In the experiments reported here, which were carried out in a low humidity atmosphere, the assembling process was mainly caused by the water evaporation from the wet stripped area that was being formed in the vicinity of the array's leading edge. After carrying on the particles from the suspension bulk toward the array's leading, the water influx, which compensates for the evaporation, presses them toward the array while poring through the array cavities. The polystyrene particles and the glass plates that we used were negatively charged and repulsed each other when are in close contact inside the suspension. The close interparticle electrostatic repulsion and the hydrodynamic influx pressure at the array's leading edge determines the dense packing of the particles in our experiments. An Alder type phase transition to form the crystalline particle array may occur as the result of the dense hexagonal packing (*32*). In competition, the close neighboring particles are attracted to each other by the lateral capillary immersion forces, which also causes dense hexagonal packing. The relation between the packing force due to the hydrodynamic pressure of the water influx and the packing force due to the lateral capillary forces depend on the pressure gradient ΔP defined in eq 2 and, thus, on the thickness of the wetting film, h_f. Our preliminary theoretical description shows that the lateral capillary part of the packing force increases with the square of the hydrodynamic part. In other words, if we attempt to accelerate the assembling process twice by increasing the evaporation rate and doubling the pressure gradient in eq 2, the result is four times increasing the strength of the lateral capillary forces. Furthermore, increasing the lateral capillary forces (i.e., decreasing h_f and increasing ΔP in eq 2) leads to an increase in the friction force between the substrate and particles. Thus, accelerating the array growth rate may suppress the rearrangement processes inside the array and may lead to a formation of smaller domains. Experimentally, we expect that depending on the balance between the hydrodynamic stream and lateral capillary forces, which can be regulated (for example, by changing ΔP, h_c, or j_e), smaller or larger highly-ordered domains can be formed onto the substrate plate.

Continuous Fabrication of 2D Particle Array

The procedure for continuous grow of 2D particle array was applied to different suspensions of particles with diameters from 79 to 2106 nm. The substrate plates were vertically dipped in the cell filled with particle suspension. Wetting films were formed due to the capillary-rise along the plates. Mono-, bi-, trilayers, etc., started to form in the vicinity of the plate-suspension-air contact line, when the plates were kept immovable. The growing of 2D particle arrays of k layers, and especially of monolayers ($k = 1$), was continuing by withdrawing the substrate plates up from the suspensions and adjusting the withdrawal rates to the rates of the arrays growth, $v_c^{(k)}$. A snap of a regularly growing monolayer is shown in Figure 4. The array growth behavior depended on the particle diameters, polydispersity of the particles in the suspension, humidity and temperature of the environment, wetting film properties, impurities in the suspension, dust on the plates, etc. The array formation procedure developed by us is general and should work without any pitfalls, if the following conditions are met: the suspension, the atmosphere around the cell, the substrate plate, and the cell itself are free from impurities; the substrate is completely wettable by the suspension; and the particles have the same diameter and do not adsorb onto the substrate.

Diameter of the Particles. The experimental data for the growth rate of monolayer particle array versus the inverse particle diameter showed a linear dependence, which was to be expected from eq 7 (*27*). This means that the total water evaporation flux per unit length at the array's leading edge, $J_{evap} = l\, j_e$, does not depend on the particle diameter. In other words, the data laying on a straight line according to eq 7 suggest the constancy of the evaporation rate. From the slope of this straight line we obtained $J_{evap} = 8.6 \times 10^{-7}$ cm^2 s^{-1}, assuming $\beta = 1$ in eq 7 – see ref. *27* for details. In an independent experiment by weighing a small vessel with water at the same experimental conditions, we measured that the evaporation rate of pure water is $j_e = 4.3 \times 10^{-6}$ cm s^{-1}. Thus, we calculated the evaporation length, namely, $l = 0.20$ cm. We also observed (using a low-magnification microscope) that the width of the wet particle arrays along the leading edge is around 0.50 cm during the array growth, which is in an acceptable agreement with the calculated l.

Polydispersity and the Domains' Size. Polydispersity of the particles influences the quality of the obtained monolayer particle array films. A reduced polydispersity is necessary, but not sufficient, to obtain large uniformly oriented domains of highly aligned particles. For example, the monolayer of 814-nm particles with polydispersity of 2.8% consisted of small no larger than 20-30 μm in diameter differently oriented domains. In contrast, the monolayer of 953-nm particles with polydispersity of 1% contained single domains as large as 2 mm wide and 5 mm long. In these cases the particles larger than the average size initiated dislocations in the array, thus forming a monolayer of differently oriented domains. Much larger particles or impurities, which can be regarded as particles strongly deviated in size, tremendously disturbed the monolayer growth and, in most of the cases, caused multilayer formation. Due to the slower growth of the multilayers (note in eq 7 that the number of layers, k, is in the denominator) voids may also appear behind the large impurities. However, the developed techniques for coating the solid surfaces by 2D particle arrays showed that the process of array formation is inherently stable and such kind of instabilities are damped with time. For example, a bilayer or even a trilayer initiated by a particle aggregate may cause a void area but, finally, the monolayer growth is restored (27). Further, the monolayer continues to grow if it does not meet another large particle.

Domain Size and the Wetting Film Thickness. Large-sized particle-crystalline 2D array as those shown in Figure 6C were obtained using the suspension of 953-nm polystyrene particles. The single-colored regions in this photograph show a single domain of highly ordered particles or, at least, a group of domains that are unidirectionally aligned. The electron microphotographs of this monolayer array confirmed the dense hexagonal packing of the particles, which has been formed during the process of array growth and is seen in Figure 4. One sees in Figure 6C that the domain size in the central microslide region is much larger than those in the peripheral regions, where the coloring shows even some disturbances. Actually, we have controlled the array growth process by observing only along the central microslide region, where the film thickness was regulated to be equal or slightly less than the diameter of the particles. The ability to form large-sized domains mainly depends on the quality of the suspension and the thickness and stability of the wetting film on the substrate plate. With high quality suspensions, which means free from impurities with monodisperse particles, we can form larger or smaller domains by controlling either the wetting film thickness at the array's leading edge or the rate of evaporation. In our experimental setup, the evaporation rate on the plate periphery was higher than that on the plate central area. Thus, the wetting film on the plate peripheral areas had a thickness smaller than that on the central areas. The particles flowing into the thinner film regions with the suspension influx started to attract each other due to the lateral capillary immersion forces thus forming small hexagonally packed aggregates. These 2D aggregates were pressed to the array's leading edge in different orientations. The further alignment of the particles was hampered due to the pressing by the capillary menisci towards the substrate plate. The particles approaching the array's leading edge through the wetting film with thickness of the order of the particle diameter (along the central plate regions) do not attract each other due to the immersion forces. They were pressed by the hydrodynamic flux one by one, thus finding free positions in the growing domains. The dislocations in this case are formed due to the slightly larger particles failed to be removed from the suspension.

The film thickness becomes of tremendous importance for the formation of uniform monolayers when the particle size decreases down to nanometer scale. For example, after spreading an amount of water on the microslide plates, we waited until the formed film starts to shrink from the periphery and, then, measured its thickness. For these centimeter-sized water films we measured a thickness of about 120 nm, which scattered with ± 5 nm for successively formed films. Unexpectedly, this value within the experimental error did not depend on the presence of sodium dodecyl sulfate and sodium chloride. The measured thickness of water films on horizontal microslides explains the experimental fact that the monolayer particle array of 144-nm particles were easily formed on an inclined microslide (25). In these

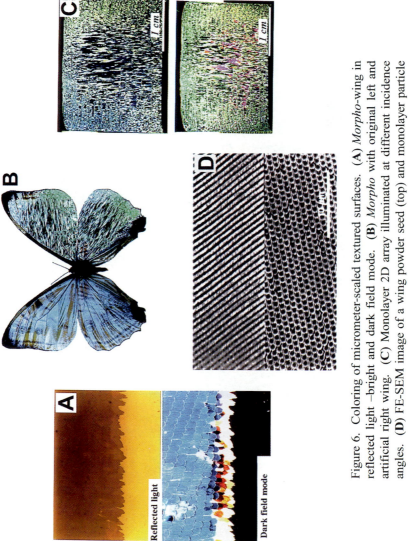

Figure 6.   Coloring of micrometer-scaled textured surfaces. (**A**) *Morpho*-wing in reflected light –bright and dark field mode. (**B**) *Morpho* with original left and artificial right wing. (**C**) Monolayer 2D array illuminated at different incidence angles. (**D**) FE-SEM image of a wing powder seed (top) and monolayer particle array.

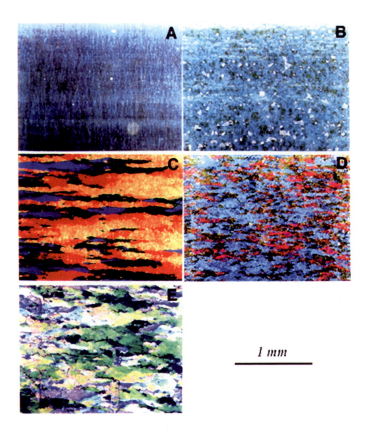

Figure 7. Microphotographs of monolayer particle array films illuminated by sun light at an incidence angle of 58° and an observation angle of 0° (i.e., normal to the array). The particle diameters are: **(A)** 309 nm, **(B)** 479 nm, **(C)** 953 nm, **(D)** 1083 nm, and **(E)** 2106 nm. (Reproduced with permission from ref. *27*. Copyright 1996 American Chemical Society.)

experiments, we succeeded to form about 1-cm^2 large homogeneous dense particle monolayer (with ellipsometric average thickness of 136 ± 2 nm and refractive index of 1.35 ± 0.01) and about 0.3-cm^2 large dense particle bilayer.

The Monolayer Particle Array as Diffraction Gratings

A dependence of the color pattern on the size of the particles is shown in Figure 7. The microphotographs were taken from a direction normal to the array (i.e., observation angle of 0°) and illuminated by sun light at an incidence angle of 58°. The particle diameters in the monolayer arrays were 309 nm (Figure 7A), 479 nm (7B), 953 nm (7C), 1083 nm (7D), and 2106 nm (7E). These microphotographs show that monolayer array from small particles, especially those with diameters less than the wavelength of visible light, exhibit uniform coloring due to interference of the light. In contrast, two-dimensional arrays from particles that have diameters larger than the wavelength of the visible light exhibit brilliant diffraction colors, which depend also on the size and orientation of the domains.

The theory of the diffraction grating has been already developed (*33*). However, the mechanical way for producing echelette or echelon gratings is very difficult and requires special equipment. The difficulties are greater as higher is the groove density. We hope that our technique for producing 2D array allows to produce good quality gratings for reflected and transmitted light in relatively cheep way. In fact, there are differences between the diffraction gratings made of stripes and particle array, which have 6-fold axial symmetry. These differences must be accounted for using the theory of diffraction of a 2D arrangement of assembled particles (*34*). For the time being, however, we just note the light dispersion from monolayer particle arrays. Figure 7, for example, illustrates a decomposition of the white light in colors (see also eq 18 in ref. *33*). For taking these pictures, the incidence angle was fixed at 58° and the observation one at 0°. The color depended on the grating period and the order of interference. Monolayer arrays of 309-nm particles in Figure 7A are dark but slightly blue due to the interference as described in ref. *17*. Using 309-nm particle monolayers we were able to observe only the blue grating spectral color when both the incidence and the observation angles were above 60°. As the size of the particles becomes higher, the decomposition of the white light in colors becomes clearly dependent on the domain orientation. Carefully observing the hexagonally packed particle monolayer (see, e.g,, the upper part of Figure 4) one can find that depending on the array orientation the grating period can have two major values: the interparticle distance (which for a densely packed monolayer is equal to the diameter of the particles) or the interparticle distance multiplied by $\sqrt{3}/2$. Taking also into account that only the first and second order interference maximums have enough intensity to be seen, one can explain the appearance of four different distinguishable colors in Figures 6C and 7C.

Comparison between Coloring of a *Morpho*-Butterfly Wing and 2D Array

Regular structure of textured surfaces often brings about an unexpected effect of color mimicking the natural beauty such as opal and butterfly wing. The colors and structure of our particle array samples were compared to those of the wings of *Morpho* butterflies. We examined the wings of three types butterflies by using optical and electron microscope as well. Two of them are beautifully colored and the electron microscope observations showed a periodicity of the surface fiber type structure of the order of 650 nm for the saturated blue *Morpho rhetenor* (*Morpho (Cypritis) rhetenor rhetenor* Cramar from Guiana) and of the order of 800 nm for the light blue *Morpho sulkowskyi* (*Morpho (Cytheritis) stoffeli stoffeli* Le Moult & Real from Peru). The third butterfly (caught evening time around Tokyo), which is almost colorless, has the same shaped structure of the wings surface but the periodicity is of the order of 10 μm. These experimental facts imply that the periodicity of the surface fine texture is an important factor for producing of brilliant natural colors as, e.g., the colors of morphos.

The colors and structure of our 953-nm particle array samples are compared to those of the wings of *Morpho sulkowskyi*. The similarities are shown in Figure 6. We chose the pair of particle monolayers and butterfly to mimic each other by coloring. The naked-eye comparison is shown in Figures 6B and 6C. The optical microscopy image of a butterfly wing in reflected light bright and dark field mode is shown in Figure 6A. Detailing the surface with a comparison of the surface periodicity are seen in the electron microphotographs in Figure 6D. Our experiments showed that for the same incidence angle of illumination the dispersed light coloring of the textured surfaces mainly depends on the texture periodicity but, also, on the orientation of the surface domains. As seen in Figure 6A the scales on the butterfly wing are unidirectionally ordered around a preferred direction like tiles on a roof, whereas the domains in the particle array (Figure 6C) are randomly oriented. This fact results in some color-pattern differences between the wings and particle arrays (Figure 6B). The main difference is that of the single coloring of the wing in comparison with color change of the monolayer array, when the incidence and the observation angles are varied. This difference may be explained by the very complicated fine structure of the *Morpho* wing (*35*) in comparison with the hexagonally packed particle.

Conclusions

We found a method for continuous formation of 2D particle array, analyzed the parameters determining the array growth, developed an experimental setup, and produced centimeter-size 2D array from small polystyrene particles. The obtained 2D particle array illuminated by white light mimic the dispersed light coloring of *Morpho*-butterfly wings. Microscopic observations showed that the similarities in the coloring of both particle array and butterfly wings are due to the same scale periodicity of the surface texture, which is of the order of or a little above the wavelength of visual light. Our findings are quite versatile for a future use of 2D particle array to mimic in some aspects, e.g., coloring, naturally textured surfaces. However, the best application of our method to fabricating particle array may be found in producing optical and electronical devices as, for example, high density digital memories.

Literature Cited

1. Deckman, J.A.; Dunsmuir, J.H. *J. Vac. Sci. Technol. B* **1983**, *1*, 1109-1112.
2. Weitz, D.A.; Gramila, T.J.; Genack, A.Z.; Gerstein, J.I. *Phys. Rev. Lett.* **1980**, 45, 355-358

3. Burch, J.M. *Nature* **1953**, 171, 889-890
4. Yoldas, B.E.; Partlow, D.P. *Applied Optics* **1984**, *23*, 1418-1424.
5. Hinz, P.; Dislich, H. *J. Non-Cryst. Solids* **1986**, *82*, 411-416.
6. Hahn, R.E. Seraphin, B.O. In *Physics of Thin Films*, Academic: New York, 1978.
7. Yablonovich, E.; Cody, G. *IEEE Tran. Electron Devices* **1982**, *ED-29*, 300-305.
8. Hayashi, S.; Kumamoto, Y.; Suzuki, T.; Hirai, T. *J. Colloid Interface Sci.* **1991**, *144*, 538-547.
9. Perrin, J. *Ann. Chim. Phys.* **1909**, *18*, 5-114.
10. Deckman, J.A.; Dunsmuir, J.H. *Appl. Phys. Lett.* **1982**, *41*, 377-379.
11. Deckman, J.A.; Dunsmuir, J.H.; Garoff, S.; McHenry, J.A.; Peiffer, D.G. *J. Vac. Sci. Technol. B* **1988**, *6*, 333.
12. Dunsmuir, J.H.; Deckman, J.A.; McHenry, J.A, **1989**, J. A. U. S. Pat.No. 4,801,476.
13. Denkov, N.D.; Velev, O.D.; Kralchevsky, P.A.; Ivanov, I.B.; Yoshimura, H.; Nagayama, K. *Langmuir* **1992**, *8*, 3183-3190.
14. Dushkin, C.D.; Yoshimura, H.; Nagayama, K. *Chem. Phys. Lett.* **1993**, *204*, 455-460.
15. Nagayama, K. *Phase Transitions* **1993**, *45*, 185.
16. Denkov, N.D.; Velev, O.D.; Kralchevsky, P.A.; Ivanov, I.B.; Yoshimura, H.; Nagayama, K. *Nature*, **1993**, *361*, 26.
17. Dushkin, C.D.; Nagayama, K.; Miwa, T.; Kralchevsky, P.A. *Langmuir*, **1994**, *151*, 79.
18. Dimitrov, A.S.; Dushkin, C.D.; Yoshimura, H.; Nagayama, K. *Langmuir*, **1994**, *10*, 432-440.
19. Kralchevsky, P.A.; Paunov, V.N.; Ivanov, I.B.; Nagayama, K. *J. Coll. Interface Sci.*, **1992**, *151*, 79-94.
20. Velev, O.D.; Denkov, N.D.; Paunov, V.N.; Kralchevsky, P.A.; Nagayama, K. *Langmuir*, **1993**, *9*, 3702.
21. Kralchevsky, P.A.; Nagayama, K. *Langmuir*, **1994**, *10*, 23-36.
22. Blake, T. D. *in Surfactants (Ed. T.F. Tadros)*, **1984**, 221, Academic Press, London.
23. Pansu, B.; Pieranski, Pi.; Pieranski, Pa. *J. Phys.* **1984**, *45*, 331-339.
24. Kralchevsky, P.A.; Ivanov, I.B.; Nikolov, A.D. *J. Colloid Interface Sci.* **1986**, *112*, 97-107.
25. Dimitrov, A.S.; Nagayama, K. *Chem. Phys. Lett.* **1995**, *243*, 462-468.
26. Adachi, E.; Dimitrov, A.S.; Nagayama, K. *Langmuir*, **1995**, *9*, 1057-1060.
27. Dimitrov, A.S.; Nagayama, K. *Langmuir*, **1996**, *12*, 1303-1311.
28. Dimitrov, A.S. *Ph.D. Thesis*, The University of Tokyo: Tokyo, 1995.
29. van Winkle, D.H.; Murray, C.A. *Phys. Rev. A* **1983**, *34*, 562-573.
30. Pieranski, P.; Strzelecki, L.; Pansu, B. *Phys Rev. Lett.* **1983**, *50*, 900-903.
31. Murray, C.A.; van Winkle, D.H. *Phys. Rev. Lett.* **1987**, *58*, 1200-1203.
32. Alder, B.J.; Hoover, H.G.; Young, D.A. *J. Chem. Phys.* **1968**, *49*, 3688-3696.
33. Born, M.; Volf, E. *Principles of Optics*, 6th ed.; Pergamon Press: New York, 1993, pp 401-414.
34. Goodwin, J.W.; Ottewill, R.H.; Parentich, A. *J. Phys. Chem.* **1980**, *84*, 1580-1586.
35. Ghiradella, H. *Applied Optics* **1991**, *30*, 3492-3500.

Chapter 29

Preparation and Characterization of Cross-Linked Hydroxypropyl Cellulose Hydrophilic Films

Cheng Qian Song[1], Morton H. Litt, and Ica Manas-Zloczower

Department of Macromolecular Science and Engineering, Case Western Reserve University, Cleveland, OH 44106

Hydroxypropyl cellulose(HPC) films with isotropic or liquid crystalline structures have been prepared by photoinitiated crosslinking of HPC in dimethylacetamide(DMAc) solutions. The equilibrium swelling degree for both kinds of films depended on the crosslinking conditions. p-Nitrophenol, indophenol sodium, and Chrome Black T were used to evaluate the permeation behavior of HPC films. The influence of penetrant size, shape and interaction with film molecules on the diffusivity was also investigated.

Fundamental studies of the transport of small organic molecules through polymeric films or membranes have been actively pursued for many years (1-6). The investigations generally included a wide variety of polymers and penetrants over a broad range of experimental conditions. Moreover, hydrophilic films largely used in biomedical applications, have also been studied extensively (7-13). However, there are no reported data on penetration studies through liquid crystalline hydrogel films. A comparison of the permeation properties of isotropic and liquid crystalline films in this study correlates the structural regularities with the permselectivity.

The permselectivity of a film depends on its structure and properties. Characteristics of the structure include the average pore size and size distribution. The molecular interaction between penetrant and film molecules is another major factor which determines the film separation efficiency.

Hydroxypropyl cellulose(HPC) is a hydrophilic polymer with a semi-rigid rod backbone. HPC films can not be used in biomedical applications because the polymer is water soluble. However, HPC can be crosslinked in both the isotropic and liquid

[1]Current address: Research and Development Center, Montell Polyolefins, 912 Appleton Road, Elkton, MD 21921

0097–6156/96/0648–0490$15.00/0
© 1996 American Chemical Society

crystalline states by using photoinitiated crosslinking (*14*). While keeping their hydrophilicity, such crosslinked HPC films can be used as hydrogel films and are considered the first example of hydrogel films which can be obtained in either isotropic or liquid crystalline states.

The HPC films we have prepared could have solute sieving properties as do dialysis or ultrafiltration membranes. Methods often used to determine permeation properties and to estimate pore size of such films are bubble pressure/solvent permeability (*15,16*), gas adsorption/desorption (*17,18*), thermoporometry (*17*), high resolution electron microscopy (*18*) and selective solute sieving (*19-21*). Of the techniques mentioned above, the solute sieving method using well-characterized pure solutes of known size and shape provided information most relevant for characterizing the films. The methodology is quite general and has been widely applied to various biological transport barriers and films (*22*).

Experimental

Materials. Food Grade HPC-LF (weight-average molecular weight 95,000) from Aqualon Company was used after drying in vacuo at 60 °C for 24 hours. Anhydrous N,N-dimethylacetamide (DMAc) (Aldrich) was used as received. Hexamethoxymethylmelamine (HMMM) (Pfaltz & Bauer Inc.), the crosslinking agent, and triphenylpyrylium trifluoromethanesulfonate (TPTS)(Lancaster Synthesis Ltd.), the cationic photocatalyst, were used without further purification. 1,4-Diazabicyclo[2.2.2]octane (DABCO) (Aldrich, used as received) was added to prevent premature crosslinking of the samples.

The penetrants, p-nitrophenol (Matheson Coleman & Bell Manufacturing Chemists), indophenol sodium (Aldrich Chemical Co., Inc.), Chrome Black T (Crompton & Knowles Corp.) and methylene blue chloride (Merck & Co., Inc.) were used without further purification.

Concentration Measurement. The UV absorption spectra of penetrants and their concentrations were determined using a Perkin-Elmer Lambda 9 UV/VIS/NIR spectrophotometer.

Preparation of Films. Isotropic and liquid crystalline HPC films were prepared from 30 wt% and 50 wt% HPC/DMAc solutions respectively. The solution compositions are listed in Table I. The films were fabricated as follows:

1. A solution of DMAc containing crosslinker (HMMM), photocationic catalyst (TPTS) and base (DABCO) was made up as needed before use. The solution was then mixed with HPC powder which had been dried in vacuo at 60 °C for 24 hours.

2. The suspension was mixed by hand for several minutes to dissolve the polymer and then was allowed to stand for one hour in order to achieve uniformity and to eliminate most of the air bubbles.

3. The solution was then poured onto a Pyrex glass plate with a spacer (0.3 mm) and an upper glass plate was clamped over the lower plate containing the polymer solution.

Table I Formula of HPC solutions for the preparation of films

Conc. (wt%)	HPC-LF (g)	DMAc (g)	HMMM[a] (g)	TPTFS[b] (g)	DABCO[c] (mg)
30	2.00	4.67	0.13	0.065	2.65
50	2.00	2.00	0.13	0.039	1.59

[a] Crosslinker: Hexamethoxymethylmelamine (HMMM).
[b] Photoinitiator: 2,4,6-Triphenylpyrylium trifluoromethanesulphonate (TPTFS).
[c] Base: 1,4-Diazabicyclo[2.2.2]octane (DABCO).

4. After standing for 30 minutes, the samples were UV irradiated (GE UV lamp, 275W; sample to lamp distance was kept at about 25 cm) for a preset time at room temperature.

5. The samples were postcured at 65 °C for various times as specified.

6. The films were demolded from the glass plates and immersed in DMAc containing 1 wt% DABCO for two hours to neutralize and remove the acid.

7. The films were then rinsed in distilled water several times and kept in distilled water until they were used.

Permeability Measurement. If a film of thickness L and area A separates two chambers containing a penetrant at different concentrations, the permeation of the penetrant from the high concentration chamber to the low concentration chamber can be calculated according to Fick's first law (*23*):

$$F=D(C_1-C_2)/L \tag{1}$$

where F is the flux of penetrant passing through the film, C_1 and C_2 are the concentrations of penetrant on either side of the film and D is the diffusion coefficient. The surface concentrations C_1 and C_2 can be correlated with the bulk concentrations in the feed cell, C_f and the receiving cell, C_r, respectively:

$$C_1=C_f S; \qquad C_2=C_r S \tag{2}$$

where S is the partition coefficient. Substituting eqn.(2) into eqn.(1) gives:

$$F=DS(C_f-C_r)/L \tag{3}$$

or,

$$F=P(C_f-C_r)/L \tag{4}$$

where P=DS is called the permeability coefficient.

The above derivation is based on the assumption of infinite cells so that the concentrations on each side of the film can be assumed constant. For finite cells, the derivation must be modified to consider the change in concentrations during the course of diffusion. By definition the flux, F, can be expressed as:

$$F=dQ/(Adt)=VdC/(Adt) \tag{5}$$

where dQ is the amount of penetrant transported during the time interval dt, A is the effective area of the film, V is the volume of the cells, and dC is the concentration change during the time interval dt. Combining equations (4) and (5) and integrating the result with the boundary condition $C_r=0$ at $t<=t_g$ (lag time) generates the following working equation for the calculation of permeability coefficients:

$$-(1/2)\ln(1-2C_r/C_o)=(t-t_g)PA/(VL) \tag{6}$$

where C_o is the initial concentration of penetrant in the feed cell.

Plotting the left hand side of equation(6) versus time, allows one to calculate the lag time, t_g, from the intercept and the permeability coefficient, P, from the slope of the line. The diffusion coefficient, D, is related to the lag time, t_g, via (24):

$$D=L^2/(6t_g) \tag{7}$$

The partition coefficient can then be calculated from permeability and diffusion coefficients.

The experimental set-up for the permeability measurement is shown in Figure 1. The feed cell and receiving cell are 500 ml containers with magnetic stirring. Vigorous stirring (500 rpm) was applied to induce turbulent flow of the fluid in order to mitigate the concentration polarization at each film-solution interface (25,26). At the concentration of penetrants used (0.01g/L), all the permeation experiments showed linear correlation between the concentration and permeation time if plotted according to equation (6). Therefore, the influence of concentration polarization is insignificant under the experimental conditions.

Results and Discussion

Swelling Properties of Films. It has been reported (14) that HPC films crosslinked at room temperature have isotropic or liquid crystalline structures depending on the initial HPC concentration. In this study, isotropic films were prepared from 30 wt% HPC/DMAc solutions, while liquid crystalline films were prepared from 50 wt% HPC/DMAc solutions. Since the swelling degree is important for hydrophilic films (27,28), the correlation of equilibrium swelling properties of the films with their preparation conditions is listed in Table II. The swelling degree is expressed by the polymer content of the swollen films. The films used in this work were crosslinked for four hours under UV radiation and postcured for half an hour at 65 °C, which is slightly less than complete cure, in order to maintain a high level of flexibility.

Effect of Penetrant Size on Permeation. In order to understand the permeation behavior of the films, small organic molecules with different molecular weights and sizes were used to characterize the permselectivity. p-Nitrophenol was dissolved in either a basic buffer solution (NaHCO$_3$+NaOH, pH=9.5) to form its salt or in an acidic buffer solution (NaAc+HAc, pH=5.6) to keep it in the neutral form. Indophenol sodium and Chrome Black T were both dissolved in the basic buffer solution to keep their salt form. Neutral indophenol has a limited solubility in water.

The structure and conformation of the penetrants were modeled using the Alchemy II molecular modeling program (Tripos Associates) as shown in Figures 2, 3 and 4. p-Nitrophenol has only one aromatic ring and is flat. Indophenol sodium has two rings connected at an angle which increases its effective cross section. Chrome Black T has an equivalent of four rings which are located in two planes tilted to each other.

As well known, an increase in the penetrant size in a series of chemically similar penetrants generally leads to a decrease in the diffusion and permeability

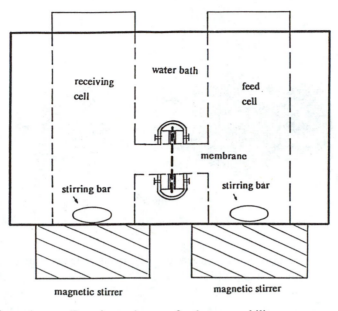

Figure 1   Experimental set-up for the permeability measurement.

Table II Preparation of films with various swelling properties in water

Crosslinking Conditions	Polymer Content (wt%) Isotropic Films[a]	Polymer Content (wt%) LC Films[b]
3.5 hours UV radiation without postcuring	13.4	20.2
4.0 hours UV radiation without postcuring	16.5	26.3
4.0 hours UV radiation with 0.5 hour postcuring[c]	25.7	40.3
4.0 hours UV radiation with 4 hours postcuring[c]	29.9	42.9
4.0 hours UV radiation with 24 hours postcuring[c]	29.4	43.1

[a] Isotropic films were crosslinked from 30 wt% HPC/DMAc solutions.
[b] LC films were crosslinked from 50 wt% HPC/DMAc solutions.
[c] Postcuring temperature was 65 °C.

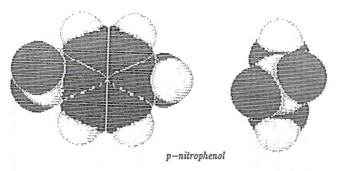

p−nitrophenol

Figure 2   Conformational structure of p-nitrophenol after energy minimization.

Indo−phenol sodium

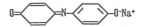

Figure 3   Conformational structure of indophenol after energy minimization.

coefficients. In addition, the more uniform the size of the film pores, the larger the decrease in the diffusion and permeability coefficients as the molecular cross-section increases. The activation energy for diffusion was also found to be proportional to the molecular dimensions of the penetrant molecule, with correlation varying between the first and second power of the cross-sectional area (*29,30*). It is understood that both the structure of the film and the size of the penetrant play an important role in transport phenomena.

Figure 5 shows the permeation of the penetrants through an isotropic I24.5 film (The letter I or LC is assigned to indicate isotropic or liquid crystalline films respectively. The number following the letter gives the polymer content by weight at the equilibrium swelling of films in distilled water). The curves are close to each other. In contrast, the permeation of the penetrants through a liquid crystalline LC40.0 film shows significant differences between the various species (Figure 6). From the plots in Figures 5 and 6 various permeation characteristics were calculated; they are listed in Table III. The permeability coefficients of p-nitrophenol, p-nitrophenol sodium and indophenol sodium through the I24.5 film are almost identical. However the diffusion coefficient decreases from the salt to the neutral form of p-nitrophenol indicating a stronger interaction between the neutral p-nitrophenol and the film molecules.

It is obvious that the difference for a given set of penetrants in both permeability and diffusion coefficients is larger for the liquid crystalline films than for the lower concentration isotropic films. The permeability coefficient decreases with the increase of the molecular interaction or the size of penetrants. The largest molecule studied, Chrome Black T, showed no penetration through either the isotropic or the liquid crystalline films after 72 hrs., though the films turned black due to absorption of the penetrant.

Effect of Penetrant shape on Permeation. In order to study the penetrant shape effect on permeation we made use of methylene blue, a tricyclic planar molecule (*31*), with a molecular weight between those of indophenol sodium and Chrome Black T. Since it is planar it is not hindered very much when passing through the films. Figures 7 and Figure 8 show methylene blue permeation through an isotropic I25.7 film and a liquid crystalline LC39.1 film respectively. Table IV gives the permeability, diffusion and partition coefficients calculated for this system. The permeation parameters measured for methylene blue are smaller than those of indophenol sodium for the isotropic films and are larger for the case of liquid crystalline films. Qualitatively, methylene blue can be considered to have comparable permeability with indophenol sodium. This implies that anisotropic penetrant molecules are oriented and move along their long dimension during the permeation process (*32*). Berens and Hopfenberg (*33*) reported that the diffusivities of n-alkane and other elongated or flattened molecules through amorphous films are higher than the diffusivities of spherical molecules of similar volume or molecular weight. Aitkin and Barrer (*34*) has even tried to establish the linear correlation of diffusivity with the product $l_1 x l_2$, where l_1 is the minimum molecular dimension of the penetrant and l_2 is the next smallest dimension perpendicular to l_1.

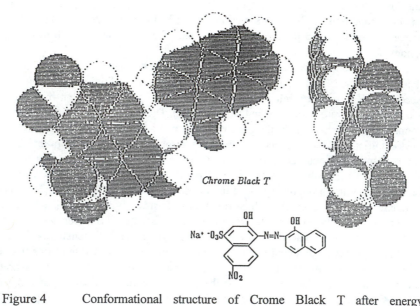

Figure 4   Conformational structure of Crome Black T after energy minimization.

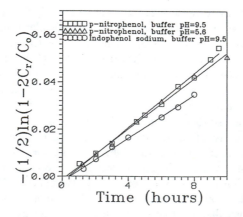

Figure 5   Permeation of penetrants through an isotropic I24.5 film.

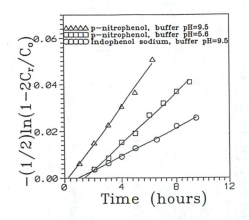

Figure 6   Permeation of penetrants through a liquid crystalline LC40.0 film.

Table III Permeability of small organic penetrants through HPC films

Penetrant	Isotropic Film(I24.5)[a]				LC Film(LC40.0)[b]			
	t_g (hr)	P (cm^2/sec)	D (cm^2/sec)	S	t_g (hr)	P (cm^2/sec)	D (cm^2/sec)	S
p-Nitrophenol pH=9.5	0.23	3.5E-6	2.9E-7	12	0.23	4.0E-6	1.6E-7	25
p-Nitrophenol pH=5.6	0.37	3.8E-6	1.8E-7	21	1.32	2.7E-6	2.8E-8	95
Indophenol sodium pH=9.5	0.44	3.0E-6	1.5E-7	20	0.89	1.5E-6	4.2E-8	35
Chrome Black T pH=9.5	No penetration after 72 hours				No penetration after 72 hours			

[a] Thickness = 0.380 mm.
[b] Thickness = 0.285 mm.

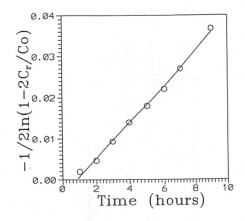

Figure 7   Permeation of methylene blue through an isotropic I25.7 film.

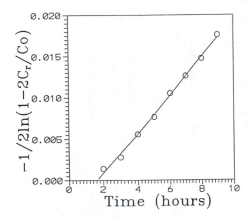

Figure 8   Permeation of methylene blue through a liquid crystalline LC39.1film.

Table IV Permeability of Methylene Blue through HPC films

Penetrant	Isotropic Film(I25.7)[a]				LC Film(LC39.1)[b]			
	t_g (hr)	P (cm^2/sec)	D (cm^2/sec)	S	t_g (hr)	P (cm^2/sec)	D (cm^2/sec)	S
Methylene Blue in 0.1 M NaCl	0.82	2.4E-6	5.4E-8	43	1.63	1.8E-6	5.7E-8	32

[a] Thickness = 0.310 mm.
[b] Thickness = 0.450 mm.

Summary and Conclusions

The swelling degrees of both isotropic and liquid crystalline HPC films can be adjusted by changing the crosslinking conditions. HPC films crosslinked in the liquid crystalline state with higher polymer content(40.0wt%) in the swollen state show better permselectivity for small organic penetrants than films crosslinked in the isotropic state with lower polymer content(24.5 wt%) in the swollen state. The diffusivity for p-nitrophenol was much lower than that for its salt, probably because of the strong interaction between the acidic phenolic hydroxyl group and the basic oxygen in the film. Diffusivities decrease with the increase of the effective size of the penetrants. Chrome Black T has no penetration through either the isotropic or the liquid crystalline films after 72 hours, probably because it is a large molecule and because it contains two phenolic hydroxyl groups.

Acknowledgments

This work was supported by a grant from the Johnson & Johnson Focused Giving Program.

Literature Cited

1.	Park, J.S.; Ruckenstein, E. *J. Appl. Polym. Sci.,* **1989**, *38*, 453.
2.	Cabasso, I.; Jagur-Grodzinski, J.; Vofsi, D. *J. Appl. Polym. Sci.,* **1974**, *18*, 2137.
3.	Fels, M.; Huang, R.Y.M. *J. Macromol. Sci., Phys.,* **1971**, *B5*, 89.
4.	Huang, R.Y.M.; Lin, V.J.C. *J. Appl. Polym. Sci.,* **1968**, *12*, 2615.
5.	Long, R.B. *Ind. Eng. Chem.,* **1965**, *4*, 445.
6.	Barrer, R.M.; Barrie, J.A.; Slater, J.J. *J. Polym. Sci.,* **1958**, *27*, 177.
7.	Yasuda, H.; Ikenberry, L.K.; Lamaze, C.E. *Makromol. Chem.,* **1969**, *125*, 108.
8.	Pusch, W. *Desalination,* **1975**, *16*, 65.
9.	Serensen, T.S.; Jensen, J.B. *J. Non-Equilib. Thermodyn.,* **1984**, *9*, 1.
10.	Misra, A.; Kroesser, F.W. *J. Polym. Symp.,* **1973**, *41*, 145.
11.	Benavente, J. *J. Non-Equilib. Thermodyn.,* **1984**, *9*, 217.
12.	Benavente, J.; Vazquez-Gonzalez, M.I. *Separation Sci. and Tech.,* **1989**, *24*, 1001.
13.	Ishihara, K.; Kobayashi, M.; Ishimaru, N.; Shinohara, I. *Polymer J.,* **1984**, *16*, 625.
14.	Song, C.Q.; Litt, M.H.; Manas-Zloczower, I. *J. Appl. Polym. Sci.,* **1991**, *42*, 2517.
15.	Capannelli, G.; Vigo, F.; Munari, S. *J. Membrane Sci.,* **1983**, *15*, 289.
16.	Munari, S.; Bottino, A.; Capannelli, G.; Moretti, P. *Desalination,* **1985**, 53, 11.
17.	Smolders, C.A.; Vugteveen, E. In *Materials Science of Synthetic Membranes,* Lloyd, D.R., Ed.; ACS, Washington, D.C., **1985**, 327.
18.	Zeman, L.; Tkacik, G. In *Materials Science of Synthetic Membranes,* Lloyd, D.R. Ed.; ACS, Washington, D.C., **1985**, 339.

19.   Schwarz, H.-H.; Bossin, E.; Fanter, D. *J. Membrane Sci.,* **1982**, *12,* 101.

20.   Kassotis, J.; Shmidt, J.; Hodgins, L.T.; Gregor, H.P. *J. Membrane Sci.,* **1985**, *22,* 61.

21.   Leypoldt, J.K. *J. Membrane Sci.,* **1987**, *31,* 289.

22.   Curry, F.E. In *Handbook of Physiology,* Renkin, E.M.; Michel, C.C., Eds., American Physiological Society, Bethesda, MD, **1984**, Chap. 8.

23.   Comyn, J. *Polymer Permeability,* Elsevier Applied Science Publishers, London and New York, **1985**, 5.

24.   Daynes, H. *Proc. Roy. Soc.,* **1920**, *A97,* 286.

25.   Jonsson, G. *Desalination,* **1984**, *51(1),* 61.

26.   Wijmans, J.G. *J. Membrane Sci.,* **1985**, *22,* 117.

27.   Drobnik, J.; Spacek, P.; Wichterle, O. *J. Biomed. Mater. Res.,* **1974**, *8,* 45.

28.   Yasuda, H.; Lamaze, C.E.; Peterlin, A. *J. Polym. Sci., Part A-2, Polym. Phys.,* **1971**, *9,* 1117.

29.   Van Krevelen, D.W. *Properties of Polymers*, 2nd Edn., Elsevier Scientific Publ. Co., New York, **1976**, Chap. 18.

30.   Rogers, C.E. In *Physics and Chemistry of the Organic Solid State,* Fox, D.; Labes, M.M.; Weissberger, A., Eds., Interscience Publ., New York, **1965**, Chap. 6.

31.   Marr III, H.G.; Stewart, J.M. *J. Chem. Soc., Chem. Comm.,* **1971**, 131.

32.   Rogers, C.E. In *Polymer Permeability,* Comyn, J., Ed., Elsevier Appl. Sci. Publ., London and New York, **1985**, Chap. 2.

33.   Berens, A.R.; Hopfenberg, H.B. *J. Membr. Sci.,* **1982**, *10,* 283.

34.   Aitkin, A.; Barrer, R.M. *Trans. Faraday Soc.,* **1955**, *51,* 116.

Chapter 30

Designing Coating Agents for Inorganic Polycrystalline Materials

A. K. Chattopadhyay[1], L. Ghaicha[1], Bai Yubai[2], G. Munger[2], and R. M. Leblanc[2]

[1]ICI Explosives Group Technical Centre, 701 Boulevard Richelieu, McMasterville, Québec J3G 6N3, Canada
[2]Centre de Recherche en Photobiophysique, Université du Québec à Trois-Rivières, Québec G9A 5H7, Canada

This paper deals with the investigation on the influence of four amphiphile monolayers viz. arachidic acid, a mixture of dioctadecyl amine and arachidic acid (3:1), carboxybetaine and dioleoyl-L-α-lecithin (DOPC) on ammonium nitrate crystallization. The study was carried out by means of surface pressure vs. area, surface potential vs. area and fluorescent microscopy. Results indicated a strong effect on ammonium nitrate crystallization in the presence of carboxybetaine and dioctadecylamine monolayers, whereas an inhibition in crystal growth was observed in the presence of DOPC.

This study involves crystallization behavior of inorganic salts below Langmuir films of organic monolayers for the purpose of investigating the cooperative effects of molecular arrays. A detailed appreciation of recognition factors of inorganic surfaces will lead us to the rational design of new surface active molecules applicable for inorganic surface coating agents, scale inhibitors and rust preventives (1-8).

Amongst various commercially available inorganic salts, ammonium nitrate probably occupies the largest portion in consumer market. In order to solve storage and transportation problems associated with ammonium nitrate (AN), there is a need to produce consistently noncaking particles, generally available in globular forms known as prill. It is an industrial challenge for the commercial manufacturers of ammonium nitrate to overcome the caking tendencies between ammonium nitrate particles and their subsequent changes in physical properties with time. This happens due to the multiple crystalline phase transitions in AN that occur with temperature changes and are accelerated in the presence of moisture. It is well known that the phase IV ↔ III transition of AN, which occurs at around 32°C, promotes interparticle bridging or caking.

Polymorphism of AN

Cubic	⇌	Tetragonal	⇌	Rhombic
I		II		III
170 - 123°C		125 - 84°C		84 - 32°C

Bipyramidal Tetragonal   ⇌   o, Rhombic

V   IV

<-18°C   32 - (-17)°C

The study of IV ↔ III transition kinetics has historically received a great deal of attention. However, no systematic study has ever been carried out to understand the effect of various additives (auxiliary host molecules) on the crystal growth pattern of AN. The use of crystal morphology as a determinant of specific interactions between crystal surfaces and auxiliary molecules has received considerable attention in the last decade. In the crystal structure of AN of phase IV, (100), (001), (110) and (011) planes are classified according to the arrangement of NH_4^+ and NO_3^- ions in their surfaces.

The o,Rhombic crystal stucture of phase IV AN is shown in figure 1. The (001) and (110) planes both consist of alternating layers of either NH_4^+ or NO_3^- ions whilst the (100) and (011) planes comprise layers containing both cations and anions. Some dyestuffs (well known as AN habit modifiers) viz. Cu-phthalocyanines bring about habit modification as a result their bipolar nature and adsorb on the (100) and (011) planes via SO_3^- and $-NH_3^+$ groups. Similarly additive molecules containing anionic groups are found to exert an effect via the (001) and (110) planes where they can adsorb directly onto the layers containing only NH_4^+.

From the previous work done on AN crystal habit modification aided by various additive molecules indicated a mechanistic relationship between the adsorption of the additive molecules and lattice matching. Rationalisation of such mechanism may provide a predictive tool for the design of surface coating agents useful for AN. The specific adsorption of additive molecules onto the crystal lattices and its effect on overall crystallization and crystal shapes provides a basis for rational molecular design of coating agents.

The process of crystallization that occurs in organisms can also be cited in this regard where it is largely influenced by the structure-activity relationships and recognition phenomena. For example, in the animal world the nature serves its purpose by dictating crystallization of calcium carbonate to occur in different morphological forms which we observe in the formation of bones, teeth, shells etc. Such variations in morphology are principally guided by the nature of proteins (host molecules) that influences the crystallization of calcium carbonate (9). The nature of association between an inorganic phase and an organic film principally determines the overall crystal growth pattern, its morphology and subsequent material property of the crystallized inorganic substance.

The roles of molecular recognition phenomenon at the interface of 3-D and 2-D inorganic crystals have been demonstrated by various authors (5-9). This provided us the impetus to carry out studies on crystallization of ammonium nitrate under the influence of organized monolayers of various amphiphiles. The major motivation behind this work was primarily to understand the nature of interactions of various organic substances on ammonium nitrate crystallization (9-19).

The Langmuir films have been widely used to provide organized surfaces in order to induce epitaxial crystallization. The studies revealed that the influence of charged monolayers on crystallization can be induced under the conditions where there is a structural and stereochemical matching between the attached crystal phase and the polar head groups constituting the organic surfaces (6-12).
In the present paper, crystallization of ammonium nitrate (AN) under compressed Langmuir monolayers of arachidic acid, dioctadecylamine, carboxybetaine and dioleoyl-L-α-lecithin (DOPC), is investigated by measuring surface pressure versus surface area, surface potential versus surface area, and using Fluorescence Microscopy and Scanning Electron Microscopy (Π- A, ΔV-A, FM and SEM).

The main purpose of the present work was to study the monolayer properties of the organic phases on AN solution substrates of different concentrations as well as to investigate the effect of the nature of monolayers in inducing crystallization and growth of AN crystals. The monolayers of amphiphile molecules can be served as useful models for elucidating structure and function of the polar moieties on their interactions with the subphase. This study provides some important indications about AN crystallization under the monolayers of amphiphiles mentioned above.

Materials and Methods

Pure grade arachidic acid (Aldrich, USA), dioctadecylamine (Applied Sciences Laboratories Inc.), Carboxybetaine, sample donated by Le Perchec Group (13), and dioleoyl-L-a-lecithin, Sigma Inc., were used as received without further purification. The molecular structures of the amphiphiles are given in figure2.

A mixture of chloroform (spectrophotometric grade from Burdick and Jackson) and ethanol (1:1) was used as the spreading solvent for the amphiphiles. Both solvents were distilled before use. Recrystallized AN was obtained from J.T.Baker Inc.(USA). To eliminate any organic impurities from the subphase, the AN solutions were filtered and washed twice with chloroform before use. The quality of the AN solution subphase was also tested prior to each run by compressing the surface in the absence of a monolayer while noting any rise in surface pressure.

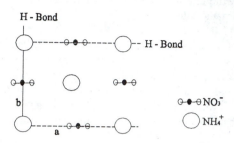

AN phase IV

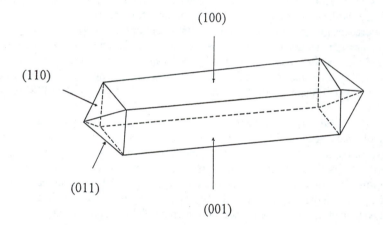

Figure 1.   Phase IV crystal structure of ammonium nitrate.

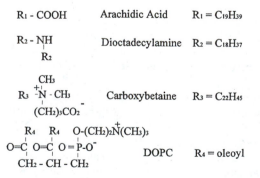

Figure 2.   Molecular structures of the amphiphiles used.

The behaviour of AN crystallization was studied by surface pressure-area and surface potential-area isotherms. The monolayers were spread on aqueous subphase containing AN, where AN concentration was varied between 0-55% by weight. The surface potential-area isotherms were obtained simultaneously with an automated Langmuir trough described elsewhere (*14*).

The surface crystallization of AN was directly observed on the Langmuir trough using fluorescence microscopy (FM). The total surface area of the trough used for this purpose was 200 mm by 15 mm and it was milled to a depth of 3 mm. The fluorescence technique described elsewhere (*15*) involves addition of a small amount of fluorescent probe to the subphase. Rhodamine B, used as a fluorescent probe, was obtained from J.T.Backer Inc. (USA). The surface crystallization was investigated by FM with 55% aqueous subphase under two different conditions:
(a) The rhodamine solution (1% of the monolayer concentration) was injected in the subphase after the monolayer was spread and stabilized.
(b) The rhodamine solution was injected in the subphase prior to spreading the monolayer.
The results obtained from both methods were similar.

SEM pictures of AN crystals grown on the Langmuir-Blodgett films of the surfactants were taken by Hitachi S-2700 Scanning Electron Microscope. The L-B films were formed by transferring the surfactant monolayers onto the silicon slides at a constant surface pressure of 20 mN/m by means of an autoadvancing Teflon barrier. Eight monolayers of each surfactant were deposited by horizontal X-type transfer from AN subphase on the silicon slides and AN crystals were allowed to grow in a desiccator.

Results and Discussion

Arachidic acid. The surface pressure-area isotherms of arachidic acid on pure water and on AN subphases at pH 4.5 are shown in figure 3. With the increased AN concentration in the aqueous subphase, the monolayers were changed from condensed state to an expanded one, particularly below the phase transition point at ~24mN/m. However, above the phase transition there was no significant change in molecular surface area of arachidic acid with the increase in AN concentration. Further increase in molecular area was also noticed with the increase in waiting time before the film compression. In fig. 4, the differences in compression behavior of arachidic acid films on 55% AN solution are shown.

These differences could probably due to the (i) increased spacing between the head groups due to increased repulsion and (ii) increase in effective head group

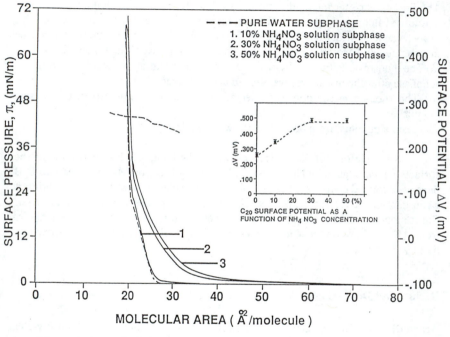

Figure 3.   Π-A and ΔV-A isotherms of arachidic acid at AN
solution - air interface.
Inset : ΔV as a function of AN concentration.
(Reproduced from reference 8b. Copyright 1996 Marcel
Dekker.)

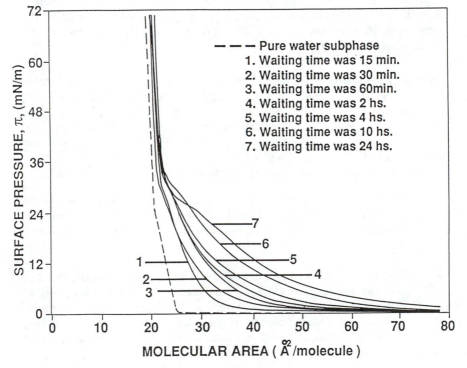

Figure 4. II-A isotherms of arachidic acid as a function of waiting time. (Reproduced from reference 8b. Copyright 1996 Marcel Dekker.)

volume due to the association of the head group moiety with NH_4^+ and NO_3^- through hydrogen bonding(*16*). The surface potential as shown in the inset of fig.3, was found to increase with increased AN concentration. The increase in surface potential of arachidic acid films with increased AN concentration also reveals that there is an increase in surface charge accumulation at the monolayer level.

The FM micrographs of AN surface crystallization taking place in the presence of arachidic acid monolayers are shown in figure 5. From these micrographs it was evident that the film compactness of arachidic acid monolayers (by viewing the monolayers at various surface pressures) had hardly any effect on surface crystallization. The FM images remained more or less similar at all stages of film compression. This reveals that the -COOH as a polar head group has minimum or no influence on AN crystallization in 2D plane.

Dioctadecylamine (DOA). DOA molecules are comprised of two long hydrophobic chains and a secondary amine functional group . Because of its poor hydrophilicity, DOA molecules do not form a stable monolayer. The improvement in their monolayer stability can often be observed by mixing the pure monolayer with some long chain carboxylic acid or alcohol. In order to achieve a stable monolayer, DOA was mixed with arachidic acid at a molar ratio of 3:1. From Π-A and ΔV-A isotherms of DOA/ arachidic acid mixed monolayers shown in figure 6, several points are apparent:

•   The Π-A profiles of the mixed monolayers in the presence of AN in the subphase are completely different from the one obtained from pure water. This indicates that the film undergoes some major changes in the presence of AN.

•   Regardless of the concentration of AN in the aqueous subphase, the Π-A isotherms exhibit two distinct discontinuities, the first one occurring at ~10mN/m and the second one at ~24mN/m. The second discontinuity can be attributed to the phase transition of pure carboxylic acid present in the film, whereas the first transition in Π-A profiles could be due to the demixing of the film in the presence of AN.

•   Contrary to the arachidic acid monolayers, the mixed monolayers show a decrease in surface potential in the presence of AN. This indicates a change in physico-chemical properties of the monolayer contributed by the -NH-groups of DOA. An opposite trend in surface potential can be attributed to the reversal of monolayer charge through protonation of amine groups of DOA and their subsequent interactions with AN (*20,21*).

The effect of DOA/arachidic acid monolayers on surface crystallization of AN was investigated by FM at different surface pressures (figure 7) with a subphase

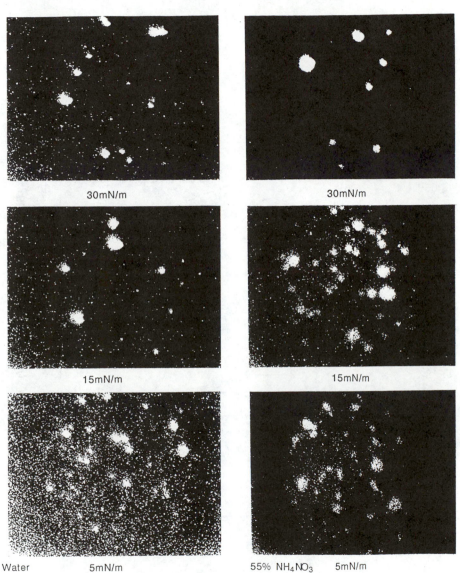

Figure 5.   FM images of arachidic acid spread on pure water and
55% AN solution at surface pressures of 5 mN/m, 15 mN/m and
30 mN/m.
(Reproduced from reference 8b. Copyright 1996 Marcel Dekker.)

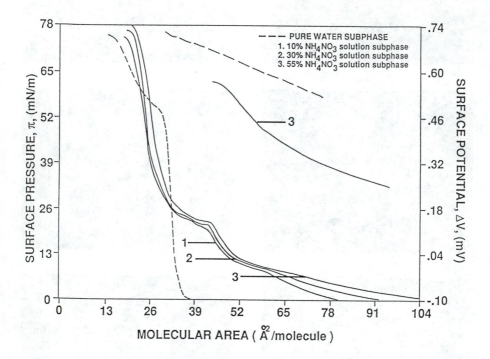

Figure 6.   II-A and ΔV-A isotherms of DOA/arachidic acid spread on pure water and 10%, 30% and 55% AN solution subphases.
(Reproduced from reference 8b. Copyright 1996 Marcel Dekker.)

20mN/m

20mN/m

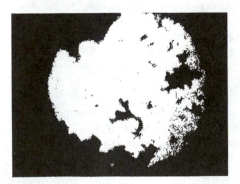

15mN/m

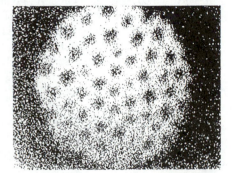

15mN/m

Water   6mN/m

55% NH$_4$NO$_3$ 6mN/m

Figure 7.   FM images of DOA/arachidic acid spread on pure water and 55% AN
solution at 6mN/m, 15mN/m and 20mN/m surface pressures.
(Reproduced from reference 8b. Copyright 1996 Marcel Dekker.)

comprising 55 wt% AN . The nucleation and subsequent growth of crystals which appeared visually as domains at the monolayer surface, took place within 15 minutes after compressing the films at a desired pressure. The observed induction time for the domains to form, which was apparently absent in the case of arachidic acid, strongly suggests that there exists a specific interaction between the DOA polar head groups and AN. The size and stability of such domains depend mostly on the monolayer compressibility. Viewed from the top, the crystals appeared to be dendritic in morphology. It must be noted in this regard that no such domains were observed with a subphase containing water only. This rules out the possibility of formation of any domains of amphiphiles caused by surface micellization.

Carboxybetaine (CB). Figure 8 shows the II-A and ΔV-A isotherms of pure carboxybetaine. The monolayer at the air-water surface has a fairly large compressibility, as implied by the fact that the surface pressure begins to rise at about 160Å^2 and collapses at about 60Å^2, giving an area of compression around 100Å^2. It can also be seen that the monolayer collapses at relatively higher pressure around 40Å^2, suggesting that a stable monolayer is formed at the air-water surface.
In the presence of AN solution subphase, however, the CB monolayers exhibited decrease in collapse pressure as well as molecular area with increased AN concentration.

From the molecular structure of this surfactant, it can be seen that the head group comprises a tether group which bears both positive and negative charges. In the presence of AN, when the monolayer is compressed, the tether group is bent and squeezed out of the subphase which can explain the large decrease in the molecular area.

Crystallization of AN under CB monolayers at a 55 wt% AN solution - air interface was observed by means of FM (figure 9). After compressing the film at a desired surface pressure, it also required a certain period of induction time for the flower-like crystal domains to be visible by FM. It should be noted that FM images of Carboxybetaine spread on pure water were similar to those of DOA/arachidic acid monolayers. From FM studies, visually a marked difference between CB and DOA/arachidic acid monolayers was observed with regard to the rate of formation of AN crystals and their growth patterns. The study revealed that CB monolayers accentuated surface crystallization of AN more than DOA/arachidic acid mixed films.

Dioleoyl-l-α-lecithin (DOPC). Despite having a zwitterionic head group, DOPC showed a different behaviour when compared with carboxybetaine. The major differences between these two molecules stems from the differences in head group size, hydrocabon chains and the nature of ionic groups. DOPC molecules comprise double hydrocarbon chains with large head groups containing phosphonate ions.
The presence of AN in the aqueous subphase expands the DOPC monolayer and

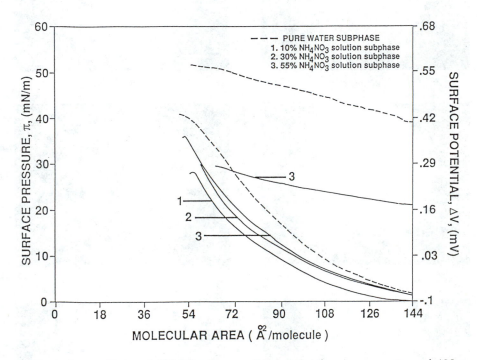

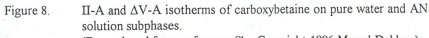

Figure 8.   Π-A and ΔV-A isotherms of carboxybetaine on pure water and AN solution subphases.
(Reproduced from reference 8b. Copyright 1996 Marcel Dekker.)

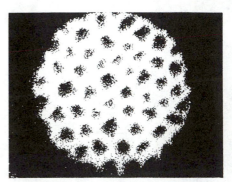

20mN/m   20mN/m

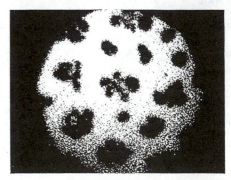

10mN/m   10mN/m

Water   5mN/m   55%NH$_4$NO$_3$ 5mN/m

Figure 9.   FM images of carboxybetaine spread on pure water and 55% AN
                solution at 5mN/m, 10mN/m and 20mN/m.
                (Reproduced from reference 8b. Copyright 1996 Marcel Dekker.)

decreases its surface potential values (fig.10). The association of AN with the polar head groups in the DOPC monolayer has expanding and fluidizing effects, as it is indicated by the enhanced compressibility of the monolayers.

FM micrographs (figure 11) exhibited no surface crystallization of AN under the compressed monolayer of DOPC. This indicates that DOPC, unlike other amphiphilic systems of the present study, inhibits AN crystallization.

The results obtained from four different amphiphilic monolayer systems indicate that the crystallization of AN is influenced by the nature of molecules present in the monolayers. The changes in the functionalities of the polar head groups produce significant differences in nucleation and growth pattern of AN crystals as observed by FM. These studies also showed that both DOA and CB with amine functionalities present in their polar head groups induce AN crystallization more favorably than arachidic acid or DOPC molecules.

The crystallization of AN in the presence of amine group is due to the stereochemical correspondence of ammonium ions and the protonated amines of the surfactant head groups. At a pH 4.5 (pH of the saturated subphase) both DOA and CB polar groups bear a positive charge. In such events, it is possible that the mode of interaction of AN at the interface occurs through the nitrate ions and the protonated amines head groups of the surfactants. This may provide a degree of stereochemical recognition between the organic-inorganic boundary layer.

For proper interpretation of the results, however, further elucidation involving molecular packing as well as structural details such as position, conformation and alignment of the head groups in relation to the disposition of AN crystals is required. DOPC is found to inhibit AN crystallization. Although DOPC contains amine functional groups, the crystal inhibition property of DOPC presumably comes from the presence of phosphonate functionalities which are known to be crystallization inhibitors and habit modifiers when present in various mineral salts (*22-23*). In spite of the widespread application of phosphonates in industrial processes, the mechanism of their action is not yet fully understood.

The SEM micrographs of AN crystals grown on L-B films of two different surfactants viz. DOA and CB (figure 12) clearly demonstrate that DOA influences the crystals to grow in needle shaped o,rhombic forms of phase IV, whereas CB modifies the pattern in the form of plates. In the case of DOA, the molecules comprising -NH_2^+- groups act as nucleating sites for AN crystals to grow in o,rhombic forms because of the charge and structural similarity. However, in the presence of CB molecules with zwitterionic head groups comprising quaternary amines and carboxylate ions provide a major effect on the habit modification of AN crystals (*16-19*). Such morphological changes in AN crystals grown under L-B films of different surface active agents revealed the importance of surface binding sites. Therefore by

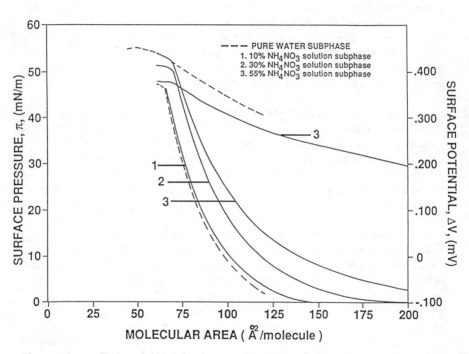

Figure 10.   Π-A and ΔV-A isotherms of DOPC on pure water and AN solution
             subphases.
             (Reproduced from reference 8b. Copyright 1996 Marcel Dekker.)

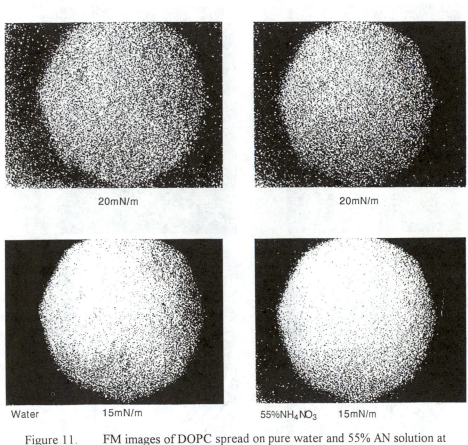

Figure 11. FM images of DOPC spread on pure water and 55% AN solution at 15mN/m and 20mN/m.
(Reproduced from reference 8b. Copyright 1996 Marcel Dekker.)

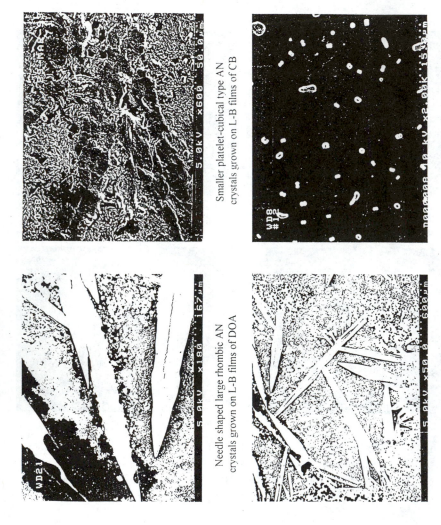

Figure 12.   SEM pictures of AN crystals grown on the L-B films of DOA and CB.

building such stereochemistry and functionality into surface active molecules strongly bound coatings and nucleating layers can be created. As it is briefly mentioned in the Introduction that AN caking occurs due to the phase transition of AN from IV - III. A coating agent, besides providing an organic boundary on top of the AN particles, must alter the habit of AN crystals on the surface of AN particles in order to prevent the outer part of the particles from undergoing phase transitions.

Acknowledgements. The authors are thankful to ICI Explosives Group Technical Centre and NSERC Canada for the financial support.

Literature Cited

1. Hamilton, A.; Lehn, J. M.; Sesslen, J. L.; J. Am.Chem.Soc., **1986**, *108*, 5158.
2. Hosseini, M. W.; Lehn, J. M.; J. Am. Chem. Soc., **1987**,*109*, 7047.
3. Hosseini, M. W.; Blaker, A. J.; Lehn, J. M.; J. Am. C. Soc., **1990**, *112*, 3896.
4. Bianconi, P. A.; Lin, J.; Strelecki, A. R.; Nature, **1991**, *349*, 6307.
5. (a) Weissbuch, I.; Addadi, L.; Lahav, M.; Leiserowitz, L.; Science, **1991**,*253*, 637. (b) Sjolin, C. J.; Agr. Food Chem.,*1972*, *20*, 895.
6. Landau, E. M.; Levanon, M.; Leiserowitz, L.; Lahav, M.; Sagiv, J.; Nature, **1985**,*318*, 353.
7. Landau, E. M.; Dopovitz-Biior, R.; Levanon, M.; Leiserowitz, L.; Lahav, M.; Sagiv, J.; Mol. Cryst. Liq., **1986**, *134*, 323.
8. (a) Landau, E. M.; Wolf, S. G.; Levanon, M.; Leiserowitz, L.; Lahav, M.; Sagiv, J.; J. Am. Chem. Soc., **1989**, *111*, 1436.
   (b) Yubai, B.; Munger, G.; Leblanc, R. M.; Ghaïcha, L.; Chattopadhyay, A. K.; J. Dispersion Sci. Technol. (in press)
9. Rajam, S.; Heywood, B. R.; Walker, J. B. A.; Mann, S.; Davey, R. J.; Birchall, J. D.; J. Chem. Soc. Faraday Trans., **1991**, *87, 5*, 727.
10. Heywood, B. R.; Rajam, S.; Mann, S.; J. Chem. Soc. Faraday Trans., **1991**, *87*, 735.
11. Heywood, B. R.; Mann, S.; Adv. Mater., **1992**, *4*, 278.
12. Mann, S.; Nature, **1988**, *332*, 6160.
13. Chevalier, Y.; Y. Storet, Y.; Pourchet, S.; Le Perchec, P.; Langmuir, **1991**, *7*, 848.
14. Dijkmans, H.; Munger, G.; Aghion, J.; Leblanc, R. M.; Can. J. Biochem. **1981**, *59*, 328.
15. Grainger, D. W.; Reichert, A.; Ringsdorf, H.; Salesse, C.; FE. Let. **1989**, *252*, 72.
16. Ghaïcha, L.; Chattopadhyay, A.K.; Tajmir-Riahi, H. A.; Langmuir, **1991**, *7*, 2007.

17. Gaïnes, G. L.; Insolubles Monolayers at the liquid-Gas Interfaces, Wiley Intersci., New York (1966).

18. Mann, S.;Heywood, B. R.; Rajam, S.; Birchall, J. D.; Proc. R. Soc. Lond. **1989**, *A423*, 457.

19. Black, S. N.; Bromley, L. A.; Cottier, D.; Davey, R.J.; Dobbs, B.; Rout, J.E.; J. Chem.Soc. Faraday Trans., **1991**, *87*, 3409.

20. Ghaïcha, L.; Leblanc, R. M.; Chattopadhyay, A.K.; Langmuir, **1993**, *9*, 288.

21. Chattopadhyay, A. K.; Shah, D.O.; Ghaïcha, L.; Langmuir, **1992**, *8*, 27.

22. Van der Leeden, M.C.; Reedijk, J.; van Rosmalen, G. M.; Estudios Geol. **1992**, *38*, 279.

23. Weijnen, M. C. P.; van Rosmalen, G. M.; Bennema, P.; J. Cryst.Gro. **1987**, *82*, 528.

Chapter 31

Amine–Quinone Polymers as Binders for Metal-Particle-Tape Formulation

Effect of Polymer Structure on Dispersion Quality and Corrosion Protection

Antony P. Chacko, Russell I. Webb, and David E. Nikles

Department of Chemistry and Center for Materials for Information Technology, University of Alabama, Tuscaloosa, AL 35487–0336

The effect of polymer structure for amine-quinone polyurethanes on the metal particle dispersion and corrosion protection in metal particle tape was studied. The polyurethanes were block copolymers containing the monomers: 2,5-bis(N-2-hydroxyethyl-N-methylamino)-1,4-benzoquinone (AQM-1), polytetrahydrofuran diol (M_n = 650), and 2,4-tolylene diisocyanate. Three amine-quinone polyurethanes containing 20, 30 or 40 weight percent AQM-1 were investigated. These polymers had a two-phase microstructure consisting of a discrete crystalline hard segment phase dispersed in a continous amorphous phase. Adsorption isotherms from THF solution showed that the polymer having 40% AQM-1 had the highest affinity for the iron particle surface. Dynamic rheological measurements on the coating formulations showed that storage modulus was lowest for the polymer containing 20% AQM-1. Magnetic remanence data were interpreted to show that the polymer containing 20 % AQM-1 had the best magnetic dispersion. The polymer containing 40% AQM-1 had the best corrosion protection. These observations were interpreted within a film formation model where the had segment partitioned onto the iron particle surface, providing a crystalline barrier against attack by oxygen or moisture.

As the information storage industry continuously endeavors to increase data storage capacity, metal particle magnetic recording tape is the leading candidate for high density storage media. Metal particle (MP) tape consists of fine iron particles dispersed in a polymeric binder on a polymer substrate. The iron particles have the highest saturation magnetization and can be made with coercivities exceeding 2000 Oe, making them the best commercially available magnetic particles (*1, 2*). Iron is inherently susceptible to corrosion and this has led to concerns about the reliability of MP tape for data archiving (*3*). The iron particle corrosion problem has been largely solved by the use of ceramic coatings

0097–6156/96/0648–0525$15.00/0

on the particle (4, 5). However, different iron pigments can give MP tape that display vastly different rates of corrosion (6). Recent data show vastly different rates of corrosion for commercial D-2 tapes from two different manufacturers (7).

MP tape is made by a continuous web coating process where a magnetic ink, consisting of iron particles dispersed in a mixture of organic solvents, is applied to a polyester base film. The ink also contains binder polymers, lubricants, and other additives that facilitate tape handling. The signal output performance of the tape depends on the quality of the magnetic dispersion (8). Magnetic particles have strong tendency to form aggregates because of their strong magnetic attraction forces. High performance media requires consistent and uniform magnetic characteristics over the entire media surface which can be only achieved by a well dispersed suspension of the magnetic particles. The ideal dispersion consists of individual iron particles coated with a binder polymer. The binder polymer magnetically separates the iron particles to minimize magnetic interactions. Magnetic interactions between the particles lead to noise on the tape. However, to maximize the signal level the volume fraction of magnetic particles should be maximized. This leads to a tradeoff between increasing the volume fraction of pigment and minimizing the interparticle magnetic interactions. A typical tape formulation contains 30 volume percent pigment.

Magnetic tape coating formulations are made by dispersing iron particles in a mixture of organic solvents. In the dispersion process the particles are subjected to high shear in a mill, which breaks up aggregates and allows a wetting binder to adsorb on the particle surface. The ideal magnetic ink would contain individual magnetic particles coated with the wetting polymer, suspended in the solvent. The use of improved dispersion techniques (high shear mixing and milling equipment) and new functionalized wetting polymers (thermoplastic polyurethanes or polyvinylchlorides) have enhanced dispersion quality (9). However, the high magnetization (σ_s = 125 to 145 emu/g) and high specific surface area (45 to 60 m^2/g) make these iron particles difficult to disperse.

Our interest in the study of long term performance of metal particle tape has led to the exploration of means to enhance the stability of MP tape. Erhan et al synthesized quinone amine polymers which was reported to have high affinity for surface of iron and displaced moisture from the surface (10, 11). We have synthesized polyurethanes, containing a 2,4-diamino-1,4-benzoquinone functional group, which has been shown to inhibit corrosion of commercial iron pigments in MP tape (12-14). These polymers were also used to replace the ceramic coatings, used in commercial particles, giving iron particles with good corrosion resistance and better magnetic properties (15). These demonstrations inspired us to do a systematic study of structure-property-processing relations of these polymers on their application in MP tape coating formulations. The amine-quinone polyurethanes have a two phase microstructure, consisting of a crystalline hard segment dispersed in an amorphous, soft segment continuous phase (13). We speculate that the amine-quinone functional group strongly adsorbs to the particle surface where the hard segment crystllizes around the particles. This provides a barrier against attack by oxygen or moisture, thus preventing corrosion. The high affinity of the amine-quinone polymers for the iron particle surface suggests that these polymers may be useful for dispersing the particles. In this report we examine the use of amine-quinone polyurethanes to disperse iron particles and make a connection between the rheology of the magnetic inks and the quality of the magnetic dispersion in tapes made from the inks.

Experimental

Materials. The amine quinone polyurethanes were synthesized by melt polymerization from 2,5-bis(N-2-hydroxyethyl-N-methylamino)-1,4-benzoquinone (AQM-1), Terathane 650 (polytetrahydrofuran), and tolylene diisocyanate. The hard segment content was increased by increasing the AQM-1 content from 20 weight percent for AQPU-13 to 30 weight percent for AQPU-16 to 40 weight percent for AQPU-15. 4-N,N-Dimethylaminopyridine was used as polymerization catalyst. The synthesis of amine quinone monomer was described elsewhere (*13, 14*).

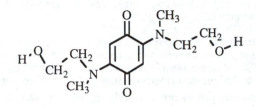

Amine-Quinone Monomer (AQM-1)

Two different commercial polyvinylchloride wetting binders, UCARMAG-536 (Union Carbide) and MR-110 (Nippon Xenon), and a commercial polyurethane, CA-271 (Morton International), were included for comparison. The magnetic pigments were a commercial ceramic coated iron having a coercivity of 1500 Oe, saturation magnetization of 122 emu/g and a specific surface area of 49 m²/g. The average particle diameter was 200 nm. The particles were shown by electron microprobe to have the approximate composition; Fe 60 mole percent, Al 2.5 mole percent, 0.3 mole percent Si, O 38 mole percent. This was consistent with iron particles protected from oxidation with an aluminosilcate ceramic coating.

Adsorption experiments. Adsorption experiments were carried out by adding a known amount of iron particles to a 10 mL solution containing a known concentration the polymer in tetrahydrofuran. The resulting mixture was shaken to ensure complete suspension of metal particles and to ensure intimate contact between metal particles and polymers. The suspension was allowed to equilibrate for a week, the particles allowed to settle, and the supernatant was filtered through a 2 micron pore-size syringe filter to completely remove the particles. The polymer concentrations were determined by measuring the absorbance of the supernatant polymer solution. From the concentration before and after adsorption, the amount of polymer adsorbed in mg polymer/g pigment was calculated.

Magnetic dispersion and coatings. The metal particle dispersions were made by milling a mixture of a the iron particles with a polymer solution in cyclohexanone with 2 mm steel beads using a reciprocating paint shaker for 6 hours. The formulation was simplified by excluding other ingredients in a typical commercial MP tape formulation such as lubricants, antistatic agents, dispersants and abrasives. The particle loading was fixed at 75 weight percent (of the total solid content) and a 20 weight percent solid content was used. For example, a typical formulation contained 7.5 g pigment, 2.5 g polymer and 40 g cyclohexanone. After milling, a portion of the dispersion was removed for rheological study and the remaining was used for coating. MP tape coatings were made by casting the dispersion onto a polyester web using a hand draw-down knife coater. The coating was simultaneously drawn through a magnetic field to longitudinally orient the iron particles. The coatings were dried in an oven at 50°C for 24 hours.

Rheological measurements. Rheological measurements were carried out on a Carri-Med model CS-100 controlled stress rheometer with a 6 cm parallel plate fixture. Constant shear stress measurements were made using the steady mode and sinusoidal oscillations were done using the dynamic mode. The gap height used was 1 mm which was much greater than diameter of the iron particle (200 nm). The samples were directly taken from the milling stage for the measurements and a small shear was applied manually before loading the sample for rheological test. Similar preshear, loading technique and rest time was used for all samples and the reproducibility was generally good.

Magnetic measurements. The magnetic hysteresis loops and isothermal and dc remanence magnetization curves were measured on a Digital Measurement Systems model 880 vibrating sample magnetometer. The remanence curves were analyzed using a technique described by Huang et al. (*16*) to obtain values of the interparticle interaction parameter α introduced by Che and Bertram (*17*). We have used the interparticle interaction field parameter as a measure of the dispersion state of the particles (*18*). In an ideal magnetic dispersion with no agglomerates the value of α would be zero. MP tape samples show negative values for α, indicative of demagnetizing interparticle interactions. The closer α is to zero , the better the magnetic dispersion.

Corrosion experiments. Test pieces (6 mm diameter) were punched from the hand drawn magnetic tape samples and subjected to accelerated aging conditions. The level of iron corrosion was determined by obtaining values for the saturation magnetization (σ_S) from magnetic hysteriesis curves. When iron corrodes it forms non magnetic oxide, thus a decrease in σ_S is a direct measure of the amount of iron corrosion. Test pieces were exposed to accelerated aging environment, either soaked in pH 2 aqueous buffer or placed in a temperature/humidity chamber at 80°C and 80% relative humidity. Periodically the samples were removed , σ_S measured then returned for further exposure. The value of σ_S after exposure were divided by the initial values of σ_S giving the relative saturation magnetization.

Results and Discussion

Adsorption isotherms. Figure 1 shows adsorption isotherms for adsorption of AQPU-13, AQPU-16, and AQPU-15 onto metal particles from tetrahydrofuran at 25°C. Plots of mg polymer adsorbed per g of particle obeyed the Langmuir relation,

$$\theta = A \times \frac{BC}{1 + BC} \tag{1}$$

equation 1, where A is the Langmuir capacity , B is Langmuir affinity, and C is the concentration of polymer solution. Values for the Langmuir affinity, Table I, for the three polymers were similar, however the Langumuir capacity increased with increasing AQM-1 content. AQPU-15, having the highest AQM-1 content, had the highest affinity.

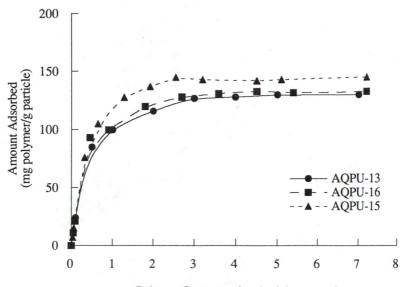

Figure 1. Adsorption isotherms for adsorption of amine-quinone polymers onto iron particles from THF solution.

Table I. Parameters from the the the fit to the Langmuir Adsorption Equation

Polymer	AQM-1 content	Capacity (A)	Affinity (B)
AQPU-13	20%	140±2	2.59±0.23
AQPU-15	40%	157±3	2.96±0.31
AQPU-16	30%	142±3	3.00±0.40

Rheology of metal particle dispersions. Rheological measurements were made on the metal particle dispersions in order to compare the different amine-quinone polymers with the commercial polymers. In Figure 2 is a plot of viscosity as a function of shear rate for metal particle dispersions made with amine-quinone polyurethanes, a commercial poly(vinylchloride) wetting binder or a commercial polyurethane. All amine-quinone polyurethanes showed similar viscosity at high shear rate. AQPU-13 had the lowest viscosity at low shear rates, suggesting a lower yield stress. The presence of a yield stress and shear-thinning, suggests the break-down of a network structure in the coating fluid.   Viscoelastic measurements were carried out to assess the magnitude of elastic contribution resulting from structure formation. The storage modulus G'(ω) was a measure of the elasticity in the system, while the loss modulus G"(ω) was a measure of the viscous response, where ω is the angular frequency at which the measurements are carried out. In Figure 3 is a plot of G' as a function of ω for the different metal particle dispersions. In all cases G' increased with increasing frequency. AQPU-13, containing the lowest AQM-1 content, showed lowest storage modulus while AQPU-15 showed the highest storage modulus. Higher values of G' for a magnetic dispersion suggests an increase in the network structure. The network

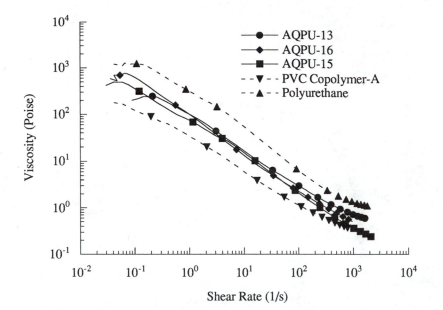

Figure 2. Plot of viscosity as a function of shear rate for the metal particle dispersions.

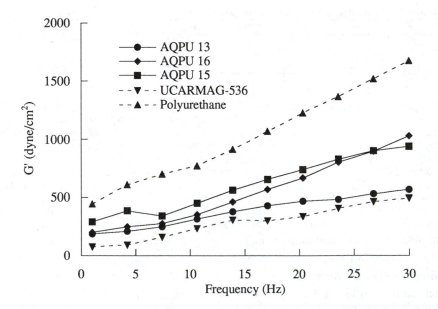

Figure 3. Plot of storage modulus, G', as a function of frequency for the metal particle dispersions.

structure arising from increased magnetic attractions between the particles in the dispersion and therefore a lower quality of dispersion. At every frequency the dispersion containing UCARMAG-536 had the lowest value of G' and therefore the best dispersion. The dispersion containing the commercial polyurethane had the highest value of G' and was therefore the worst dispersion. For the amine-quinone polyurethane the dispersion containing AQPU-13 had the lowest value of G', while AQPU-15 had the highest the value of G'. Therefore AQPU-13 gave the best dispersion and the quality of the dispersion approached that provided by the commercial wetting binder UCARMAG-536.

Dispersion Quality in Magnetic Coatings. A major source of media noise is attributed to the interparticle interactions which arises from the magnetic particle aggregates or agglomerates (*19*). These particle interactions have been analyzed through the use of magnetic remanance curves and using a model by Che and Bertram (*17*). The interparticle interaction parameter, α, extracted from magnetic remanance curves measured on the magnetic coatings prepared from various dispersions are tabulated in Table II. For an ideal dispersion where particles are separated, the interparticle interaction fields are negligible and values of α are expected to be zero. As the particles come into contact due to insufficient steric repulsion, the interaction field increases and the alpha deviates from zero. The higher the absolute value of α, the poorer the dispersion quality. As expected, the values of α for all the amine-quinone polymers were negative, indicating that the interparticle interaction were demagnetizing in nature. AQPU-13 had the lowest absolute of α, suggesting that it had the best magnetic dispersion. AQPU-15 had the highest absolute values of α, suggesting that this dispersion was the worst. The remanence measurements on the tape correlated well with the rheological measurements on the coating fluids. By both measures, AQPU-13 gave the best dispersion, and the quality of the dispersion decreased in the order AQPU-16 > AQPU-15.

Table II.   Interparticle interaction parameter obtained from magnetic remanence measurements

Binder Polymer	α
AQPU-13	-0.117
AQPU-16	-0.128
AQPU-15	-0.136

The quality of dispersion generally dependent on the amount of polymer adsorption and conformation of the adsorbed layer on particle surfaces. Although AQPU-15 had the highest adsorption, AQPU-13 found to have better dispersion from both rheological measurements on dispersion and magnetic measurements on the tape. This may be attributed to the differences in the conformation of adsorbed polymer chains on the metal particle surface. Because of the affinity of amine-quinone functional group to iron surfaces, the hard segments containing this monomer adsorbed strongly to the metal particle surface. The adsorption of these polymers was be interpreted in terms of segregated train-tail conformation in which hard segment preferentially anchors to the particle surface and the excluded soft segment blocks extend away from the surface as tails. Longer tails from the longer soft segment chains in AQPU-13 were more suitable for sterically blocking the magnetic attraction between the particles in the dipsersion.

Corrosion Protection. In a comparative accelerated corrosion study, the MP tape coatings were exposed to pH 2 buffer and the amount of corrosion, as measured by the relative saturation magnetization was determined as a function of exposure time, Figure 4. The MP tape made with commercial polyvinyl chloride binders completely corroded after about 2 hr exposure to pH 2 buffer. There was no loss in saturation magnetization for the tape made with amine quinone polymers even after 16 hour exposure. Among the amine quinone polyurethanes AQPU-15 showed the best corrosion protection while AQPU-13 the least. The higher affinity of AQPU-15 for the metal particle surface leads to better coverage by amine-quinone functional group, providing a better protective barrier than AQPU-13. This trend was also seen in samples exposed to 80°C and 80% relative humidity, Figure 5, where AQPU-15 had the best corrosion protection. Transmission electron microscope studies on these amine-quinone polyurethanes suggest a higher hard segment segregation and hard segment crystallite size with AQPU-15 (Nikles, D.E.; Chacko, A.P.; Webb, R. I. The University of Alabama, unpublished data). The hard segment may be crystallizing around particle surface and blocking moisture absorption, which leads to reduced corrosion. The corrosion inhibition of steel by quinone monomers and oligomers has been recently reported (20-21). The affinity of the amine-quinone functional group to iron may arise from the interaction between this electron rich molecule and the vacant d-orbitals in iron. The nature of the interaction between amine quinone polymer and the iron surface is currently under investigation .

Conclusions

The composition of amine-quinone polyurethane has a strong influence on the performance of the magnetic coatings made with these polymers. The amount of polymers adsorbed on the particle surface increases with amine-quinone monomer content. The polymer with highest amine-quinone content gave a better corrosion protection. The dispersion quality is governed by the nature of conformation of the adsorbed polymer chains. The polymer having the lowest amine-quinone content, and therefore the highest soft segment content gave the best dispersions. There was a good correlation between the elasticity of the magnetic ink, as measured by G', and the quality of the magnetic dispersion in tape made from these inks, as measured by the interparticle interaction parameter, α.

Acknowledgments

We thank Dr. Peter C. Clark, Department of Mineral Engineering, for the use of his rheometer. This project was supported in part by the Department of Defence Advanced Research Project under a grant administered by the National Storage Industry Consortium.

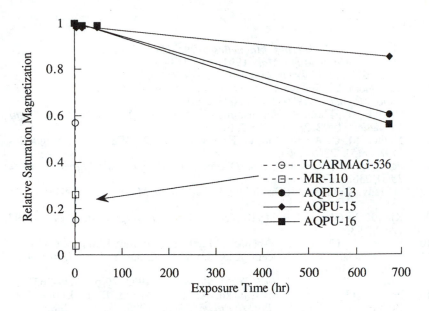

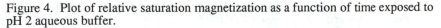

Figure 4. Plot of relative saturation magnetization as a function of time exposed to pH 2 aqueous buffer.

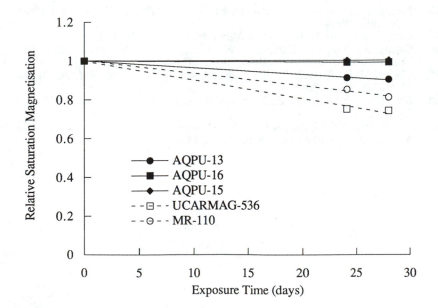

Figure 5. Plot of relative saturation magnetization as a function of time exposed to 80°C and 80% relative humidity.

Literature Cited

1. Richter, H. J. *IEEE Trans. Magnetics* **1993,** *29(5)*, 2185-2201.
2. Bate, G. *J.Magn. Magn. Mat.* **1991,** *100*, 413-424.
3. Speliotis, D. *IEEE Trans. Magnetics* **1990,** *26(1)*, 124-126.
4. Yamamoto, Y.; Sumiya, K.; Miyake, A.; Kishimoto, M.; Taniguchi, T. *IEEE Trans. Magnetics* **1990,** *26(5)*, 2098-2100.
5. Okazaki, Y.; Hara, K.; Kawashima, T.; Sato, A.; Hirano, T. *IEEE Trans. Magnetics* **1992,** *28(5)*, 2365-2367.
6. Mathur, M. C. A.; Hudson, G. F.; Hackett, L. D. *IEEE Tran. Magnetics*, **1992,** *28(5)*, 2362-2364.
7. Parker, M. R.; Venkataram, S.; DeSmet, D. *IEEE Trans. Magnetics*, **1992,** *28(5)*, 2368-2370.
8. O'Grady, K.; Gilson, R.G.; Hobby, P. C. *J.Magn. Magn. Mat.* **1991,** *95*, 341-355.
9. Kim, K. J.; Glasgow, P. D.; Kolycheck, E. G. *J. Magn. Magn. Mat.* **1993,** *120*, 87-93.
10. Kaleem, K.; Chertok, F.; Erhan, S. *Prog. Org. Coatings*, **1987,** *15*, 63-71.
11. Nithianandam, V. S.; Kaleem, K.; Chertok, F.; Erhan, S. *J. Appl. Polym. Sci* **1991,** *42*, 2893-2897.
12. Nikles , D.E.; Liang. J. *IEEE Trans. Magnetics.* **1993,** *29(6)*, 3649-3651.
13. Nikles, D.E.; Cain, J. L.; Chacko, A.P.; and Webb; R. I.; Liang, J.-L., Belmore. K. *J. Poly. Sci. Part A, Polym. Chem.* **1995,** *33*, 2881-2886.
14. Nikles, D.E.; Cain, J.; Chacko, A.P.; Webb, R. I.;. Liang, J.; Belmore. K. *Polymer Materilas Encyclopedia* , CRC Press., in press.
15. Nikles, D.E.; Cain, J. L.; Chacko, A. P.; Webb, R. I. *IEEE Trans. Magnetics* **1994,** *30(6)*, 4068-4070.
16. Huang, P.; Harrell, J. W.; Parker, M. R. *IEEE Trans. Magnetics* . **1994,** *30(6)*, 4002-4004.
17. Che, X.; Bertram, H. N. *J.Magn. Magn. Mat.* **1992,** *116*, 121.
18. Cheng, S.; Fan, H.; Harrell, J. W.; Lane, A. M.; Nikles, D. E. *IEEE Transactions on Magnetics* **1994,** *30(6)*, 4071-4073.
19. Clarke, M. D.; Bissell, P. R.; Chantrell, R.W.; Gilson, R. *J.Magn. Magn. Mat.* **1991,** *95*, 17-26.
20. Slavcheva, E; Sokolava, E.; Raicheva, S. *J. Electroanal. Chem.* **1993,** *360*, 271-282.
21. Muralidharan, S.; Phani, K.L.N.; Pitchumani, S.; Ravichandran, S.; Iyer, S. V.K. *J. Electrochem. Soc.* **1995,** *142(5)*, 1478-1483.

INDEXES

Author Index

Affiliation Index

Subject Index

Bestsellers from ACS Books

The ACS Style Guide: A Manual for Authors and Editors
Edited by Janet S. Dodd
264 pp; clothbound ISBN 0–8412–0917–0; paperback ISBN 0–8412–0943–X

Understanding Chemical Patents: A Guide for the Inventor
By John T. Maynard and Howard M. Peters
184 pp; clothbound ISBN 0–8412–1997–4; paperback ISBN 0–8412–1998–2

Chemical Activities (student and teacher editions)
By Christie L. Borgford and Lee R. Summerlin
330 pp; spiralbound ISBN 0–8412–1417–4; teacher ed. ISBN 0–8412–1416–6

Chemical Demonstrations: A Sourcebook for Teachers,
Volumes 1 and 2, Second Edition
Volume 1 by Lee R. Summerlin and James L. Ealy, Jr.;
Vol. 1, 198 pp; spiralbound ISBN 0–8412–1481–6;
Volume 2 by Lee R. Summerlin, Christie L. Borgford, and Julie B. Ealy
Vol. 2, 234 pp; spiralbound ISBN 0–8412–1535–9

Chemistry and Crime: From Sherlock Holmes to Today's Courtroom
Edited by Samuel M. Gerber
135 pp; clothbound ISBN 0–8412–0784–4; paperback ISBN 0–8412–0785–2

Writing the Laboratory Notebook
By Howard M. Kanare
145 pp; clothbound ISBN 0–8412–0906–5; paperback ISBN 0–8412–0933–2

Developing a Chemical Hygiene Plan
By Jay A. Young, Warren K. Kingsley, and George H. Wahl, Jr.
paperback ISBN 0–8412–1876–5

Introduction to Microwave Sample Preparation: Theory and Practice
Edited by H. M. Kingston and Lois B. Jassie
263 pp; clothbound ISBN 0–8412–1450–6

Principles of Environmental Sampling
Edited by Lawrence H. Keith
ACS Professional Reference Book; 458 pp;
clothbound ISBN 0–8412–1173–6; paperback ISBN 0–8412–1437–9

Biotechnology and Materials Science: Chemistry for the Future
Edited by Mary L. Good (Jacqueline K. Barton, Associate Editor)
135 pp; clothbound ISBN 0–8412–1472–7; paperback ISBN 0–8412–1473–5

For further information and a free catalog of ACS books, contact:
American Chemical Society
Customer Service & Sales
1155 16th Street, NW, Washington, DC 20036
Telephone 800–227–5558